Topics in Biodiversity and Conservation

Volume 21

Springer's book series, Topics in Biodiversity and Conservation, brings together some of the most exciting and topical papers in biodiversity and conservation research. The result is a series of useful themed collections covering issues such as the diversity and conservation of specific habitats or groups of organisms, and the key dilemma of resource use versus conservation.

Based on Springer's popular journal, Biodiversity and Conservation, the series provides access to selected peer-reviewed papers which represent the cutting edge of current research to provide a valuable overview of progress in each topic addressed. With their diversity of case studies and depth of investigation, these collections will be of particular interest for courses including biodiversity and/or conservation issues, and to advanced students and researchers working in related fields.

Dipayan Das • Payel Paul
Monoranjan Chowdhury
Editors

Plant Diversity and Conservation Strategies of the Himalayan Biodiversity Hotspot

Springer

Editors
Dipayan Das
Department of Microbiology
Royal School of Biosciences
The Assam Royal Global University
Guwahati, Assam, India

Payel Paul
Department of Botany
North Bengal St. Xavier's College
Jalpaiguri, West Bengal, India

Monoranjan Chowdhury
Department of Botany
University of North Bengal
Siliguri, West Bengal, India

ISSN 1875-1288 ISSN 1875-1296 (electronic)
Topics in Biodiversity and Conservation
ISBN 978-3-032-21696-0 ISBN 978-3-032-21697-7 (eBook)
https://doi.org/10.1007/978-3-032-21697-7

This Springer imprint is published by the registered company Springer Nature Switzerland AG
The registered company address is: Gewerbestrasse 11, 6330 Cham, Switzerland

Foreword

It is with great enthusiasm and appreciation that I present this foreword for the book *Plant Diversity and Conservation Strategies of the Himalayan Biodiversity Hotspot*, edited by Dr. Dipayan Das, Dr. Payel Paul, and Prof. Monoranjan Chowdhury. This book, comprising 20 insightful chapters contributed by esteemed researchers from across the globe, is a significant scholarly endeavor dedicated to one of the most ecologically rich yet fragile ecosystems on our planet, the Himalayas. The volume sits firmly within the continuum of earlier regional floristic works and complements previous contributions of the series "Topics in Biodiversity and Conservation," thereby providing continuity and context for ongoing botanical research in the region.

The Himalayan region, recognized as one of the world's foremost biodiversity hotspots, harbors an extraordinary variety of plant species, including many endemic and threatened taxa. The chapters in this volume delve into the floristic diversity, vegetation patterns, and unique ecological aspects of the region, offering valuable perspectives on conservation challenges and sustainable management strategies. The contributions cover a broad spectrum of topics, ranging from the distribution of orchids and medicinal plants to the impact of climate change, invasive species, and anthropogenic pressures on the fragile Himalayan ecosystem. The outcome of the book state that each plant species supports a range of other organisms that are often host-restricted, particularly fungi and insects. Consequently, the loss of a single plant species may lead to the loss of multiple dependent organisms, some of which may possess significant scientific or applied value, including novel bioactive compounds or potential applications as biological control agents.

This book is particularly relevant at a time when global biodiversity is under unprecedented stress due to climate change and human activities. The discussions on conservation policies, ecotourism, ex situ conservation strategies, and protected area management provide practical insights into preserving the unique phytodiversity of the Himalayas. Furthermore, the ethnobotanical perspectives discussed in several chapters indicates the deep interconnection between local communities and their natural environment, emphasizing the need for inclusive and sustainable conservation approaches.

I commend the editors and contributors for their dedication to this important cause and am confident that this volume will serve as a valuable resource for researchers, conservationists, policymakers, and students engaged in biodiversity conservation. It is my sincere hope that this book will inspire further research and conservation initiatives aimed at protecting the Himalayan biodiversity for future generations.

Former Professor of Botany, University of Delhi, Delhi, India

Former Vice-Chancellor, Mansarovar Global University, Bhopal, India

Arun K. Pandey

Preface

The Himalayas, a majestic mountain range spanning several nations, stand as one of the most significant biodiversity hotspots on Earth. This vast expanse of towering peaks, deep valleys, and diverse ecosystems harbors an extraordinary wealth of plant life, many of which are endemic, rare, or medicinally significant. The region is not only a cradle of unique flora but also serves as a lifeline for millions of people, providing essential ecosystem services such as water regulation, soil stabilization, and climate moderation. The Himalayan biodiversity supports traditional knowledge systems, sustains livelihoods, and plays a crucial role in cultural and spiritual practices among indigenous communities. This fragile ecosystem is increasingly under threat due to climate change, habitat destruction, invasive species, and unsustainable human activities, leading to alarming rates of biodiversity loss. As the ecological balance of the Himalayas is disrupted, the repercussions extend far beyond regional boundaries, impacting global climate patterns and biodiversity networks.

In this context, the present volume, *Plant Diversity and Conservation Strategies of the Himalayan Biodiversity Hotspot*, serves as an essential compendium that brings together diverse perspectives on the floristic wealth of the region and the imperative need for conservation strategies to safeguard its ecological integrity. By integrating scientific research with traditional knowledge and conservation policy insights, this book not only reveals the rich botanical heritage of the Himalayas but also offers pragmatic solutions for preserving it. Through its comprehensive approach, the volume aims to foster awareness, promote sustainable conservation efforts, and contribute to the global discourse on biodiversity protection in mountainous ecosystems. It indicates the urgent need for collaborative action among scientists, conservationists, policymakers, and local communities to ensure that the Himalayas continue to thrive as a sanctuary of biodiversity for generations to come.

The editor presents this book as a collective effort of distinguished scholars, researchers, and conservationists across the world. The chapters explore the multifaceted dimensions of plant diversity in the Himalayas, assessing the distribution of floral species, ecological dynamics, and the pressing challenges that threaten biodiversity. It presents strategic conservation measures to ensure the long-term sustainability of these invaluable diverse ecosystems on various altitudinal rages of Himalayas.

The book begins with an overarching discussion on the Himalayas' significance in global biodiversity, highlighting the region's evolutionary history, unique ecological landscapes, and role as one of the world's biodiversity hotspots. It further delves into the intricate patterns of floristic diversity and vegetation dynamics, offering insights into endemic species, rare plant groups, and the delicate balance of high-altitude ecosystems. Special emphasis is given to orchids, rhododendrons, maples, medicinal plants, and wild relatives of various domesticated crops, which hold immense ecological and economic value while facing significant conservation challenges.

The sustainable use and management of plant resources emerge as a critical theme throughout this volume. The discussion extends to ethnobotanical knowledge systems, where indigenous communities have long relied on native flora for medicine, nutrition, and cultural practices. The intersection of traditional knowledge with modern conservation strategies provides a compelling case for integrating local wisdom with scientific advancements to promote biodiversity conservation.

Recognizing the numerous threats to Himalayan biodiversity, this book indicates the impact of climate change, habitat fragmentation, and anthropogenic activities on Himalayan vegetation. The ongoing shifts in temperature and precipitation patterns have led to habitat loss, species migration, and ecosystem instability. The expansion of human settlements, unsustainable land use, and invasive alien species exacerbate biodiversity decline. Several chapters propose adaptive and preventive strategies, including sustainable ecotourism, protected area management, and community-led conservation initiatives, to mitigate these impacts.

From a policy perspective, this volume provides a critical evaluation of existing conservation frameworks, identifying gaps in biodiversity governance and recommending policy refinements for enhanced environmental protection. It indicates the importance of ex situ and in situ conservation approaches, advocating for a more integrated approach to ecosystem management, aligns scientific research with community participation and legislative support.

Beyond its academic value, this book holds significance for a broad audience. For researchers, it provides a comprehensive reference on Himalayan plant diversity, conservation strategies, and ecological challenges. Students will find valuable insights into biodiversity conservation principles, ethnobotanical applications, and environmental policy discussions. Conservation practitioners and policymakers can benefit from its recommendations for sustainable resource management, while nature enthusiasts and environmental advocates will gain a deeper appreciation of the Himalayas' irreplaceable ecological heritage.

Through this volume, the editors aimed to foster awareness, inspire action, and contribute to the global discourse on biodiversity conservation. The Himalayas are not just a geographical marvel but a vital ecological treasure that sustains millions of lives and supports countless species. Protecting this fragile ecosystem is not only a regional responsibility but a global imperative.

We express our sincere gratitude to all contributing authors for their valuable research and dedication to biodiversity conservation. We also extend our appreciation to the reviewers for their constructive feedback and to Springer for facilitating the publication of this important work.

It is our hope that this book will inspire further research and conservation efforts, ensuring that the Himalayas continue to thrive as a sanctuary of biodiversity for generations to come.

Guwahati, Assam, India — Dipayan Das
Jalpaiguri, West Bengal, India — Payel Paul
Siliguri, West Bengal, India — Monoranjan Chowdhury

Acknowledgments

The successful completion of this volume, *Plant Diversity and Conservation Strategies of the Himalayan Biodiversity Hotspot*, under the esteemed book series Topics in Biodiversity and Conservation, would not have been possible without the support and contributions of eminent scientists and professors from different institutions.

First and foremost, we express our sincere gratitude to all the contributing authors for their valuable chapters, insightful research, and dedication to advancing knowledge on plant diversity and conservation in the Himalayan region. Their expertise and commitment have significantly enriched the quality of this book.

We extend our deep appreciation to the reviewers (Prof. A.K. Pandey, Dr. Swarnendu Roy, Prof. S.C. Roy, Prof. M.M. Sardesai, Dr. Sourish Dasgupta, Dr. Birson Ingti, Dr. Pankaj Losan Sharma, Dr. Tanushree Sarkar, and Dr. Asit Ray) for their meticulous evaluation and constructive feedback, which have greatly enhanced the clarity and scientific rigor of the chapters. Their time and effort in refining this work are truly commendable.

Our heartfelt thanks go to Springer and the editorial team of the Topics in Biodiversity and Conservation series for their unwavering support, guidance, and professionalism throughout the publication process. Their efforts in ensuring the highest standards of academic publishing are deeply appreciated.

Dr. Dipayan Das would like to thank The Assam Royal Global University, Dr. Payel Paul would like to thank North Bengal St. Xavier's College, and Prof. Monoranjan Chowdhury would like to thank the University of North Bengal for continuous support and motivation and for providing the necessary infrastructure.

We also acknowledge the invaluable support of various research institutions, universities, and funding agencies that facilitated the research endeavors of the contributing authors. Their commitment to biodiversity conservation and scientific research has played a crucial role in shaping the discussions presented in this volume.

Finally, we are grateful to our families, friends, and colleagues for their encouragement and patience during the preparation of this book. Their constant support has been a source of motivation throughout this journey.

We hope that this volume will serve as a valuable resource for researchers, conservationists, and policymakers dedicated to preserving the rich plant diversity of the Himalayan Biodiversity Hotspot.

Dipayan Das
Payel Paul
Monoranjan Chowdhury

Contents

Contributors

Puja Acharya Sikkim Regional Centre, G.B. Pant National Institute of Himalayan Environment, Pangthang, Gangtok, Sikkim, India

Arjun Adhikari Department of Botany, Madhab Choudhury College, Barpeta, Assam, India

Mrinalini Agarwal Department of Biotechnology, Graphic Era (Deemed to be University), Uttarakhand, India

Pandey Akansha Department of Botany & Pharmacognosy, Captain Srinivasa Murthy Central Ayurveda Research Institute (CSM CARI), Chennai, Tamil Nadu, India

Gyanendro Singh Amom Regional Ayurveda Research Centre (RARC), 1st Floor, CMO Building, District Hospital Colony, Dimapur, Nagaland, India

Chandan Barman Department of Botany, Gour Banga University, Malda, West Bengal, India

Basabi Barua Hazra Department of Botany, Triveni Devi Bhalotia College, Raniganj, West Bengal, India

Debahooti Baruah Department of Botany, Tezpur College, Tezpur, Assam, India

Partha Pratim Baruah Department of Botany, Gauhati University, Guwahati, Assam, India

Jeewan Kumar Bastola Department of Arts and Humanities, Samtse College of Education, Royal University of Bhutan, Samtse, Bhutan

Rajneesh Bhardwaj School of Agriculture, Graphic Era Hill University, Dehradun, Uttarakhand, India

Ganga Datt Bhatt Department of Plant Resources, Plant Research Center, Dhangadhi, Kailali, Sudurpashchim Province, Nepal

Rajib Biswas Department of Conservation Biology, Durgapur Government College, Durgapur, West Bengal, India

Devanjal Bora Botany Section, Central Ayurveda Research Institute, CCRAS, Ministry of AYUSH, Guwahati, Assam, India

Bishmita Boruah Department of Botany, Gauhati University, Guwahati, Assam, India

Monoranjan Chowdhury Department of Botany, University of North Bengal, Siliguri, West Bengal, India

Deep Shekhar Das Taxonomy and Biosystematics Laboratory, Department of Botany, University of Calcutta, Kolkata, West Bengal, India

Dipayan Das Department of Microbiology, Royal School of Biosciences, The Assam Royal Global University, Guwahati, Assam, India

Chandra Ghosh Tea Taxonomy and Ecology Laboratory, Department of Tea Science, University of North Bengal, Raja Rammohunpur, Darjeeling, West Bengal, India

Sourish Das Gupta Department of Botany, North Bengal St. Xavier's College, Rajganj, West Bengal, India

Soumitra Halder Department of Botany, Visva-Bharati, Santiniketan, West Bengal, India

Nidaa Harun Department of Botany, Faculty of Life Sciences, University of Okara, Okara, Pakistan

Pabitra Jha Tribhuvan University, Pokhara, Kaski, Gandaki Province, Nepal

Nikhilesh Joardar Department of Microbiology and Immunology, The University of North Carolina at Chapel Hill, School of Medicine, Chapel Hill, NC, USA

Department of Bioengineering, School of Engineering and Technology, Dr. Vishwanath Karad MIT World Peace University, Pune, Maharashtra, India

Namini Joshi Department of Biotechnology, Graphic Era (Deemed to be University), Uttarakhand, India

Himamani Kalita Department of Botany, Gauhati University, Guwahati, Assam, India

Rowndel Khwairakpam School of Agriculture, Graphic Era Hill University, Dehradun, Uttarakhand, India

Sapna Avinash Kondalkar Regional Ayurveda Research Institute, CCRAS, Gwalior, Madhya Pradesh, India

Pardeep Kumar Department of Botany, Mata Gujri College, Fatehgarh Sahib, Punjab, India

Anupam Kundu Applied Phycology Laboratory, Department of Botany, Visva-Bharati (Central University), Santiniketan, West Bengal, India

Surya Kumar Maharjan Tribhuvan University, Institute of Forestry, Hetauda Campus, Hetauda, Makawanpur, Bagmati Province, Nepal

Debabrata Maity Taxonomy and Biosystematics Laboratory, Department of Botany, University of Calcutta, Kolkata, West Bengal, India

Piyush Mathur Department of Botany, Institute of Science, Banaras Hindu University, Varanasi, Uttar Pradesh, India

Achintya Mitra Regional Ayurveda Research Institute, Gangtok, Sikkim, India

Amitava Nayek Department of Botany, Triveni Devi Bhalotia College, Raniganj, West Bengal, India

Arvind Singh Negi School of Agriculture, Graphic Era Hill University, Dehradun, Uttarakhand, India

Shivani Negi Department of Biotechnology, Graphic Era (Deemed to be University), Uttarakhand, India

Vikram S. Negi Department of Botany and Microbiology, HNB Garhwal University, B.G.R. Campus, Pauri Garhwal, Uttarakhand, India

Tej Kumar Nepal Department of Prevention & Education, Anti-Corruption Commission, Thimphu, Bhutan

Krishna Nikhil Regional Ayurveda Research Centre (RARC), 1st Floor, CMO Building, District Hospital Colony, Dimapur, Nagaland, India

Aaratrik Pal Taxonomy of Angiosperms and Biosystematics Laboratory, Department of Botany, University of North Bengal, Raja Rammohunpur, Darjeeling, West Bengal, India

Biswajit Paul Botany Section, Central Ayurveda Research Institute, CCRAS, Ministry of AYUSH, Guwahati, Assam, India

Payel Paul Department of Botany, North Bengal St. Xavier's College, Jalpaiguri, West Bengal, India

Adharsika Pradhan Sikkim Regional Centre, G.B. Pant National Institute of Himalayan Environment, Pangthang, Gangtok, Sikkim, India

Shouvik Pramanik Applied Phycology Laboratory, Department of Botany, Visva-Bharati (Central University), Santiniketan, West Bengal, India

Akshita Raj Department of Biotechnology, Graphic Era (Deemed to be University), Uttarakhand, India

Balwant Rawat School of Agriculture, Graphic Era Hill University, Dehradun, Uttarakhand, India

Janhvi Mishra Rawat Department of Biotechnology, Graphic Era (Deemed to be University), Uttarakhand, India

Sandeep Rawat Sikkim Regional Centre, G.B. Pant National Institute of Himalayan Environment, Pangthang, Gangtok, Sikkim, India

Anindya Sundar Ray Ethnopharmacology Laboratory, Department of Biological Sciences and Microbiology, Biswa Bangla Biswabidyalay (University), Bolpur, Shibpur, Birbhum, West Bengal, India

Sourav Roy Applied Phycology Laboratory, Department of Botany, Visva-Bharati, Santiniketan, West Bengal, India

Kingsuk Sarkar Regional Ayurveda Research Institute, Gangtok, India

Kushankur Sarkar Taxonomy of Angiosperms and Biosystematics Laboratory, Department of Botany, University of North Bengal, Raja Rammohunpur, Darjeeling, West Bengal, India

Sukanya Sarma Department of Botany, Salbari College, Baksa, Assam, India

Animesh Sen Botany Section, Regional Ayurveda Research Institute, CCRAS, Ministry of AYUSH, Gangtok, Sikkim, India

Alok Sharma CCRAS- Regional Ayurveda Research Institute, Gwalior, Madhya Pradesh, India

Nagayya Shiddamallayya Medicinal Plants Research (MPR) Section, Central Council for Research in Ayurvedic Sciences (CCRAS), Janakpuri, New Delhi, India

Dinesh Singh Department of Botany, Pt. L. M. S. Campus, Sri Dev Suman Uttarakhand University, Rishikesh, Uttarakhand, India

Bhagat Suberi Department of Forestry College of Natural Resources, Royal University of Bhutan, Lobesa, Punakha, Bhutan

Saini Sultana Tea Taxonomy and Ecology Laboratory, Department of Tea Science, University of North Bengal, Raja Rammohunpur, Darjeeling, West Bengal, India

Shaheena Umbreen Department of Botany, Faculty of Life Sciences, University of Okara, Okara, Pakistan

Walsan Kalarikkal Vishnu Department of Botany & Pharmacognosy, Captain Srinivasa Murthy Central Ayurveda Research Institute (CSM CARI), Chennai, Tamil Nadu, India

Rajendra Yonzone Department of Botany, Victoria Institution (College), Baithakkhana, Kolkata, West Bengal, India

Abbreviations

2-iP	N6-(2-isopentenyl) Adenine
°C	Degree Celsius
ACAP	Annapurna Conservation Area Project
Arc GIS 9.2	Area Reference Coordinate Geographic Information System 9.2
asl	Above Sea Level
BGCI	Botanic Gardens Conservation International
BRs	Biosphere Reserves
CABI	Centre for Agriculture and Biosciences International
CBD	Convention on Biological Diversity
CBC	Community-Based Conservation
CbT	Community-Based Tourism
CFP	Community Forest Program
CI	Cultural Importance Index
CITES	Convention on International Trade in Endangered Species of Wild Fauna and Flora
cm	Centimeter
CR	Critically Endangered
CV	Cultural Value
DD	Data Deficient
DNA	Deoxyribonucleic Acid
DN	Digital Number
DPR	Department of Plant Resources
EbA	Ecosystem-based Adaptation
EIAs	Environmental Impact Assessments
EN	Endangered
ERDAS	Earth Resources Data Analysis System
EST-SSR	Expressed Sequence Tag-Simple Sequence Repeat
EX	Extinct
GBPNIHE	G.B. Pant National Institute of Himalayan Environment

GCPs	Ground Control Points
GIS	Geographic Information System
GLOFs	Glacial Lake Outburst Floods
GoN	Government of Nepal
GPS	Global Positioning System
gSSR	Genomic Simple Sequence Repeat
HAWS	High-Altitude Wetlands
HKPL	Hindukush-Karakoram Pamir Landscapes
HWI	Himalayan Wetlands Initiative
ICIMOD	International Centre for Integrated Mountain Development
IHR	Indian Himalayan Region
IBA	Indole-3-Butyric Acid
IPCC	Intergovernmental Panel on Climate Change
IRS-1A	Indian Remote Sensing Satellite-1A
IRS-1B	Indian Remote Sensing Satellite-1B
IRS-1C	Indian Remote Sensing Satellite-1C
IRS-1D	Indian Remote Sensing Satellite-1D
ISRO	Indian Space Research Organization
ISSG	Invasive Species Specialist Group
IUCN	International Union for Conservation of Nature
KL	Kangchenjunga Landscape
Km	Kilometer
KSL	Kailash Sacred Landscapes
Landsat TM-1990	Landsat Thematic Mapper-1990
Landsat-8 SR 2015	Landsat-8 Surface Reflectance 2015
LC	Least Concern
LTP	Lata-Tolma-Phagti
LULC	Land Use/Land Cover
m	Meter
MAPs	Medicinal and Aromatic Plants
masl	Meter Above Sea Level
MBRs	Mountain Biosphere Reserves
MBT	Main Boundary Thrust
MCT	Main Central Thrust
MFT	Main Frontal Thrust
MoEF & CC	Ministry of Environment, Forests and Climate Change
MoFSC	Ministry of Forests and Soil Conservation
MT	Metric Tons
NAPCC	National Action Plan on Climate Change
NDBR	Nanda Devi Biosphere Reserve
NDNP	Nanda Devi National Park
NE	Not Evaluated
NEI	North-Eastern India
NGOs	Non-Governmental Organizations

NMPB	National Medicinal Plants Board
NMSHE	National Mission for Sustaining the Himalayan Ecosystem
NP	National Park
NRCS	Natural Resources Conservation Service
NRSC	National Remote Sensing Centre
NT	Near Threatened
NTFP	Non-Timber Forest Product
OMF	Orchid Mycorrhizal Fungi
PA	Protected Area
PIC	Polymorphism Information Content
POWO	Plants of the World Online
PSK	Pindari-Sunderdhunga-Kafni
RFC	Relative Frequency of Citation
RI	Relative Importance Index
RS	Remote Sensing
SAC	Space Application Center
SARS-CoV-2	Severe Acute Respiratory Syndrome Coronavirus 2
SCoT	Start Codon Targeted
sp.	Species
SSC	Species Survival Commission
SSR	Simple Sequence Repeat
STD	South Tibetan Detachment System
SWM	Solid Waste Management
TAL	Terai Arc Landscape
TAR	Tibetan Autonomous Region
TEK	Traditional Ecological Knowledge
UK	United Kingdom of Great Britain and Northern Ireland
UNESCO-MAB	United Nations Educational, Scientific and Cultural Organization - Man and the Biosphere Programme
UNESCO's	United Nations Educational, Scientific and Cultural Organization's
UR	Use Report
USA	United States of America
USD	United States Dollar
UV	Ultra Violet
Uv	Use Value
VU	Vulnerable
WFO	World Flora Online
WHO	World Health Organization
WNBR	World Network of Biosphere Reserves
WWF	World Wide Fund for Nature
w.r.t.	With Respect To
μM	Micromolar

Part I
Himalaya and Its Diversity

Chapter 1
The Himalayas and its Position in Global Plant Biodiversity

Walsan Kalarikkal Vishnu, Pandey Akansha, Krishna Nikhil, Gyanendro Singh Amom, and Nagayya Shiddamallayya

Abstract The Himalayas are included as one of the 36 global biodiversity hotspots due to its luxuriant endemic biodiversity and the threat of habitat loss. Its orogeny is the result of a profound collision between the Indian and Eurasian plates, which generated its uplift creating spatial variations in altitudinal and climatic factors across latitudinal and longitudinal orientations. The region is broadly divided latitudinally into western, central, and eastern Himalayas owing to the spectrum of socio-cultural, climatic, and biogeographical heterogeneity. Various factors—preformative (pre-Himalayan geographical isolation), formative (establishment of land connections and plant migrations, biogeographic heterogeneity of the land masses, and Himalayan uplift) and post-formative (altitudinal and climatic variations)—have contributed towards the prolific biodiversity and endemism in the region. General vegetation and forest profiles exhibit noticeable latitudinal, longitudinal, and altitudinal gradients. The Himalayan region is characterised by high plant biodiversity. Species richness and endemism gradually decrease from the eastern to western Himalayas. Several studies have documented high endemic species richness and the presence of critically endangered, endangered, vulnerable flora in the biodiversity-rich landscapes of the Himalayas. The unique flora, species richness and endemism highlight the importance of immediate and sustained conservation efforts and natural resource management in the Himalayas.

Keywords Himalayas · Himalayan orogeny · Species richness · Endemism · Plant biodiversity

---

W. K. Vishnu (✉) · P. Akansha
Department of Botany and Pharmacognosy, Captain Srinivasa Murthy Central Ayurveda Research Institute (CSM CARI), Chennai, Tamil Nadu, India
e-mail: vishnu.walsan@ccras.nic.in

K. Nikhil · G. S. Amom
Regional Ayurveda Research Centre (RARC), Dimapur, Nagaland, India

N. Shiddamallayya
Medicinal Plants Research (MPR) Section, Central Council for Research in Ayurvedic Sciences (CCRAS), New Delhi, India

D. Das et al. (eds.), *Plant Diversity and Conservation Strategies of the Himalayan Biodiversity Hotspot*, Topics in Biodiversity and Conservation 21,
https://doi.org/10.1007/978-3-032-21697-7_1

1.1 Introduction

The Himalayan region is a geographically diverse and biodiversity-rich mountain province, which separates the Indian subcontinent from the Tibetan plateau, and thereby from the rest of Eurasia (Bahadur et al. 2018). It is 2400 km long and 300–400 km wide, with altitudes ranging from around 500 m in the foothills of the Indo-Gangetic plain to around 8000 m in the Greater Himalayas (Kayastha and Kayastha 2020). It hosts several mountains including the Mount Everest, and is the source of several prime rivers, including the Ganges, Brahmaputra and Indus. The ranges have profoundly affected the climate and shaped the livelihood of the communities in the large drainage basin of the rivers, making it culturally, biologically and geographically significant (Kayastha and Kayastha 2020).

1.2 Orogeny and Spatial Classification of the Himalayas

Himalayan orogeny is a geologically recent and ongoing phenomenon. The ancient supercontinent Gondwanaland which covered present-day India was split to form Western and Eastern Gondwana around 180 Mya (Chatterjee and Scotese 1999). Around 120 Mya, another tectonic event separated the present-day India-Madagascar-Seychelles fragment from Western Gondwana. This fragment migrated northward, demounting Madagascar and Seychelles fragments during around 80–80 and 65 Mya respectively (Manish and Pandit 2018). The northbound Indian plate subsequently collided with Eurasia at around 55–50 Mya, progressing towards the west till 40 Mya (van Hinsbergen et al. 2012). This profound collision and the subsequent geologic faulting led to the Himalayan orogeny.

The geological fault thus occurring has given rise to different fault lines, south of the tectonic suture zone (Indus-Tsangpo suture zone) (DiPietro and Pogue 2004; Singh and Patel 2022). The Himalayas can be stratigraphically divided into different yet juxtaposed litho-tectonic zones, bounded by these major fault lines (DiPietro and Pogue 2004) as Tethyan Himalaya, Greater Himalaya/Higher Himalaya, Lesser Himalaya and the Sub-Himalaya from north to south (Fig. 1.1).

While the Himalayas are stratified geologically from north to south, it has a climatic, cultural and socio-economic gradient from west to east. The initial settlements in the Himalayas have a gradient of Afghan-Iranian, Indo-Aryan, Tibeto-Burman, South-East Asian and indigenous tribal influences from west to east (Karan and Mather 1976). A gradual change in climatic behaviour is also observed from west to east, owing to the increasing proximity of the ranges in the east to the tropical region (Apollo et al. 2021) and other factors, which have been discussed in Sect. 1.3.1. Therefore, the Himalayan arc has most often been broadly classified into western, central and eastern Himalayas based on these cohesive association of geographical patterns and human occupancy (Burrard and Hayden 1908; Dhital 2015; Andrejczuk 2016; Dwivedi et al. 2024) (Fig. 1.2). But, the geographical

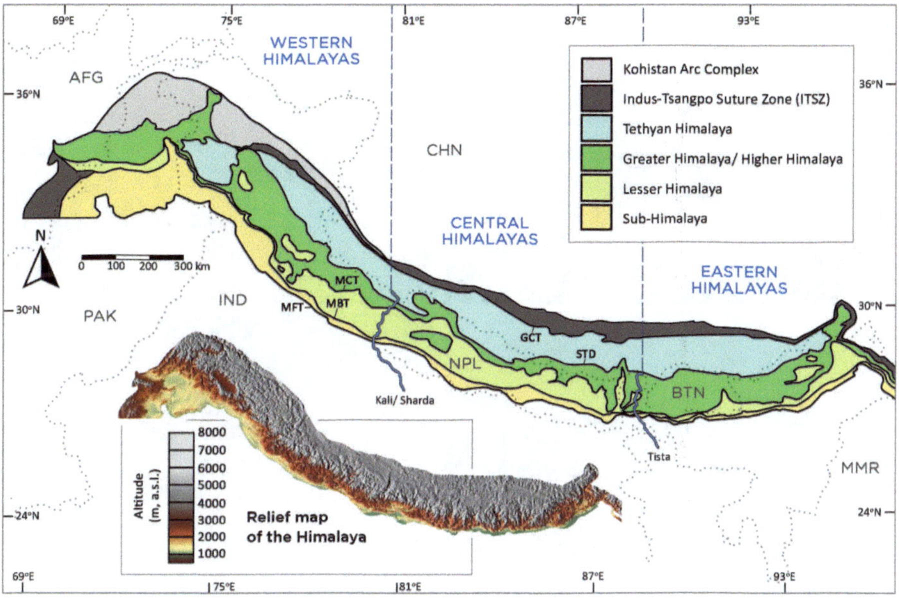

Fig. 1.1 Generalised geological map of the Himalayan region showing latitudinal and longitudinal stratification. Major fault lines from north to south are South Tibetan Detachment System (STD), Main Central Thrust (MCT), Main Boundary Thrust (MBT), and Main Frontal Thrust (MFT). Countries have been represented by their ISO 3166-1 alpha-3 codes. (Illustrated by VWK; Relief map adapted from Braxmeier (2024))

boundaries of this tripartite classification vary among authors. We have considered the region of the Himalayas to the west of river Kali as the western Himalayas. This includes the Himalayas in Afghanistan and Pakistan, the states and/or union territories of Jammu & Kashmir, Ladakh, Himachal Pradesh and Uttarakhand of India, and the corresponding areas in Tibetan Autonomous Region (TAR). The Himalayan region between the rivers Kali and Tista is the central Himalayan region, comprising the areas of Nepal, western Sikkim and corresponding areas in TAR. The area towards the east of Tista River comprises the Eastern Himalayas, which includes the rest of Sikkim, Bhutan, the Himalayas in North-Eastern India and northern Myanmar, and the corresponding areas in TAR.

1.3 Biodiversity in the Himalayas: Global, Geological and Geographical Perspectives

The Himalayas are characterised by a wealthy biodiversity. The region is placed among the 36 global biodiversity hotspots due to its high endemism and the looming threat of habitat loss. Though the geographical range and borders of the hotspots have been revised and regrouped over the years, the region has been a regular

Fig. 1.2 Variations in the vegetation and landscape of the Himalayas. Tropical evergreen forest (**a**) and the transition zone to temperate pine forests (**b**) in upper Dibang valley, Arunachal Pradesh; (**c**) Pine Forest on the way to Kalimpong, West Bengal; Highland landscape resembling cold tundra near Tsomgo lake (**d**) and on the way to Nathula pass (**e**) in Sikkim; Nival zone of Mishmi hills seen from Arunachal Pradesh. (Photographs by NK (**a–c** and **f**) and VWK (**d** and **e**))

inclusion in the list of global biodiversity hotspots since its inception (Myers 1988, 1990; Myers et al. 2000; Mittermeier et al. 2011). The Himalayan hotspot houses several ecoregions prioritised for conservation under the "Global 200 ecoregions" (Olson and Dinerstein 1998), and 330 of important bird areas on the earth (Pandit et al. 2014). The region forms the centre of diversity for many plants, such as *Meconopsis*, *Potentilla*, *Primula* and *Sorbus*, making it a critical hotspot in the perspective of global plant biodiversity (Pandit and Kumar 2013; Pandit et al. 2014).

Globally, the highest levels of endemism are often concentrated in island biodiversity hotspots, such as New Caledonia, Polynesia–Micronesia, the Caribbean Islands, and the East Melanesian Islands. However, according to Kier et al. (2009), the Himalayan hotspot stands out as one of the most significant endemic areas among mainland regions (Table 1.1). In their study, the region was ranked among the top 27 out of the 90 recognised global terrestrial biogeographic regions, distinguished by exceptionally high plant endemism and severe habitat loss. These 27 regions are considered the most critical worldwide, where extraordinary species uniqueness coincides with intense threats, making them top priorities for global conservation.

1.3.1 Drivers of Biodiversity and Endemism in the Himalayas

Several factors have contributed towards the proficient plant biodiversity and high endemism in the Himalayas. Based on the geological period of influence, the factors can be broadly divided into pre-formative (before the collision of Indian and Eurasian plates), formative (during the period of collision and active Himalayan uplift), and post-formative (after the period of active Himalayan uplift to present).

1.3.1.1 Pre-Formative Factors

Pre-Himalayan Geographical Isolation

The Himalayas possess relatively high endemism for a non-island geographical region. Some authors suggest that the geographical isolation of the Indian plate after its separation from Gondwanaland for around 45 million years has contributed to the development of this high endemic biodiversity (Manish and Pandit 2018). This speculation is substantiated by fossil evidence showing the diversification of angiosperms in this period (Srivastava 2011). Srivastava (2011) also suggested that the upper Cretaceous-Palaeogene flora of the Indian plate, which was devoid of Dipterocarpacean members, may be regarded as the original stock for the present-day Indian flora. Land connections established after the collision were responsible for migrations and further enhancement of biodiversity.

1.3.1.2 Formative Factors

Establishment of Land Connections and Biogeographic Heterogeneity

The first migration wave, which occurred after the collision of Indian and Eurasian plates ensued a biodiversity boom in the Himalayas, owing to its position at the boundary of two continental plates, creating an ecotone with diverse flora and fauna. The region was favourable to migration after the collision mainly due to two reasons. Firstly, there was a biological vacuum formed due to extinctions during Cretaceous-Tertiary volcanism events during around 65 Mya (Manish and Pandit 2018). Secondly, due to the replacement of Tethyan marine deposition with terrestrial

Table 1.1 Endemism in mainland ecoregions among global biodiversity hotspots

Sl. No.	Biodiversity hotspot	Location	% Plant endemism[a]	Total endemism score[b]
1	Cape floristic region	Africa	69.00	771.4
2	Atlantic forest	South America	40.00	300
3	Tropical Andes	South America	50.00	300
4	Tumbes-Choco-Magdalena	South America	25.00	220.9
5	Mountains of Southwest China	East Asia	29.17	200
6	Western Ghats	South Asia	51.54[c]	190
7	Southwest Australia	Southeast Asia and Asia-Pacific	52.92	180
8	Guinean forests of West Africa	Africa	20.00	176
9	Sundaland	Southeast Asia and Asia-Pacific	60.00	129
10	California Floristic Province	North and Central America	60.89	110
11	Mesoamerica	North and Central America	17.30	100
12	Madrean pine–oak woodlands	North and Central America	75.00	95
13	Maputaland–Pondoland–Albany	Africa	23.46	84.6
14	Succulent Karoo	Africa	38.37	67
15	Forests of East Australia	Southeast Asia and Asia-Pacific	25.97	60
16	Himalaya	South Asia	31.60	60
17	Eastern Afromontane	Africa	31.01	56.5
18	Caucasus	Europe and Central Asia	25.00	50
19	Indo-Burma	South Asia	51.85	41.5
20	Irano-Anatolian	Europe and Central Asia	41.67	40
21	Mediterranean Basin	Europe and Central Asia	52.00	34
22	Cerrado	South America	44.00	26.2
23	Mountains of Central Asia	Europe and Central Asia	27.27	25
24	Chilean winter rainfall and Valdivian forests	South America	50.28	20
25	Coastal forests of eastern Africa	Africa	43.75	NA[d]
26	Horn of Africa	Africa	55.00	NA[d]
27	North American coastal plain	North and Central America	–	NA[d]

[a]As per Mittermeier et al. (2011)
[b]As per Kier et al. (2009)
[c]Includes data from Sri Lanka also
[d]Ecoregions congruent with the hotspots not represented by Kier et al. (2009)

sedimentation, connectivity was established between the newly joined continental plates (Mehrotra et al. 2005). These favourable conditions aided in the large-scale migration of plants initially flora with tropical affinities into the Himalayan landforms (Pandit et al. 2014). Initial migrants were Sino-Japanese and Malayan plants, such as *Dipterocarpus*, *Shorea*, rosewoods (*Dalbergia*), crape myrtles (*Lagerstroemia*), nutmegs (*Myristica*) and tropical almonds (*Terminalia*) (Singh and Singh 1987). This first wave of voracious migrations lasted for around 30 million years shaping its biodiversity, till it slowed down because of the dissection of connectivity due to the growing Himalayan uplift (Manish and Pandit 2018).

Himalayan Uplift

Geophysical development of the Himalayas after the collision event was mainly characterised by elevation episodes and the onset of monsoons during the period from around 45 Mya to 5 Mya (Manish and Pandit 2018). During this period, an elevational climatic (thereby regional) gradient (tropical, temperate and alpine regions) was established and temperate plants migrated to the Himalayas, ensuing a second migration wave, boosting its biodiversity through new corridors (Tabata 2008). The immigrant taxa in the second wave included high-altitude plants like alder (*Alnus*), birch (*Betula*), cedar (*Cedrus*), fir (*Abies*), larch (*Larix*), spruce (*Picea*) and willow (*Salix*), along with other primitive alpine plants (Mehrotra et al. 2005; Tabata 2008).

1.3.1.3 Post-Formative Factors

Altitudinal and Climatic Variations

Climate and geology are important determinants of mountain biodiversity around the world (Antonelli et al. 2018). The Himalayas exhibit an extraordinary diversity in climatic (temperature and precipitation) and altitudinal factors. A gradual change in climatic factors is observed from west to east, ranging from a subtropical climate in the foothills of the western Himalayas, through a tropical climate in the central Himalayas to a subequatorial climate in the eastern Himalayas (Andrejczuk 2016). The Himalayas also accommodate almost the entire spectra of altitudinal variation in climate, ranging from humid subequatorial, through cold arctic to snow-clad nival zones, over a distance of around 100 km (Andrejczuk 2016). Additionally, the orography profoundly affects the precipitation, and thereby the biodiversity (Singh and Kumar 1997). Precipitation markedly increases from the west (around 500 mm in Afghanistan) to the east (around 3000 mm in Sikkim, India). Climatic fluctuations, including cooling and drying, are more pronounced in the west than in the more tropical east (Rana et al. 2019). Consequently, plant vegetation tends to be more diverse from western to eastern Himalayas (Padmawathe et al. 2004). The Himalayas are also notably more biodiverse in the southern slopes than northern slopes, especially in the central Himalayas, as evidenced by the arid Tibetan plateau. Himalayas pose a massive rain shadow on the northern side, as well as generate *foehn* winds during the monsoon season, which generate a dry belt in the northern leeward side of the Himalayas (Wang et al. 2019).

1.4 Generalised Profile of Vegetation and Plant Biodiversity in the Himalayas

The Himalayan vegetation exhibits spatial and altitudinal variations (Fig. 1.2). This section presents the altitudinal vegetation profile of the Himalayas, with a focus on higher vascular plants. Pandit et al. (2014) observed that western Himalayan floral elements generally have Euro Caucasian-Mediterranean affinities, while vegetation in the eastern Himalayas generally has Sino-Japanese, North American, and Indo-Malayan affinities.

Himalayan foothills (upto 1000–1200 m.a.s.l.) are generally represented by tropical semi-deciduous forests dominated by sal (*Shorea robusta*) and *Terminalia* sp. (Pandit et al. 2014). The region exhibits spatial variation in vegetation from sub-tropical deciduous forests with scrub in the Kashmir Himalayas to tropical evergreen and semi-evergreen forests in the Arunachal Himalayas (Andrejczuk 2016). In addition to the characteristic *Shorea-Terminalia* forests, horn scrub vegetation dominated by *Acacia* sp. and *Zizyphus* sp., steppe vegetation dominated by *Artemisia* sp., and sub-tropical semi-desert vegetation comprising *Capparis* sp., *Calotropis* and grasses are also found in the western Himalayas. Tropical bamboos, represented by *Dendrocalamus* and *Arundinaria* are seen in the central Himalayas. Evergreen and semi-evergreen forests, consisting of *Ficus*, *Syzygium* etc. are seen in the eastern Himalayas (Rau 1974).

Above the foothills (from around 1000–1200 m.a.s.l. to 2000–2200 m.a.s.l.), the forests are generally represented by subtropical longleaf Indian pine/chir pine (*Pinus roxburghii*) and oak (*Quercus* sp.) (Pandit et al. 2014). *Rhododendrons* begin at around 1500 m. Oak-*Rhododendron* (*Quercus incana—Rhododendron arboretum*) forests, with Acer and Aesculus are profuse in the west (Rau 1974). Mixed evergreen forests are also seen in this zone from central Himalayas to the east, comprising of *Alnus*, *Prunus* etc., with *Quercus* sp. dominating the upper belt.

Subtropical and tropical forests transition into temperate forests at higher altitudes (from around 2000–2200 m.a.s.l. to 3500–3800 m.a.s.l.) (Andrejczuk 2016). This region is generally dominated by conifers such as blue pine (*Pinus wallichiana*), deodar (*Cedrus deodara*), spruce (*Picea*) and fir (*Abies*), with plenty of *Rhododendron*-oak forests, *Acer* and *Aesculus* (Pandit et al. 2014). Andrejczuk (2016) observed a decrease in the population of conifers with an increase in the population of deciduous trees in the temperate region. Another notable trend is the domination of *C. deodara* and *Larix griffithiana* in western and eastern landscapes respectively. *Rhododendrons* are also profuse in this region (Rau 1974).

In the sub-alpine zone (from around 3500–3800 m.a.s.l. to 4200–4600 m. a.s.l.), the woody vegetation transitions to a rich meadow flora (Andrejczuk 2016). The treelines of a considerable portion of the Himalayan arc are seen in this region, comprising birch (*Betula utilis*), fir trees (*Abies pindrow*, *A. densa*, *A. spectabilis* etc.), and different species of *Rhododendrons*. Though *Rhododendrons* dominate the shrublands and meadows in this zone, they decline towards the west, being absent in the Kashmir Himalayas (Rau 1974). Notably, the domination of birches increases from east to west (Andrejczuk 2016; Singh et al. 2020). Bushes of *Juniperus communis* and *J. wallichiana*, and other plants such as *Anemone*, *Corydalis*, *Epilobium*, *Gentiana*, *Geranium*, *Meconopsis*, *Polygonum*, *Potentilla*, *Primula*, *Ranunculus* and *Saxifraga* among others are also found in this zone (Rau 1974; Andrejczuk 2016; Manish and Pandit 2018).

The alpine zone (from around 4200–4600 m.a.s.l. to around 5000 m.a.s.l.) is primarily represented by dry alpine scrubs of *Juniperus communis*, dwarf *Rhododendron* and *Cassiope* communities (Pandit et al. 2014). General vegetation limits are at an altitude of 4200 (in the west) to 4600 m.a.s.l. (in the east), though considerable variations occur (Andrejczuk 2016). The Himalayas often exhibit unusually high treelines. *Juniperus tibetica* forms a treeline in Tibet at 4900 m.a.s.l. (Singh et al. 2020). *Rhododendrons* make associations at high altitudes. *Rhododendron anthopogon* and *Berberis* sp. are seen at 4800 m.a.s.l. at the Kumaon Himalayas in the west, while associations of *Rhododendron anthopogon*, *R. nivale*, *Ephedra* and others are seen at above 4500 m.a.s.l. at the North-East Frontier Himalayas in the far east. Other plants observed in this zone include *Aconitum*, *Androsace*, *Aquilegia*, *Arenaria*, *Aster*, *Astragalus*, *Bergenia*, *Corydalis*, *Draba*, *Leontopodium*, *Lloydia*, *Paraquilegia Potentilla*, *Primula*, *Rheum*, *Saussurea*, *Saxifraga*, *Sedum*, *Sibbaldia* and *Trollius* (Rau 1974).

The Himalayas are usually characterised by a nival zone or zone of perpetual snow, ice and bare rocks by altitudes of around 5000 m.a.s.l. Even though the region is highly inhospitable to plant communities, certain interesting cold tolerance plant species have been sampled from this zone. *Eremogone bryophylla* was sampled between 6100 and 6200 m.a.s.l. by Wollaston during the first British Reconnaissance Expedition to Mount Everest (1921) (Dentant 2018). Al-Shehbaz (2000) described a new species, *Lepidostemon everestianus* from a sample collected from 6400 m.a.s.l. during the 1935 British expedition to the Mount Everest, led by Eric E. Shipton (Al-Shehbaz 2000). Species of *Leontopodium* and *Sedum* were recorded from altitudes of 6000 m.a.s.l. (Rau 1974).

1.5 Plant Species Richness in the Himalayas

Species richness, the count of distinct species within a particular area or sample is a commonly used index for assessing plant diversity in a region. It helps in the planning and management of conservation-related policies and programs (Fedor and Zvaríková 2019). Plant diversity in the Himalayan region has been a topic of

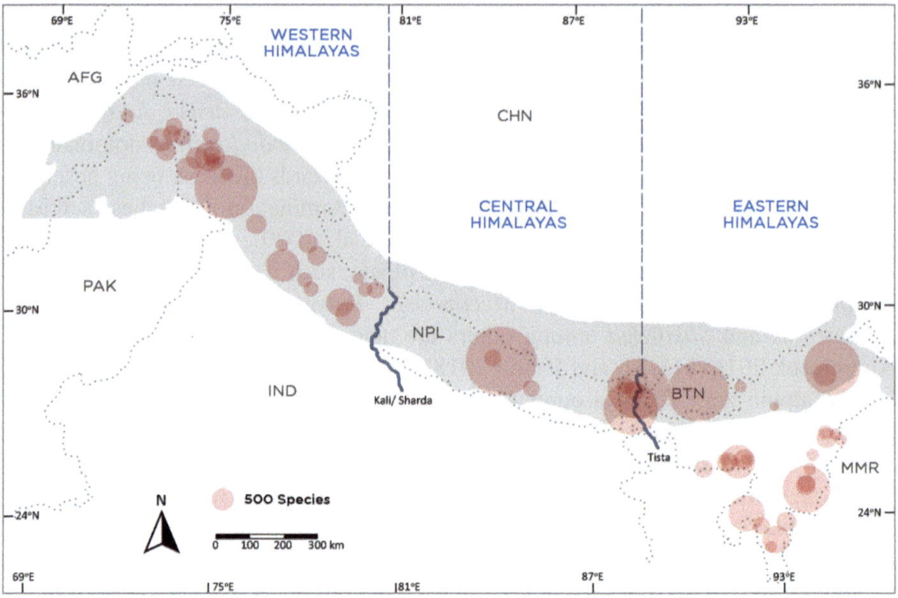

Fig. 1.3 Bubble map representing species richness across the Himalayan region. Representation is based on available literature and is not exhaustive. Countries have been represented by their ISO 3166-1 alpha-3 codes. (Illustrated by VWK)

research for many decades. Indian Himalayas alone host about 18,440 species of plants (Mir et al. 2020). Rana et al. (2019) observed that species richness generally declines three times from the eastern to western Himalayas. Quantitatively, the scaled density of species richness was highest in the eastern (5.30) Himalayas, followed by the central (4.31) and western (2.07) Himalayas (Joppa et al. 2013). It was also observed that while species richness gradually declines with elevation in the western Himalayas, tree richness peaks at around 1000 m.a.s.l., and shrub richness peaks at around 2000 m.a.s.l. in the eastern Himalayas. Though there is a growing interest to study the floristics of the Himalayan region, the currently available data on species richness of the Himalayas is non-exhaustive due to uneven terrain, extreme climatic conditions, and geopolitical restrictions, thereby making interpretations difficult (Fig. 1.3).

1.5.1 Plant Species Richness in the Western Himalayas

Plant richness of the western Himalayas has been comprehensively documented in many regional floras (Collett 1921; Nair 1977; Osmaston 1978; Naithani 1984; Aswal and Mehrotra 1994; Dhaliwal and Sharma 1999; Singh and Rawat 2000; Kaur and Sharma 2004; Singh 2006, 2018).

Naran Valley of Pakistan, located in the Irano-Turanian region is a critical botanical transition zone which harbours 198 plant species (Khan et al. 2014b). In independent studies, 252, 354, 154, 114 and 42 plant species were enumerated respectively from the Thandiani sub-forests division (Khan et al. 2016), Manoor valley (Rahman et al. 2019), Kaghan valley (Hashim et al. 2023), Yakhtangay hills (Ullah et al. 2022), and Nandiar valley (Haq et al. 2010) of Pakistan.

Floristic studies and regional floras from Kashmir (union territories Jammu & Kashmir, and Ladakh; erstwhile state of Jammu & Kashmir) reported 879 (species of) angiosperms, 1065 algae (Kaloo and Amin 2020), 420 bryophytes (Ismail et al. 2020), 424 lichens (Khare et al. 2020), 207 pteridophytes, and 41 gymnosperms (Dar and Dar 2020). In a recent study, Dar and Khuroo (2020) enumerated 5056 taxa of angiosperms. Floristic studies from Overa Wildlife Sanctuary (Aijaz et al. 2022), Tangmarg forest division (Altaf et al. 2022), Dachigam National Park (Haq et al. 2021) and Gurez valley (Haq et al. 2023a) enumerated 37, 74, 190 (diversity index of 3.09) and 155 plant species respectively. 364 vascular plants were identified from Gulmarg Wildlife Sanctuary (Wani and Pant 2023). Studies have also covered Keran valley (183 sp.) (Haq et al. 2019), Shawilks mountain range (116 sp.; 46 families) (Haq et al. 2023b), and Poonch valley (430 sp.) (Khan et al. 2014a).

Himachal Pradesh is floristically rich. The first volume of the state flora enumerates 1202 species belonging to 85 families from *Ranunculaceae* to *Caprifoliaceae* (Chowdhery and Wadhwa 1984). Masoodi and Sundriyal (2020) documented 811 species of non-timber forest produced from the mountains of Himachal Pradesh. Studies were carried out from Bhabha Valley (313 sp.) (Chawla et al. 2008), Dhauladhar mountains (255 sp.) (Ahmad et al. 2020), Shimla water catchment sanctuary (476 sp. of vascular plants) (Rana et al. 2022), Shikari Devi alpine pasture (62 sp.) (Kumar and Chauhan 2024), Chandi beat of Majathal Wildlife Sanctuary (97 genera) (Kumari et al. 2023), and Sangla valley (320 sp. of angiosperms) (Sharma et al. 2014) in Himachal Pradesh.

Several floristic studies have been carried out in Uttarakhand with major emphasis on vascular plants and lower plants being relatively understudied (Bargali et al. 2022). The state flora documented 854 angiosperms and 19 gymnosperms (Pusalkar and Srivastava 2018). Uniyal et al. (2007) documented 4700 species (1523 genera) of plants from the state. Rawat et al. (2016) documented 651 spermatophytes (444 genera) from the montane zone in western Ramganga valley. Pteridophyte diversity was assessed from Mussoorie Dehradun region (112 sp.) (Stewart 1942), Valley of Flowers National Park (30 sp.) (Wadhwa et al. 2024), Chakrata (103 sp.) (Kumar and Raizada 1989), and Nanda Devi National Park (78 sp.) (Samant et al. 2005). 237 species of orchids, including two endemic taxa, were documented from Gori valley (Jalal et al. 2018).

1.5.2 Plant Species Richness in the Central Himalayas

A recent updated floristic study from Nepal reveals the presence of 6653 species of angiosperms, 28 gymnosperms, 534 pteridophytes, 1150 bryophytes, 2025 fungi, 771 lichens and 807 algae (Kunwar et al. 2010). Rajbhandari and Rai (2017) initially listed 5309 species of flowering plants, which was subsequently expanded (Rajbhandari and Rai 2019) with 1600 additional species. Region-specific studies from the Nepal Himalayas include enumeration of 189 taxa from Daman-Simbhanjyang Area (Thapa Magar and Chaudhary 2022), 198 orchid species from Annapurna conservation area (Tiruwa et al. 2024) and 229 species of trees from lower Kanchenjungha Singhalila ridge. A plant diversity assessment revealed higher species richness in areas of low erosion and high anthropogenic pressure in mid-elevational forests of the central Himalayas (Khera et al. 2001).

1.5.3 Plant Species Richness in the Eastern Himalayas

Bhutan had botanically been a terra incognita for a very long period (Rau 1974). Rana et al. (2019) stated that Bhutan has the highest angiosperm diversity in the Himalayas with 5452 species.

In Sikkim, Kangchenjunga Landscape (KL) is among the richest areas in terms of species diversity, though the studies are not exhaustive. A recent systematic review of KL plant richness mentions 5198 species spread over 216 families, with *Orchidaceae* (704 species) being the most diverse family (Kandel et al. 2019). *Pinaceae* (10 species) was recorded as the most species-rich family among gymnosperms. KL-Darjeeling and KL-Bhutan regions were also reported as species-rich areas, with Darjeeling alone contributing to 4000 species distributed among 160 families of flowering plants (Kandel et al. 2019). A detailed vegetation analysis among the tree ecotone species along the Tuksam-Dzongri transect in Sikkim reported 267 species belonging to 81 families (Kumar and Pandey 2023).

Several species richness studies have been carried out in Arunachal Pradesh. Rana et al. (2019) enumerated 4039 species of angiosperms, while Ambrish (2013) estimated around 5000 plant species from the ecologically significant mountain ranges of Arunachal Pradesh. 482 plant species, including 421 angiosperm species belonging to 129 families were enumerated by Saikia et al. (2017) from Himalayan forests in the state. 57 species of trees were identified by diversity assessment using remote sensing satellite images from West Kameng and Tawang districts (Kumar et al. 2019). Recent studies identified high species diversity within certain genera such as *Ficus* (21 species in Pakke Wildlife Sanctuary) and *Musa* (21 species from different regions) in Arunachal Pradesh (Ranibala et al. 2018; Parbo et al. 2024).

Floristic data of Dibru Saikhowa National Park in Assam was generated as a part of the M.Phil. work of Gogoi (1997), and a comprehensive flora of the region was generated by Baishya and Bora (2002).

Though the boundaries of the Himalayas do not include the Indian north-eastern hill states to the south of Assam and Arunachal Pradesh, there exists a significant phytogeographical continuity due to similarity in biomes, terrains, plant migration and distribution (Pandit et al. 2014). Therefore, a description of species richness in the eastern Himalayas would be incomplete without the inclusion of these states.

A recent study from Garo hills in Meghalaya has enumerated 257 plant species belonging to 83 families from Siju Wildlife Sanctuary (Roy et al. 2014). Other studies from the state have enumerated 287 woody species from 6 districts (Lasushe et al. 2024), 35 plant species from west Khasi hills (Jyndiang et al. 2021), and 30 tree species from east Khasi hills (Dkhar and Wani 2021). Floristic studies from Swer (Mishra et al. 2004), Ialong and Railang (Upadhaya et al. 2003) sacred groves have enumerated 63 angiosperm families, 82 species and 80 species (across 62 genera) respectively. 395 plant species were recorded from Khloo Paiu Ram Pyrthai, Urkhla, and Khloo Langdoh sacred grooves (Jamir and Pandey 2003). 1300 species belonging to 645 genera have been enumerated from 82 community-managed forests of Khasi hills (Mir et al. 2022).

The Flora of Mizoram includes 884 plant species belonging to 90 angiosperm families (Singh et al. 2002). A total of 67 and 63 different plant species were documented from disturbed and undisturbed areas of semi-evergreen forest of Lunglei district (Lalfakawma et al. 2009). Floristic studies from Pala wetland reserve forest (Nohro and Jayakumar 2020), Murlen National Park (Sengupta and Sekhar Dash 2023), Reiek forest in Mamit district (Lalzarzovi and Lalnuntluanga 2022), and Lalsavunga park (Ao et al. 2023) enumerated 65 (from 31 families), 321 (from 73 families) and 213 plant species, and 41 tree species respectively.

Nagaland hosts proficient plant diversity in Mt. Saramati, Patkai and Barial hill ranges. Floristic studies from Tuensang district (Panda 2012), Fakim Wildlife Sanctuary (Ao et al. 2020), sub-tropical forest belt in Mokokchung (Ao et al. 2021), and riparian vegetation along Dzücha river of Kohima (Keretsu et al. 2022) enumerates 30 species (from *Ericaceae* family), 60 species (27 families), 118 species (84 genera, 43 families), and 31 species respectively. Family and genus-specific enumerations from the state include the documentation of 378 species of rare and endangered orchids (Changkija et al. 1994) and 14 wild species of *Musa* (Deb et al. 2023).

The Flora of Manipur documented around 3000 higher plants belonging to 205 families, with highest representations in *Orchidaceae* (249 sp.), *Poaceae* (200 sp.) and *Fabaceae* (168 sp.) respectively (Singh et al. 2000). Floristic studies from Kangchup hills (Kikim et al. 2023), Langol hills (Shalini 2021), and Senapati district (Meetei et al. 2017) have documented 195 plant species (145 genera), 282 plant species (210 genera and 82 families), and 43 tree species respectively. In another study, the species richness index was found to be highest in Konthoujam Lairembi sacred grove, when compared to other sacred groves in Manipur (Ashalata Devi et al. 2005).

Deb (1981) recorded 1545 species (862 genera and 193 families) from Tripura. Majumdar et al. (2012) enumerated 144 species (103 genera, 46 families) from the moist deciduous forests in Tripura.

1.6 Plant Endemism in the Himalayas

An early assessment of plant endemism by Chatterjee (1939) suggested 3165 endemic species in the Himalayas, representing 46% of all endemic species in India and Burma. Later reassessments (Irwin and Narasimhan 2011; Irwin et al. 2021) suggested the Himalayas contribute to 12% of all endemic genera in India.

A recent study by Tiwari et al. (2024) recorded a total of 1076 endemic species of plants from the Indian Himalayan Region (IHR), with eastern Himalayas accommodating the larger fraction of endemic plants. Eastern (scaled endemism density of 4.07; scaled to 100,000 plant species and 1000 km^2) and central (3.43) Himalayas, were included in the identified endemism regions of the world (Joppa et al. 2013), while western Himalayas (1.83) were not included. As per an assessment by the Botanical Survey of India, there is proficient plant endemism in eastern (466 taxa), central (314 taxa) and western (297 taxa) Himalayas (Singh et al. 2015). Eight pteridophytes and four gymnosperms are also reportedly endemic to the Himalayas (Chadha et al. 2008; Singh and Srivastava 2013).

1.6.1 Plant Endemism in the Western Himalayas

Singh et al. (2015) enumerated 297 taxa of endemic plants from the western Himalayas. A study on the endemism of *Poaceae* in the western Himalayas reported 13 species which were strictly endemic to the region (Yadav et al. 2020). The genus *Kashmiria* (*Plantaginaceae*) is distributed in Western Himalayan subdivisions (Irwin et al. 2021).

153 species of plants in the Kashmir Himalayas are endemic (Mir et al. 2020). In a study conducted among high altitude threatened medicinal plants in Himachal Pradesh, 44 species were seen to be critically endangered, endangered, vulnerable or endemic (Kumar et al. 2017). According to Bargali et al. (2022), Uttarakhand hosts 93 species of flowering plants, 15 species of ferns, and 18 species of gymnosperms which are endemic to the state. The western Himalayan endemic genera *Ivanjohnstonia* (*Boraginaceae*) and *Pseudodanthonia* (*Poaceae*) were reported from Uttarakhand (Irwin et al. 2021).

1.6.2 Plant Endemism in the Central Himalayas

Shrestha and Joshi (1996) reported 246 species of endemic plants from Nepal. Later, the Department of Plant Resources (DPR), Ministry of Forests and Environment, Government of Nepal (GoN) updated the number to 289 (DPR 2013). After recent revisions, currently DPR, GoN estimates the number of endemic species in Nepal to be 312 across 126 genera and 46 families (Rajbhandari and Rai 2017). Among angiosperms, *Apiaceae* is the largest family contributing to endemism in Nepal (Tiwari et al. 2019). Pradhan and Shrestha (2021) recognised 30 species of endemic bryophytes endemic to the region. *Asplenium pseudofugax, Bolbitis andreisii* and *Polystichum annapurnicola* are the known species of endemic pteridophytes of Nepal (Kandel 2020).

1.6.3 Plant Endemism in the Eastern Himalayas

Eastern Himalayan region harbours the greatest number of endemic species in Himalayan ranges as per the available floristic reports. In a recent study by Tiwari et al. (2024), Arunachal Pradesh, Meghalaya and Sikkim were identified as the states with the greatest endemic plant richness in the IHR. Irwin et al. (2021) reported the eastern Himalayan endemic genus *Parakaempferia* (*Zingiberaceae*) from Arunachal Pradesh, Assam and Meghalaya.

Tiwari et al. (2024) report 294 endemic plants from Arunachal Pradesh. Pal (2013) documented 37 plant species (out of 170) as endemic in the Lower Subansiri District, Arunachal Pradesh. The remarkable endemic biodiversity of the state can be attributed to its unique phytogeographical position sharing Burma-Sino-Malaysian floras. Notably, Arunachal Pradesh shares endemic species common with Assam (9 sp.), Meghalaya (17 sp.) and Sikkim Himalayas (21 sp.). The endemic genus *Stapletonia* (*Poaceae*) is restricted to Arunachal Pradesh (Irwin et al. 2021).

Tiwari et al. (2024) reported 213 endemic plants from Meghalaya. In the Flora of Jowai, Balakrishnan (1981) identified 51 species endemics to Meghalaya. Mir et al. (2019) recorded that 548 plant taxa distributed in Meghalaya are endemic to the eastern Himalayas or Indo-Burma, with 115 species being restricted to Meghalaya. Lasushe et al. (2022) reported that 70 out of 385 threatened plant species from Meghalaya are endemic to the eastern Himalayas or Indo-Burma, with 24 being restricted to Meghalaya.

A total of 212 endemic plants were reported from Sikkim by Tiwari et al. (2024). Ghosh et al. (2022) reported 93 flowering plant taxa restricted to Sikkim. The endemic genus *Brachycaulos* (*Rosaceae*) is also restricted to Sikkim (Irwin et al. 2021). Deb (1981) identified that a total of 36 species endemic to the eastern Himalayas are distributed in Tripura. Two species share restricted distribution between Tripura, Mizoram and Chittagong, and sixteen taxa are distributed only in Assam and Tripura. State forest report of Mizoram has reported 27 species (16 families) as endemic to Mizoram (Singh et al. 2016). Mao and Gogoi (2016) have enumerated 21 taxa endemic to Nagaland in the Flora of Dziiko/Dzukou valley. Darjeeling hosts 405 dicot and 76 monocot species that are endemics to the Himalayas, among which 29 dicots and 76 monocots are exclusive to Darjeeling (Bhujel 1996; Nirola and Das 2017).

The Flora of Bhutan reported 212 species endemics to Bhutan (Grierson and Long 1983). After reports on the extended distribution records of several endemic plants across political boundaries, the total endemics are now estimated at 144 species (Yangzom and Thomas 2017).

1.7 Conclusions

The Himalayas have deep-rooted cultural importance beyond their floristic and endemic significance. It carved the livelihood of a large number of inhabitants in the drainage basins of its rivers across geographical and political boundaries. Settlements in the Himalayan foothills have co-evolved with the hills, depending on them for numerous ecosystem and biodiversity services. Amid all the cultural, geographical and biological significance, it is well known that the Himalayas are under threat and there is an immediate need for taking significant conservation measures. Climate change, deforestation, unplanned development, overexploitation, invasive species and hunting are some of the age-old yet still relevant factors, which are posing threats to the Himalayan biodiversity. Gaps in the understanding of floristics in several Himalayan regions also need attention. A clear floristic picture is important for designing sustainable conservation strategies and natural resource management.

Acknowledgements The authors acknowledge the National Medicinal Plants Board (NMPB), Ministry of AYUSH, Government of India for the funding. Sincere thanks are due to Rabinarayan Acharya, MD, PhD (Director General, CCRAS, New Delhi), Chitra Subramani, PhD (Assistant Director In-charge, CSMCARI, Chennai), and Vilas Gangurde, MD (Research Officer In-charge, RARC Dimapur) for their encouragement and support.

Conflict of Interest The authors express no conflicting or competing interests.

Funding Information This study was supported by the National Medicinal Plants Board (NMPB), Ministry of AYUSH, Government of India (Grant No. F.No.Z.18017/187/CSS/R&D/AS-01/2021-22-NMPB-IV A, dated 21.01.2022).

References

Ahmad M, Uniyal SK, Batish DR et al (2020) Patterns of plant communities along vertical gradient in Dhauladhar Mountains in lesser Himalayas in North-Western India. Sci Total Environ 716: 136919. https://doi.org/10.1016/j.scitotenv.2020.136919

Aijaz S, Ahmad K, Qaisar K et al (2022) Floristic diversity under different habitats in Overa wildlife sanctuary of Jammu & Kashmir, India. Indian Forester 148:1104. https://doi.org/10.36808/if/2022/v148i11/164125

Al-Shehbaz IA (2000) *Lepidostemon* (Brassicaceae) is no longer monotypic. Novon 10:329. https://doi.org/10.2307/3392977

Altaf A, Haq SM, Shabnum N, Jan HA (2022) Comparative assessment of Phyto diversity in Tangmarg Forest division in Kashmir Himalaya, India. Acta Ecol Sin 42:609–615. https://doi.org/10.1016/j.chnaes.2021.04.009

Ambrish K (2013) Floristic diversity of Arunachal Pradesh upper Subansiri District. Bishen Singh Mahendra Pal Singh, Dehradun

Andrejczuk W (2016) Himalaje: Szkic Fizycznogeograficzny—Biota, Piętra Roślinne I Krajobrazy. Acta Geogr Silesiana 24:29–49

Antonelli A, Kissling WD, Flantua SGA et al (2018) Geological and climatic influences on mountain biodiversity. Nat Geosci 11:718–725. https://doi.org/10.1038/s41561-018-0236-z

Ao A, Changkija S, Tripathi SK (2020) Species diversity, population structure, and regeneration status of trees in Fakim Wildlife Sanctuary, Nagaland, Northeast India. Biodiversitas J Biol Divers 21(6):2777–2785. https://doi.org/10.13057/biodiv/d210654

Ao A, Changkija S, Tripathi SK (2021) Stand structure, community composition and tree species diversity of sub-tropical forest of Nagaland, Northeast India. Trop Ecol 62:549–562. https://doi.org/10.1007/s42965-021-00170-5

Ao A, Mere K, Tripathi SK (2023) Stand structure, regeneration potential and biomass carbon stock of subtropical Forest of Mizoram, Northeast India. Indian J Ecol 50:1972–1979. https://doi.org/10.55362/IJE/2023/4162

Apollo M, Andreychouk V, Mostowska J et al (2021) Himalayan nature-based tourism: potential, regional diversity, nature conservation and touristic load. Int J Conserv Sci 12:1527–1546

Ashalata Devi K, Khan ML, Tripathi RS (2005) Sacred groves of Manipur, Northeast India: biodiversity value, status and strategies for their conservation. Biodivers Conserv 14:1541–1582. https://doi.org/10.1007/s10531-004-0530-5

Aswal BS, Mehrotra BN (1994) Flora of Lahaul-Spiti. Bishen Singh Mahendra Pal Singh, Dehradun

Bahadur B, Krishnamurthy KV, Pullaiah T (2018) Introduction. In: Ethnobotany of India, Volume 4: Western and Central Himalaya. Apple Academic Press, New York, p 496

Baishya AK, Bora PJ (2002) Dibru—Saikhowa biosphere reserve. In: Singh NP, Singh KP (eds) Floristic diversity and conservation strategies in India. Botanical Survey of India, Calcutta, pp 2413–2456

Balakrishnan NP (1981) Flora of Jowai and Vicinity, Meghalaya: a contribution towards a detailed knowledge of the flora of the Northeastern Region of India. Botanical Survey of India, Howrah

Bargali H, Kumar A, Singh P (2022) Plant studies in Uttarakhand, Western Himalaya–a comprehensive review. Trees For People 8:100203. https://doi.org/10.1016/j.tfp.2022.100203

Bhujel R (1996) Studies on the dicotyledonous flora of Darjeeling district. University of North Bengal

Braxmeier H (2024) Relief Map. https://maps-for-free.com/. Accessed 18 Nov 2024

Burrard SG, Hayden HH (1908) A sketch of the geography and geology of the Himalaya Mountains and Tibet. Superindentent Government Printing, India, Culcutta

Chadha J, Chander H, Kumar B (2008) Endemic pteridophytes of India: distribution and threat status. In: Rawat GS (ed) Special habitats and threatened plants of India. Wildlife Institute of India, Dehradun

Changkija S, Kumar Y, Gurung PB (1994) Orchids of Nagaland. Department of Forests, Ecology, Environment & Wildlife, Govt. of Nagaland, Kohima

Chatterjee D (1939) Studies on the endemic flora of India and Burma. J R Asiat Soc Bengal 5:19–67

Chatterjee S, Scotese CR (1999) The breakup of Gondwana and the evolution and biogeography of the Indian plate. Pinsa 65:397–425

Chawla A, Rajkumar S, Singh KN et al (2008) Plant species diversity along an altitudinal gradient of Bhabha Valley in western Himalaya. J Mt Sci 5:157–177. https://doi.org/10.1007/s11629-008-0079-y

Chowdhery HJ, Wadhwa BM (1984) Flora of Himachal Pradesh, vol 1. Botanical Survey of India, Howrah

Collett H (1921) Flora simlensis: a handbook of the flowering plants of Simla and the neighbourhood. Thacker, Spink & Co, Calcutta and Simla
Dar AR, Dar GH (2020) Diversity of gymnosperms in Jammu and Kashmir State. In: Biodiversity of the Himalaya: Jammu and Kashmir State. Springer Singapore, Singapore, pp 449–466
Dar GH, Khuroo AA (2020) An updated taxonomic checklist of angiosperms in Jammu and Kashmir state. In: Biodiversity of the Himalaya: Jammu and Kashmir State. Springer Singapore, Singapore, pp 467–519
Deb DB (1981) The flora of Tripura State, vol 1. Today and Tomorrow's Printers and Publishers, New Delhi
Deb CR, Tiatemsu P, Vupru TN, Paul A (2023) Diversity and distribution of wild Musa in Nagaland, India. Open J For 13:315–337. https://doi.org/10.4236/ojf.2023.133019
Dentant C (2018) The highest vascular plants on earth. Alp Bot 128:97–106. https://doi.org/10.1007/s00035-018-0208-3
Dhaliwal DS, Sharma M (1999) Flora of Kullu district, Himachal Pradesh. Bishen Singh Mahendra Pal Singh, Dehradun
Dhital MR (2015) Geology of the Nepal Himalaya. Springer, Cham
DiPietro JA, Pogue KR (2004) Tectonostratigraphic subdivisions of the Himalaya: a view from the west. Tectonics 23(5):119. https://doi.org/10.1029/2003TC001554
Dkhar DE, Wani AM (2021) Assessment of plant biodiversity at 3 different elevation of protected forest of subtropical pine forest of Meghalaya. Int J Curr Microbiol Appl Sci 10:157–169. https://doi.org/10.20546/ijcmas.2021.1011.019
DPR (2013) Plants of Nepal: fact sheet. Department of Plant Resources (DPR), Kathmandu
Dwivedi S, Singh V, Mahra K et al (2024) Functional foods in the northwestern Himalayan region of India and their significance: a healthy dietary tradition of Uttarakhand and Himachal Pradesh. J Ethn Foods 11:20. https://doi.org/10.1186/s42779-024-00236-4
Fedor P, Zvaríková M (2019) Biodiversity indices. In: Fath B (ed) Encyclopedia of ecology. Elsevier, Amsterdam, pp 337–346
Ghosh J, Das DS, Midday M et al (2022) The status of endemic angiosperms of Sikkim (India). Pleione 16:271–277. https://doi.org/10.26679/Pleione16.3.2022.271-277
Gogoi AB (1997) The study of the herbaceous plants of Tinsukia sub-div. of Dibrugarh District, Assam with ref. to their taxonomy & utilization. Gauhati University, Assam
Grierson AJC, Long DG (1983) Flora of Bhutan: including a record of plants from Sikkim. Royal Botanic Garden, Edinburgh
Haq FU, Ahmad H, Alam M et al (2010) Species diversity of vascular plants of Nandiar Valley, Western Himalayas, Pakistan. Pakistan J Bot 42:213–229
Haq SM, Malik AH, Khuroo AA, Rashid I (2019) Floristic composition and biological spectrum of Keran – a remote valley of northwestern Himalaya. Acta Ecol Sin 39:372–379. https://doi.org/10.1016/j.chnaes.2018.12.001
Haq SM, Shah AA, Yaqoob U, Hassan M (2021) Floristic quality assessment index of the Dagwan stream in Dachigam National Park of Kashmir Himalaya. Proc Natl Acad Sci India Sect B Biol Sci 91:657–664. https://doi.org/10.1007/s40011-021-01247-w
Haq SM, Khoja AA, Lone FA et al (2023a) Floristic composition, life history traits and phytogeographic distribution of forest vegetation in the Western Himalaya. Front For Glob Chang 6:1–17. https://doi.org/10.3389/ffgc.2023.1169085
Haq SM, Waheed M, Bussmann RW, Arshad F (2023b) Vegetation composition and ecological characteristics of the forest in the Shawilks Mountain range from Western Himalayas. Acta Ecol Sin 43:798–809. https://doi.org/10.1016/j.chnaes.2022.10.008
Hashim M, Bushra E, Dasti AA et al (2023) Exploration of species diversity and vegetation pattern in temperate conifer forests along altitudinal gradients in the Western Himalayas. Front For Glob Chang 6:1–14. https://doi.org/10.3389/ffgc.2023.1195491
Irwin SJ, Narasimhan D (2011) Endemic genera of angiosperms in India: a review. Rheedea 21:87–105

Irwin SJ, Narasimhan D, Rekha G (2021) Reassessment of endemic angiosperm genera in India. Rheedea 31:267. https://doi.org/10.22244/rheedea.2021.31.04.04

Ismail Z, Khuroo AA, Bhat MY et al (2020) An updated checklist of bryophytes in Jammu and Kashmir state. In: Dar GH, Khuroo AA (eds) Biodiversity of the Himalaya: Jammu and Kashmir State. Springer, Singapore, pp 379–414

Jalal JS, Kala A, Bisht S et al (2018) Orchids of Gori valley. Wildlife Institute of India, Dehradun

Jamir SA, Pandey HN (2003) Vascular plant diversity in the sacred groves of Jaintia Hills in Northeast India. Biodivers Conserv 12:1497–1510. https://doi.org/10.1023/A:1023682228549

Joppa LN, Visconti P, Jenkins CN, Pimm SL (2013) Achieving the convention on biological diversity's goals for plant conservation. Science 341:1100–1103. https://doi.org/10.1126/science.1241706

Jyndiang P, Lal EP, Khare N, Sen S, Wani AM (2021) Phytodiversity of threatened ethno medicinal plant species in West Khasihills District of Meghalaya, India. Int J Curr Microbiol Appl Sci 10(12):139–146. https://doi.org/10.20546/ijcmas.2021.1012.015

Kaloo ZA, Amin S (2020) Algal diversity in Jammu and Kashmir State. In: Dar GH, Khuroo AA (eds) Biodiversity of the Himalaya: Jammu and Kashmir State. Springer Singapore, Singapore, pp 285–317

Kandel DR (2020) Pteridophytes of Nepal. In: Siwakoti M, Jha PK, Rajbhandary S, Rai SK (eds) Plant diversity of Nepal. Botanical Society of Nepal, Kathmandu, pp 10–22

Kandel P, Chettri N, Chaudhary RP et al (2019) Plant diversity of the Kangchenjunga landscape, eastern Himalayas. Plant Divers 41:153–165. https://doi.org/10.1016/j.pld.2019.04.006

Karan PP, Mather C (1976) Art and geography: patterns in the Himalaya. Ann Assoc Am Geogr 66: 487–512. https://doi.org/10.1111/j.1467-8306.1976.tb01106.x

Kaur H, Sharma M (2004) Flora of Sirmaur, Himachal Pradesh. Bishen Singh Mahendra Pal Singh, Dehradun

Kayastha RB, Kayastha R (2020) Glacio-Hydrological Degree-Day Model (GDM) useful for the Himalayan River basins. In: Himalayan weather and climate and their impact on the environment. Springer, Cham, pp 379–398

Keretsu K, Singh MR, Temjen W (2022) Herbaceous riparian ecotonal species along the Dzücha River, Kohima, Nagaland and their ethnomedicinal uses. Agric Sci Dig 45(5):891–896. https://doi.org/10.18805/ag.D-5558

Khan MA, Khan MA, Hussain M, Mujtaba G (2014a) Plant diversity and conservation status of Himalayan region Poonch Valley Azad Kashmir (Pakistan). Pak J Pharm Sci 27:1215–1239

Khan SM, Page S, Ahmad H, Harper D (2014b) Ethno-ecological importance of plant biodiversity in mountain ecosystems with special emphasis on indicator species of a Himalayan Valley in the northern Pakistan. Ecol Indic 37:175–185. https://doi.org/10.1016/j.ecolind.2013.09.012

Khan W, Khan SM, Ahmad H et al (2016) Vegetation mapping and multivariate approach to indicator species of a forest ecosystem: a case study from the Thandiani sub Forests Division (TsFD) in the Western Himalayas. Ecol Indic 71:336–351. https://doi.org/10.1016/j.ecolind.2016.06.059

Khare R, Upreti DK, Haq MU, Behera BC (2020) Diversity of lichens in Jammu and Kashmir state. In: Dar GH, Khuroo AA (eds) Biodiversity of the Himalaya: Jammu and Kashmir state. Springer Singapore, Singapore, pp 343–377

Khera N, Kumar A, Ram J, Tewari A (2001) Plant biodiversity assessment in relation to disturbances in mid-elevational forest of central Himalaya, India. Trop Ecol 42:83–95

Kier G, Kreft H, Tien ML et al (2009) A global assessment of endemism and species richness across Island and mainland regions. Proc Natl Acad Sci USA 106:9322–9327. https://doi.org/10.1073/pnas.0810306106

Kikim A, Khan MR, Thokchom A (2023) Plant diversity assessment in sub tropical mixed forests of Kangchup Hills, Manipur. Int J Sci Res 12:25–27

Kumar A, Chauhan H (2024) Plant diversity status in the alpine pasture of Shikari Devi, district Mandi, Himachal Pradesh. Ann Arid Zone 63:53–60. https://doi.org/10.59512/aaz.2024.63.1.6

Kumar D, Pandey A (2023) Vegetation structure along an elevation gradient at the treeline ecotone of eastern Himalayan forests in Sikkim. In: Singh SP, Reshi ZA, Joshi R (eds) Ecology of Himalayan treeline ecotone. Springer Nature Singapore, Singapore, pp 247–265

Kumar S, Raizada MB (1989) Survey and distribution of fern-allies and ferns of Chakrata Forest Division. Uttar Pradesh. Indian For 115:578–589

Kumar P, Manikandan R, Srivastava SK, Panwar GS (2017) Conservation status and uses of high altitude threatened m edicinal plants, Himachal Pradesh. Phytotaxonomy 16:115–121

Kumar A, Kishore BSPC, Saikia P et al (2019) Tree diversity assessment and above ground forests biomass estimation using SAR remote sensing: a case study of higher altitude vegetation of North-East Himalayas, India. Phys Chem Earth Parts A/B/C 111:53–64. https://doi.org/10.1016/j.pce.2019.03.007

Kumari K, Verma RK, Kumar R, Thakur A (2023) Study on arboreal floristic diversity, phytosociology and conservation strategy of endangered species in Chandi beat of Majathal wildlife sanctuary, Himachal Pradesh, Western Himalaya. Int J Environ Clim Chang 13:3263–3280. https://doi.org/10.9734/ijecc/2023/v13i102994

Kunwar RM, Shrestha K, Dhungana SK et al (2010) Floral biodiversity of Nepal an update. J Nat Hist Mus 25:295–311

Lalfakawma L, Sahoo UK, Roy S et al (2009) Community composition and tree population structure in undisturbed and disturbed tropical semi-evergreen forest stands of North-East India. Appl Ecol Environ Res 7:303–318. https://doi.org/10.15666/aeer/0704_303318

Lalzarzovi ST, Lalnuntluanga (2022) Plant species diversity in a tropical semi-evergreen forest in Mizoram (northeastern India): assessing the effectiveness of community conservation. J Threat Taxa 14:21055–21067. https://doi.org/10.11609/jott.7549.14.5.21055-21067

Lasushe K, Mir AH, Singh PP et al (2022) A comprehensive checklist of threatened plants of Meghalaya, Northeast India. J Asia-Pacific Biodivers 15:435–441. https://doi.org/10.1016/j.japb.2022.03.008

Lasushe KU, Singh PP, Chaudhary KL et al (2024) Diversity, biomass and carbon stock of Woody species in community reserves of Meghalaya, Northeast India. Int J Ecol Environ Sci 50:757–769. https://doi.org/10.55863/ijees.2024.0200

Majumdar K, Shankar U, Datta BK (2012) Tree species diversity and stand structure along major community types in lowland primary and secondary moist deciduous forests in Tripura, Northeast India. J For Res 23:553–568. https://doi.org/10.1007/s11676-012-0295-8

Manish K, Pandit MK (2018) Geophysical upheavals and evolutionary diversification of plant species in the Himalaya. PeerJ 6:e5919. https://doi.org/10.7717/peerj.5919

Mao AA, Gogoi R (2016) Flora of Dziiko/Dzukou Valley. Botanical Survey of India, Kolkata

Masoodi HUR, Sundriyal RC (2020) Richness of non-timber forest products in Himalayan communities—diversity, distribution, use pattern and conservation status. J Ethnobiol Ethnomed 16:56. https://doi.org/10.1186/s13002-020-00405-0

Meetei S, Das A, Singh E (2017) Tree species composition and diversity in subtropical forest of Manipur, North-East India. Indian For 143:1169–1176

Mehrotra RC, Liu XQ, Li CS et al (2005) Comparison of the tertiary flora of Southwest China and Northeast India and its significance in the antiquity of the modern Himalayan flora. Rev Palaeobot Palynol 135:145–163. https://doi.org/10.1016/j.revpalbo.2005.03.004

Mir AH, Upadhaya K, Roy DK et al (2019) A comprehensive checklist of endemic flora of Meghalaya, India. J Threat Taxa 11:14527–14561. https://doi.org/10.11609/jott.4605.11.12.14527-14561

Mir AH, Tyub S, Kamili AN (2020) Ecology, distribution mapping and conservation implications of four critically endangered endemic plants of Kashmir Himalaya. Saudi J Biol Sci 27:2380–2389. https://doi.org/10.1016/j.sjbs.2020.05.006

Mir AH, Sarma K, Upadhaya K (2022) Assessing the effectiveness of community managed forests for plant diversity conservation in Meghalaya, Northeast India. Plant Divers 44:243–254. https://doi.org/10.1016/j.pld.2021.11.010

Mishra BP, Tripathi OP, Tripathi RS, Pandey HN (2004) Effects of anthropogenic disturbance on plant diversity and community structure of a sacred grove in Meghalaya, Northeast India. Biodivers Conserv 13:421–436. https://doi.org/10.1023/B:BIOC.0000006509.31571.a0

Mittermeier RA, Turner WR, Larsen FW et al (2011) Global biodiversity conservation: the critical role of hotspots. In: Zachos FE, Habel JC (eds) Biodiversity hotspots: distribution and protection of conservation priority areas. Springer Berlin Heidelberg, Berlin, pp 3–22

Myers N (1988) Threatened biotas: "Hot spots" in tropical forests. Environmentalist 8:187–208. https://doi.org/10.1007/BF02240252

Myers N (1990) The biodiversity challenge: expanded hot-spots analysis. Environmentalist 10: 243–256. https://doi.org/10.1007/BF02239720

Myers N, Mittermeier RA, Mittermeier CG et al (2000) Biodiversity hotspots for conservation priorities. Nature 403:853–858. https://doi.org/10.1038/35002501

Nair NC (1977) Flora of Bashahr Himalayas. International Bioscience Publishers, Madras

Naithani BD (1984) Flora of Chamoli, vol 1. Botanical Survey of India, Kolkata

Nirola S, Das AP (2017) Endemic monocot flora of Darjeeling Himalaya, West Bengal, India. Pleione 11:116–124

Nohro S, Jayakumar S (2020) Tree species diversity and composition of the Pala Wetland Reserve Forest, Mizoram, Indo-Burma hotspot, India. Biocatal Agric Biotechnol 23:101474. https://doi.org/10.1016/j.bcab.2019.101474

Olson DM, Dinerstein E (1998) The global 200: a representation approach to conserving the Earth's most biologically valuable ecoregions. Conserv Biol 12:502–515. https://doi.org/10.1046/j.1523-1739.1998.012003502.x

Osmaston AE (1978) A forest flora for kumaon. Bishen Singh Mahendra Pal Singh, Dehradun

Padmawathe R, Qureshi Q, Rawat GS (2004) Effects of selective logging on vascular epiphyte diversity in a moist lowland forest of eastern Himalaya, India. Biol Conserv 119:81–92. https://doi.org/10.1016/j.biocon.2003.10.024

Pal GD (2013) Flora of lower Subansiri District, Arunachal Pradesh (India), vol 1. Botanical Survey of India, Kolkata

Panda S (2012) Checklist of Ericaceae in Tuensang District of Nagaland, India with special reference to Mt. Saramati. J Threat Taxa 4:3454–3461. https://doi.org/10.11609/JoTT.o2939.3454-61

Pandit MK, Kumar V (2013) Land-use change and conservation challenges in the Indian Himalaya. In: Sodhi NS, Gibson L, Raven PH (eds) Conservation biology – voices from the tropics. Wiley, Oxford, pp 121–133

Pandit MK, Manish K, Koh LP (2014) Dancing on the roof of the world: ecological transformation of the Himalayan landscape. Bioscience 64:980–992. https://doi.org/10.1093/biosci/biu152

Parbo D, Kumar A, Devi A et al (2024) Diversity of fig species and their ecological services in Pakke wildlife sanctuary, Arunachal Pradesh, India. J Wildl Biodivers 8:55–80

Pradhan N, Shrestha P (2021) A handbook of the bryophytes of Nepal, vol 1. Department of Plant Resources, National Herbarium and Plant Laboratories, Godawari

Pusalkar P, Srivastava SK (2018) Flora of Uttarakhand (Vol. 1). Botanical Survey of India, Kolkata

Rahman IU, Afzal A, Iqbal Z et al (2019) Role of multivariate approaches in floristic diversity of Manoor Valley (Himalayan region), Pakistan. Appl Ecol Environ Res 17:1475–1498. https://doi.org/10.15666/aeer/1702_14751498

Rajbhandari KR, Rai SK (2017) A handbook of the flowering plants of Nepal, vol 1. Department of Plant Resources, Ministry of Forests and Environment, Government of Nepal, Kathmandu

Rajbhandari KR, Rai SK (2019) A handbook of the flowering plants of Nepal, vol 2. Department of Plant Resources, Ministry of Forests and Environment, Government of Nepal, Kathmandu

Rana SK, Price TD, Qian H (2019) Plant species richness across the Himalaya driven by evolutionary history and current climate. Ecosphere 10:e02945. https://doi.org/10.1002/ecs2.2945

Rana D, Kapoor KS, Bhatt A, Samant SS (2022) Community structure and species diversity of forest vegetation in a protected area of Western Himalayan region of India. Environ Dev Sustain 24:14011–14032. https://doi.org/10.1007/s10668-021-02020-9

Ranibala G, Singh SS, Khan ML (2018) Species diversity and distribution of Musa species (Musaceae) in Arunachal Pradesh, north eastern India. Ann Plant Sci 7:2342–2346

Rau MA (1974) Vegetation and phytogeography of the Himalaya. In: Mani MS (ed) Ecology and biogeography in India. Springer, Dordrecht, pp 247–280

Rawat DS, Tiwari JK, Tiwari P, Singh H (2016) Floristic diversity of montane zone of Western Ramganga Valley, Uttarakhand, India. J Econ Taxon Bot 40:104–125

Roy DK, Talukdar AD, Choudhury MD, Sinha BK (2014) Diversity in angiosperm flora of Siju wildlife sanctuary, south Garo Hills district of Meghalaya, India. Indian J Plant Sci 3:87–101

Saikia P, Deka J, Bharali S et al (2017) Plant diversity patterns and conservation status of eastern Himalayan forests in Arunachal Pradesh, Northeast India. For Ecosyst 4:28. https://doi.org/10.1186/s40663-017-0117-8

Samant SS, Joshi HC, Arya SC, Pant S (2005) Diversity, distribution and conservation of pteridophytes in Nanda Devi biosphere reserve, west Himalaya. Indian Fern J 22:100–111

Sengupta R, Sekhar Dash S (2023) A contribution to the flora of Murlen National Park, Mizoram, a part of Indo-Burma biodiversity hotspot in India. Nelumbo 65:252–277. https://doi.org/10.20324/nelumbo/v65/2023/173057

Shalini L (2021) Plant diversity assessment in a sub-tropical forest ecosystem of Manipur, Northeast India. Int J Mod Eng Technol Sci 3:1446–1464

Sharma P, Rana JC, Devi U et al (2014) Floristic diversity and distribution pattern of plant communities along altitudinal gradient in Sangla Valley, northwest Himalaya. Sci World J 2014:1–11. https://doi.org/10.1155/2014/264878

Shrestha TB, Joshi RM (1996) Rare, endemic and endangered plants of Nepal. WWF Nepal Program, Kathmandu

Singh H (2006) Flora of Chamba District, Himachal Pradesh. Bishen Singh Mahendra Pal Singh, Dehradun

Singh PB (2018) Flora of Mandi district Himachal Pradesh: north west Himalaya. Bishen Singh Mahendra Pal Singh, Dehradun

Singh P, Kumar N (1997) Effect of orography on precipitation in the western Himalayan region. J Hydrol 199:183–206. https://doi.org/10.1016/S0022-1694(96)03222-2

Singh P, Patel RC (2022) Miocene development of the main boundary thrust and Ramgarh thrust, and exhumation of lesser Himalayan rocks of the Kumaun-Garhwal region, NW-Himalaya (India): insights from fission track thermochronology. J Asian Earth Sci 224:104987. https://doi.org/10.1016/j.jseaes.2021.104987

Singh SK, Rawat GS (2000) Flora of Great Himalayan National Park, Himachal Pradesh. Bishen Singh Mahendra Pal Singh, Dehradun

Singh JS, Singh SP (1987) Forest vegetation of the Himalaya. Bot Rev 53:80–192. https://doi.org/10.1007/BF02858183

Singh NP, Srivastava RC (2013) Gymnosperms of India: a check list. Botanical Survey of India, Kolkata

Singh NP, Chauhan AS, Mondal MS (2000) Flora of Manipur, vol 1. Botanical Survey of India, Calcutta, West Bengal

Singh GP, Singh KP, Singh DK (2002) Flora of Mizoram. Botanical Survey of India, Ministry of Environment and Forests, Kolkata

Singh P, Karthigeyan K, Lakshminarasimhan P et al (2015) Endemic vascular plants of India. Botanical Survey of India, Kolkata

Singh OP, De SK, Cajee L (2016) State of environment report of Mizoram 2016. Department of Environment, Forests and Climate Change. Government of Mizoram, Mizoram

Singh SP, Gumber S, Singh RD, Singh G (2020) How many tree species are in the Himalayan treelines and how are they distributed? Trop Ecol 61:317–327. https://doi.org/10.1007/s42965-020-00093-7

Srivastava R (2011) Indian upper cretaceous-tertiary flora before collision of Indian plate: a reappraisal of central and western Indian flora. Mem Geol Soc India 77:281–292

Stewart RR (1942) The ferns of Mussoorie and Dehradun. Royal Botanical Garden, Calcutta

Tabata H (2008) On the Himalayan uplift and Himalayan corridors. Himal J Sci 2:256–257. https://doi.org/10.3126/hjs.v2i4.934

Thapa Magar G, Chaudhary S (2022) Wild floristic diversity of daman-Simbhanjyang area, Makwanpur District, Central Nepal. Int J Appl Sci Biotechnol 10:112–123. https://doi.org/10.3126/ijasbt.v10i2.46188

Tiruwa BL, Neupane BD, Kadariya R et al (2024) Diversity of orchids in terms of their distribution, uses and conservation in Annapurna conservation area of Nepal. Am J Plant Sci 15:422–440. https://doi.org/10.4236/ajps.2024.156030

Tiwari A, Uprety Y, Rana SK (2019) Plant endemism in the Nepal Himalayas and phytogeographical implications. Plant Divers 41:174–182. https://doi.org/10.1016/j.pld.2019.04.004

Tiwari H, Sekar KC, Pandey A et al (2024) Diversity, distribution and need of urgent conservation of endemic plants in Himalaya. Biodivers Conserv 33:2285–2303. https://doi.org/10.1007/s10531-024-02815-y

Ullah H, Khan SM, Jaremko M et al (2022) Vegetation assessments under the influence of environmental variables from the Yakhtangay Hill of the Hindu-Himalayan range, North Western Pakistan. Sci Rep 12:20973. https://doi.org/10.1038/s41598-022-21097-4

Uniyal BP, Sharma JR, Chaudhary U, Singh DK (2007) Flowering plants of Uttarakhand (A checklist). Bishen Singh Mahendra Pal Singh, Dehradun

Upadhaya K, Pandey HN, Law PS, Tripathi RS (2003) Tree diversity in sacred groves of the Jaintia hills in Meghalaya, Northeast India. Biodivers Conserv 12:583–597. https://doi.org/10.1023/A:1022401012824

van Hinsbergen DJJ, Lippert PC, Dupont-Nivet G et al (2012) Greater India Basin hypothesis and a two-stage Cenozoic collision between India and Asia. Proc Natl Acad Sci 109:7659–7664. https://doi.org/10.1073/pnas.1117262109

Wadhwa BM, Rao RR, Hajra PK (2024) Botany of the valley of flowers National Park and its environs. Nelumbo 29:129–175. https://doi.org/10.20324/nelumbo/v29/1987/74686

Wang Y, Yang K, Zhou X et al (2019) The formation of a Dry-Belt in the north side of central Himalaya Mountains. Geophys Res Lett 46:2993–3000. https://doi.org/10.1029/2018GL081061

Wani ZA, Pant S (2023) Assessment of floristic diversity and community characteristics of Gulmarg Wildlife sanctuary, Kashmir Himalaya. Geol Ecol Landscapes 9(1):296–316. https://doi.org/10.1080/24749508.2023.2196767

Yadav R, Tripathi S, Jaiswal S et al (2020) Endemic species of Poaceae from Western Himalaya: diversity and distribution. Pleione 14:301. https://doi.org/10.26679/Pleione.14.2.2020.301-309

Yangzom R, Thomas MB (2017) Endemic plants of Bhutan – a preliminary checklist. In: Bhutan Biodivers. Specim. Portal, Natl. Herb. Natl. Biodivers. Centre, Minist. Agric. Serbithang, Thimphu, Bhutan. http://bhutanbiodiversity.net/checklists/checklist.php?cl=8&emode=0. Accessed 13 Nov 2024

Chapter 2
The Himalayas: One of the World's Biodiversity Hotspots

Nidaa Harun and Shaheena Umbreen

Abstract One of the most important hotspots for biodiversity in the world, the Himalayas includes Nepal, India, Bhutan, China and Pakistan. They are home to a multitude of endemic plants and animals that are especially suited to the diverse geoclimatic conditions of the area. This chapter examines the ecological relevance of the Himalayas, which are home to a wide variety of climate zones, seasonal variations, and complicated topography that support abundant biodiversity. These varied habitats, which range from subtropical forests to alpine meadows, support a variety of species worldwide and provide vital ecosystem services, such as resources and a supply of fresh water, to millions of people. However, this natural richness is seriously threatened by human activity and climate change. This study highlights the pressing need for sustainable measures to safeguard the environmental health of the Himalayas by looking at patterns of biodiversity, conservation requirements, and local reliance on these resources. The results offer important new information about how protecting the Himalayas might help local livelihoods and international efforts to conserve biodiversity.

Keywords Geoclimatic conditions · Himalayas · Alpine meadows

2.1 Introduction

Historically, the region's isolation and complex geography have contributed to its rich biodiversity (Pullaiah 2024a, b). The Himalayas are the most important biological region in the world, spanning five countries: Nepal, India, Bhutan, China, and Pakistan. Many of the remarkable varieties of flora and wildlife found in this vast mountain range are indigenous to the area. The Himalayas' diverse variety of climatic conditions and intricate terrain, which differ throughout the area, are responsible for its tremendous biodiversity (section "Altitudinal and Climatic

N. Harun (✉) · S. Umbreen
Department of Botany, Faculty of Life Sciences, University of Okara, Okara, Pakistan
e-mail: nidaadr@uo.edu.pk

D. Das et al. (eds.), *Plant Diversity and Conservation Strategies of the Himalayan Biodiversity Hotspot*, Topics in Biodiversity and Conservation 21,
https://doi.org/10.1007/978-3-032-21697-7_2

Variations" in Chap. 1). The Himalayas have become a major focus for ecological and environmental research, as their unique position as a natural barrier also serves as a reservoir for biodiversity. Over millions of years, the region's isolation due to its complex geography has allowed for the evolution of numerous unique species, contributing to its status as one of the world's primary biodiversity reserves (Pullaiah 2024a, b).

For generations, the indigenous Himalayan people have relied on the biological resources and ecosystem services offered by Himalayan ecosystems for their subsistence (Dar and Khuroo 2020a, b, c). Natural resources can be found in a range of forest types, semi-wild regions, and other ecosystems (Miller 1999). The type and location of the bioresources themselves, as well as the people who use them, determine how these resources are used, and whether in a sustainable way (Bandyopadhyay and Parveen 2004; Pradhan and Badola 2008). Despite having rich biological resources, Himalayan people are still deprived, probably due to the challenging high-altitude environment they live in. The researchers feel that a complete research and evaluation of Himalayan bioresource utilisation trends and dynamics would serve as a good platform for evaluating poverty reduction in the Himalayan region. Naturally, this would be a precondition for the long-term sustainable use of Himalayan biodiversity and appropriate conservation planning (Oli and Gupta 2008; Badola 2009).

The four world's main rivers originate in the Himalayas: the Indus, the Ganges, the Brahmaputra, and the Mekong. These "separated feeders," which stream life from their events through arteries in every direction, supply fresh water to most of Asia, which is essential for all life on the Earth. These vital habitats are simply irreplaceable in the role they provide to the global economy, especially when taking into account the effects of these rivers on the wider remaining region of Asia Pacific, which currently accounts for over half of the worldwide population, almost 40 percent of the global gross domestic product (GDP), and 3.96 billion in 2013 which is about 55% of the global total.

2.2 Geographical Features

The Himalayas are a collection of mountain ranges spreading five nations and occupying a land area of over 600,000 km^2. Three primary sections can be used to group the full range:

- **The Western Himalayas**: This region includes parts of India, Pakistan, and Afghanistan.
- **The Central Himalayas**: This section spans across Nepal and the northern Indian states.
- **The Eastern Himalayas**: Including Bhutan and parts of India, this region is known for its heavy rainfall and dense forests (Fig. 2.1).

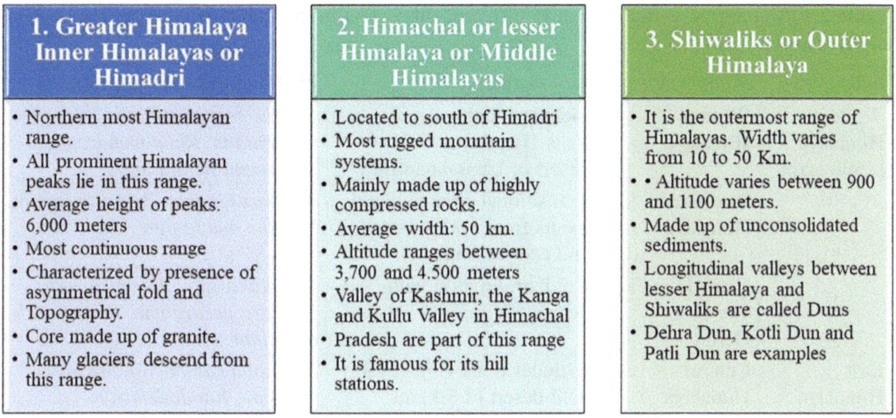

Fig. 2.1 Regional division of the Himalayas (Valdiya 2015)

This division corresponds with broader classifications in Chap. 1 (Sect. 1.2), which also highlighted their socio-cultural, climatic, and biogeographical heterogeneity.

2.3 Regional Division of the Himalayas

One of the most powerful and geographically intricate mountain ranges in the world, the Himalayas stretch over 2400 km through five countries: Nepal, India, Bhutan, China and Pakistan. This range is not only significant for its towering peaks but also for its diverse ecological zones, which are categorised into three primary regions: the Great Himalayas, the Lesser Himalayas, and the Outer Himalayas. Chapter 1 (Sects. 1.2 and "Himalayan Uplift") provided a foundation for this division by describing the geological formation, orogeny, and tectonic uplift processes that shaped the Himalayas The region's climate and biodiversity are significantly shaped by each of these zones (Table 2.1).

2.3.1 The Great Himalayas (Inner Himalayas)

The Great Himalayas, also known as the Inner Himalayas, represent the highest and most formidable section of the Himalayan range. This zone is characterised by its towering peaks, such as Everest (8848 m), Kanchenjunga (8586 m), and Nanga Parbat (8126 m), among others. The Great Himalayas extend from the west to the northwest range's northernmost edge, providing a continuous band of high mountains (Fig. 2.2) (Pullaiah 2024a, b).

Table 2.1 Number of endemic taxa in different subcenters in the Indian Himalayan region (Dar and Khuroo 2020a, b, c)

Subcentre	Microcentre	Key features	Important taxa (Examples)
Western Himalaya: [~500 taxa]	Western Himalaya	Includes Garhwal and trans-Himalayan cold desert of Uttarakhand	*Cedrus deodara*, *Pinus wallichiana*, *Rhododendron arboreum*, *Betula utilis*
	Northwest Himalaya	Transitional zone with elements from both western and central zones	*Juniperus indica*, *Berberis lycium*, *Daphne mucronata*
	Trans-Himalayan Cold Desert	Cold, arid region with xerophytic flora	*Caragana* spp., *Artemisia* spp., *Ephedra gerardiana*, *Saussurea obvallata*
East Himalaya: [~544 taxa]	Central Himalaya (Sikkim–Darjeeling)	Includes trans-Himalayan cold desert of Sikkim	*Magnolia campbellii*, *Michelia doltsopa*, *Rhododendron falconeri*, *Meconopsis* spp.
	Eastern Himalaya	Richest in biodiversity; includes Arunachal Pradesh	*Rhododendron* spp., *Dendrobium* spp., *Taxus wallichiana*, *Paris polyphylla*, *Panax* spp.

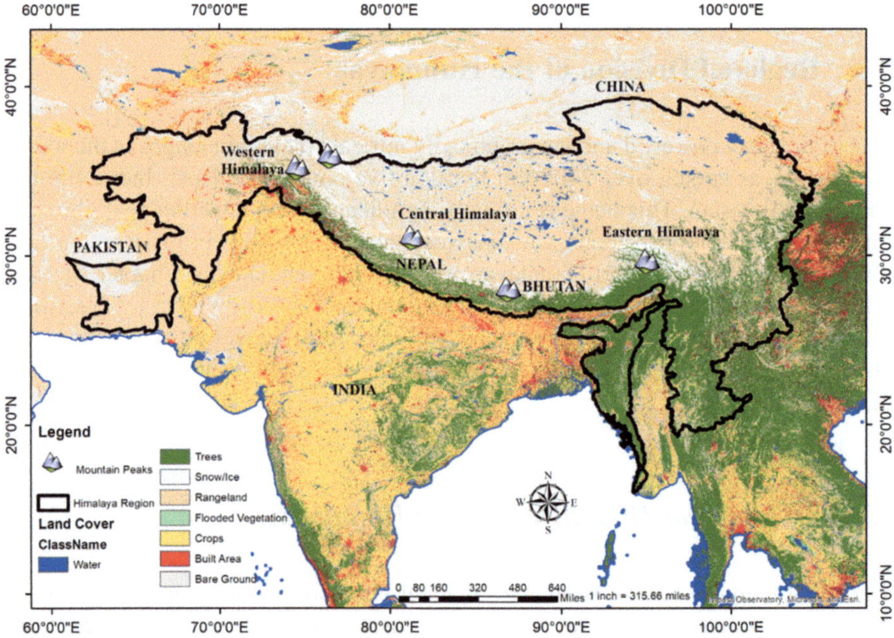

Fig. 2.2 Geographical map of Himalayan, showing its division into various regions. (GIS Arc software v10.7.1)

2.3.1.1 Climatic Conditions and Biodiversity

The Great Himalayas experience some of the most extreme climatic conditions on the Earth, with temperatures often plunging below freezing, especially at higher altitudes. The region is mostly covered in ice and snow across the year, and it boasts the highest concentrations of the glaciers outside the polar regions. These glaciers are vital for the hydrology of the entire region, feeding major river systems such as the Ganges, Brahmaputra, and Indus (Gansser 1964). Biodiversity in the Great Himalayas is adapted to these harsh conditions. Vegetation is sparse, with hardy species such as alpine shrubs and grasses dominating the landscape. The tree line, which marks the upper limit of tree growth, occurs at about 4000 m, beyond which only the most resilient plants can survive. Animal life in this zone includes iconic species including the Himalayas tahr (*Hemitragus jemlahicus*), the snow leopard (*Panther auncia*), and several ibex and wild sheep species. The animals keep changing to the cold, thin air and rugged terrain (Jackson and Ahlborn 1988).

2.3.1.2 Geological Significance

The Great Himalayas represent one of the most geologically dynamic regions on Earth, formed as a result of the ongoing collision between the Indian and Eurasian tectonic plates. This orogeny began approximately 50 million years ago during the Eocene epoch, not fifty thousand as often misinterpreted, and continues today with tectonic uplift rates averaging around 5 mm per year (Pullaiah 2024a, b). This persistent tectonic activity contributes not only to the vertical growth of the mountain range but also to frequent seismic events, making the Himalayas one of the most earthquake-prone regions globally.

The geological dynamism of the Himalayas plays a crucial role in shaping the region's topographic heterogeneity, which in turn supports its exceptional ecological and biodiversity richness. The continuous folding, faulting, and uplifting processes give rise to new landforms, river systems, and microhabitats, influencing patterns of species evolution, migration, and endemism.

However, the same tectonic forces that foster biological diversity also pose significant risks. High-magnitude earthquakes, landslides, and glacial lake outburst floods (GLOFs) are frequent in this seismically active belt, threatening both human populations and fragile ecosystems. The interplay between geological evolution and ecological processes makes the Himalayas a living laboratory for studying mountain-building phenomena, landscape development, and their impacts on biodiversity.

In essence, the geological history and ongoing activity of the Himalayas are not only central to the understanding of plate tectonics and mountain formation but are also fundamental drivers of biogeographical diversity and environmental vulnerability in the region.

2.3.2 The Lesser Himalayas (Middle Himalayas)

The Lesser Himalayas, or Middle Himalayas, form a transitional mountain belt between the Great Himalayas and the Siwalik Hills, with elevations typically ranging from 3700 to 4500 m. This zone is marked by rugged terrain, deep valleys, and steep forested slopes, contrasting sharply with the snow-covered peaks of the higher ranges.

Geologically composed of metamorphic and sedimentary rocks, the region lies along active thrust zones such as the Main Central Thrust (MCT), making it prone to landslides and erosion. These processes contribute to the formation of diverse microhabitats, supporting rich biodiversity, including temperate and subalpine forests. Its biodiversity and forest types echo the altitudinal zonation patterns outlined earlier in Chap. 1 (Sect. 1.4). *Rhododendron* and oak forests are examples of vegetation profiled in both chapters.

Ecologically, the Lesser Himalayas act as a biogeographic bridge, linking tropical lowlands with alpine zones and sustaining both wildlife corridors and human livelihoods dependent on agriculture, forestry, and medicinal plants.

2.3.2.1 Climate and Biodiversity

The climate of the Lesser Himalayas is generally milder and less extreme compared to that of the Great Himalayas. While snowfall can occur at higher elevations during the winter months, the region's moderate climate fosters a rich diversity of flora and fauna. Vegetation ranges from temperate broadleaf forests to coniferous forests, featuring species such as *Rhododendron* spp., *Pinus* spp., *Cedrus* spp. and other native plants that dominate the landscape (Pullaiah 2024a, b).

Among these, *Rhododendron* species are particularly prominent and ecologically significant. These flowering shrubs and trees, known for their vibrant blooms, often form dense understory layers in temperate forests. They not only add striking seasonal colour but also play an important role in supporting pollinators such as bees and birds. In traditional Himalayan cultures, *Rhododendron arboreum* (locally known as 'Burans' or 'Lali Gurans') holds medicinal and cultural value, with its flowers used in herbal remedies and local beverages.

This diverse vegetation provides critical habitats for several iconic and ecologically important animal species, including the Himalayan black bear (*Ursus thibetanus*), red panda (*Ailurus fulgens*), and the Himalayan monal (*Lophophorus impejanus*), which is also the national bird of Nepal. The biodiversity of the Lesser Himalayas is especially notable due to the complex interplay of altitudinal gradients, microclimatic zones, and varied topography, all within relatively short spatial distances (Gurung 2012).

Furthermore, the region acts as an ecological bridge between tropical lowlands and alpine zones, supporting both endemic and migratory species. Conservation efforts in the Lesser Himalayas are increasingly important, as anthropogenic pressures and climate change continue to threaten these fragile ecosystems.

2.3.2.2 Cultural and Economic Importance

The Lesser Himalayas hold substantial cultural and economic significance for the millions of people residing within their bounds. This region supports diverse traditional mountain livelihoods that are closely adapted to its terrain and resources. Key economic activities include agriculture, forestry, and tourism, which together form the backbone of the local economy.

Terrace farming is widely practiced on the steep mountain slopes, enabling the cultivation of staple crops such as rice, maize, and millet. These agroecosystems not only ensure food security but also reflect deep-rooted indigenous knowledge of soil and water conservation.

The forests of the Lesser Himalayas are vital to both household economies and regional markets. They provide essential resources including fuelwood, timber, fodder, medicinal plants, and non-timber forest products (NTFPs). Many of these, such as *Nardostachys jatamansi* and *Picrorhiza kurroa*, have significant value in traditional medicine and pharmaceutical industries.

Moreover, the scenic landscapes and rich cultural heritage of the region, including pilgrimage sites and traditional architecture, make it a growing hub for eco-tourism and cultural tourism, providing alternate income streams and promoting conservation awareness (Singh 2017).

In essence, the Lesser Himalayas are not only a biodiversity-rich ecological zone but also a cultural landscape where nature and livelihoods are intricately linked.

2.3.3 The Outer Himalayas (Shiwalik Hills)

With elevations ranging from 900 to 1200 m, the Outer Himalayas, commonly referred to as the Shiwalik Hills, constitute the lowest and southernmost range of the Himalayan system. The plains of northern India and Nepal give way to a succession of low, rocky hills that define this zone (Table 2.2).

2.3.3.1 Climate and Vegetation

The Outer Himalayas, also known as the Shiwalik Hills, experience a generally warm climate characterised by hot summers and mild winters. During the monsoon season, the region receives substantial rainfall, which sustains its lush vegetation and contributes to the overall greenness of the landscape throughout much of the year.

The Shiwalik Hills are predominantly covered with dense forests, particularly dominated by *Shorea robusta* (sal) and *Tectona grandis* (teak), along with other hardy and moisture-loving species that thrive in the region's subtropical conditions (Pullaiah 2024a, b). These deciduous forests form a crucial ecological zone that supports a diverse range of wildlife.

Table 2.2 Biogeographic zones and phytogeographical regions with land area coverage, no. of species and endemic species (Bhat et al. 2021a, b, c; Davis et al. 1995; Watson et al. 2011)

Phytogeographical regions	Biogeographic zones	Province	States covering	% of total area	No. of Spp.	Endemic Species
Western Himalayan	Trans-Himalaya	Ladakh mountains	Regions of Jammu & Kashmir	3.2	5000	430
		Plateau Tibetan	Upper parts of Himachal Pradesh	2.3	1500–1600	70
	Himalayan	Northwest Himalaya	Himachal Pradesh, Uttarakhand	1.5	3500	800
Eastern Himalayan		Central Himalayan	Sikkim	0.2	5500	100
		East Himalayan	Arunachal Pradesh	2.4	3000	285

The fauna of the Outer Himalayas includes several large and ecologically significant mammals, such as the Asian elephant (*Elephas maximus*), leopard (*Panthera pardus*), and various species of deer, including chital (*Axis axis*) and sambar (*Rusa unicolor*), as well as different species of monkeys like the rhesus macaque (*Macaca mulatta*) and Hanuman langur (*Semnopithecus entellus*). These forests serve as vital wildlife corridors and breeding grounds, playing a key role in regional biodiversity conservation.

2.3.3.2 Geological Formation and Soil

The Shiwalik Mountains, forming the southernmost and geologically youngest range of the Himalayas, owe their origin to the ongoing tectonic convergence between the Indian and Eurasian plates. This active collision has not only led to the uplift of the Great and Lesser Himalayas but continues to push the Shiwaliks upward, contributing to their fragile and dynamic geomorphology (Gansser 1964).

Geologically, the region is composed of poorly consolidated alluvium, conglomerates, sandstones, and siltstones, making the soils loose, coarse-textured, and highly susceptible to erosion. Frequent seismic activity, combined with intense monsoonal rainfall and steep slopes, has sculpted a landscape of rugged hills, unstable slopes, and deep gorges, prone to landslides and flash floods.

Despite this instability, the region supports a range of ecological niches, from rocky outcrops and dry scrublands to fertile valleys and river terraces, which host rich biodiversity. The continuous tectonic uplift and erosion not only shape the terrain but also foster the development of mosaic habitats, contributing to the ecological complexity of the Himalayan foothills.

2.4 Climate and Weather Patterns

2.4.1 Climate Zones

The Himalayas encompass a broad spectrum of climatic zones, ranging from subtropical conditions at the lower elevations to alpine and even nival (perpetual snow) climates at the highest altitudes. This reflects the climatic and altitudinal gradients discussed in Chap. 1 (section "Altitudinal and Climatic Variations"), which explained how rainfall increases eastward and biodiversity richness follows these patterns. This dramatic altitudinal variation over short horizontal distances results in distinct ecological zones, each supporting a unique array of plant and animal life. Subtropical forests in the foothills gradually transition into temperate broadleaf and coniferous forests, followed by alpine meadows and cold desert conditions near the snow line. This complex climatic gradient contributes to the Himalayas being one of the most biodiverse mountain ranges in the world (Fig. 2.3) (Pullaiah 2024a, b). The Himalayan climate can be divided into several distinct zones:

2.4.1.1 Temperate Forests

Located predominantly within the Lesser Himalayas, the temperate forests span mid-elevations and are characterised by high biodiversity and a mix of broadleaf hardwoods and coniferous species. Common vegetation includes *Quercus* spp. (oaks), *Aesculus indica*, *Cedrus deodara* and *Abies pindrow*. These forests provide critical habitat for a wide range of fauna, including mammals, birds, and insects, and offer key ecosystem services such as carbon sequestration, water regulation, and soil stabilisation (Gurung 2012).

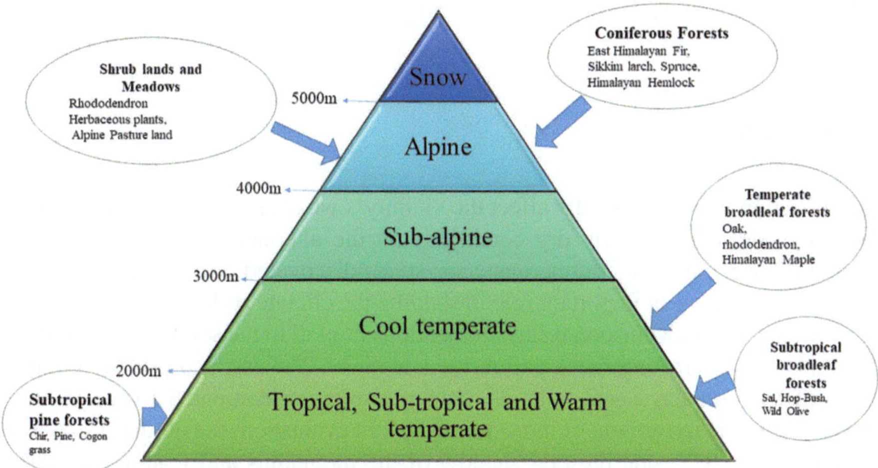

Fig. 2.3 Different vegetation zones under multiple climate zones

2.4.1.2 Subtropical Forests

The subtropical zone, occupying the lower slopes and foothills, experiences warmer temperatures and higher rainfall. Vegetation here includes dense evergreen and deciduous forests, dominated by *Shorea robusta* (sal), *Terminalia* spp., and various bamboo and fern species. This region supports high plant and animal diversity, including numerous endemic and threatened species, and plays an essential role in agriculture, water recharge, and cultural heritage (Singh 2017).

2.4.1.3 Alpine Region

Above the treeline, generally at elevations exceeding 3500 m, the alpine zone is marked by cold temperatures, intense solar radiation, and short growing seasons. Vegetation is sparse and dominated by hardy herbaceous plants, shrubs, and cushion-forming species like *Rhododendron anthopogon* and *Meconopsis* spp. This region is critical habitat for charismatic and endangered species, including the Snow Leopard (*Panthera uncia*) and Himalayan Ibex (*Capra sibirica*), adapted to the harsh alpine environment (Pullaiah 2024a, b).

2.4.1.4 Nival (Arctic-Like) Zone

The highest reaches of the Himalayas, including glaciated peaks like Mount Everest, fall under the nival or arctic-like zone. Found above 5500 m, this zone is permanently covered by snow and ice, with minimal biological activity. Temperatures remain well below freezing for most of the year, and only a few extremophile organisms, such as cold-adapted lichens, mosses, and microfauna, survive here. This zone is crucial for cryospheric processes, influencing regional hydrology and global climate systems (Dar and Khuroo 2020a, b, c).

2.4.2 Monsoon Impact

The monsoon winds drastically affect the vicinity's weather, with heavy rains on the southern side and relatively dry conditions on the alternative side. This climatic sample is essential for the numerous ecosystems discovered within the Himalayas. The monsoon plays an important role in shaping the climate styles of the Himalayas. The Indian summer Monsoon brings tremendous rainfall to the southern facets of the mountain range, creating moist and fertile situations. The east and primary components of the Himalayas get hold of more precipitation compared to the western phase, due to the movement of monsoon winds. In evaluation, the northern slopes of the Himalayas lie inside the rain shadow of the mountains and feature a far drier climate. In addition to the summer monsoon, other wind systems influence the Himalayan climate, such as the Monsoon of Westerlies and East Asian Winter.

The combination of these wind systems and the region's complex topography contributes to dramatic climatic variation over relatively short distances (Dar and Khuroo 2020a, b, c; Gurung 2012).

2.4.3 *Seasonal Variability*

Seasonal patterns in the Himalayas are shaped by a complex interplay of altitude, topography, and monsoonal influence. These factors cause significant variation in temperature, precipitation, and phenology across different elevations and regions. While general seasonal divisions apply, their intensity and duration vary widely from the subtropical foothills to the alpine and glaciated zones.

The Himalayan year can broadly be divided into four climatic seasons:

- ***Spring (March to May)***:
  Spring marks a transition from the cold winter to warmer conditions, particularly in the lower and middle elevations. Temperatures become mild and vegetation begins to regenerate. This is a peak period for flowering, with forests displaying vibrant blooms of *Rhododendron*, *Primula* and *Magnolia* species. It is also a favoured time for eco-tourism and biodiversity research.
- ***Summer/Monsoon (June to September)***:
  Dominated by the Southwest Monsoon, this season brings intense rainfall, particularly to the Eastern and Central Himalayas. It fosters rapid vegetative growth, replenishes rivers and aquifers, and sustains agricultural cycles. However, it also triggers landslides, floods, and erosion, especially in geologically fragile zones.
- ***Autumn (October to November)***:
  A post-monsoon period characterised by clear skies, moderate temperatures, and minimal precipitation. It is widely regarded as one of the best seasons for trekking, field research, and cultural festivals. Forests display striking autumn foliage, and wildlife activity is prominent before the onset of winter.
- ***Winter (December to February)***:
  Winter brings cold, dry conditions to the Himalayas, with heavy snowfall at higher elevations. Subzero temperatures are common above 3000 m, and alpine zones enter a phase of ecological dormancy. In contrast, lower elevations experience a relatively mild winter, allowing continued human activity and some agricultural operations.

2.5 Geoclimatic Features and Their Impact on the Ecosystem

The Himalayas are characterised by steep altitudinal gradients, complex topography, and strong climatic variability, which together generate a multitude of microclimates across relatively short distances. Chapter 1 (Sect. 1.3.1) outlined similar drivers, noting how pre-formative, formative, and post-formative geological events created ecotones and microhabitats. For instance, south-facing slopes in Himalayan valleys receive more solar radiation, making them warmer and drier, while north-facing slopes remain cooler and moister, favouring different vegetation types (An et al. 2014a, b).

These microclimatic differences, combined with elevation ranging from 900 m to over 8500 m, support a remarkable diversity of ecosystems from tropical forests and temperate woodlands at lower elevations to alpine meadows, tundra, and glaciated zones at higher altitudes. The result is a highly heterogeneous landscape, contributing to exceptional ecological diversity and species endemism throughout the Himalayan arc.

2.5.1 Leeward and Windward Slopes

The southern (windward) slopes of the Himalayas face the prevailing southwest monsoon winds and receive high annual precipitation. These moist conditions promote the growth of dense subtropical and temperate forests, rich in both floral and faunal biodiversity. In contrast, the northern (leeward) slopes, shielded from monsoonal rains, remain comparatively dry and are dominated by species adapted to xeric and subalpine conditions (Dar and Khuroo 2020a, b, c). This contrast in moisture availability contributes to the zonal distribution of vegetation and the development of distinct ecological niches on either side of the range.

2.5.2 Floristic Regions

The Himalayan flora is biogeographically influenced by two major floristic kingdoms: the Paleotropical and the Holarctic. The Eastern Himalayas belong to the Paleotropical Kingdom, reflecting affinities with Southeast Asian tropical flora, whereas the Western Himalayas fall under the Holarctic Kingdom, which is closely related to temperate Eurasian floras (Takhtajan 1986).

These divisions are not merely historical classifications, they signify deep evolutionary and ecological patterns in plant distribution, shaped by climatic history, geological isolation, and topographic heterogeneity. This biogeographical dichotomy underscores the Himalayas' role as a confluence zone for tropical and temperate biodiversity.

2.6 Himalayan Biodiversity Hotspots

A biodiversity hotspot is considered a biogeographic region that is rich in biodiversity reservoirs that need to be protected from human interventions. To be a biodiversity hotspot a region must have lost 70% of its principal vegetation and at least

0.5% or 1500 endemic species of plants. These areas serve as the shelter for locally endemic species. There are about 35 biodiversity hotspots in the world which supports to protect world's 60% of endemic plants and vertebrates approximately (Maiti and Maiti 2017). These are the areas within the landscape that have exceptional value for the organisms or ecosystem process. These areas are of significant value for some species with regard to some vital biological activities like reproduction, overwintering, and resources (Lindenmayer and Franklin 2013).

The Himalayan biodiversity hotspot is the reservoir of diverse ecosystems ranging from alluvial grasslands and subtropical broadleaf forests to alpine meadows above the tree line. Vascular plants have been recorded at the height of 6000 m along with the numerous large birds and mammals, including vultures, tigers, elephants, rhinos and wild water buffalo. Zachos and Habel (2011) stated that the Himalayan has one of the world's grand mountain chains with a great range of biodiversity.

2.7 Floristic Composition of Himalayas

With around 750 species, the *Orchidaceae* is the largest family of flower-producing plants in the Himalaya Hotspot. Between 5000 and 6000 m is where a zone of rock and ice begins. Vascular plants have been recorded at some of the highest elevations on Earth despite these harsh conditions. Cushion plants were found at elevations above 6200 m, while *Ermania himalayensis* scree plants, which belong to the *Brassicaceae* family, were recorded at 6300 m on the slopes of Mount Kamet in the northwestern Himalayan region (Körner 2003; Chettri et al. 2023). The following are some of the Himalayan genera' geographic affinities: (Table 2.3).

Table 2.3 Geographic affinities of some of the Himalayan genera

Geographic affinities	Representative plant genera	References
Cosmopolitan	*Cyperus, Euphorbia, Cuscuta, Drosera, Juncus, Liparis, Luzula, Liparis, Plantago, Polygonum, Solanum, Polygonum, Poa* etc.	Dar and Khuroo (2020a, b, c)
Tropical	*Desmodium, Musa, Rotal, Strobilanthes, Saccharum, Sida, Ficus, Crotalaria,* etc.	Dar and Khuroo (2020a, b, c)
Central Asian	*Allardia, Aphragmus, Christolea, Desideria, Lindelofia, Leptorhabdos, Lipschitziella, Mandragora, Paraquilegia, Pegaeophyton, Phaeonychium, Thylacospermum, Anthogonium, Atelanthera, Lasiocaryum, Perovskia, Lasiocaryum*, etc.	Dar and Khuroo (2020a, b, c)
Sino-Himalayan	*Dobinea, Echinacanthus, Herpetospermum, Holboellia, Lamiophlomis, Leycesteria, Lithostegia, Meconopsis, Microula, Monochosorum, Neillia, Oreocomopsis, Omphalogramma, Onosma, Orinus, Toricellia, Ponerorchis, Prinsepia, Pseudostachyum,*	Dar and Khuroo (2020a, b, c)

(continued)

Table 2.3 (continued)

Geographic affinities	Representative plant genera	References
	Drepanostachyum, Pterocyclus, Risleya, Roscoea, Sinocrassula, Omphalogramm, Maharanga, Merillopanax, Soroseris, Spongiocarpella, Lamiophlomis, Decaisnea, Chaetoseris etc.	
East Asian	*Hemiphragma, Hydrangea, Leptodermis, Lasionotus, Neottianthe, Oreorchis, Phryma, Phyllostachys, Cephalotaxus, Cardiocrinum, Cleyera, Tricyrtis, Phryma, Neottianthe, Boenninghausenia, Gamblea, Gastrochilus, Aucuba* etc.	Dar and Khuroo (2020a, b, c)
Southeast Asian	*Buchnania, Clarkella, Colquhounia, Corallodiscus, Cranitome, Dittoceras, Engelhardtia, Exbucklandia, Falconeria, Lasiococca, Leucosceptrum, Loxostigma, Monomeria, Papilionanthe, Phaialacanthus, Pleione, Poikilospermum, Sabia, Sarcopyramis, Streptolirion, Boeica, Caulokaempferia* etc	Dar and Khuroo (2020a, b, c)
Eurasian	*Biebersteinia, Lamium, Lomatogonium, Paeonia, Loxostigma, Falconeria, Natsiatum, Saxifraga, Sedum, Engelhardtia, Taucheria, Cranitome, Colquhounia, Patrinia, Peucedanum, Atropa, Carum, Rheum, Saussurea, Lathraea* etc.	Dar and Khuroo (2020a, b, c)

2.7.1 Forest Types and Endemic Plant Species

The flora of the Himalayas includes tropical forests, subtropical forests, temperate forests and alpine meadows. Each forest type supports different plant communities, ranging from broadleaf trees to conifers and alpine shrubs (Pullaiah 2024a, b). About one-third of Himalayan flora is native to the region. Out of the 10,000 species of flowering plants present in the Himalayas, nearly 3160 (31.6%) are native (Myers et al. 2000; Pullaiah 2024a, b). Six families, viz. *Tetracentraceae*, *Hamamelidaceae*, *Circaeasterceae*, *Butomaceae*, *Brachycaulaceae* and *Stachyuraceae*, and nearly 71 genera, including *Actinocarya*, Arcyosperma, *Brachycaulos*, *Bryocarpum*, *Bulleyia*, *Bhutanthera*, *Catamixis*, *Cyananthus*, *Cyathopus*, *Dodecadenia*, *Edgaria*, *Eurycarpus*, *Ivanjohnstonia*, *Kedarnatha*, *Ladakiella*, *Lalldhowjia*, *Cautleya*, *Kashmiria*, *Cavea*, *Circaeaster*, *Lomatogoniopsis*, *Milula*, *Nardostachys*, *Neopicrorhiza*, *Oreosolon*, *Orthoraphium*, *Dolomiaea*, *Platystemma*, *Parakaempferia*, *Paroxygraphis*, *Parrotiopsis*, *Pycnoplinthus*, *Picrorhiza*, *Piptanthus*, *Pycnoplinthopsis*, *Roylea*, *Risleya*, *Setulocarya*, *Sphaerosacme*, *Solms-laubachia*, *Jaeschkea*, *Stapletonia*, *Stelbanthus*, *Stracheya*, *Tibetia*, *Pseudodanthonia*, *Tordyliopsis*, *Megacodon*, *Trachydium*, etc. are endemic to the Himalaya. Of these, nearly one-third genera are monotypic (Uniyal and Mathur 1994; Rana and Ranade 2009). The Himalayas are home to numerous endemic plant species, including the Himalayan blue poppy (*Meconopsis* spp.) and various species of *Rhododendron*, and *Calamus*. *Calamus pseudoerectus*, a climbing rattan palm that was recently discovered in the Eastern Himalayas, is characterised by its distinctive morphological features, which include cirrate leaves, densely spinous petioles, and tubular sheaths with light brown spines. The subtropical forests of Darjeeling, India, especially the Dalkajhar forest near Bagdogra, are home to this species. Road development, the growth of tea plantations, and deforestation for resources and tourists are some of the activities endangering its habitat. Because of its small population and restricted geographic range, the species is listed as endangered and protection is advised. Local communities make use of its tender shoots, and wildlife, particularly elephants, eats its fruits (Mondal and Chowdhury 2020). *Rhododendron*s, which thrive at elevations between 1600 and 3600 m, are found in 36 species and 45 varieties in the Sikkim and Darjeeling Himalayas. These plants are important for their ecological and cultural significance in addition to their biodiversity. Threatened species like *Rhododendron niveum* and *Rhododendron dalhousiae* have been successfully propagated through conservation efforts employing biotechnological techniques, such as tissue culture, which has helped to preserve them in specialised conservation parks (Pradhan and Lachungpa 1990). These plants (Table 2.4) are adapted to the specific environmental conditions of their respective regions (Shrestha 1997a, b; Dar and Khuroo 2020a, b, c).

Table 2.4 Taxonomic diversity of various groups in the Himalayan ecosystem

Taxonomic group	Estimated no. of species	Representative species	% of Himalayan flora	Ecological significance/ Notes	References
Monera (Bacteria)	Data unavailable	*Escherichia coli*, *Pseudomonas* spp., *Bacillus* spp.	Unknown	Vital for soil nutrient cycling; species-specific diversity poorly studied	Bhat et al. (2021a, b, c)
Protista (Protozoa)	Data unavailable	*Entamoeba histolytica*, *Giardia lamblia*	Unknown	Common waterborne parasites; regional biodiversity data limited.	Sharma and Singh (2020a, b)
Fungi (excluding lichens)	~150,000	*Aspergillus* spp., *Penicillium* spp., *Rhizopus* spp.	~5%	Important for nutrient recycling and traditional healing practices.	Kumar et al. (2019a, b)
Lichens	~1159	*Parmotrema* spp., *Cladonia* spp., *Usnea* spp.	~3.7%	Bioindicators of pollution; prevalent in alpine regions.	Joshi and Upreti (2015a, b)

(continued)

Table 2.4 (continued)

Taxonomic group	Estimated no. of species	Representative species	% of Himalayan flora	Ecological significance/ Notes	References
Algae	~500	*Chlorella* spp., *Spirogyra* spp., *Ulva* spp.	~0.5%	Found in glacial and high-altitude water bodies	Sharma et al. (2017a, b, c)
Bryophytes	~2000	*Marchantia polymorpha*, *Sphagnum* spp.	~1%	Support wet-land function and soil moisture retention	Bansal and Chandra (2020)
Pteridophytes	~600	*Dryopteris* spp., *Adiantum* spp.	~0.6%	Thrive in shady, moist habitats; vital to forest understorey	Rana et al. (2009)
Gymnosperms	~50	*Pinus wallichiana*, *Cedrus deodara*, *Taxus baccata*	~0.1%	Dominate conifer zones; medicinal and timber values	Negi et al. (2018a, b)
Angiosperms	~10,000	*Rhododendron* spp., *Primula* spp., *Saussurea obvallata* (Brahma Kamal)	~30%	Most dominant group; rich in medicinal, ornamental, and cultural value	Gaur et al. (2019a, b, c)

2.7.2 Faunal Diversity of the Himalayas

The Himalayas are one of the most significant global biodiversity hotspots because of their remarkable biodiversity. Numerous endemic species are buoyed by the region's diverse range of temperatures and stark elevation differences. The fauna of the Himalayas is mostly composed of two types of animals: vertebrates and invertebrates. These groups are essential to maintaining the overall biological equilibrium and viability of the ecosystems in the region. Because of their distinctive adaptations to the region's high altitude, severe climate, and isolation, many Himalayan species are endangered and in need of conservation.

2.7.2.1 Invertebrates

Invertebrates make up a significant fraction of the biodiversity found in the Himalaya. The main group of invertebrates in the region is called Arthropoda, and it consists of insects, arachnids, and crustaceans. It is estimated that the Himalayas are home to over 15,000 species of arthropods, including popular ones like the Asian honeybee (*Apis cerana*), which is essential for pollination, and the swallowtail butterfly (*Papilioma chaon*), which is required to maintain the ecosystem's diversity. Arthropods, especially bees, beetles, and butterflies, are necessary for pollination, which impacts the capacity of flowering plants to proliferate. They also help with the cycling of nutrients and the management of pests because they are scavengers and decomposers (Kumar and Dar 2020a, b).

Moreover, an important component of the ecosystem in the Himalayan region is molluscs and annelids. For instance, molluscs that filter water, such as *Lymnaea stagnalis* and *Pila globosa*, are found in the aquatic habitats of the Himalayas and are essential to controlling the quality of the water. For the purpose of breaking down organic matter and aerating the soil, annelids like *Eisenia fetida* and *Lumbricus terrestris* are crucial. The Himalayan woods and grasslands depend on the nitrogen cycle that these organisms foster, which also increases soil fertility (Sharma and Khuroo 2020a, b).

One of the distinguishing characteristics of Himalayan invertebrates is the presence of rotifers and nematodes in the soil and freshwater bodies. Rotifers such as *Brachionus plicatilis* thrive in the cold water of high-altitude lakes, contributing to the nutrient cycle and serving as food for other aquatic organisms (Negi et al. 2020a, b). Nematodes interact with the microbial communities in the soil and encourage decomposition, which improves the soil's overall health. This is notably true with *Ascaris lumbricoides* and *Caenorhabditis elegans*. At different elevations in the Himalayas, their existence contributes to the soils' stability and fertility (Sharma et al. 2018a, b).

2.7.2.2 Vertebrates

There is an extensive diversity of vertebrates in the Himalayas. The rivers and high-altitude lakes in the area are frequently home to fish, particularly cold-water species like *Schizothorax* spp. (Snow Trout) and *Salmo trutta* (Brown Trout). Because these animals may fish for food and a means of sustenance, they are essential to the way of life of the local communities. They have adapted to the tough, low-oxygen surroundings. Fish populations in the Himalayan freshwater habitats are directly impacted by pollution, habitat degradation, and climate change. These ecosystems are extremely sensitive to environmental changes (Bhat et al. 2021a, b, c).

According to eminent scientist Dhan Bahadur Gurung, it is the first endemic fish species in Bhutan and unique to the entire globe. The nation where the fish is native is honoured by the name *Parachiloglanis bhutanensis*, given to the fish. The village of Khaling, through which the creek where it was found flows, is mentioned in the common name, Khaling Torrent Catfish. The catfish was found by the researchers around 1 km east of the Khaling settlement. The indigenous fish has a flat body and a large, blunt, rounded head that rises to a point that is vertically aligned with the eye at a roughly 45° angle from the snout. From the tip of its nose to the end of its tail, the fish's length is 101 mm.

The fish was also noted by researchers to have other characteristics including thick, leathery fins. One of the physical components that makes it an unusual fish is the ventral aspect of the mouth and snout, which mimics the form of a bell or hammer. The back flap of tissue that connects the maxillary barbel to the mouth is what gives this form (Gurung 2017).

Despite their diminutive numbers, reptiles and amphibians are crucial for maintaining the balance of ecosystems. Significant bioindicators are amphibians, such as *Bufo melanostictus* (Asian Common Toad) and *Nanorana* spp., because of their sensitivity to environmental changes. Because they prey on smaller species, reptiles like *Varanus bengalensis*, commonly known as the *Bengal Monitor*, and *Najanaja*, often known as the Indian Cobra, plays a crucial role in preserving a healthy ecology (Dar and Khuroo 2020a, b, c). The Himalayas are home to a wide range of amphibians and reptiles, such as the Himalayan pit viper (*Gloydius himalayanus*) and the Kashmir frog (*Bufo viridis*). Due to the particular climatic conditions in the area, these species have adapted (Shrestha 1997a, b).

The remote jungles of northeastern India's Arunachal Pradesh state are home to *Megophrys ancrae*, a newly discovered species of frog. The distinctive feature of this species, which is frequently called the "horned frog," is its long, horn-like upper eyebrows. Members of the Megophryidae family are litter frogs, also known as *Megophrys* frogs. Native to Southeast Asia, this family ranges from the eastern foothills of the Himalayas to the Greater Sunda Islands, Indonesia, and the Philippines (Inger and Colwell 2014). According to Frost et al. (2014), the Megophryidae family included around 180 species distributed throughout nine genera as of 2014.

These frogs have become famous for possessing exceptional camouflage that allows them to seamlessly integrate into their surroundings. Some species have skin folds that mimic leaf veins because of how well they blend in, and horned frogs' head projections help them imitate fallen leaves. This adaptation helps the frog evade predators by confusing or discouraging them with its unique features (Nabhitabhata et al. 2014).

The Himalayas are home to an astounding 700 different kinds of birds, demonstrating the region's unparalleled avian diversity. Famous species include the Himalayan vulture, *Gyps himalayensis*, and the White-bellied Redstart, *Phoenicurus erythrogastrus*. Numerous of these species migrate, covering great distances to reproduce or find food within the Himalayan region. Birds perform ecological tasks like pollination and seed distribution in addition to adding to the area's aesthetic appeal. Some species are culturally and symbolically significant to the people living in the area, such as the national bird of Nepal, the Himalayan Monal (*Lophophoru simpejanus*) (Gaur et al. 2019a, b, c).

The Snow Leopard (*Panthera uncia*) is a flagship species for conservation among the *Himalayan mammals*. As an apex species, the elusive predator represents the well-being of the Himalayan ecosystem. Other notable mammals are the Red Panda (*Ailurus fulgens*), which is endangered due to habitat degradation and poaching, and the Himalayan Musk Deer (*Moschus chrysogaster*), which is hunted for its prized musk. Because they keep other species' populations under check and because their decrease can cause ecological imbalances, these mammals are essential to the health of ecosystems (Negi et al. 2020a, b).

2.7.2.3 Important Fauna of the Himalayan Region

The Himalayan region, renowned for its rich biodiversity, is home to a variety of remarkable animal species (Table 2.5). Below is an overview of notable fauna found in this region.

Red Panda (*Ailurus fulgens*) primarily found in temperate woods with thick bamboo undergrowth; the red panda is a symbol of the Himalayan region. It is found in Nepal, northern India, Bhutan and southwestern China. Because to habitat loss, hunting, and climate change, this arboreal mammal is in risk of extinction. Because they keep bamboo forests in balance, red pandas are essential to their ecology. To protect their populations, conservation measures are essential (Negi et al. 2020a, b).

Snow Leopard (*Panthera uncia*) known as the "ghost of the mountains," the snow leopard lives between 3000 and 5500 m.a.s.l. in the Himalayan alpine and subalpine regions. This apex predator regulates herbivore populations, which is a vital ecological function. Because of habitat loss, poaching, and conflicts between people and wildlife, it is endangered. The survival of this elusive species depends on conservation initiatives, such as habitat preservation and anti-poaching tactics (Jackson and Ahlborn 1988).

The Eurasian Lynx (*Lynx lynx*) thick-furred and suited to frigid climes, the Eurasian lynx can be found in the upper elevations of the Himalayas. It maintains the balance of the prey population by feeding on small animals and birds as a solitary and cunning predator. Its survival is threatened by habitat fragmentation and hunting, underscoring the necessity of conservation efforts (Dar and Khuroo 2020a, b, c).

The Tibetan wolf (*Canis lupus chanco*) the Himalayan rocky areas and high-altitude grasslands are home to the Tibetan wolf. This grey wolf subspecies is an apex predator that feeds on herbivores such as yaks and blue sheep. Measures to lessen human-wolf interactions are necessary because the main threats to its population are habitat degradation and retaliatory kills by herders (Negi et al. 2020a, b).

Red Fox (*Vulpes vulpes*) the red fox is found all over the Himalayas and thrives in a variety of environments, including alpine tundra and woodlands. This versatile species contributes to seed distribution and pest control. Hunting and habitat loss can affect local populations even when they are reasonably prevalent (Shrestha 1997a, b).

Tibetan Fox (*Vulpes ferrilata*) Native to the high-altitude Himalayan plains, the Tibetan fox has adapted admirably to the hostile, dry climate. Pikas and other small mammals make up the majority of its diet. Despite its constant population, future problems may arise from habitat damage caused by livestock grazing (Dar and Khuroo 2020a, b, c).

Bucerotidae (*Buceros bicornis*) the lower Himalayan woods are home to a number of hornbill species, including the Great Hornbill. They are frugivorous birds that are essential to the spread of seeds and the regrowth of forests. Their existence is seriously threatened by habitat loss brought on by hunting and deforestation. For Himalayan forests to remain ecologically healthy, hornbill populations must be protected.

Table 2.5 Faunal diversity in the Himalayas with taxonomic groups and ecological significance

Taxonomic group	Estimated no. of species	Representative species	% representation in region	Ecological role and significance	References
Invertebrates					
Platyhelminthes (Flatworms)	Data unavailable	*Fasciola hepatica*, *Schistosoma* spp.	Unknown	Parasitic helminths in freshwater ecosystems; impact livestock and human health.	Sharma et al. (2018a, b)
Rotifera (Wheel animals)	Data unavailable	*Brachionus plicatilis*, *Philodina* spp.	Unknown	Key plankton in high-altitude water bodies; help in nutrient cycling and food web stability.	Gaur et al. (2019a, b, c)
Nematoda (Roundworms)	Data unavailable	*Ascaris lumbricoides*, *Caenorhabditis elegans*	Unknown	Regulate microbial populations and contribute to decomposition in soil.	Negi et al. (2020a, b)
Annelida (Segmented worms)	Data unavailable	*Eisenia fetida*, *Lumbricus terrestris*	Unknown	Soil engineers, promoting aeration and organic matter turnover.	Pandian (2019a, b)
Arthropoda	~15,000	*Apis cerana* (Honey bee), *Papilioma chaon* (Swallowtail), *Hippodamia variegata* (Ladybird beetle)	~45%	Most diverse group; crucial for pollination, pest regulation, and ecosystem resilience.	Kumar and Dar (2020a, b)
Mollusca	~500	*Pila globosa*, *Lymnaea stagnalis*	~1.5%	Maintain aquatic ecosystem balance; serve as bioindicators of pollution.	Sharma and Khuroo (2020a, b)
Vertebrates					
Fishes	~200	*Schizothorax* spp., *Salmo trutta* (Brown trout)	~0.5%	Cold-water species; important for alpine river and lake ecology and local fisheries.	Bhat et al. (2021a, b, c)
Amphibians	~50	*Bufo melanostictus*, *Nanorana* spp.	~0.1%	Sensitive to environmental stress; serve as early indicators of ecosystem change.	Gaur et al. (2019a, b, c)
Reptiles	~100	*Varanus bengalensis*, *Naja naja* (Cobra)	~0.3%	Regulate pest populations and serve vital links in trophic chains.	Sharma et al. (2017a, b, c)
Birds	~700	*Phoenicurus erythrogastrus* (Redstart), *Gyps himalayensis* (Himalayan Vulture)	~2.5%	Rich avifaunal diversity, many endemic; essential for eco-tourism and seed dispersal.	Dar and Khuroo (2020a, b, c)
Mammals (Wild)	~100	*Panthera uncia* (Snow Leopard), *Moschus chrysogaster* (Himalayan Musk Deer)	~1%	Charismatic megafauna; critical to high-altitude ecosystem health and global conservation priorities.	Negi et al. (2020a, b)

Tiger (*Panthera tigris*) One of the most recognisable animals in the area is the Bengal tiger, which is located in the foothills of the Himalayas. As solitary predators, tigers regulate prey populations, which is an important function they do for the ecology. According to Dinerstein et al. (2017), conservation activities are essential to ensuring their survival since they face risks from habitat loss and poaching.

Sloth bear (*Melursus ursinus*) which is indigenous to the Indian subcontinent, is well-known for both its shaggy coat and its unusual eating technique, which involves sucking insects with its lips. Many forested areas, including sections of the Himalayan foothills, are home to this bear (Sukumar 2003).

Long-tailed Marmot (*Marmota caudata*) in the Himalayan highlands, the long-tailed marmot can be found. This ground-dwelling rat hibernates in the winter and is well suited to the hard alpine environment. It influences plant communities and soil structure, which contributes to its role in the ecosystem (Ruppert et al. 2016).

Yellow-throated Marten (*Martes flavigula*) This medium-sized mammal, which lives in the forests of the Himalayas, is easily recognised by its distinctive yellow throat patch. The perceptive yellow-throated marten eats a variety of small animals and insects (Gurung et al. 2020).

Long-eared Hedgehog (*Hemiechinus auritus*) is the long-eared hedgehog, which can be found in the lower Himalayan regions, is identified by its long ears and spiky coat. It feeds on small invertebrates like worms and insects and is nocturnal (Yakovlev et al. 2018).

Chinkara (*Gazella bennettii*) also called the Indian gazelle, is found in the arid and semi-arid regions of the Himalayas. It is well known for moving quickly and gracefully, and it does well in dry settings. The species is at risk from hunting and the destruction of its natural habitat (Bhardwaj et al. 2019).

Blackbuck (*Antilope cervicapra*) is unique in that they are native to the Indian subcontinent and are incredibly beautiful. Massive, swirling horns are used to identify males. Although it is not typically found there, it inhabits the grasslands and slopes of the Himalayas (Sankaran et al. 2008).

Baya Weaver (*Ploceus philippinus*) well-known small bird for its intricate nest-building methods resides in the Himalayan foothills. The creation of elaborate hanging nests to attract females is a vital part of the reproductive strategy employed by male baya weavers.

2.8 Conclusions

The Himalayas, one of the world's most critical biodiversity hotspots, play a pivotal role in sustaining global ecological stability and supporting millions of livelihoods across South and Central Asia. The region's exceptional species richness and endemism are primarily shaped by its complex topography, elevation gradients, and diverse microclimates, which together create a mosaic of unique ecosystems ranging from subtropical forests to alpine tundra.

However, the ecological integrity of the Himalayas is increasingly under threat from a range of anthropogenic pressures, including unsustainable resource extraction, habitat fragmentation, infrastructure development, and climate change. These drivers are accelerating the loss of biodiversity, destabilising ecosystem functions, and undermining the socioeconomic resilience of local communities. To address these challenges, there is an urgent need for comprehensive and integrated conservation strategies. These should prioritise: Habitat protection and restoration in ecologically sensitive zones; Sustainable natural resource management, balancing ecological limits with human needs; Legislative and policy frameworks that enforce environmental safeguards and incentivise conservation.

Equally essential is the active engagement of local communities, whose traditional knowledge, dependence on natural resources, and lived experiences are invaluable for the success of conservation initiatives. Empowering these communities through participatory governance, capacity building, and benefit-sharing mechanisms can foster long-term stewardship of Himalayan ecosystems. Protecting the Himalayas requires a multifaceted approach—combining robust scientific research, climate-adaptive development planning, and cross-border cooperation. This chapter underscores the need for an integrated strategy that places ecological conservation, cultural preservation, and sustainable development at the core of regional and global agendas. Only through such coordinated and forward-looking efforts can we ensure the long-term resilience, ecological functionality, and heritage value of the Himalayas for present and future generations.

Acknowledgements The authors express their sincere gratitude to the Faculty of Life Sciences, University of Okara, for providing the necessary facilities and support to carry out this work. We also thank our colleagues and peers for their valuable suggestions and encouragement during the preparation of this article.

References

An Z, Kutzbach JE, Prell WL, Porter SC (2014a) Evolution of Asian monsoons and phased uplift of the Himalaya–Tibetan plateau since Late Miocene times. Nature 411:62–66

An L, Zhang Q, Yu Z, Zhang J (2014b) Microclimate variability and ecological diversity in the Himalayas. J Mount Sci 11(3):785–798

Badola R (2009) Mountain biodiversity and livelihood security: case of the Indian Himalayas. TERI Press, New Delhi

Bandyopadhyay J, Parveen S (2004) Conservation of Himalayan biodiversity: legal and policy imperatives. Mount Res Develop 24(3):208–211

Bansal P, Chandra S (2020) Bryophyte diversity and conservation in the Indian Himalayas. Biodivers Conserv 29(2):123–134

Bhardwaj M, Arora R, Mishra R (2019) Conservation status of Chinkara (*Gazella bennettii*) in the semi-arid regions of India. Wildlife Bio 25(2):101–110

Bhat A, Sharma R, Puri P, Bhau B (2021a) Diversity and conservation status of endemic flora in the Indian Himalayas. Biodivers Conserv 30(4):943–960

Bhat SA, Wani KA, Rafiq R (2021b) Freshwater fish diversity in high-altitude Himalayan lakes. Aqua Biodiv J 14(2):233–245

Bhat SA, Wani KA, Rafiq R (2021c) Soil microbial diversity and function in Himalayan agro-ecosystems. Environ Microbial Rep 18(3):201–209
Chettri N, Shakya B, Thapa R, Sharma E (2023) Himalaya: mountains of life—biodiversity in the eastern Himalayas. Conservation and Sustainable Use Report, ICIMOD
Dar GH, Khuroo AA (2020a) An updated checklist of flowering plants of the Indian Himalayan region: with conservation status and distribution range. Biodivers Conserv 29(5):1121–1146
Dar GH, Khuroo AA (2020b) Biodiversity of the Indian Himalayas: endemism, challenges, and conservation management. J Himalayan Eco Sustain Devlop 13(2):23–39
Dar GH, Khuroo AA (2020c) Avifaunal diversity and conservation in the Indian Himalayas. J Ornithol Ecol 22(4):195–210
Davis SD, Heywood VH, Hamilton AC (1995) Centres of plant diversity: a guide and strategy for their conservation. WWF/IUCN, Cambridge
Dinerstein E, Loucks C, Wikramanayake E (2017) The status of Bengal tiger populations in the Himalayan foothills. Ecol Evolution 8(3):355–366
Frost D, Grant T, Faivovich J, Bain R, Haas A, Haddad C, de Sá R, Channing A, Wilkinson M, Donnellan S (2014) The amphibian tree of life. Bullet Am Mus Nat Hist 297(1):1–370
Gansser A (1964) The geology of the Himalayas. Interscience Publishers, New York
Gaur RD, Tewari LM, Purohit AN, Tiwari U (2019a) Endemic and endangered plants of the Western Himalaya. Biodivers Conserv 8(4):355–374
Gaur RD, Tiwari JK, Purohit AN (2019b) Ecological roles of microinvertebrates in Himalayan aquatic ecosystems. Himal Biosci Rev 9(1):42–56
Gaur RD, Tiwari JK, Purohit AN (2019c) Flora of the Indian Himalaya: biodiversity and conservation concerns. Botanical Survey of India Bulletin 61(1):112–129
Gurung H (2012) The geography of Nepal and Himalayas, 3rd edn. Himal Books
Gurung DB (2017) New species of Khaling torrent catfish discovered in Bhutan. J Bhutanese Biodiver 6(1):45–52
Gurung DB, Arora R, Rawat GS (2020) Ecology and behavior of yellow-throated Marten in the Himalayas. Himalayan J Ecol 15(3):203–211
Inger RF, Colwell RK (2014) Species of horned frogs in the family Megophryidae in Southeast Asia. J Trop Ecol 12(4):65–81
Jackson R, Ahlborn G (1988) Observations on the ecology of snow leopard in West Nepal. In: Proceedings of the fifth international snow leopard symposium, pp 65–87
Joshi S, Upreti DK (2015a) Diversity and distribution of lichen species in the Indian Himalayas. Environ Sci J 30(2):123–134
Joshi Y, Upreti DK (2015b) Lichen diversity in the Indian Himalaya: a bioindication perspective. Environ Monitor Assess 187(9):583
Körner C (2003) Alpine plant life: functional plant ecology of High Mountain ecosystems. Springer. https://doi.org/10.1007/978-3-642-18970-8
Kumar R, Dar MA (2020a) Arthropod diversity in the Himalayan landscape: patterns and implications. J Entomol Stud 13(3):145–160
Kumar S, Dar GH (2020b) Arthropods and invertebrates in the Himalayas: a study of ecological significance. Himalayan J Bio 12(2):85–99
Kumar M, Sharma A, Prasad R (2019a) Fungi and their ethnomedicinal uses in the Himalayas. Mycosphere 10(5):931–945
Kumar A, Singh R, Reddy MS (2019b) Role of fungi in nutrient recycling in the Himalayan ecosystems. Myco Res 15(2):56–67
Lindenmayer DB, Franklin JF (2013) Conserving forest biodiversity: a comprehensive multiscaled approach. Island Press, Washington, DC
Maiti P, Maiti P (2017) Biodiversity: perception, peril and preservation, 2nd edn. PHI Learning, New Delh
Miller KR (1999) Living in the environment: principles, connections, and solutions, 11th edn. Brooks/Cole Publishing Company

Mondal AK, Chowdhury S (2020) Plant resources and their utilization in the eastern Himalayas. Springer

Myers N, Mittermeier RA, Mittermeier CG, da Fonseca GAB, Kent J (2000) Biodiversity hotspots for conservation priorities. Nature 403(6772):853–858. https://doi.org/10.1038/35002501

Nabhitabhata J, Chan-Ard T, Frost DR (2014) Amphibians and reptiles of Southeast Asia and the Himalayas: a field guide. Southeast Asian J Herpetol 7(3):121–150

Negi CS, Bhatt A, Rawat DS (2018a) Diversity and conservation status of gymnosperms in the Indian Himalayas. Fores Ecol Manag 412(1):234–248

Negi GCS, Singh SP, Kuniyal JC (2018b) Coniferous forests of the western Himalaya: ecology, conservation and sustainable use. Himalayan J Forest 28(3):211–226

Negi CS, Rawal RS, Chaturvedi P (2020a) Endangered vertebrate species in the Himalayas and their conservation status. J Conserv Bio 18(2):165–178

Negi GCS, Singh SP, Kuniyal JC (2020b) Soil nematode and microbial biodiversity in high-altitude regions of the Indian Himalaya. Mount Ecosystem J 18(2):87–98

Oli KP, Gupta R (2008) Sustainable use of biodiversity in the Indian Himalayas. IUCN India Country Office

Pandian TJ (2019a) Annelids and soil fertility in montane regions. Ind J Soil Biol 6(3):112–121

Pandian TJ (2019b) Reproduction and development in Annelida. CRC Press

Pradhan S, Badola R (2008) Biodiversity and livelihood: case study of sustainable use of resources in the Indian Himalayas. Int J Sustain Dev World Ecol 15(3):214–220

Pradhan UC, Lachungpa ST (1990) Sikkim-Himalayan *Rhododendrons*. Primulaceae Books

Pullaiah T (2024a) Biodiversity of the Himalayas: conservation and sustainability. Springer

Pullaiah T (2024b) Global biodiversity, volume 3: selected regions and countries. CRC Press, Boca Raton

Rana TS, Ranade SA (2016) Diversity of pteridophytes in the Indian Himalayan region. Fern Gazette 20(1):45–59

Rana TS, Ranade SA (2009) Endemic flora of the Indian Himalayas and its conservation status. Curr Sci 97(5):667–671

Ruppert N, Byers DA, Arndt J (2016) Impact of marmots on plant communities in the alpine zone. J Alpine Ecol 10(1):135–147

Sankaran R, Rahmani AR, Manakadan R (2008) Blackbuck and its grassland habitat: a study on conservation and management. Conserv Sci J 12(2):78–89

Sharma K, Khuroo AA (2020a) Freshwater mollusks of the Himalaya: ecological roles and threats. J Limnol 31(2):101–118

Sharma P, Khuroo AA (2020b) Soil invertebrates and their role in nutrient cycling in the Himalayan forests. Himalayan Environ J 13(1):95–110

Sharma KP, Singh A (2020a) Protozoan diversity in freshwater bodies of the Indian Himalayas. Inter J Protozool 8(1):24–31

Sharma P, Singh V (2020b) Protozoan diversity in Himalayan water bodies. J Parasitol 9(4): 155–167

Sharma S, Negi V, Rawat S (2017a) Algae diversity in high-altitude lakes of the Himalayas. J Phycol 14(2):235–244

Sharma R, Verma A, Mehta P (2017b) Freshwater algal flora of high-altitude Himalayan lakes. J Phycol Res 29(2):77–84

Sharma R, Verma A, Mehta P (2017c) Reptilian diversity and conservation threats in the Himalayan foothills. Herpetol Conserv Notes 29(1):75–84

Sharma E, Chettri N, Tse-ring K (2018a) Himalayan biodiversity in the changing world. Springer

Sharma V, Tiwari R, Bhatnagar M (2018b) Helminth parasites of livestock and freshwater systems in the Himalayas. J Parasitol Res 16(4):223–233

Shrestha BB (1997a) Amphibians and reptiles of the Himalayas: their ecological and conservation significance. Environ Conserv J 21(1):85–92

Shrestha BB (1997b) Conservation of endemic species in the Himalayas: biodiversity hotspot and beyond. Kathmandu Univ Sci Eng Technol 6(3):87–100

Singh P (2017) Himalayan ecology and environment. Bishen Singh Mahendra Pal Singh Publishers, Dehradun

Sukumar R (2003) The sloth bear in the Indian subcontinent. Int J Mammalian Biol 68(2):225–235

Takhtajan A (1986) Floristic regions of the world. University of California Press, Berkeley

Uniyal BP, Mathur K (1994) Endemic flora of the Himalayas. Bishen Singh Mahendra Pal Singh

Valdiya KS (2015) The making of India: geodynamic evolution. Springer, pp 241–243

Watson J, Dudley N, Segan DB (2011) The performance and potential of protected areas. Nature 474(7353):254–258

Yakovlev A, Kuznetsov O, Yershov O (2018) Ecology and adaptation of long-eared hedgehogs in Himalayan habitats. J Mammal 42(4):565–573

Zachos FE, Habel JC (2011) Biodiversity hotspots: distribution and protection of conservation priority areas. Springer, Heidelberg

Chapter 3
Floristic and Fungal Diversity and Vegetation Pattern of the Central Himalayas

Surya Kumar Maharjan, Pabitra Jha, and Ganga Datt Bhatt

Abstract The Central Himalayas, where elevation varies from few hundred meters in the south to the highest peak of the world in the north, presents one of the longest elevational gradients in the world. Besides, it lies at the crossroads of the Indomalayan and the Palearctic realms and the Western and Eastern Himalayas. Owing to this elevational gradient and the associated changes in temperature, precipitation, topography, and soil conditions, and the unique geographic position, the Central Himalayas is a home to an incredibly rich floristic diversity. In this chapter, we have made an effort to enrich the information compiled from the published and unpublished literature with observations made during our works/ explorations in the region over the past few decades to present comprehensive summaries of floristic and fungal diversity and vegetation patterns in the Central Himalayas. In addition, we have also made an effort to document major conservation issues being faced by the biodiversity of the Central Himalayas. Finally, we concluded the chapter with the conservation initiatives being undertaken in response to those conservation issues in the Central Himalayas.

Keywords Biodiversity · Conservation · Elevational gradient · Nepal Himalayas · Terrestrial plant diversity

S. K. Maharjan (✉)
Tribhuvan University, Institute of Forestry, Hetauda Campus, Makawanpur, Bagmati Province, Nepal
e-mail: surya.maharjan@hc.tu.edu.np

P. Jha
Tribhuvan University, Institute of Forestry, Pokhara Campus, Kaski, Gandaki Province, Nepal

G. D. Bhatt
Department of Plant Resources, Plant Research Center, Kailali, Sudurpashchim Province, Nepal

D. Das et al. (eds.), *Plant Diversity and Conservation Strategies of the Himalayan Biodiversity Hotspot*, Topics in Biodiversity and Conservation 21,
https://doi.org/10.1007/978-3-032-21697-7_3

3.1 Introduction

3.1.1 The Himalayas and Its Longitudinal Division

The Himalayas is a mountain range in Asia that separates the Indo-Gangetic Plain from the Tibetan Plateau (Gurung 2021). Geologically, the Himalayas is the youngest mountain range that began forming between 45 and 55 million years ago due to the convergence and collision of the continental crust of two tectonic plates (Chap. 1, Sects. 1.2 and "Himalayan Uplift"), namely, the Indian Plate and the Eurasian Plate (Valdiya 1984; Pandit et al. 2014). It extends from Nanga Parbat in the west to Namcha Barwa in the east spanning over five countries in Asia, namely from east to west, Nepal, India, Bhutan, China and Pakistan (Fig. 3.1).

The Himalayas is made up of three parallel mountain ranges from north to south often referred to as the Greater Himalayas, the Lesser Himalayas, and the Outer Himalayas. The Greater Himalayas is the northernmost and the highest mountain range of the Himalayas with an average elevation of 6000 m.a.s.l. This range is primarily composed of granites, gneisses and other metamorphic rocks supporting its characteristic massive height and rugged terrain (Dhital 2015a). Its peaks are permanently covered in snow and are known for their glacial activities, which supply fresh water to the numerous rivers of the region including, the Indus, the Ganges, and the Brahmaputra. To the north of the Greater Himalayas is the Trans-Himalayan region (referred to as a fourth mountain range—the Tethys or Trans or Tibetan Himalayas by Singh and Pusalkar (2020)). This region is characterised by high plateaus, wide valleys, and barren landscapes. Its high elevation, which often exceeds 4000 m.a.s.l., makes it a challenging environment for floral and faunal habitation. Situated to the south of the Greater Himalayas, the Lesser Himalayas is also known as the Mahabharat range. This range is characterised by lower height and less rugged terrain. It is primarily composed of sedimentary rocks like limestone, shale, and sandstone (Dhital 2015b). Its peaks are seasonally covered in snow. It is known for its lush forests and fertile valleys between the Lesser and Greater Himalayas. The Outer Himalayas also known as the Siwalik range is the southernmost mountain range of the Himalayas. This range is characterised by lower height and the presence of round hills with foothills. It is primarily composed of unconsolidated sediments like sand, gravel and clay, which makes it geologically fragile (Dhital 2015c). It is known for its wide fertile Doon valleys between the Siwalik hills and the Lesser Himalayas and the Bhabar region that functions as a water recharge zone for the southern Indo-Gangetic Plain. To the south of the Outer Himalayas is the Indo-Gangetic Plain also known as Terai. The fertile Terai that encompasses northern regions of the Indian subcontinent, including most of modern-day northern and eastern India, most of eastern Pakistan, virtually all of Bangladesh and the southern plains of Nepal (Valdiya 2016) is known as the food basket of the Himalayas (Koshal 2014).

Singh (1971) has divided the Himalayas into eight longitudinal sections. However, for the purpose of this chapter, we divided the Himalayas into three

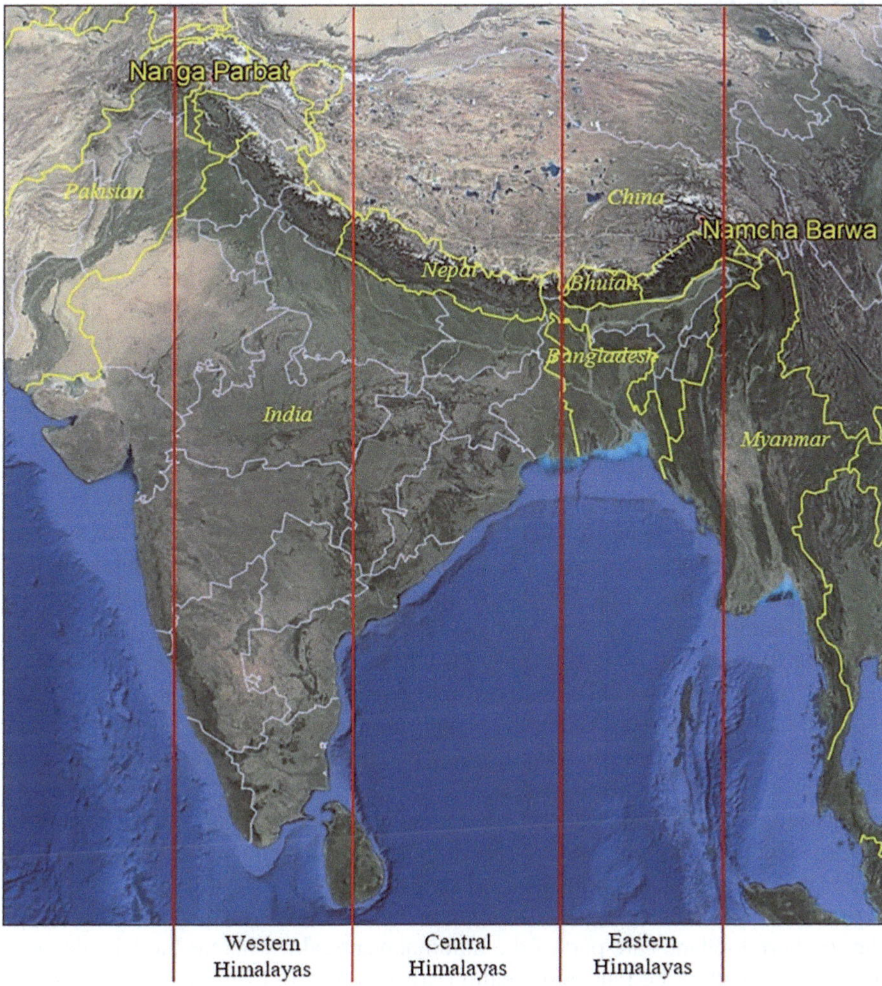

Fig. 3.1 Longitudinal division of the Himalayas. Snow-covered mountains in the upper middle portion of the image represent the Himalayas. Lines in yellow colour represent country boundaries and those in red colour are used for longitudinal sectioning of the Himalayas. (Source: Image clipped from the Googleearth.org accessed on 11th Sept 2024)

longitudinal sections i.e., Western, Central and Eastern Himalayas as has been done by Chatterjee (1964) but with some modifications. We called the western section that stretches between the Indus River in the west and the Mahakali River in the east—the Western Himalayas. The central section that stretches between the Mahakali River in the west and the Tista River in the east—the Central Himalayas. Finally, the eastern section that stretches between the Tista River in the west and the Brahmaputra River in the east—the Eastern Himalayas (Fig. 3.1).

3.1.2 The Central Himalayas and Its Uniqueness

The major part of the Central Himalayas except for the extreme eastern part, which lies in Sikkim, India, falls in Nepal. Therefore, it is often referred to as Nepal Himalayas. Since eight of the 14 peaks over 8000 m.a.s.l. namely, Mt. Everest (8849 m.a.s.l.), Kanchenjunga (8586 m.a.s.l.), Lhotse (8516 m.a.s.l.), Makalu (8485 m.a.s.l.), Cho-Oyu (8188 m.a.s.l.), Dhaulagiri I (8167 m.a.s.l.), Manaslu (8163 m.a.s.l.), and Annapurna (8091 m.a.s.l.) are part of this section (Peakbagger 2024), the Himalayas attains maximum height in this section. The elevation of the Central Himalayas varies from a few hundred meters in the foothills of the Outer Himalayas in the south to 8849 m.a.s.l., the highest peak of the world, in the north presenting one of the longest elevational gradients in the world. Temperature decreases more or less linearly along this gradient. Generally, a fall of 1 °C in the mean annual temperature can be observed at a rise of 270 m in elevation, the rate of decrease often being higher at higher elevations above 1500 m.a.s.l. (Singh and Singh 1987).

The precipitation pattern of the Central Himalayas is governed by the summer monsoon that deposits a considerable amount of precipitation on the southern slopes of the Lesser and the Outer Himalayas on the windward side. Since the summer monsoon enters from the east and migrates west, the amount of precipitation decreases from east to west, but the contribution of winter precipitation (thanks to winter westerlies) to the total annual precipitation increases (Singh and Singh 1987). In general, the topography becomes steeper and the soil becomes shallower as we climb upward and northward along the Central Himalayan gradient (Maharjan et al. 2021).

The Central Himalayas, situated at the junction of the Indomalayan and Palearctic realms, exhibit remarkable biogeographic richness. As elaborated in Chap. 2 (Sect. 2.5.2), this unique positioning fosters significant species overlap and transitional vegetation zones, enhancing both endemism and ecological resilience in the region. The southern lowlands are part of the Indomalayan realm whereas the Himalayas in the north fall within the Palearctic realm (Udvardy 1975). Consequently, the region benefits from the flora and fauna of both regions, resulting in a rich variety of species.

Thanks to the elevational gradient and the associated changes in temperature, precipitation, topography, and soil conditions, and the unique geographic position, the Central Himalayas is a hotspot of floral and faunal diversity. In this background, this chapter intends to summarise the floristic and fungal diversity and vegetation pattern of the Central Himalayas, and document major conservation issues being faced by the biodiversity of the Central Himalayas and initiatives being undertaken in response to those conservation issues in the Central Himalayas.

3.2 Insights from Literature and First-Hand Observations

In this chapter, the information compiled from the published and unpublished literature available to the authors was enriched with the observations made by the authors during their works/explorations in the region over the past few decades to present comprehensive summaries of floristic and fungal diversity and vegetation patterns in the Central Himalayas, and major conservation issues being faced by the biodiversity of the Central Himalayas and initiatives being undertaken in response to those conservation issues in the Central Himalayas.

3.3 Insights and Observations

3.3.1 The Central Himalayas: A Treasure Trove of Floristic Diversity

The Nepal Himalayas and the Sikkim Himalayas are part of the Central Himalayas. Thanks to their geographic proximity and similar climatic conditions, both regions show considerable overlap in their floristic and fungal diversity. In this chapter, we focused mainly on the floristic and fungal diversity of the Nepal Himalayas. The Nepal Himalayas exhibits exceptional floristic and fungal diversity, largely due to its position at the confluence of the Indomalayan and Palearctic realms. This biodiversity richness aligns with global evaluations of Himalayan flora, where the Central Himalayas is recognised for its high species density and significant levels of endemism (Chap. 1, Sect. 1.5, Table 1.1). Furthermore, Chap. 2 (Sect. 2.5) highlights how the region's complex terrain and ecological gradients give rise to unique microhabitats that promote species diversification and evolutionary processes. Besides, the species migrates from the western and eastern parts of the Himalayas, making it further richer in floristic and fungal diversity. The floristic and fungal diversity of the Nepal Himalayas can be divided into the following groups.

3.3.1.1 Algae

Algae are the diverse group of non-flowering plants having simple reproductive organs and lacking developed embryos after fertilization. Algae lack true roots, stems or leaves and are found in a wide variety of shapes, colours and sizes, ranging from microscopic phytoplanktons to large seaweeds. It grows in almost all kinds of habitats, be it freshwater springs, rivers and ponds or hypersaline lagoons and salt lakes. We could not find the most recent data on the algal richness of the Nepal Himalayas in published form. According to the National Herbarium & Plant Laboratories, Nepal database, there are a total of 1101 taxa of algae including 1001 algae species reported by Prasad (2013) found in Nepal. Of these, 29 species (12 species of

Chlorophyta, 9 species of Cyanophyta, 4 species of Bacillariophyta, 2 species each of Rhodophyta and Euglenophyta) are endemic to Nepal (Rai and Ghimire 2020).

3.3.1.2 Fungi

Fungi are members of the group of eukaryotic organisms that include microorganisms like yeasts, rusts, smuts, mildews, molds, and mushrooms. According to the fungal diversity update made by (Kost and Adhikari 2015; Adhikari 2016, 2021, 2023) there are a total of 2470 species of fungi found in Nepal, of which, 131 species are endemic to Nepal. Of the 2470 species of fungi found in Nepal, 1291 species are wild mushrooms, of which 159 species are edible, 74 species have some medicinal value, and 100 species are poisonous (Devkota and Aryal 2020).

3.3.1.3 Lichens

Lichens are mutualistic associations comprising a dominant fungus that provides the overall shape and one or more algae, bacteria, or other incorporated fungi (Hawksworth and Grube 2020). They can survive extreme cold and dry conditions. They remain dormant when conditions are adverse and recover quickly when conditions become favourable. The Nepal Himalayas is rich in lichen diversity. It holds the record for the highest elevation for the growth of lichens at 7400 m.a.s.l. in the Makalu-south wall area of eastern Nepal (Baniya et al. 2010; Sochting 2015). Particularly, the sub-alpine zone of the Nepal Himalayas (3000–4000 m.a.s.l.) with 249 species of lichens is known for its rich lichen diversity (Baniya 2024). A recent checklist shows that there are 1129 taxa including 1078 species and 51 infra-specific categories of lichens found in Nepal (Baniya et al. 2022).

3.3.1.4 Bryophytes

Bryophytes are the diverse group of plants after angiosperms and the oldest ever known living land plants on the Earth. Bryophytes have long been recognised as very important components of almost all the Earth's major ecosystems, and this is particularly true for Nepal, where the wide range of geology, topography and climate create ideal conditions for a wide variety of bryophytes. So far, a total of 1217 species of bryophytes have been reported from Nepal, of which 31 species are endemic to Nepal (Pradhan 2020).

3.3.1.5 Pteridophytes

Pteridophytes (ferns and ferns-allies) are non-flowering vascular plants (Fig. 3.2). So far, a total of 583 taxa (including 550 species and 33 subspecies) of pteridophytes

Fig. 3.2 *Woodwardia unigemmata* photographed from Phulchoki, Lalitpur, Bagmati Province, Central Nepal. (Photo credit: Ganga Datt Bhatt)

have been reported from Nepal, of which 3 species are endemic to Nepal (Fraser-Jenkins et al. 2015; Fraser-Jenkins and Kandel 2019; Kandel and Fraser-Jenkins 2020). Though pteridophytes are found in tropical to sub-alpine zones, the maximum species richness is observed at ca. 2000 m.a.s.l. in the Nepal Himalayas (Bhattarai et al. 2004).

3.3.1.6 Gymnosperms

Gymnosperms are seed-bearing plants with an ovule not enclosed in a carpel, as with angiosperms. This group consists of trees and shrubs. The largest group of gymnosperms is the conifers, which includes cone-bearing trees such as pine, yew, cedars, etc. (Fig. 3.3). Gymnosperms are important plants in the flora and vegetation of the Nepal Himalayas. Gymnosperms, especially conifers, play an important role in the forests of the Nepal Himalayas. A total of 26 taxa including 24 species and 2 infra-species of gymnosperms are recorded from Nepal (Shrestha et al. 2018; Pandey et al. 2020; Rajbhandari et al. 2020). Though gymnosperms are found from 201 to 5200 m.a.s.l., the species richness of gymnosperms peaks at 3300 m.a.s.l. in the Nepal Himalayas (Shrestha et al. 2018; Pandey et al. 2020).

3.3.1.7 Angiosperms

Angiosperms (flowering plants) are the largest and the most diverse group within the plant kingdom (Fig. 3.4). The systemic study of the flowering plants in the Nepal

Fig. 3.3 *Pinus roxburghii* forest photographed from Patan, Baitadi, Sudurpashchim Province, Western Nepal. (Photo credit: Ganga Datt Bhatt)

Fig. 3.4 *Rhododendron arboreum* photographed from Milke area, Sankhuwasabha, Koshi Province, Eastern Nepal. (Photo credit: Ganga Datt Bhatt)

Himalayas started as early as the early nineteenth century and till date the study is continuously going on. The number of flowering plant species in Nepal can vary depending on the source (Shrestha et al. 2022). Based on the National Herbarium and Plant Laboratories database, a total of 5619 species of flowering plants have been reported from Nepal, of which 348 taxa are endemic to Nepal. This number is lower than the estimated total of flowering plant species in the Nepal Himalayas, which is approximately 6500 species (Stearn 1978).

3.3.2 Vegetation Pattern of the Central Himalayas

Vegetation along the Central Himalayas varies notably with its varying elevation, precipitation, topography, and soil conditions. Various researchers have classified vegetation of the Central Himalayas in many ways but based on studies confined to the narrow sections of the Central Himalayas e.g., Offen et al. (2021) confined to the Sikkim Himalayas; Shrestha (1982) confined to Karnali region in North-West Nepal; Shrestha et al. (2005) confined to Trans-Himalayan Zone in North-West Nepal. In this chapter, we have made an effort to review information available in the published and unpublished literature and enrich them with our observations to come up with more comprehensive vegetation types of the Central Himalayas, which are as follows:

3.3.2.1 Tropical Deciduous Forests

These forests are confined to the foothills of the Himalayas, especially at elevations below 1000 m.a.s.l., where mean annual precipitation ranges from 1500 to 2000 mm. They are typically dominated by a semi-deciduous Dipterocarp-*Shorea robusta* (locally known as Sal), a valuable timber species. Other associated tree species include *Terminalia alata*, *Adina cordifolia*, *Lagerstroemia parviflora*, *Aegle marmelos*, *Madhuca longifolia*, *Butea monosperma*, and *Syzygium cumini*. Unlike other areas, the areas along the water courses (riverine habitats) are dominated by *Dalbergia sissoo, Senegalia catechu* and *Bombax ceiba* also referred to as the tropical deciduous riverine forests (Rokaya et al. 2024). These tropical deciduous forests are often interspersed with tropical grasslands. These grasslands are dominated by *Saccharum spontaneum*. Other associated grass species include *Phragmites karka*, *Narenga porphyrocoma*, *Themeda arundinacea* and *Arundo donax*. These grasslands provide open spaces that support different species of flora and fauna, creating a rich mosaic of habitats. The foothills of the Himalayas with this interspersion of the tropical deciduous forests, tropical deciduous riverine forests and the tropical grasslands are home to numerous renowned protected areas in Asia e.g., Shuklaphanta National Park, Bardia National Park, Chitwan National Park, Parsa National Park, etc. in Nepal; and Kishanpur Wildlife Sanctuary, Dudhwa National Park, Katarniaghat Wildlife Sanctuary, Suhelva Wildlife Sanctuary, Valmiki Tiger Reserve etc. in India (Wikramanayake et al. 2011).

3.3.2.2 Subtropical Forests

Subtropical forests are confined to elevations between 1000 and 2000 m.a.s.l. These forests feature tree species like *Schima wallichii*, *Castanopsis* sp. (together forming Schima-Castanopsis forests in moist hill slopes), *Pinus roxburghii* (found mostly on the south-facing slopes), and *Alnus nepalensis* found along the watercourses on the steeper slopes (Rokaya et al. 2024).

3.3.2.3 Temperate Forests

With increasing elevation and decreasing precipitation, the subtropical forests give way to the temperate forests, especially at elevations between 2000 and 3000 m.a.s.l. These forests feature temperate broadleaf tree species such as *Quercus* sp., *Aesculus indica*, *Juglans regia, Acer* sp., *Magnolia campbellii*, and *Rhododendron arboreum* (found predominantly on north and west facing slopes), and coniferous tree species such as *Pinus wallichiana* (found mostly on the south-facing slopes), *Abies pindrow* and *Picea smithiana* (found mainly on north and west facing slopes, Rokaya et al. 2024). These species are prominently discussed in Chap. 2 (Sect. 2.4.1.1), where temperate forests of the Lesser Himalayas are recognised for their biodiversity and ecological services including carbon sequestration and soil stabilisation.

3.3.2.4 Subalpine Forests

The sub-alpine forests, typically found at elevations between 3000 and 4000 m.a.s.l., serve as a transitional zone between the dense forests of lower elevations and the treeless alpine scrublands and meadows of higher elevations. These forests feature a mix of coniferous tree species such as *Abies spectabilis*, *Pinus wallichiana*, *Picea smithiana* and *Juniperus* sp. and combinations of broadleaf tree species such as *Rhododendron* sp., *Betula utilis*, and *Quercus semecarpifolia* (Rokaya et al. 2024).

3.3.2.5 Alpine Scrublands and Meadows

In the Central Himalayas, the tree line elevation typically ranges from around 3400 to 4500 m.a.s.l. (Gaire et al. 2023). Typically, tree growth ceases above the treeline; however, cold-adapted woody shrubs and grasses can thrive up to elevations of 5000 m.a.s.l., giving rise to alpine scrublands and meadows (TISC 2002). Chapter 1 complements this by documenting key species such as *Rhododendron anthopogon*, *Berberis*, *Primula*, and *Saussurea* in these high-altitude zones, emphasising their ecological adaptations and elevational ranges (Chap. 1, Sect. 1.4). The alpine scrublands feature woody shrub species like *Caragana* sp., *Rhododendron* sp., *Berberis* sp. and *Juniperus* sp. Whereas the alpine meadows feature grass species like *Carex* sp., *Kobresia* sp., *Calamogrostis* sp., *Agrotis micantha* and *Festuca* sp. (Rokaya et al. 2024).

3.3.2.6 Nival Zone

Nival zone refers to the cold mountain tops of the Greater Himalayas above 5000 m. a.s.l. that remain covered by snow in most of the seasons. This zone has no evidence of important vegetation except for scattered carpets of mosses like *Bryum* sp. and lichens like *Gyalideas cutellaris and Stereocaulon* sp. (Rokaya et al. 2024).

3.3.2.7 Trans-Himalayan Vegetation

The region north of the Greater Himalayas that is not exposed to the monsoon rains is called the Trans-Himalayan zone. It includes high plateaus and wide (inner) valleys. These high plateaus and valleys with elevations above 3600 m.a.s.l. have cold and rather arid desertic climates. This region features dwarf perennial herbs like *Carex* sp., *Deyeuxia* sp., *Poa* sp. and dwarf shrubs like *Juniperus* sp., *Caragana* sp., *Rhododendron* sp., *Berberis* sp., and *Lonicera* sp. (Shrestha et al. 2005).

3.3.3 Conservation Issues in the Central Himalayas

The Central Himalayas, a treasure trove of floristic and fungal diversity, is facing numerous conservation issues, which are as follows:

3.3.3.1 Land Use Land Cover Change

The steadily increasing population in the Central Himalayas is leading to the conversion of forests, shrublands and grasslands for cultivation, settlements, and infrastructure development (Chapagain 2016). For instance, the Nepalese part of the Kailash Sacred Landscape, an important transboundary landscape in the Central Himalayas, has experienced 9% loss in forest cover and a 12% gain in cropland between 1990 and 2009 (Uddin et al. 2015). In recent decades, the expansion of cropland and urbanisation has been identified as key factors causing land use land cover change in Nepal (Ning et al. 2023).

3.3.3.2 Biological Invasion

Studies (Lamsal et al. 2018; Shrestha and Shrestha 2021) have shown that the number of naturalised and invasive alien plant species is continuously increasing in the Central Himalayas. Besides, it has been predicted that the intensity of such biological invasion is likely to increase in the Central Himalayas with an increase in tourist numbers (Bhattarai et al. 2014) and ongoing climate change (Shrestha and

Shrestha 2021). According to (Shrestha and Shrestha 2021), 182 alien flowering plant species have naturalised in the Nepal Himalayas, of which 27 species are reported to be invasive. Of the 27 species that are reported to be invasive, *Mikania micrantha*, *Lantana camara* and *Chromolaena odorata* have been identified to be the most problematic invasive alien plant species in Nepal (Dhungana et al. 2024).

3.3.3.3 Climate Change

Average temperature has been reported to rise by 1.7 °C in Central Himalayas between 1973 and 2009, which is comparatively higher than the estimated average rise of 1.14 °C for the Himalayas (Khanal 2010). Further aggravating the situation, higher elevation areas in the Central Himalayas are experiencing a more rapid rate of warming compared to lower elevation areas i.e., elevation-dependent warming (Pandey et al. 2024). As a result, treelines (KC and Ghimire 2015) and montane plant species (Gaire et al. 2014) are being forced to shift upslope; glaciers are retreating at alarming rates (Sigdel et al. 2020) and homogenisation of vegetation has occurred in more than 67% of the Himalayas (Xu et al. 2024).

3.3.3.4 Unregulated Harvesting and Trade of Non-Timber Forest Products (NTFPs)

The Central Himalayas harbours rich diversity of wild plant species with non-timber values such as edible food (e.g., *Morchella conica*, *Zanthoxylum armatum*), fuel-wood (e.g., *Juniperus* sp., *Betula utilis*), fodder (e.g., *Litsea monopetala*, *Ficus* sp.), resin (e.g., *Pinus roxburghii*), gums (e.g., *Shorea robusta*, *Senegalia catechu*), and medicinal values (e.g., *Ophiocordyceps sinensis*, *Nardostachys jatamansi*). In recent times, with an increase in access to the global market, the demand for NTFPs has increased (Bashyal and Roberts 2024). Consequently, NTFPs are being over-harvested as a cash crop and for export. Such unregulated harvesting of NTFPs is likely to pose a significant threat to both biodiversity and the livelihoods of local communities (Ghimire et al. 2021).

3.3.3.5 Migration

According to the National Population and Housing Census 2021, 2.2 million Nepalese are staying abroad. Besides, the number of youths migrating from rural to urban areas for employment and education is steadily increasing in Nepal (Basnet et al. 2020; KC 2020). As a result, agricultural lands in rural areas are being abandoned and the interaction between rural households and forests is changing, in turn having consequences on the rural environment and livelihoods (KC et al. 2021; Chhetri et al. 2023).

3.3.3.6 Tourism

The Himalayas is a dream destination for mountaineers, hikers, and trekkers. The region is observing a steady increase in the number of foreign tourists post-COVID (Highlights Tourism 2024). The Himalayan environment, especially that of high mountain areas, sensitive to outside influences may be disturbed provided the tourism is not regulated.

3.3.4 Conservation Initiatives in the Central Himalayas

In response to the above-mentioned challenges, various conservation initiatives have been initiated. Some of the notable efforts are as follows:

3.3.4.1 Community-Led Forest Management

Nepal's revolutionary Forest Act 1993 allowed Nepal's national forests to be handed over to the local community groups, referred to as community forest user groups, for their protection and management. Currently, there are 23,682 community forest user groups protecting and managing 2.5 million ha of forests in Nepal (MoFE 2024). As a result of this community-led forest management, Nepal has nearly doubled its forest cover between 1992 (3.88 million ha of forests) and 2016 (6.63 million ha of forests, Smith et al. 2023).

3.3.4.2 Establishment of Network of Protected Areas

The Central Himalayas are home to several protected areas. Some of the notable protected areas in the region include: Sagarmatha National Park—a home to Mt. Everest and a UNESCO World Heritage Site; Chitwan National Park—Nepal's first national park established in 1973 and a UNESCO World Heritage Site; Annapurna Conservation Area—the largest conservation area in Nepal; Khangchendzonga National Park—a UNESCO World Heritage Site; Singalila National Park—famous for its panoramic views of the Himalayas; Neora Valley National Park—famous for its rich biodiversity and dense forests. These protected areas are playing a crucial role in preserving the region's rich biodiversity and diverse vegetation types.

3.3.4.3 Transboundary Cooperation

Transboundary cooperation is crucial in addressing shared environmental challenges such as climate change, biodiversity loss, and the overexploitation of critical natural resources since these challenges often span multiple countries and regions. The Greater Terai Arc Landscape (TAL) is a transboundary conservation initiative that aims to connect 14 Nepalese and Indian protected areas in the Outer Himalayas (Siwalik hills), the adjoining Bhabar areas and the Terai plains between the Yamuna River in the west and the Bagmati River in the east. By linking these 14 protected areas through biological corridors, TAL intends to address the challenges of biodiversity loss and the overexploitation of critical natural resources (MoFSC 2015).

Nepal and India have been actively collaborating to combat the trafficking of plant resources and wildlife. This collaboration includes strengthening the capacity of law enforcement agencies including forest departments, customs, police, and para-military forces; harmonising legislation and judicial procedures; and implementing legal enforcement mechanisms like the Convention on International Trade in Endangered Species of Wild Fauna and Flora (CITES) to tackle challenges related to the illegal trade of plant species and wildlife (Chisholm 2009).

Similarly, the Central Himalayas is a part of the regional initiative called the International Centre for Integrated Mountain Development (ICIMOD) that promotes transboundary landscape management in the Hindu Kush Himalayan Region. This initiative has prioritised six transboundary landscapes in the Hindu Kush Himalayan Region, of which three landscapes namely, Kailash, Everest, and Kangchenjunga (Gurung et al. 2019) are in the Central Himalayas, to achieve integrated ecosystem management to conserve biodiversity and ensure the flow of ecosystem goods and services across the countries in the region (Gurung et al. 2019). Meanwhile, to address the interconnected challenges of biodiversity loss, impaired ecosystem functionality, and the subsequent decline in ecosystem services exacerbated by climate warming and anthropogenic pressures in the Hindu Kush Himalayan Region, Kattel (2022) has proposed a resilience management framework, advocating for a collaborative approach wherein scientific expertise and societal engagement converge with national governments and multilateral organisations in the region.

3.3.4.4 Sustainable Practices

In response to immense tourism pressure in the Central Himalayas, efforts are being made to promote sustainable tourism. For instance, community-based tourism (CbT) initiatives support capacity building of the locals (providing training, ranging from hospitality management to guiding tours) and setting up sustainable tourism infrastructures like community homestays, which can generate economic benefits to local communities while providing authentic experiences to the tourists. CbT initiatives are also making important strides in promoting responsible travel that includes but is not limited to low-impact trekking, use of renewable energy, and going local.

Responsible travel helps minimise the environmental footprint of tourism and maximise contributions to socio-economic well-being of the locals. One of the noteworthy CbT initiatives in the Central Himalayas is the Annapurna Conservation Area Project (ACAP) in Nepal, which has meaningfully contributed to both conservation efforts and local livelihoods (Basnyat and Kafle 2022). Ecosystem-based adaptation (EbA) practices are also being implemented in this region to reduce the adverse impacts of climate change by using naturally available ecosystem services and aims to complement the sustainable development goal (Munang et al. 2013). Ecosystem conservation is one of the main focuses of the EbA practices (Chaudhary et al. 2021).

3.4 Conclusions

Being at the crossroads of the Indomalayan and the Palearctic realms and the Western and the Eastern Himalayas, the Central Himalayas with one of the longest elevational gradients in the world boasts an incredibly rich floristic and fungal diversity. It is home to more than 1001 species of algae; 2470 species of fungi; 1078 species of lichens; 1217 species of bryophytes; 550 species of pteridophytes; 24 species of gymnosperms; and 5619 species of angiosperms. The vegetation pattern of the Central Himalayas is quite diverse and varies considerably with varying elevation, precipitation, topography, and soil conditions. The major vegetation types of the Central Himalayas include tropical deciduous forests (up to 1000 m.a.s.l), subtropical forests (1000–2000 m.a.s.l.), temperate forests (2000–3000 m.a.s.l), sub-alpine forests (3000–4000 m.a.s.l.), alpine scrublands and meadows (treeline to 5000 masl), nival zone (above 5000 m.a.s.l.) and Trans-Himalayan vegetation. The Central Himalayas is facing numerous conservation issues such as land use land cover change, biological invasion, climate change, unregulated harvesting and trade of NTFPs, migration, and tourism. In response to these conservation challenges, several conservation initiatives including community-led forest management, establishment of network of protected areas, transboundary cooperation, and sustainable practices such as community-based tourism and ecosystem-based adaptation are underway in the Central Himalayas.

Acknowledgements The authors express their sincere gratitude to the Institute of Forestry, Tribhuvan University, and the Department of Plant Resources, Plant Research Center, for providing the facilities, logistical support, and encouragement essential for this study. We are also thankful to local communities of the Central Himalayas for their cooperation and assistance during field surveys.

References

Adhikari MK (2016) Revised checklist to the Mycotaxa proposed from Nepal. Bul Dept Pl Res 38: 1–11

Adhikari MK (2021) New record of two parasitic fungi on Malva sylvestris L. from Nepal. J Plant Res 19:12–17

Adhikari MK (2023) New record of fungi *Cerotelium malvicola* (Speg.) Dietel (Uredinales) parasitic on hibiscus species from Nepal. J Plant Res 21:1–4. https://doi.org/10.3126/bdpr.v21i1.57189

Baniya CB (2024) Lichens of Nepal. In: Rokaya MB, Sigdel SR (eds) Flora and vegetation of Nepal. Springer Nature Switzerland, Cham, pp 219–237

Baniya CB, Solhøy T, Gauslaa Y, Palmer MW (2010) The elevation gradient of lichen species richness in Nepal. Lichenologist 42:83–96. https://doi.org/10.1017/S0024282909008627

Baniya CB, Tamang R, K.C. A (2022) Lichens of Nepal (checklist). National Herbarium and Plant Laboratories, Department of Plant Resources, Government of Nepal, Lalitpur

Bashyal R, Roberts DL (2024) A systematic survey of online trade in the caterpillar fungus Ophiocordyceps sinensis. Oryx 58:29–37. https://doi.org/10.1017/S003060532300131X

Basnet N, Timmerman MC, van der Linden J (2020) The education-to-work transition of young rural—urban migrants in Kathmandu, Nepal. J Appl Youth Stud 3:331–344. https://doi.org/10.1007/s43151-020-00027-z

Basnyat S, Kafle T (2022) Community-based tourism experiences in Nepal: a path to sustainable tourism in the post-COVID-19 era. In: Costa RA, Brandão F, Breda Z, Costa C (eds) Planning and managing the experience economy in tourism. IGI Global Scientific Publishing, pp 176–188

Bhattarai KR, Vetaas OR, Grytnes JA (2004) Fern species richness along a central Himalayan elevational gradient, Nepal. J Biogeogr 31:389–400. https://doi.org/10.1046/j.0305-0270.2003.01013.x

Bhattarai KR, Måren IE, Subedi SC (2014) Biodiversity and invasibility: distribution patterns of invasive plant species in the Himalayas, Nepal. J Mt Sci 11:688–696. https://doi.org/10.1007/s11629-013-2821-3

Chapagain PS (2016) Changing population, landuse and environment in the Nepal Himalayas. J Hydrol Meteorol 8:66–71. https://doi.org/10.3126/jhm.v8i1.15574

Chatterjee SP (1964) Fifty years of science in India: progress of geography, 1st edn. Indian Science Congress Association, Calcutta

Chaudhary S, Adhikari BR, Chaudhary P et al (2021) Ecosystem-based adaptation (EbA) in the Hindu Kush Himalaya: status, Progress and challenges. In: Mukherjee M, Shaw R (eds) Ecosystem-based disaster and climate resilience. Springer Nature Singapore, pp 29–51

Chhetri R, Yokying P, Smith A et al (2023) Forest, agriculture, and migration: contemplating the future of forestry and agriculture in the middle-hills of Nepal. J Peasant Stud 50:411–433. https://doi.org/10.1080/03066150.2021.1978983

Chisholm H (2009) Regional co-operation for biodiversity conservation: lessons learnt from around the world and the way forward for South Asia

Devkota S, Aryal HP (2020) Wild mushrooms of Nepal. In: Siwakoti M, Jha PK, Rajbhandary S, Rai SK (eds) Plant diversity in Nepal. Botanical Society of Nepal, Kathmandu, pp 41–54

Dhital MR (2015a) Introduction to higher Himalaya. In: Dhital MR (ed) Geology of the Nepal Himalaya: regional perspective of the classic collided Orogen. Springer International Publishing Switzerland, Cham, pp 197–211

Dhital MR (2015b) Introduction to lesser Himalaya. In: Dhital MR (ed) Geology of the Nepal Himalaya: regional perspective of the classic collided Orogen. Springer International Publishing Switzerland, Cham, pp 65–80

Dhital MR (2015c) Introduction to Siwaliks. In: Dhital MR (ed) Geology of the Nepal Himalaya: regional perspective of the classic collided Orogen. Springer International Publishing Switzerland, Cham, pp 371–384

Dhungana S, Yuangyai N, Sinutok S (2024) Impact, management, and use of invasive alien plant species in Nepal's protected area: a systematic review. J Ecol Environ 48:182–195. https://doi.org/10.5141/jee.23.081

Fraser-Jenkins CR, Kandel DR (2019) Ferns and fern-allies of Nepal - Volume 2. National Herbarium and Plant Laboratories, Department of Plant Resources, Ministry of Forests and Soil Conservation, Kathmandu, Nepal

Fraser-Jenkins CR, Kandel DR, Pariyar S (2015) Ferns and fern-allies of Nepal - Volume 1. National Herbarium and Plant Laboratories, Department of Plant Resources, Ministry of Forests and Soil Conservation, Kathmandu, Nepal

Gaire NP, Koirala M, Bhuju DR, Borgaonkar HP (2014) Treeline dynamics with climate change at the Central Nepal Himalaya. Clim Past 10:1277–1290. https://doi.org/10.5194/cp-10-1277-2014

Gaire NP, Fan Z-X, Chhetri PK et al (2023) Treeline dynamics in Nepal Himalaya in a response to complexity of factors. In: Singh SP, Reshi ZA, Joshi R (eds) Ecology of Himalayan treeline ecotone. Springer Nature Singapore, pp 519–563

Ghimire P, Pudasaini N, Lamichhane U (2021) Status, prospects and challenges for non-timber forest products conservation in Nepal: a critical review. Grassroots J Nat Resour 4:1–16. https://doi.org/10.33002/nr2581.6853.040101

Gurung SB (2021) Mountain peaks of Nepal Himalaya. J Tour Himal Adven 3:63–73. https://doi.org/10.3126/jtha.v3i1.39118

Gurung J, Chettri N, Sharma E et al (2019) Evolution of a transboundary landscape approach in the Hindu Kush Himalaya: key learnings from the Kangchenjunga landscape. Global Ecol Conserv 17:e00599. https://doi.org/10.1016/j.gecco.2019.e00599

Hawksworth DL, Grube M (2020) Lichens redefined as complex ecosystems. New Phytol 227: 1281–1283. https://doi.org/10.1111/nph.16630

Highlights Tourism (2024) Nepal welcomes record numbers amidst encouraging tourism trends. https://highlightstourism.com/2024/04/01/nepal-welcomes-record-numbers-amidst-encouraging-tourism-trends/

Kandel DR, Fraser-Jenkins CR (2020) Ferns and fern-allies of Nepal - Volume 3. National Herbarium and Plant Laboratories, Department of Plant Resources, Ministry of Forests and Soil Conservation, government of Nepal, Nepal

Kattel GR (2022) Climate warming in the Himalayas threatens biodiversity, ecosystem functioning and ecosystem services in the 21st century: is there a better solution? Biodivers Conserv 31: 2017–2044. https://doi.org/10.1007/s10531-022-02417-6

KC S (2020) Internal migration in Nepal. In: Bell M, Bernard A, Charles-Edwards E, Zhu Y (eds) Internal migration in the countries of Asia: a cross-national comparison. Springer Nature Switzerland, Cham, pp 249–267

KC A, Ghimire A (2015) High-altitude plants in era of climate change: a case of Nepal Himalayas. In: Öztürk M, Hakeem KR, Faridah-Hanum I, Efe R (eds) Climate change impacts on high-altitude ecosystems. Springer International Publishing Switzerland, pp 177–187

KC B, Race D, Fisher R, Jackson W (2021) Changing rural livelihoods and Forest use transition in the Middle Hills of Nepal. Small-Scale For 20:479–501. https://doi.org/10.1007/s11842-021-09477-6

Khanal S (2010) Climatic warming in the Himalaya during recent decades. Master's Project, Yale University - Yale School of Forestry & Environmental Science

Koshal AK (2014) Changing current scenario of Rice-wheat system in indo-Gangetic plain region of India. Int J Sci Res Publ 4:1–13

Kost G, Adhikari MK (2015) Mycota. In: Miehe G, Pendry C, Chaudhary RP (eds) Nepal: an introduction to the natural history, ecology and human environment of the Himalayas: a companion to the Flora of Nepal. Royal Botanic Garden Edinburgh, Edinburgh, pp 203–210

Lamsal P, Kumar L, Aryal A, Atreya K (2018) Invasive alien plant species dynamics in the Himalayan region under climate change. Ambio 47:697–710. https://doi.org/10.1007/s13280-018-1017-z

Maharjan SK, Sterck FJ, Dhakal BP et al (2021) Functional traits shape tree species distribution in the Himalayas. J Ecol 109:3818–3834. https://doi.org/10.1111/1365-2745.13759
MoFE (2024) Annual progress report, 2023/24. Kathmandu
MoFSC (2015) Strategy and action plan 2015–2025, Terai arc landscape, Nepal, Ministry of Forests and Soil Conservation, Kathmandu
Munang R, Thiaw I, Alverson K et al (2013) Climate change and ecosystem-based adaptation: a new pragmatic approach to buffering climate change impacts. Curr Opin Environ Sustain 5:67–71. https://doi.org/10.1016/j.cosust.2012.12.001
Ning C, Subedi R, Hao L (2023) Land use/cover change, fragmentation, and driving factors in Nepal in the last 25 years. Sustainability 15:6957. https://doi.org/10.3390/su15086957
Offen IA, Bobrowski M, Jahan R et al (2021) Vegetation-environment relationships, species richness, and climate change impacts along the elevational gradient in the Sikkim Himalaya - general findings and preliminary results from a reconnaissance field trip. In: Hartmann M, Barclary MVL, Weipert J (eds) Biodiversität und Naturausstattungim Himalaya VII. Verein der Freunde & Förderer des Naturkundemuseums Erfurt e.V, pp 9–40
Pandey B, Nepal N, Tripathi S et al (2020) Distribution pattern of gymnosperms' richness in Nepal: effect of environmental constrains along elevational gradients. Plants 9:1–13. https://doi.org/10.3390/plants9050625
Pandey P, Ali SN, Bhardwaj A et al (2024) Pattern and imprints of elevation-dependent warming on central Himalayan cryosphere as revealed by earth observation datasets. Result Earth Sci 2:1–16. https://doi.org/10.1016/j.rines.2024.100021
Pandit MK, Manish K, Koh LP (2014) Dancing on the roof of the world: ecological transformation of the Himalayan landscape. BioScience 64:980–992. https://doi.org/10.1093/biosci/biu152
Peakbagger (2024) World 8000-meter peaks. In: Peakbagger.com. https://www.peakbagger.com/list.aspx?lid=200. Accessed 24 Sept 2024
Pradhan N (2020) Bryoflora of Nepal. In: Siwakoti M, Jha PK, Rajbhandary S, Rai SK (eds) Plant diversity in Nepal. Botanical Society of Nepal, Kathmandu, pp 62–70
Prasad V (2013) Biodiversity: algae. In: Jha PK, Neupane FP, Shrestha ML, Khanal IP (eds) Biological diversity and conservation. Nepal Academy of Science and Technology, Lalitpur, pp 97–103
Rai SK, Ghimire N (2020) Algal explorations in Nepal. In: Siwakoti M, Jha PK, Rajbhandary S, Rai SK (eds) Plant diversity in Nepal. Botanical Society of Nepal, Kathmandu, pp 16–40
Rajbhandari KR, Joshi L, Chhetri R, Khatri S (2020) A handbook of the gymnosperms of Nepal. National Herbarium and Plant Laboratories, Department of Plant Resources, Lalitpur, Nepal
Rokaya MB, Parajuli B, Timsina B (2024) Vegetation and forest in Nepal. In: Rokaya MB, Sigdel SR (eds) Flora and vegetation of Nepal. Springer Nature Switzerland, Cham, pp 37–88
Shrestha TB (1982) Ecology and vegetation of north-west Nepal (Karnali region). Royal Nepal Academy, Kathmandu
Shrestha BB, Shrestha KK (2021) Invasions of alien plant species in Nepal: patterns and process. In: Pullaiah T, Ielmini MR (eds) Invasive alien species: observations and issues from around the world (volume 2: issues and invasions in Asia and the Pacific region). Wiley, pp 168–183
Shrestha MR, Rokaya MB, Ghimire SK (2005) Vegetation pattern of trans-Himalayan zone in the north-west Nepal. Nepal J Plant Sci 1:129–135
Shrestha KK, Bhattarai S, Bhandari P (2018) Handbook of flowering plants of Nepal (volume 1. Gymnosperms and angiosperms: Cycadaceae - Betulaceae). Scientific Publishers, Jodhpur, India
Shrestha KK, Bhandari P, Bhattarai S (2022) Plants of Nepal (gymnosperms and angiosperms). Heritage Publishers and Distributors Pvt. Ltd., Kathmandu
Sigdel SR, Zhang H, Zhu H et al (2020) Retreating glacier and advancing forest over the past 200 years in the Central Himalayas. J Geophys Res Biogeosci 125:1–14. https://doi.org/10.1029/2020JG005751
Singh RL (1971) India: a regional geography. National Geographical Society of India

Singh DK, Pusalkar PK (2020) Floristic diversity of the Indian Himalaya. In: Dar GK, Khuroo AA (eds) Biodiversity of the Himalaya: Jammu and Kashmir State. Springer Nature Singapore Pte Ltd, pp 93–126

Singh JS, Singh SP (1987) Forest vegetation of the Himalaya. Bot Rev 53:80–192. https://doi.org/10.1007/BF02858183

Smith AC, Hurni K, Fox J, Van Den Hoek J (2023) Community forest management led to rapid local forest gain in Nepal: a 29-year mixed methods retrospective case study. Land Use Policy 126:106526

Sochting U (2015) Lichens Nepal. In: Miehe G, Pendry C, Chaudhary RP (eds) Nepal: an introduction to the natural history, ecology and human environment of the Himalayas: a companion to the flora of Nepal. Royal Botanic Garden Edinburgh, Edinburgh, pp 191–199

Stearn WT (1978) Introduction. In: Hara H, Stearn WT, Williams LHJ (eds) An enumeration of the flowering plants of Nepal. Trustees of the British Museum (Natural History), London, pp 7–13

TISC (2002) Forest and vegetation types of Nepal. Tree improvement and Silviculture component (TISC), natural resource management sector assistance Programme, Ministry of Forests and Soil Conservation, Nepal

Uddin K, Chaudhary S, Chettri N et al (2015) The changing land cover and fragmenting forest on the roof of the world: a case study in Nepal's Kailash sacred landscape. Landsc Urban Plan 141: 1–10. https://doi.org/10.1016/j.landurbplan.2015.04.003

Udvardy MDF (1975) A classification of the biogeographical provinces of the world. IUCN Occasional Paper no. 18 45

Valdiya KS (1984) Evolution of the Himalaya. Techtonophysics 105:229–248. https://doi.org/10.1016/0040-1951(84)90205-1

Valdiya KS (2016) Indo-gangetic plains: evolution and later developments. In: Valdiya KS (ed) The making of India: geodynamic evolution, 2nd edn. Springer International Publishing Switzerland, Cham, pp 723–745

Wikramanayake E, Dinerstein E, Seidensticker J et al (2011) A landscape-based conservation strategy to double the wild tiger population. Conserv Lett 4:219–227. https://doi.org/10.1111/j.1755-263X.2010.00162.x

Xu J, Wang X, Liu C et al (2024) Widespread homogenization in vegetation activities along the elevational gradients across the Himalaya over the past 40 years. Sci Total Environ 953:176179. https://doi.org/10.1016/J.SCITOTENV.2024.176179

Chapter 4
Orchid Diversity of Indian Himalayan Region

Biswajit Paul, Animesh Sen, and Devanjal Bora

Abstract The Indian Himalayan region is renowned for its rich floristic diversity, including the *Orchidaceae* family, one of the largest families of flowering plants. This chapter examines the diversity and currently accepted orchid species of the region. A total of 1014 orchid species, spread across 129 genera, have been reported, with *Bulbophyllum* and *Calanthe* representing the highest numbers of epiphytic and geophytic species, respectively. The eastern part of the region exhibits greater diversity and is enriched with more epiphytic orchids, primarily due to its dense forest cover. In contrast, the western region, characterised by a dry and cool climate, is dominated by geophytic orchids. Among the orchids found in this region, 65 species are classified under various threatened categories by the IUCN. Despite numerous threats to their habitat, the region continues to yield new species and distributional records, further enriching its orchid diversity on an ongoing basis.

Keywords Eastern Himalaya · Western Himalaya · North East India · Conservation · *Orchidaceae*

Supplementary Information The online version contains supplementary material available at https://doi.org/10.1007/978-3-032-21697-7_4.

B. Paul (✉) · D. Bora
Botany Section, Central Ayurveda Research Institute, CCRAS, Ministry of AYUSH, Guwahati, Assam, India
e-mail: biswajitpaul@ccras.nic.in

A. Sen
Botany Section, Regional Ayurveda Research Institute, CCRAS, Ministry of AYUSH, Gangtok, Sikkim, India

D. Das et al. (eds.), *Plant Diversity and Conservation Strategies of the Himalayan Biodiversity Hotspot*, Topics in Biodiversity and Conservation 21,
https://doi.org/10.1007/978-3-032-21697-7_4

4.1 Introduction

Orchids, belonging to the family *Orchidaceae*, are among the most diverse and captivating flowering plants. The family comprises approximately 30,000 species across more than 1000 genera, making it one of the largest and most varied plant families in the world (Dressler 1981; Misra 2007; Kindlmann et al. 2023). *Orchidaceae* alone contributes roughly 40% of the world's monocotyledons (Khasim et al. 2020). Known for their complex and highly specialised flowers, orchids exhibit remarkable adaptations to their environments, particularly in terms of pollination strategies and habitat specialisation (Arditti and Ghani 2000). Orchids display various growth forms, including epiphytic, lithophytic, terrestrial, and mycoheterotrophic species, thriving in a wide range of environments, from tropical rainforests and temperate forests to arid deserts and alpine regions. While epiphytes grow on trees and lithophytes on rocks, extracting moisture and nutrients from the air and surrounding organic matter, terrestrial orchids thrive in soil, and mycoheterotrophs associate with mycorrhizal fungi (Dressler 1993; Singh et al. 2019).

In India, the oldest record of orchids appears in the *Charak Samhita* written by Charak (A.D. 100), where the medicinal properties of some indigenous herbs, including orchids, are mentioned. However, the scientific study of orchids in India dates back to the seventeenth century. Linnaeus (1753), in his *Species Plantarum*, included Indian orchids, and other notable contributors such as William Roxburgh (1832), Griffith (1851), Lindley (1857, 1859), and Atkinson (1882) made significant contributions to the documentation and description of Indian orchids.

The Indian Himalayan Region (IHR) is a biodiversity hotspot, home to approximately 8000 species of flowering plants, and is recognised as one of the world's key biodiversity areas (Sharma and Samant 2017). The *Orchidaceae* family contributes about 10% to the Indian flora, with 155 genera and 1263 species. The northeastern region of India, particularly the Eastern Himalaya, hosts around 856 species, demonstrating a rich diversity (De et al. 2016). In contrast, the diversity of orchids decreases from the Eastern to the Western Himalaya, with the latter harboring approximately 239 taxa across 72 genera (Singh Jalal and Jayanthi 2015).

Several important studies on orchids of the IHR have been published over the past centuries, including *The Orchids of the Sikkim Himalayas* by George King and Robert Pantling (1898), *The Orchids of North-Western Himalaya* by J.F. Duthie (1906), *Flora of British India* (1888–1890), and *A Century of Indian Orchids* (1895) by Sir J. D. Hooker. Orchids are of significant economic value, with many species having medicinal, ethnobotanical, ornamental, and horticultural importance (Mulgaonkar and Shraddha 2015; Rahamtulla et al. 2020a). Moreover, the orchids of the IHR are of conservation interest due to their ecological roles and the increasing threats they face (De and Medhi 2014).

Although numerous publications, such as articles, handbooks, and pictorial guides, provide information on the orchid flora of different regions within the IHR, distribution records and taxonomic updates are continually being refined.

Therefore, this chapter aims to provide a comprehensive list of orchids in the IHR, including their currently accepted names, habitats, and distributions.

4.2 Indian Himalayan Region

The Himalayas, one of the largest mountain ranges in the world, extend across several South Asian countries. A majority of the Himalayan region lies within India, where it forms a natural barrier along the northern boundary of the Indian subcontinent (Khuroo et al. 2021; Roy and Purohit 2018). The Indian Himalayan Region (IHR) is situated between 27°38′ N latitude and 72°89′ E longitude, covering approximately 16.2% of India's total land area. The region stretches over 2500 km in length and varies in width from 350 to 150 km, running from west to east. The altitude of the IHR ranges from 200 m.a.s.l to over 8000 m.a.s.l. (Roy and Purohit 2018). This spatial and elevational framework closely aligns with the longitudinal and altitudinal divisions of the Himalayas discussed in Chap. 3 (Sect. 3.1.1), which explains the progression from the Outer to the Greater Himalayas and its impact on vegetation zonation. Similarly, Chap. 1 (Sect. 1.2) highlights how altitudinal gradients shape ecological zones and floral community structure across the Himalayan belt. Administratively, the IHR is divided into 12 states and union territories: Jammu & Kashmir, Ladakh, Himachal Pradesh, Uttarakhand, Sikkim, Assam, Arunachal Pradesh, Meghalaya, Nagaland, Manipur, Mizoram, Tripura, and the hill areas of West Bengal (Fig. 4.1).

This region is characterised by a diverse range of altitudes, climates, weather conditions, and soil and forest types, ranging from tropical forests in the foothills to

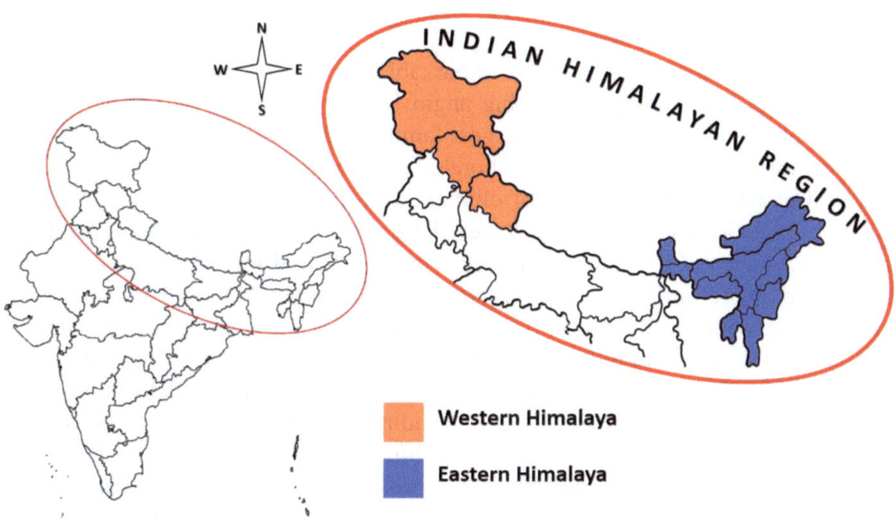

Fig. 4.1 The Indian Himalayan Region

alpine meadows near the highest peaks. This geographical and climatic diversity supports a rich variety of flora and fauna, including orchids. The floristic diversity of the IHR includes 8000 species of angiosperms, 44 species of gymnosperms, 600 species of pteridophytes, 1736 species of bryophytes, 1159 species of lichens, and 6900 species of fungi (Singh and Hajra 1996; Samant et al. 2007; Wani et al. 2022).

The IHR is typically divided into two major subregions: the Eastern Himalaya and the Western Himalaya based on geo-climatic differences and biodiversity patterns. The Eastern Himalayan region, often referred to as the "Cradle of Flowering Plants" (Takhtajan 1969), encompasses the seven northeastern states of India (Arunachal Pradesh, Assam, Meghalaya, Mizoram, Manipur, Nagaland, and Tripura), as well as Sikkim and the Darjeeling district of West Bengal. This area forms a unique phytogeographical zone, renowned for its exceptional floristic diversity, which is the highest in the Indian subcontinent. Characterised by high precipitation and a humid climate, the Eastern Himalayas are notably richer in species diversity and endemism compared to the Western Himalayas. This region is primarily inhabited by broad-leaved species, with common taxa including *Rhododendrons*, magnolias, laurels, tree ferns, and orchids (Chowdhery 2015).

In contrast, the Western Himalayas, which include the states and union territories of Jammu & Kashmir, Himachal Pradesh, Uttarakhand, and Ladakh, differ markedly from the Eastern Himalayas in terms of climate and vegetation. The Western Himalayas experience a cooler, drier climate, with lower rainfall, which decreases further towards the north-western regions, particularly in areas such as Ladakh. This region is predominantly covered by coniferous forests and is less conducive for orchid diversity (Rau 1975; Singh Jalal and Jayanthi 2015).

4.3 Orchid Diversity

The Indian Himalayan Region holds remarkable floristic diversity, encompassing a wide range of plant groups, including angiosperms, gymnosperms, pteridophytes, and bryophytes. *Orchidaceae*, the largest family of flowering plants in this region, accounts for over 75% of India's total orchid species (Fig. 4.2) (Singh and Pusalkar 2020). The IHR hosts a variety of climatic and forest ecosystems, ranging from 300 to over 5000 m in altitude. At lower elevations (300–1500 m), the tropical and subtropical zones of the Indian Himalayas provide humid forest environments that support epiphytic orchids such as *Dendrobium*, *Bulbophyllum*, and *Vanda*. In the mid-altitude range (1500–3000 m), temperate climates allow for both epiphytic and terrestrial orchids, including genera like *Coelogyne* and *Cymbidium*. At higher elevations (3000–5000 + m), cold alpine conditions favor terrestrial orchids such as *Spiranthes* and *Habenaria*. These altitudinal distribution patterns of orchids closely reflect the vegetation zonation outlined in Chap. 3 (Sect. 3.3.2), which describes the progression from tropical forests to alpine meadows. Notably, the high-altitude habitats that support alpine orchids also accommodate cold-adapted

Fig. 4.2 Some epiphytic orchids of the Indian Himalayan Region (**a**) *Coelogyne fuscescens*, (**b**) *Bulbophyllum hirtum*, (**c**) *Coelogyne lancilabia*, (**d**) *Dendrobium moschatum*, (**e**) *Pleione praecox*, (**f**) *Dendrobium longicornu*, (**g**) *Pleione hookeriana*, (**h**) *Eria coronaria*, (**i**) *Acampe praemorsa*

species such as *Saussurea* and *Primula*, as documented in Chap. 1 (Sect. 1.4), underscoring shared ecological adaptations to extreme alpine conditions.

Factors such as aspect, slope, and vegetation contribute to the creation of microhabitats that support unique orchid species. These microhabitats make orchids highly sensitive to climate change, which could impact their survival and distribution. The varying microclimates at different altitude levels significantly influence orchid distribution. For example, the orientation of slopes north-facing slopes being cooler and shadier and south-facing slopes being warmer and drier creates environments that support orchids with different moisture and temperature requirements. The dense tropical and subtropical forests in the foothills of the Eastern Himalayas, due to high annual rainfall and a humid climate, create ideal conditions for epiphytic orchids. The dense canopy and high biodiversity in these forests offer a moist and shaded environment that supports these orchids. In contrast, north-facing slopes in the region tend to be cooler and moister, favoring orchids that require lower temperatures, while south-facing slopes, being sunnier and drier, support more drought-tolerant species. Altitudinal diversity, varied forest types, and the region's complex topography contribute to the IHR's rich orchid diversity (Samant et al. 2002; Chowdhery 2015; Singh et al. 2019; Kumar et al. 2022).

Orchid distributions across the Eastern Himalayan region have been described in several studies. These include reports from northeastern India (De and Medhi 2014; Pal and Singh 2016; Kumar et al. 2022), Darjeeling (Yonzone et al. 2012; Rahamtulla et al. 2020b), Assam (Gogoi et al. 2015), Meghalaya (Kataki 1986), Manipur (Ghatak and Devi 1986; Kumar and Kumar 2005; Sadananda et al. 2019), Arunachal Pradesh (Chowdhery 1998; Hegde 2020; Nyorak 2023), Nagaland (Hynniewta et al. 2000; Deb et al. 2021), and Sikkim (Lucksom et al. 2011; Singh et al. 2020). Additionally, earlier explorations and studies have documented the orchid distribution in the Western Himalayan region. For example, Collett (1902) reported 38 orchid species across 18 genera from Shimla and the adjacent hills; Duthie (1906) documented 173 orchid species from 43 genera across the entire northwestern Himalayas, with illustrations for 53 species; Blatter (1928) described 17 orchid species from 7 genera in Kashmir; Nair (1977) recorded 43 species from 19 genera in the Bashahr Himalaya; Raizada et al. (1981) reported 89 species in 35 genera from Mussoorie; Deva and Naithani (1986) documented 239 species in 74 genera from the northwestern Himalayas; Pangtey et al. (1991) described 192 species from 61 genera in the Kumaun Himalaya; Uniyal et al. (2007) recorded 242 species in 71 genera from Uttarakhand, with Jalal et al. (2008) noting 237 species across 73 genera; and Vij et al. (2013) reported 85 species in 44 genera from Himachal Pradesh (Supplementary Table 4.1).

Based on an analysis of previous publications and multiple databases, including *Plants of the World Online* (POWO 2024), *The World Flora Online* (WFO 2024), and the e-Plant Checklist of India and e-Flora of India by the Botanical Survey of India (BSI 2024), along with validation of currently accepted botanical names, the Indian Himalayan Region (IHR) is reported to host 1014 orchid species across 129 genera. Among these, 59% (599 species) belong to 61 genera of epiphytic orchids (Fig. 4.3), 37% (373 species) are geophytic, distributed across 64 genera, and the remaining 4% (42 species) occupy mixed habitats, such as epiphytic/lithophytic and epiphytic/geophytic environments. The dominant genera in the IHR include *Bulbophyllum* (113 species), *Dendrobium* (99 species), *Coelogyne* (56 species), *Liparis* (39 species), *Oberonia* (39 species), *Calanthe* (30 species), *Cymbidium* (27 species), *Habenaria* (26 species), and *Herminium* (26 species), with

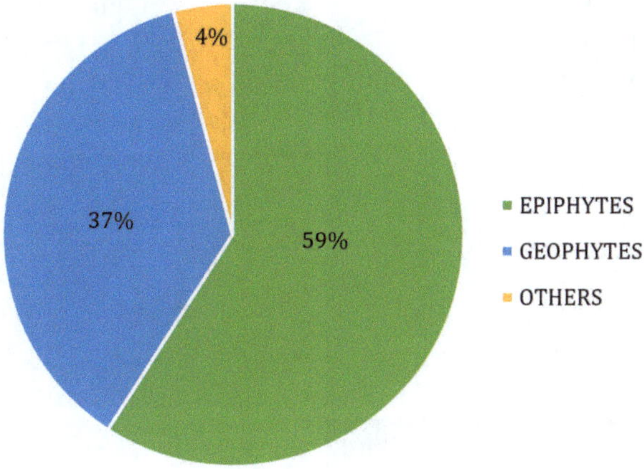

Fig. 4.3 Habitat of orchids in the Indian Himalayan Region

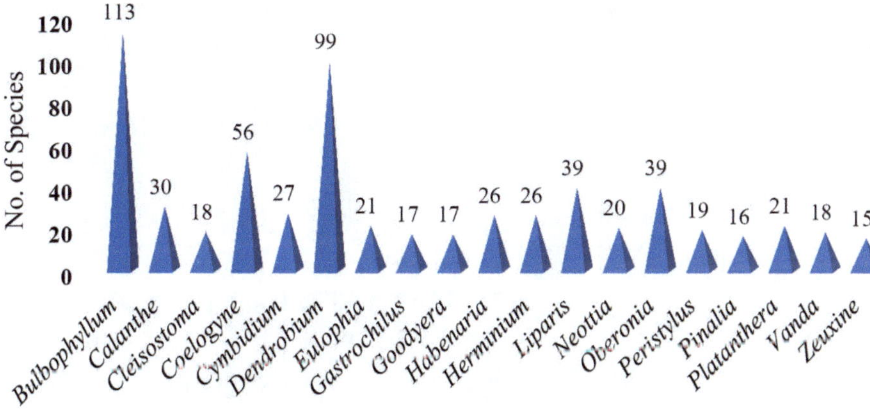

Fig. 4.4 Major orchids (Genus-wise) in the Indian Himalayan Region

Bulbophyllum and *Calanthe* representing the highest numbers of epiphytic and geophytic species, respectively (Fig. 4.4).

The Eastern Himalayan region, being the most floristically diverse part of the IHR, is home to a total of 985 orchid species spread across 127 genera. Of these, 60% are epiphytic (Fig. 4.2), which reflects the dense forest cover and humid conditions of this region. The dominant orchid genera in the Eastern Himalayan region include *Bulbophyllum*, *Dendrobium*, *Coelogyne*, *Liparis*, *Oberonia*, *Calanthe*, *Cymbidium*, *Herminium*, *Habenaria*, and *Platanthera*.

In contrast, the Western Himalayan region, despite its harsh climatic and weather conditions, supports a significant number of orchid species, totaling 262 species across 65 genera. The majority of these orchids (52%) are geophytic (Fig. 4.5), which is consistent with the dry and cold geo-climatic conditions of the region. The dominant orchid genera in the Western Himalayan region include *Dendrobium*, *Bulbophyllum*, *Habenaria*, *Herminium*, *Liparis*, *Eulophia*, *Neottia*, *Oberonia*, and *Nervilia*.

Fig. 4.5 Some geophytic orchids of the Indian Himalayan Region (**a**) *Aphyllorchis alpine*, (**b**) *Calanthe puberula*, (**c**) *Crepidium acuminatum*, (**d**) *Crepidium purpureum*, (**e**) *Cyrtosia lindleyana*, (**f**) *Goodyera repens*, (**g**) *Habenaria arietina*, (**h**) *Habenaria dentata*, (**i**) *Zeuxine affinis*

4.4 Growing Diversity/New Species or Distribution Reports

Despite the continuously growing threats to orchid habitats, new species continue to emerge in the Indian Himalayan Region (IHR). An analysis of publications from the last 5 years reveals a total of 36 new reports, of which 12 are new species and the remainder are new distributional records for various orchids. The Eastern Himalayan region has accounted for all of the new reports during the past 5 years (2019–2023), with the highest number of 17 reports-coming from the state of Arunachal Pradesh alone. This underscores the rich biodiversity of the Eastern Himalayan region.

Despite the continuously growing threats to orchid habitats, new species and distribution records continue to emerge from the Indian Himalayan Region (IHR), particularly from the Eastern Himalayas. In 2019, eight orchid species such as *Epipogium japonicum*, *Gastrochilus platycalcaratus*, *Lecanorchis taiwaniana*, *Oberonia costeriana*, *Porpax grandiflora*, *Sarcoglyphis mirabilis*, *Thrixspermum japonicum*, and *Zeuxine parvifolia* were reported as new distribution records. Additionally, two new species, *Rhomboda monensis*, and *Thrixspermum changlangensis*, were discovered in Nagaland and Arunachal Pradesh, respectively (Mao et al. 2020). In 2020, three new species *Cymbidium viride*, *Spathoglottis arunachalensis*, and *Stereochilus arunachalensis* and four new distribution records, *Dendrobium tsangianum*, *Gastrochilus changjiangensis*, *Phalaenopsis braceana*, and *Stereochilus erinaceus* were reported from the Eastern Himalayas (Mao et al. 2021). In 2021, one new species, *Bulbophyllum karbianglongensis* from Assam, along with six new distribution records *Bulbophyllum reflexipetalum*, *Ceratostylis siamensis*, *Cyrtosia integra*, *Gastrochilus yei*, *Phalaenopsis wilsonii*, and *Spathoglottis affinis* were documented (Mao et al. 2022). This was followed in 2022 by the discovery of three new species *Gastrochilus pseudocalceolaris*, *Schoenorchis mishmensis*, and *Thrixspermum patkaiensis* and two distribution records *Calanthe lamellosa* and *Goodyera alveolate* from the Eastern Himalayas (Mao et al. 2023). In 2023, three additional new species *Platanthera dambungensis*, *Platanthera lachungensis*, and *Tainia epiphytica* and four new distribution records *Anoectochilus medogensis*, *Bulbophyllum nigrescens*, *Gastrodia bambu*, and *Trachoma rhopalorrhachis* were reported from the region (Mao et al. 2024).

These continual discoveries underscore the dynamic and underexplored biodiversity of the Eastern Himalayas. They also parallel the floristic richness and endemism observed in the Central Himalayas (Chap. 3, Sect. 3.3.1), highlighting the region's importance as a center of plant evolution and speciation. However, this biodiversity is increasingly threatened. Habitat loss, fragmentation, and climate change affecting orchid diversity in the IHR mirror the broader ecological threats detailed in Chap. 2 (Sects. 2.2 and 2.6), which emphasise the critical need for timely conservation measures. Together, these findings reaffirm the necessity of sustained taxonomic and ecological research to protect and understand the unique floristic wealth of the Himalayan ecosystem.

4.5 Endangered Orchids

The International Union for Conservation of Nature (IUCN) Red List is the world's most comprehensive database of threatened species, recognised for its standardised criteria used to assess and categorise species and their associated threats (Rodrigues et al. 2006; Brummitt et al. 2015; IUCN 2024). Upon reviewing the orchid species of the Indian Himalayan Region (IHR) in the IUCN database (IUCN 2024), we identified a total of 65 species categorised under various IUCN threat categories. Of these, 37 species are classified as threatened (21, 12 and 4 species are critically endangered, endangered and vulnerable respectively). The remaining 1, 21 and 6 species are categorised as near threatened, least concern and data deficient respectively (Fig. 4.6). The genus *Paphiopedilum* contains the highest number of species in this region, with seven species, all of which are categorised as least concern. Five species of *Cypripedium* and four species of *Bulbophyllum* and *Cheirostylis* are listed under the critically endangered category. Among the endangered species, *Gastrodia* and *Neottia* show the highest number, with three species each.

4.6 Threats to Orchid Diversity

Although the family *Orchidaceae* comprises almost 10% of the world's angiosperms and is found on every continent except Antarctica, it is one of the most threatened plant families due to several factors, including climate change, deforestation, and anthropogenic activities. Global forest cover is declining primarily due to industrial development, timber exploitation, and agricultural expansion. A study indicates that by 2010 nearly 25% of endemic species in the Indian Himalayan Region may be lost due to deforestation if the current trend continues (Pandit et al. 2007). Unsustainable logging practices and resulting natural disasters have become a growing concern, posing serious threats not only to humans but also to wild orchids. In deforested regions, orchids face extreme heat and dryness, leading to their gradual decline (Besi et al. 2023).

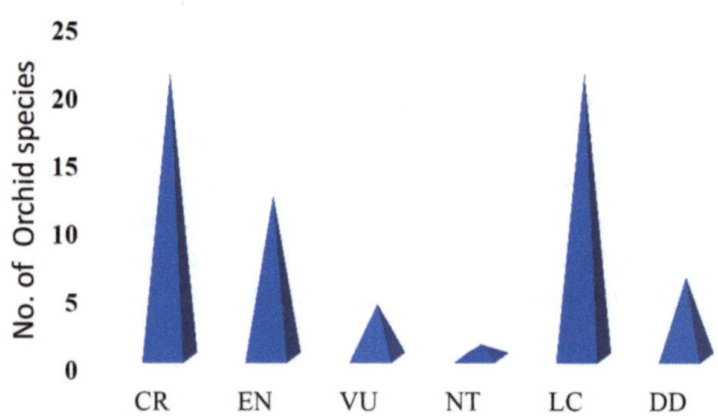

Fig. 4.6 Endangered orchids as per the IUCN categories

Many orchid species, especially those in the Western Himalayas, are rare and threatened due to habitat loss and fragmentation caused by human activities such as land development, dam construction, road building, overgrazing, and forest fires. Some orchids are restricted to specific habitats or elevations, making them particularly vulnerable to environmental changes. Others are rare due to their limited geographic distribution, specific habitat requirements, or naturally low populations. Additionally, the decline in insect pollinators, which are essential for orchid reproduction, further threatens their survival (Jalal 2012). Many orchids depend on the specific biological associations, resulting in small, isolated populations that are more susceptible to extinction risks. For instance, many orchids require specific fungi for germination, growth, and nutrient uptake. Orchid distribution and dynamics are shaped by ecological factors and orchid mycorrhizal fungi (OMF), especially *Rhizoctonia* groups which influence germination, survival, and reproduction. OMF composition changes impact orchid fitness. Network analysis reveals orchid, OMF interaction patterns, highlighting specificity, coevolution, and ecological roles, informing efficient conservation strategies through deeper microbial ecology understanding (Chakraborty et al. 2023; Li et al. 2021). Studies on *Gastrodia elata* show fungi needed for seed germination differ from those supporting later growth. Moreover, fungi aiding vegetative growth can inhibit seed germination, indicating stage-specific mycorrhizal roles (Liu et al. 2010). The fungi *Tulasnella* primarily inhabit orchid roots and stems and are highly sensitive to environmental changes, suggesting a strong reliance on their orchid hosts and surrounding plant species. This complex, coevolved nutrient relationship is difficult to restore once disrupted, posing challenges for orchid growers. However, certain substrate mixtures such as peat-based blends, coconut shells, bark shavings, native soil, or combinations of pinecone scales, mosses, and humus can help reestablish symbiosis between *Tulasnella* and transplanted or cultured orchids grown outside their natural environment (Li et al. 2021; Kaur et al. 2018). Additionally, some orchids have complex pollination systems that rely on particular pollinators, thus restricting their distribution to areas where those pollinators are present. Epiphytic orchids, in particular, depend on host plants, further limiting their distribution to these specific hosts. These specialised interactions make orchids vulnerable to environmental changes and other threats, thus increasing their extinction risk (Nilsson 1992; Santos 2000; Cozzolino and Widmer 2005; Swarts and Dixon 2009).

Tourism and recreational activities also pose significant threats to orchids worldwide, with species increasingly impacted by development, recreational activities, and the collection of wild plants in tourist areas. These activities often lead to population declines and are expected to affect more species over time, due to a growing recognition of their impact, expanding tourism infrastructure, and delays in adding species to threat assessments. The combined effects of these threats create a "threat syndrome," which necessitates integrated management approaches. The expansion of tourism infrastructure typically draws more visitors, exacerbating environmental damage and increasing the risk of wild orchid collection (Wraith and Pickering 2017).

Singh et al. (2021) reported that *Dactylorhiza hatagirea*, an endangered medicinal orchid, has a limited distribution in the Indian Western Himalayas due to its low population density, short life cycle, poor seed germination rate, specific microhabitat requirements, and high commercial demand. Sharma et al. (2017) described how overexploitation, habitat loss, and climate change pose threats to 15 native orchid species in the Himalayan region, which may lead to a rapid decrease in their populations. In Northeast India, microclimatic changes have placed several orchid species under threat, necessitating the development of more strategic conservation protocols (Kumar et al. 2022).

Habitat loss (deforestation), urbanisation, climate change, the exploitation of orchids for medicinal purposes, smuggling and illegal trading, and natural calamities are major threats to orchid diversity in the Indian Himalayan Region (Chowdhery 2015; Barman and Devadas 2013; Rahamtulla et al. 2020a; Kindlmann et al. 2023).

4.7 Conservation Status

Orchid conservation focuses on maintaining natural habitats, promoting biodiversity, and ensuring the survival of rare orchid species. Conservation efforts have been implemented for years through both in-situ and ex-situ approaches, such as orchid seed gene banks, cryopreservation, breeding methods, in-vitro conservation, propagation, protective laws, and awareness activities (Medhi et al. 2012; De et al. 2016; Mutum et al. 2023). The Indian Himalayan Region (IHR) has a large number of in-situ conservation areas, including national parks and wildlife sanctuaries. Additionally, some states, such as Arunachal Pradesh, Sikkim, and West Bengal, have designated orchid-rich habitats as "orchid sanctuaries."

In terms of ex-situ conservation, nearly all states in the region host orchidaria run by both government and private entities. Notable orchidaria include the ICAR-National Research Centre for Orchids in Pakyong, Sikkim; Loyed Botanical Garden in Darjeeling, West Bengal; National Orchidarium in Shillong, Meghalaya; and Khonghampat Orchidarium in Imphal, Manipur, among others (Pal and Nagrare 2006; Ram et al. 2011). Modern approaches to ex-situ conservation, such as asymbiotic seed germination, have been reported by Gogoi et al. (2012), who successfully germinated and acclimatised seeds of *Cymbidium eburneum*, a rare epiphytic orchid, in various basal media.

International organisations, such as the International Union for Conservation of Nature (IUCN), the Convention on International Trade in Endangered Species of Wild Fauna and Flora (CITES), the Species Survival Commission (SSC), and the Orchid Specialist Group (OSG), are working to protect orchid diversity through the formulation of action plans and regulations. Species are protected globally by CITES Appendix I (which includes certain orchid species) and Appendix II (which includes the entire *Orchidaceae* family). Nationally, biodiversity conservation is supported by the Indian Forest Act (1927), Schedule VI of the Wildlife (Protection) Act (1972), the Forest (Conservation) Act (1980), the Environment (Protection) Act (1986), the

Biodiversity Act (2002), and the Biodiversity Rule (2004) (De and Medhi 2014; Chowdhery 2015; De et al. 2016).

In Arunachal Pradesh, the largest holder of orchid diversity in the IHR, extensive research and development efforts have been focused on orchids, including the exploration of orchid flora, conservation of germplasm through rescue operations at state orchid centres, and the preservation of orchids from felled trees and disturbed forest areas. These initiatives also involve the development of herbariums, breeding and hybrid development, aseptic culture, and micropropagation of ornamental and rare, endangered, and threatened (RET) orchid species. These efforts have been ongoing for the past 48 years. Additionally, cultivation and farming technologies have been transferred through training programmes for entrepreneurs and women's self-help groups, fostering a thriving orchid industry in the state (Hegde 2020).

Several orchid species are used by tribal groups, local communities, and pharmaceutical industries for their medicinal properties. Therefore, raising awareness among these communities is crucial for the sustainable use and conservation of orchids. Furthermore, artificial propagation, along with in-situ and ex-situ conservation methods, may be adopted for commercially important species to meet demand without jeopardising their survival (Chowdhery 2015).

4.8 Conclusions

The analysis of published studies and relevant databases indicates that the Indian Himalayan Region (IHR) is home to 1014 orchid species across 129 genera. Of these, 59% are epiphytic, 37% are geophytic, and 4% thrive in mixed habitats. Dominant genera include *Bulbophyllum*, *Dendrobium*, *Coelogyne*, and *Liparis*. The Eastern Himalayas, the most species-rich part of the region, boast 985 orchid species, with 60% being epiphytic. In contrast, the Western Himalayas, despite harsher climates, support 262 species, of which 52% are geophytic, adapted to the drier, colder conditions. The Eastern Himalayas' rich orchid diversity is largely attributed to its favourable climatic conditions, including dense forests, high rainfall, and humidity, while the Western Himalayas experience lower rainfall and cooler, drier climates. This study also highlights the region's growing orchid diversity, with 36 new species and distribution reports emerging over the last 5 years, predominantly from Arunachal Pradesh, underscoring the Eastern Himalayas' exceptional biodiversity. However, the IHR faces significant conservation challenges, with 65 orchid species now listed on the IUCN Red List, 21 of which are critically endangered. Habitat loss, climate change, and anthropogenic activities pose serious threats to orchid populations.

To ensure the survival of these unique species, conservation strategies must prioritise the protection of microhabitats, ecological restoration, and the expansion of both in-situ and ex-situ conservation efforts. Public awareness and collaborative conservation initiatives are essential in mitigating these threats and safeguarding the orchid diversity of the IHR for future generations. The preservation of this remarkable botanical heritage is crucial not only for ecological balance but also for maintaining the cultural and medicinal significance of orchids in the region.

Acknowledgements The authors sincerely acknowledge the Central Council for Research in Ayurvedic Sciences (CCRAS), Ministry of AYUSH, for providing institutional support and necessary facilities to carry out this study. Special thanks are extended to colleagues and technical staff of the Central Ayurveda Research Institute, Guwahati, and the Regional Ayurveda Research Institute, Gangtok, for their assistance and cooperation during the research work.

References

Arditti J, Ghani AKT (2000) Numerical and physical properties of orchid seeds and their biological implications. New Phytol 145(3):367–421

Atkinson ET (1882) The Himalayan Districts of Northwest Provinces of India. Cosmo Publication, New Delhi

Barman D, Devadas R (2013) Climate change on orchid population and conservation strategies: a review. J Crop Weed 9(2):1–2

Besi EE, Mustafa M, Yong CS, Go R (2023) Deforestation impacts on diversity of orchids with inference on the conservation initiatives: Malaysia case study. Bot Rev 89(4):386–420

Blatter E (1928) Beautiful flowers of Kashmir. Staples and Staples Ltd., Westminster

Brummitt NA, Bachman SP, Griffiths-Lee J, Lutz M, Moat JF, Farjon A, Donaldson JS, Hilton-Taylor C, Meagher TR, Albuquerque S, Aletrari E (2015) Green plants in the red: a baseline global assessment for the IUCN sampled red list index for plants. PLoS One 10(8):e0135152

BSI (2024) e-Plant checklist of India and e-Flora of India. https://efloraindia.bsi.gov.in. Accessed 7 Aug 2024

Chakraborty O, Agrawala DK, Chakraborty AP (2023) Studies on orchidoid mycorrhizae and mycobionts, associated with orchid plants as plant growth promoters and stimulators in seed germination. In: Mathur P, Kapoor R, Roy S (eds) Microbial symbionts and plant health: trends and applications for changing climate. Rhizosphere biology. Springer, Singapore. https://doi.org/10.1007/978-981-99-0030-5_16

Chowdhery HJ (1998) Orchid flora of Arunachal Pradesh. Bisen Singh Mahendra Pal Singh, Dheradun

Chowdhery HJ (2015) Orchids of India: diversity and status. In: Mandal VP (ed) Proceedings of the National Conference on new and emerging trends in biosystematics and taxonomy. Vidya Prasarak Mandal, Thane, pp 3–10

Collett H (1902) Flora Simlensis. Thacker, Spink and Co., Shimla, pp 289–508

Cozzolino S, Widmer A (2005) Orchid diversity: an evolutionary consequence of deception? Trends Ecol Evol 20(9):487–494

De LC, Medhi RP (2014) Diversity and conservation of rare and endemic orchids of north east India-a review. Indian J Hill Farm 27(1):138–153

De LC, Rao AN, Singh DR (2016) Endangered orchids and their conservation in north east India. In: Purkayastha J (ed) Bioprospecting of indigenous bioresources of north-east India. Springer, Singapore. https://doi.org/10.1007/978-981-10-0620-3_5

Deb CR, Longchar TB, Kamba J, Jakha HY (2021) Wild orchid resources of Nagaland, India: updated status. Pleione 15(2):113–122

Deva S, Naithani HB (1986) The orchid flora of north-west Himalaya. Print & Media Associates, New Delhi

Dressler RL (1981) The orchids, natural history and classification. Harvard University Press, Cambridge, United Kingdom

Dressler RL (1993) Phylogeny and classification of the orchid family. Cambridge University Press, Cambridge, United Kingdom

Duthie JF (1906) The orchids of the north-western Himalaya. Ann Roy Bot Gard 9:81–211

Ghatak J, Devi RKJ (1986) Orchids of Manipur. In: Vij SP (ed) Biology, conservation and culture of orchids. Affiliated East West Press Pvt. Ltd., New Delhi

Gogoi K, Kumaria S, Tandon P (2012) Ex situ conservation of *Cymbidium eburneum* Lindl: a threatened and vulnerable orchid, by asymbiotic seed germination. 3 Biotech 2:337–343

Gogoi K, Das R, Yonzone R (2015) Orchids of Assam, north east India-an annotated checklist. Int J Pharm Life Sci 6(1):4123–4156

Griffith W (1851) Notulae ad plantas Asiaticas, vol 3. Bishop's College Press, Calcutta

Hegde SN (2020) Status of orchid diversity, conservation, and research and development in Arunachal Pradesh: an overview. In: Khasim S, Hegde S, González-Arnao M, Thammasiri K (eds) Orchid biology: recent trends & challenges. Springer, Singapore, pp 223–235

Hooker JD (1888–1890) *Orchidaceae.* In: Flora of British India. Reeve & Co. Ltd., Ashford, Kent 5:667–658 & 6:1–198

Hooker JD (1895) A century of Indian orchids, vol 5. Annals of the Royal Botanic Garden, Calcutta

Hynniewta TM, Kalaki SK, Wadhwa BM (2000) Orchids of Nagaland. BSI, Calcutta

IUCN (2024) The IUCN red list of threatened species. http://www.iucnredlist.org. Accessed 7 Dec 2024

Jalal JS (2012) Status, threats and conservation strategies for orchids of western Himalaya, India. J Threat Taxa 4(15):3401–3409

Jalal JS, Kumar P, Rawat GS, Pangtey YPS (2008) *Orchidaceae*, Uttarakhand, Western Himalaya, India. Checklist 4(3):304–320

Kataki SK (1986) Orchids of Meghalaya. Meghalaya Forest Department, Shillong

Kaur J, Poff KE, Sharma J (2018) A rare temperate terrestrial orchid selects similar Tulasnella taxa in ex situ and in situ environments. Plant Ecol 219:45–55

Khasim SM, Hegde SN, González-Arnao MT, Thammasiri K (eds) (2020) Orchid biology: recent trends & challenges. Springer, Singapore

Khuroo AA, Ahmad R, Hamid M, Ruquia G, Malik AH, Rashid I (2021) Alien flora of Indian Himalayan region: taxonomic, biogeographical and ecological assessment. In: Singh L, Ranjan V (eds) New vistas in Indian flora, vol 2. Bishen Singh Mahendra Pal Singh, Uttarakhand, India, pp 497–518

Kindlmann P, Kull T, McCormick M (2023) The distribution and diversity of orchids. Diversity 15(7):810

King G, Pantling R (1898) Orchids of Sikkim Himalaya. Ann Roy Bot Gard 8:1–342

Kumar CS, Kumar PS (2005) An orchid digest of Manipur, northeastern India. Rheedea-Kerala 15(1):1

Kumar S, Devi RS, Choudhury R, Mahapatra M, Biswal SK, Kaur N, Tudu J, Rath SK (2022) Orchid diversity, conservation, and sustainability in northeastern India. In: Earth systems protection and sustainability. Springer, pp 111–139

Li T, Wu S, Yang W, Selosse MA, Gao J (2021) How mycorrhizal associations influence orchid distribution and population dynamics. Front Plant Sci 12:647114

Lindley J (1857) Contributions to the orchidology of India- 1. J Linn Soc 1:170–190

Lindley J (1859) Contributions to the orchidology of India- 2. J Linn Soc 3:1–63

Linnaeus C (1753) Species Plantarum. Laurentius Salvius, Stockholm

Liu H, Luo Y, Liu H (2010) Studies of mycorrhizal fungi of Chinese orchids and their role in orchid conservation in China—a review. Bot Rev 76:241–262. https://doi.org/10.1007/s12229-010-9045-9

Lucksom SZ, Arrawatia ML, Tambe S (2011) The orchid diversity in Sikkim and effect of change of environment on the distribution of native orchids in Sikkim Himalaya, India. In: Biodiversity of Sikkim: exploring and conserving a global hotspot. Information and Public Relations Dept, Govt of Sikkim, pp 125–150

Mao AA, Dash SS, Kumar S (eds) (2020) Plant discoveries 2019. BSI, Kolkata, India

Mao AA, Dash SS, Kumar S (eds) (2021) Plant discoveries 2020. BSI, Kolkata, India
Mao AA, Agrawala DK, Dash SS, Shukla AN, Mukherjee S (eds) (2022) Plant discoveries 2021. BSI, Kolkata, India
Mao AA, Agrawala DK, Mukherjee S (eds) (2023) Plant discoveries 2022. BSI, Kolkata, India
Mao AA, Agrawala DK, Dash SS, Shukla AN, Mukherjee S (eds) (2024) Plant discoveries 2023. BSI, Kolkata, India
Medhi RP, Chakraborti S, Rampal M (2012) Orchid biodiversity in India: conservation and utilization. Ind J Genet 72(2):148–156
Misra S (2007) Orchids of India. Bishen Singh Mahendra Pal Singh, Dehra Dun, India
Mulgaonkar MS, Shraddha R (2015) Ethnobotany of orchids: a review. In: Mandal VP (ed) Proceedings of the National Conference on new and emerging trends in biosystematics and taxonomy. Vidya Prasarak Mandal, Thane, pp 59–61
Mutum RD, Chanu NM, Khanganba TN, Thongam B, Das S (2023) In vitro conservation and propagation of endangered ethno-medicinal orchids from the northeast region of India. In: Jha S, Halder M (eds) Medicinal plants: biodiversity, biotechnology and conservation. Sustainable development and biodiversity, vol 33. Springer, Singapore. https://doi.org/10.1007/978-981-19-9936-9_20
Nair NC (1977) Flora of Bashahr Himalaya. International Bioscience Publishers, Hissar, Haryana
Nilsson LA (1992) Orchid pollination biology. Trends Ecol Evol 7(8):255–259
Nyorak J (2023) Orchid flora of Arunachal Pradesh (India) - a compilation. Pleione 17(1):50–81. https://doi.org/10.26679/Pleione.17.1.2023.050-081
Pal R, Nagrare VS (2006) Orchid diversity of India: its conservation and sustainable utilization. National Research Centre for Orchids (ICAR), Pakyong, Sikkim, India
Pal R, Singh DR (2016) Endemic orchids of Northeast India. In: Purkayastha J (ed) Bioprospecting of indigenous bioresources of north-east India. Springer, pp 31–44
Pandit MK, Sodhi NS, Koh LP, Bhaskar A, Brook BW (2007) Unreported yet massive deforestation driving loss of endemic biodiversity in Indian Himalaya. Biodivers Conserv 16:153–163
Pangtey YPS, Samant SS, Rawat GS (1991) Orchids of Kumaon Himalaya. Bishen Singh Mahendra Pal Singh, Dehradun
POWO (2024) Plants of the World Online. https://powo.science.kew.org. Accessed 7 Dec 2024
Rahamtulla M, Pradhan UC, Roy AK, Rampilla V, Khasim SM (2020a) Ethnomedicinal aspects of some orchids from Darjeeling Himalaya, India. In: Khasim S, Hegde S, González-Arnao M, Thammasiri K (eds) Orchid biology: recent trends & challenges. Springer, Singapore, pp 451–472
Rahamtulla M, Roy AK, Khasim SM (2020b) Orchid diversity in Darjeeling Himalaya, India: present status and conservation. In: Khasim S, Hegde S, González-Arnao M, Thammasiri K (eds) Orchid biology: recent trends & challenges. Springer, Singapore. https://doi.org/10.1007/978-981-32-9456-1_9
Raizada MB, Naithani HB, Saxena HO (1981) Orchids of Mussoorie. Bishen Singh Mahendra Pal Singh, Dehradun
Ram RB, Lata R, Meena ML (2011) Conservation of floral biodiversity of Himalayan Mountain regions with special reference to orchids. Asian Agri-History 15(3):231–241
Rau MA (1975) High altitude flowering plants of west Himalaya. White Lotus Press, Calcutta
Rodrigues AS, Pilgrim JD, Lamoreux JF, Hoffmann M, Brooks TM (2006) The value of the IUCN red list for conservation. Trends Ecol Evol 21(2):71–76
Roxburgh W (1832) In: Caray W (ed) Flora Indica, vol 3. Serrampore, pp 450–488
Roy AB, Purohit R (2018) The Himalayas: evolution through collision. In: Indian shield: Precambrian evolution and Phanerozoic reconstitution. Elsevier, Amsterdam, The Netherlands, pp 311–327
Sadananda K, Thongam B, Konsam B (2019) Orchid species of Manipur and its commercial aspect. Int J Pharm Life Sci 10(7):6302–6314
Samant SS, Joshi HC, Arya SC (2002) Studies on the structure, composition and changes of vegetation in Nanda Devi biosphere reserve of west Himalaya. In: Sharma et al (eds) Biosphere

reserves in India and their management. Ministry of Environment and Forests, New Delhi and Kerala Forest Research Institute, Peechi, Kerala, pp 133–139

Samant PS, Singh M, Lal M, Singh A, Sharma A, Bhandari S (2007) Medicinal plants in Himachal Pradesh, north western Himalaya, India. The Int J Biodiver Sci Manag 3(4):234–251

Santos FD (2000) Orchid preference for host tree genera in a Nicaraguan tropical rain forest. Selbyana 21:25–29

Sharma P, Samant SS (2017) Diversity, distribution, indigenous uses and conservation of orchids in Parvati Valley of Kullu District, Himachal Pradesh, North Western Himalaya. J Biodivers Endanger Species 5:1–5

Sharma A, Samant SS, Bhandari S, Butola JS (2017) Diversity, distribution, and conservation status of orchids along an altitudinal gradient in Himachal Pradesh, North Western Himalaya. J Orchid Soc India 31:23–32

Singh DK, Hajra PK (1996) Floristic diversity. In: Changing perspectives of biodiversity status in the Himalaya. British Council, New Delhi, pp 23–38

Singh Jalal J, Jayanthi J (2015) An annotated checklist of the orchids of western Himalaya, India. Lankesteriana 15(1):07–50

Singh DK, Pusalkar PK (2020) Floristic diversity of the Indian Himalaya. In: Dar GH, Khuroo AA (eds) Biodiversity of the Himalaya: Jammu and Kashmir State. Springer, pp 93–126

Singh SK, Agrawala DK, Jalal JS, Dash SS, Mao AA, Singh P (2019) Orchids of India: a pictorial guide, botanical survey of India, Ministry of Environment, Forests & Climate Change, India

Singh P, Singh AK, Pamarthi RK, Tomar A, Pal R, De LC, Rai D, Singh DR (2020) Orchid database of Sikkim and Darjeeling, West Bengal. India J Pharm Phytochem 9(4S):162–166

Singh L, Bhatt ID, Negi VS, Nandi SK, Rawal RS, Bisht AK (2021) Population status, threats, and conservation options of the orchid Dactylorhiza hatagirea in Indian Western Himalaya. Reg Environ Chang 21:1–6

Swarts ND, Dixon KW (2009) Terrestrial orchid conservation in the age of extinction. Ann Bot 104(3):543–556

Uniyal BP, Sharma JR, Choudhery U, Singh DK (2007) Flowering plants of Uttarakhand: a checklist. Bishen Singh Mahendra Pal Singh, Dehradun

Vij SP, Verma J, Kumar CS (2013) Orchids of Himachal Pradesh. Bishen Singh Mahendra Pal Singh, Dehradun

Wani SA, Ahmad R, Gulzar R, Rashid I, Malik AH, Khuroo AA (2022) Diversity, distribution and drivers of alien flora in the Indian Himalayan region. Global Ecol Conserv 38:e02246

WFO (2024) An online flora of all known plants. https://www.worldfloraonline.org. Accessed 7 Dec 2024

Wraith J, Pickering C (2017) Tourism and recreation a global threat to orchids. Biodivers Conserv 26:3407–3420

Yonzone R, Lama D, Bhujel RB, Rai S (2012) Orchid species diversity of Darjeeling Himalaya of India. Int J Pharm Life Sci 3(3):1533–1550

Chapter 5
Diversity of Threatened Medicinal Plants in the Himalayan Region: Livelihood, Marketing and Conservation Practices

Sandeep Rawat, Puja Acharya, Adharsika Pradhan, Animesh Sen, and Vikram S. Negi

Abstract The Himalayan region, a biodiversity hotspot, is home to over 18,940 plant species, many of which possess significant medicinal properties. These plants have been integral to the traditional healthcare systems of local communities, providing essential resources for both health and livelihood. However, the region faces critical threats from over-exploitation, habitat destruction, and climate change, leading to the classification of numerous species as threatened or endangered. This review explores the diversity of ethno-medicinal plants in the Himalaya; their cultural and economic importance and traditional knowledge related to these species. The IUCN red list lists 114 species under different threat categories used for multiple ethno-medicinal purposes. Among different threat categories, 23 species are described under Critically Endangered, 30 under Endangered, 44 under Vulnerable and 17 under Near Threatened category. Among the different plant families, *Lauraceae* (15 species), *Fabaceae* (10 species), *Orchidaceae* (10 species), *Musaceae* (5 species), *Apocynaceae* (4 species) and *Ranunculaceae* (4 species) were dominant families of threatened medicinal plants of Himalaya. The review highlights the growing global demand for herbal products and emphasises the need for sustainable harvesting practices and cultivation initiatives to alleviate pressure on wild populations. Practical strategies for conservation, including community

Supplementary Information The online version contains supplementary material available at https://doi.org/10.1007/978-3-032-21697-7_5.

S. Rawat (✉) · P. Acharya · A. Pradhan
Sikkim Regional Centre, G.B. Pant National Institute of Himalayan Environment, Gangtok, Sikkim, India

A. Sen
Regional Ayurveda Research Institute, Gangtok, Sikkim, India

V. S. Negi
Department of Botany and Microbiology, HNB Garhwal University, Pauri Garhwal, Uttarakhand, India

D. Das et al. (eds.), *Plant Diversity and Conservation Strategies of the Himalayan Biodiversity Hotspot*, Topics in Biodiversity and Conservation 21,
https://doi.org/10.1007/978-3-032-21697-7_5

engagement and market regulation, are discussed to ensure the sustainability of medicinal plant resources for future generations.

Keywords IUCN red list · Himalayan biodiversity · Consecration efforts · Ethno-medicines · Market demand

5.1 Introduction

The Himalaya is among the most extensive mountain chains in the world stretching across the region from the Indus Valley beyond Nanga Parbat (west) to Namcha Barwa (East) across Nepal, India, Bhutan, China, Pakistan, and Afghanistan, with total area coverage of 2500 km (Shrestha et al. 2015). This geographical overview aligns with the descriptions provided in previous chapters (Chap. 1, Sect. 1.1 and Chap. 4, Sect. 4.1). The region, along with adjacent Karakoram and Tibetan plateau, is the largest elevated landscape of the Earth. It was formed by a tectonic collision occurred about 45 million years ago and subsequent thrusting of the Indian plate with the Eurasian plate. The Himalayan range is characterised by varying altitudes, slopes and aspects as well as succeeding different climates resulting in extensive ecological diversity of the region (Jain 2014). The Himalaya is the transition zone between the Palaearctic and Indo-Malayan realms having influenced both the region in its biodiversity (Morrone 2015). Topographic complexity, climatic variations, and sharp altitudinal gradients in the Himalaya support the great biological diversity (Thakur and Chawla 2019). Extreme environmental conditions in the Himalaya forced plants to accumulate various secondary metabolites, which often may have medicinal properties to combat several ailments. These medicinal and aromatic plants are considered an important resource for pharmaceuticals, cosmetic and herbal industries (Kala et al. 2006; Meke et al. 2017).

The Himalaya is exceptionally rich in biodiversity, harbouring over 18,940 plant species, which constitute approximately 10% of the world's total plant diversity (Singh and Hajra 1996; Singh and Pusalkar 2020). Among these, more than 3160 species and 61 genera are endemic to the region. The Himalaya is amongst the most extensive mountain chains of the world stretching, covering the region from the Indus Valley beyond Nanga Parba. This high level of biodiversity and endemism is further supported by observations in Chap. 1 (Sect. 1.1), which highlights the region's significant endemic species richness and the presence of critically endangered, endangered, and vulnerable flora across its biologically diverse landscapes. Chapter 2 (Sect. 2.2) also reinforces this view, describing the Himalaya as home to a wide array of endemic plants and animals. There are 2000 species of bryophytes, 1159 species of lichens, and 7020 species of fungi. Besides this, more than 980 birds (15 endemic), 300 mammals (12 endemic), 175 reptiles (50 endemic), 105 amphibians (40 endemic), and 218 fishes are present in Himalaya (Rawal et al. 2013). Thus, the region is considered as a biodiversity hotspot due to the presence of rich diversity, endemism, and critical ecosystems.

Higher plants have played essential roles in the lives of the Himalayan tribal peoples as they provide forest products for food and medicine. Since ancient times, many wild and domesticated plants have been used as healing agents. In recent years, medicinal plants have become more popular as natural ingredients for cosmetics and herbal remedies (Ashraf et al. 2024). Himalaya harbours a rich diversity of wild relatives of several crop plants and represents three sub-centres of plant origin (viz., Western Himalaya, Eastern Himalaya and North Eastern Region), while only a total 8 sub-centres are present in the world. Among these, the Western Himalaya sub-centres contribute 125 species, the Eastern Himalaya represents 82 species and North Eastern Region represents 132 species of wild relatives (Rawal et al. 2021). More than 675 wild edible plant species were reported from the Himalaya (Samant and Dhar 1997). Additionally, Himalaya harbours 683 species of fodder plants (Samant et al. 1998), 155 species of sacred plants (Samant and Pant 2003), and 118 species of essential oil-yielding medicinal plants (Samant and Palni 2000). A large number of plants with medicinal and ethno-botanic properties are present in the region.

While the Himalaya has been recognised as a biodiversity hotspot, the plant diversity, particularly medicinal and aromatic plants are declining continuously. Overharvesting of medicinal plants, especially high-value plant species for commercial use has become a prime concern. In this review, the IUCN threat status of different ethno-medicinal plants has been summarised. Furthermore, market potential, value chain, possibilities of commercial cultivation and livelihood opportunities have been discussed. Further, research gaps for conservation and commercialisation have been identified based on overall information.

5.2 Diversity of Medicinal Plants in the Himalaya

Generally, plants with medicinal benefits have been considered essential resources of the healthcare system and sources of income for low-income communities. WHO has estimated that for primary healthcare needs, 80% of the population in developing countries is dependent on medicinal plants (WHO 2002). Globally, all the traditional societies use medicinal plants for their healthcare needs. It has been estimated that more than 50,000 species of medicinal plants are used for various purposes across the globe (Hamilton 2004). Direct and indirect market account for 15–30% of the total income of poorer households in various parts of the developing world (Hamilton 2004). In Asia, over 15,000 species are used in different medicinal systems (Negi et al. 2010). Only in India, 65% of people living in rural areas rely on traditional medicine (Timmermans 2003). Particularly, communities residing in the mid-altitudes to highlands have a great dependency on medicinal for the subsistence economy of the people (Sati 2013).

The distribution of medicinal plants in the Himalaya is associated with cultural practices. Besides the traditional Ayurvedic medicine system, regional or local traditional systems are practiced in the region. Buddhist populations residing in parts of Ladakh, Himachal Pradesh Sikkim and Arunachal Pradesh have a significant impact on traditional Tibetan Pharmacopoeia (Kumar and Mishra 2012). Many species have found a place in the Indian Traditional medicine systems such as, Ayurveda, Homeopathy, Unani, Sowa-Rigpa and Sidhha. Traditional practices are available in every part of Himalaya and outside the Himalaya. In a study, a comprehensive database revealed the presence of 1123 plant species (78% species native to Himalaya) with ethno-medicinal uses in the Jammu & Kashmir state located in West Himalaya (Tali et al. 2019). A list of 1127 medicinal plants has been published from the State of Uttarakhand (Kumar et al. 2018). Many traditional medicine systems such as Amchi, Jhankri, Ayurveda, Tibetan, and Lama are practiced in Sikkim (Sherpa et al. 2015), and around 490 medicinal plant species are found their habitat in Sikkim (Panda and Misra 2010). However, more than 100 species of plants are traditionally used by rural and tribal peoples in Arunachal Pradesh (Perme et al. 2015).

Similarly, there is a broad diversity of medicinal plants available in the region, and as per an estimate, the Indian part of the Himalaya is home to more than 1750 medicinal plants. It includes 1685 species of angiosperms, 51 pteridophytes, and 12 gymnosperms (Samant et al. 1998). Many plants are described in traditional and ethno-medicinal systems and thus number varied among different studies. Therefore, other literature on documentation of medicinal plants from Himalaya used by various communities, geographic regions, valleys, districts or states exhibited variations in numbers (Samant Pant et al. 2007; Khanday et al. 2018). Few more publications reported the documentation of ethnobotany of medicinal plants used by various ethnic and tribal communities based on large-scale household surveys (people who live together in a single home) (Panmei et al. 2019).

5.3 Importance of Medicinal Plants in Traditional Systems

Early for various medical purposes, human societies relied on plants and plant-derived products. Indigenous knowledge systems are the foundation of traditional herbal remedies. These cognitive processes are deeply rooted in the local community's way of life and are essential in making decisions about using medicinal plant resources. Ethnic communities in the Himalayan region show their own culture and traditional knowledge system. Diverse ethnic communities and rural people of the Himalaya used various ethno-medicinal plants in myriad ways. Traditional herbal medicine is well recognised in the Indian Himalayan region and plays a role in healthcare and partial livelihood in rural areas of the region (Sen et al. 2011; Kala 2005). Interestingly, the traditional system of healthcare is still practiced by some of the traditional herbal practitioners (Phondani et al. 2016). In certain areas, these folk medical prescriptions are endemic and have survived through ages from one generation to next through word of mouth. The therapeutic applications in different traditional systems originated from centuries-old practices and regional traditions. The difference in culture, history and philosophy from region to region and country to country create the differences in their practices.

Traditional medicine is a broad term and practices vary from person to person. Thus, it can be classified into scholarly medicine system, folk medicine, and shamanistic medicine system (Connor and Samuel 2001). Among these, Folk and shamanistic medicine systems are considered as ethno-medicinal systems. Folk medicines are practiced in a limited number of populations with similar cultures and transmitted orally as general knowledge. However, shamanistic medicine is a spiritual form of medicine used along with herbs (Sherpa et al. 2015). The biodiversity and related traditional knowledge in the Himalayan region are at significant risk of extinction in the present day because of the growing drive for economic growth and the ensuing socio-cultural change (Raut et al. 2012).

5.4 Use of Medicinal Plants in Modern Medicine

Historically, natural compounds and their structural derivatives have played a crucial role in pharmacotherapy. Beyond the biologically active molecules derived from plants that are directly used as therapeutic drugs, many other natural bioactive substances have served as valuable "leads" or template molecules for the developing or semi-synthesis of new drugs. Modern markets of life-saving and popular drugs are filled with many drugs originating from plant sources, such as artemisinin, paclitaxel, vinblastine, vincristine, camptothecin, piperidines, phyllanthins, curcumines, withanolides, bacosides, icrosides, podophyllotoxin, and most of them has established landmark in the area of drug discovery started from Ayurvedic practices. Recently, various scientific and technological advancements, such as engineering techniques, genome mining, improved analytical tools, and progress in microbial culture, have addressed these challenges and opened up new opportunities (Newman et al. 2000). These advancements are reflected in the complexity of bioassays employed in medicinal plant drug development. They also provide detailed 'mode of action' information at the molecular level quickly and accurately (Bohlin and Bruhn 1999; Bahar et al. 2007). As a result, there has been a renewed focus on natural compounds as promising therapeutic candidates, particularly for combating antimicrobial resistance (Boakye et al. 2019).

Among the potential drug molecules, secondary metabolites like alkaloids and 80 molecules are in clinical use among the 10,000 discovered alkaloids (Zhang et al. 2021). Among these, berberine isolated from *Coptis chinensis* and *Berberis* sp. is used for inflammatory disorders; reserpine from *Rauvolfia verticillata* used to treat hypertension (Pan et al. 2013); ephedrine from *Ephedra sinica* is used for asthma; camptothecin from *Camptotheca acuminata* and vincristine from *Catharanthus roseus* used for cancers (Glade et al. 2010). Thus, the traditionally used plant needs systematic screening and clinical studies for medicinal purposes. The isolation and characterisation of pharmacologically active compounds from medicinal plants remain ongoing. In recent years, drug discovery methods have also been used to standardise herbal medicines and identify analytical marker compounds.

5.5 Threat to Medicinal Plants of Himalaya

Many of these species as well as their environments in the Himalaya are endangered by excessive exploitation, loss of habitat by deforestation, over-grazing, and climatic uncertainties (Shivali and Namrata 2022). Of these, approximately 15,000 species are at risk of extinction due to over-harvesting and habitat loss, with 20% of their wild resources already nearly depleted because of the growing human population and increased plant consumption (Baricevic et al. 2015). Even though, this concern has been recognised for decades, the risk of medicinal plant extinction has increased due to the rapid loss of species and destruction of habitats globally.

These species are rapidly declining, primarily due to factors such as population growth, urbanisation, habitat fragmentation, and the rising reliance of large global populations on finite natural resources (Woodruff 2001). Due to human activities, it is estimated that the rate of species extinction has increased significantly (Bradshaw and Brook 2014; Ganie et al. 2019). A total of 26,840 from 96,951 species evaluated for various threat categories are considered threatened (Ganie et al. 2019); and multiple risks include pollution, industrialisation, biological invasions, overexploitation, habitat loss or degradation, and accelerated climate change (Adla et al. 2022).

5.6 Threatened Medicinal Plants in the Himalaya

Among the ethno-medicinal plants of Himalaya, 114 species were identified as threatened under different categories of IUCN (2022). Species such as, *Aconitum chasmanthum, Amomum riwatchii, Bazzania bhutanica, Beilschmiedia assamica, Beilschmiedia brandisii Bulbophyllum arunachalense, Cheirostylis sherriffii, Cinnamomum champokianum, Diplomeris josephi, Gentiana kurrooa, Gymnocladus assamicus, Ilex khasiana, Lilium polyphyllum, Machilus parviflora, Meconopsis bhutanica, Meizotropis pellita, Musa argentii, Musa mannii, Musa nagalandiana, Nardostachys jatamansi, Neottia confuse, Ormosia robusta, Rhomboda arunachalensis*, and *Saussurea costus* have found place under critically endangered category. Among these, some species like *Saussurea costus, Nardostachys jatamansi*, and *Gentiana kurroa* have great industrial demand for medicinal products and thus may have high illegal exploitation from natural habitat.

Similarly, *Aconitum heterophyllum, Andrewsianthus ferrugineus, Angelica glauca, Atropa acuminate, Bistorta griersonii, Bulbophyllum trongsaense, Coptis teeta, Cypripedium himalaicum, Dactylorhiza hatagirea, Dillenia mansonii, Grewia nagensium*, *Gymnocladus burmanicus*, *Hoya bhutanica*, *Ilex venulosa*, *Illicium griffithii*, *Litsea assamica*, *Litsea khasyana*, *Meconopsis superba*, *Musa chunii*, *Neopicrorhiza minima*, *Nepenthes khasiana*, *Picrorhiza kurroa*, *Pittosporum eriocarpum*, *Podophyllum hexandrum*, *Semecarpus subpanduriformis*, *Taxus contorta*, *Taxus wallichiana*, *Trillium govanianum*, and *Trillium tschonoskii* have been recorded as endangered medicinal plant species of the Himalaya. Among these, *Aconitum heterophyllum*, *Angelica glauca*, *Dactylorhiza hatagirea*, *Podophyllum hexandrum*, *Taxus wallichiana*, *Taxus contorta*, and *Trillium govanianum* are highly traded species (Fig. 5.1). Also, 44 species have been categorised under vulnerable and 17 under near threatened categories of IUCN (Fig. 5.2). The detailed description of Himalayan medicinal plants under different threat categories is given in Supplementary Table 5.1.

Among these species, one species *Bazzania bhutanica* (family *Lepidoziaceae*) belongs to Bryophyte. While 09 species (*Abies spectabilis*, *Taxus contorta*, *Taxus wallichiana*, *Ephedra gerardiana*, *Cupressus cashmeriana*, *Pinus gerardiana*, *Cephalotaxus mannii, Cycas pectinata*, and *Picea likiangensis*) are from Gymnosperm. Monocotyledon represents 23 species and Dicotyledon represents 81 species (Fig. 5.3). Most of the Gymnosperm species found in the Himalayan region have high demand in the market due to medicinal, aroma, ornamental and food preparations.

Among the different plant taxonomic families, *Lauraceae* (15 species), *Fabaceae* (10 species), *Orchidaceae* (10 species), *Musaceae* (5 species), *Apocynaceae* (4 species), *Ranunculaceae* (4 species), *Gentianaceae* (3 species), *Melanthiaceae* (3 species), *Pinaceae* (3 species), *Taxaceae* (3 species) were recorded as dominant plant families of threatened plants (Fig. 5.4). This aligns with Chap. 4 (Sect. 4.1), which focuses on "Orchid Diversity of the Indian Himalayan Region," documenting 1014 orchid species across 129 genera, including 65 listed under IUCN threat categories. The chapter also highlights the medicinal, ornamental, and ecological importance of orchids and the urgent need for their conservation. Among these, *Orchidaceae* and *Musaceae* belong to Monocotyledon, while, *Pinaceae* and *Taxaceae* belong to Gymnosperm group of plants.

5.7 Livelihood Improvement Using Medicinal Plants

The rural households in the Himalaya collect from the wild and sell the medical plants to supplement their income. Recent investigations revealed that hundreds of tons of roots, rhizomes, tubers, leaves, and other products are traded annually for millions of USD. According to the estimates, the harvesting of medicinal plants was an important part of community livelihood goals, contributing 3–44% (on average, 12%) of annual household income (Olsen and Larsen 2003). Due to a lack of awareness or ignorance, many threatened medicinal plants get over-exploited and face the threat of extinction. At the moment, the worldwide market of medicinal and aromatic plants is estimated at 300 million USD and is increasing at a rate of 15–25%

Fig. 5.1 Representative photographs of a few important threatened medicinal plants of Himalaya. Species are represented as (**a**) *Allium roylei,* (**b**) *Aconitum heterophyllum,* (**c**) *Dactylorhiza hatagirea,* (**d**) *Aconitum violaceum,* (**e**) *Malaxis muscifera,* (**f**) *Picrorhiza kurroa,* (**g**) *Podophyllum hexandrum,* (**h**) *Lilium polyphyllum* and (**i**) *Trillium govanianum*

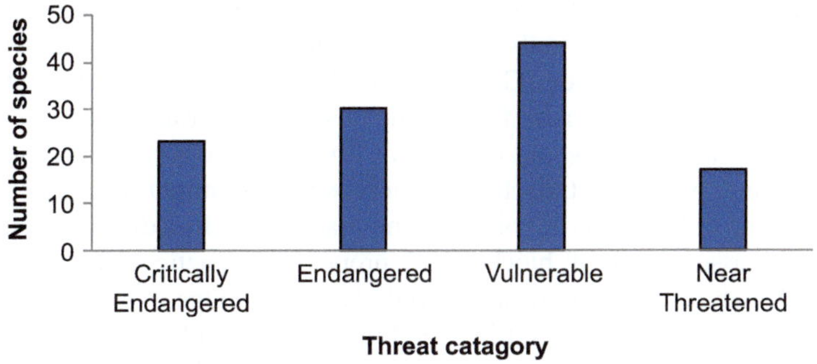

Fig. 5.2 Number of medicinal plant species present in the Himalaya under different threatened categories of IUCN

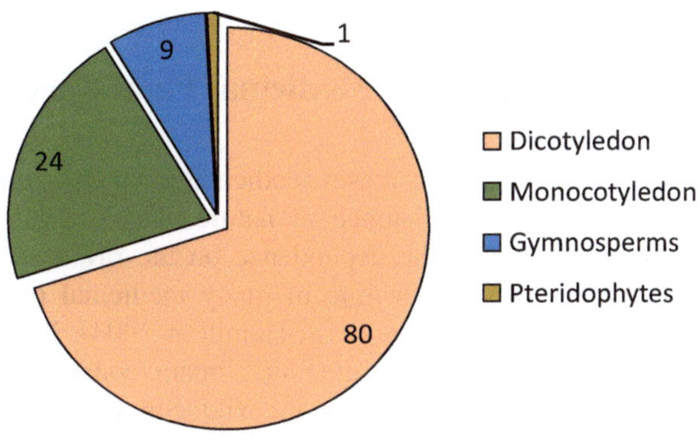

Fig. 5.3 Different taxonomic groups of threatened medicinal plant species present in the Himalaya

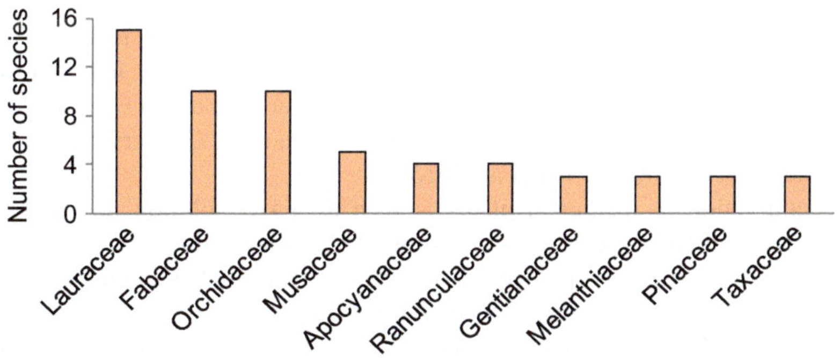

Fig. 5.4 Major families of threatened medicinal plants present in the Himalaya

annually. It has been estimated that it will reach 5 trillion USD by 2050 (Volenzo and Odiyo 2020).

India is involved in the export and import of 115 million tone plant commodities for medicinal uses and it has a value of more than 300 million USD had in 2015. China exported and imported 225 million tons of medicinal plants which had value of 1.2 trillion USD in 2015 (Volenzo and Odiyo 2020). A need for a large quantity of high-quality herbal raw materials has been estimated due to the rising demand for herbal medications worldwide (Gautam et al. 2023). The medicinal plants market normally is dependent on many factors such as raw material availability, product manufacturing cost, potency and efficacy of drug, bioprospecting, acceptability of consumers, and patenting trajectories. Bioprospecting might increase economic benefits among stakeholders involved in the marketing of medicinal plants, which involves the assessment of biochemical properties, genetic characteristics, and commercial viability of biological material. The knowledge existing or owned by local communities (particularly tribal or Indigenous peoples) is the fundamental requirement for bioprospecting or the production of useful products, such as functional foods, herbal medicine, cosmetics, food flavours, and fragrances (Volenzo and Odiyo 2020).

5.8 Cultivation of Medicinal Plant Species

Rapid globalisation increases medicinal plant demand across the world and creates pressure on their existence in nature. The growing market demand for herbal medicine increases the dependence on the raw material harvested from the wild and there is a swift decline of many medicinal plant species has been reported (Sangma and Manohara 2018; Hamilton 2004). Further, the high market demand for medicinal plants, over-grazing, climatic constraints, and other factors have led to the threat to many plants from the wild (Singh et al. 2021; Kumar and Rawat 2022). The ongoing use of a number of therapeutic plant species from restricted habitats has resulted decline in their population, which ultimately resulted in the loss of their existence. For example, Negi et al. (2019) reported a population decline of threatened medicinal plants from the Nanda Devi Biosphere Reserve (NDBR). Conservation and cultivation of medicinal and aromatic plants are important because 21% fall under the endangered category of the International Union for Conservation of Nature (Schippmann et al. 2006; Chen et al. 2016).

The cultivation of medicinal plants is seen as a key solution for meeting the growing demand for large-scale production of plant-based drugs and herbal remedies, while also helping to reduce harvest pressure on wild populations (Lambert et al. 1997). Therefore, conservation initiatives for medicinal plants require immediate attention, and bringing the wild species into commercial cultivation can reduce the risk of extinction and lessen the strain on the natural population (Lambert et al. 1997; Schippmann et al. 2002; Negi et al. 2019). Lack of cultivation practices for these species remains always a problem for growers along with quality assurance. Thus, collection from the wild is preferred among suppliers. It is estimated that around 90% of medicinal plants in use are sourced from the wild, with 70% of this collection involving harmful harvesting methods (Goraya and Ved 2017).

The Indian Himalaya is incredibly rich in source of medicinal plants but only limited success in tapping the potential of these plants has been achieved because of insufficient awareness among farmers and a lack of suitable cultivation/agronomic practices from these plants. Some medicinal plants are grown on home farms, while others are cultivated as field crops. However, the number of medicinal plant species being cultivated intensively for commercial production varies across countries. In India, 20 species are cultivated, while it is 40 in Hungary, and 100 in Germany. Despite China's long history of using medicinal plants, it cultivates only about 250 species. The domestication of a species for cultivation is complex evolutionary processes where human use of plant species results in morphological and physiological changes that differentiate domesticated varieties from their wild ancestors.

Several strategies can be used to preserve medicinal plants. It may include: (i) farmers should be encouraged to cultivate medicinal plants on their private property through healthcare farming, rather than taking advantage of natural populations (Kuniyal et al. 2015); (ii) herbal gardens at schools and other public facilities can be developed and used as germplasm repositories; (iii) It is important to encourage the planting of medicinal plants in "Van Panchayat" (community forest) and forest land so that locals can collect sustainably (iv) modern technologies such as in vitro production of planting material and secondary metabolites could be used for faster multiplication and large scale production (Rawat et al. 2018). Currently, fewer than 20 species are being cultivated in various regions of the country (Uniyal et al. 2000). But farmers can generate approximately Rs 0.12 million/ha annual income through the cultivation of higher altitude plants (Rao and Saxena 1994; Phondani et al. 2016). However, bringing medicinal plants into cultivation causes altered behavioural responses in terms of morphology, biomass, chemical composition, and productivity due to altered environmental conditions (Bhatt et al. 2012; Rawat et al. 2017). Although farmers in India are growing medicinal plants and giving them a lot of attention, many of them are still semi-domesticated. The impact of farming on the genetic diversity of crop plant species and forest tree species populations in India has been the subject of numerous researches (Bahulikar et al. 2004; Shaanker et al. 2004; Bodare et al. 2013). According to estimates, growing medicinal plants has the potential to yield significant financial benefits for farmers (Nautiyal 1995; Rao and Saxena 1994).

The cultivation of medicinal plants reduces the pressure of the dependency on the collection from the wild, which will also be beneficial for the upliftment of the rural economy (Hamilton 2004). In the Nanda Devi Biosphere of the Central Himalaya, medicinal plants like *Allium humile* and *Allium stracheyi* make up 70% of the total land area dedicated to medicinal plant cultivation (Negi et al. 2019). Among various cultivation of medicinal plants, *Carum carvi* shows the highest economic returns in the Garhwal Himalaya (Nautiyal et al. 2001). In the entire Himalayan region, rural and traditional communities are utilising medicinal and aromatic plants for healthcare and economic benefits. Five promising medicinal and aromatic herbs e.g., Rosemary (*Rosmarinus officinalis*), Satawar (*Asparagus racemosus*), Ban-tulsi (*Ocimum basilicum*), Tagar (*Valeriana jatamansi*), and Chamomile (*Matricaria chamomilla*) were chosen for cultivation on farmers' fields in specific village clusters, including Dharaunj, Mudiyani, and Gumod, through participatory action research methods. As a result, 120 farmers initiated the cultivation of these selected medicinal plants species across 14 hectares of land. To build the farmers' capacity, a series of meetings were held involving farmers, traders, and other stakeholders (Phondani et al. 2016). Many threatened medicinal plants like *Saussurea costus*, *Swertia chirayta*, *Dactylorrhiza hatagirea*, *Aconitum heterophyllum*, *Picrorhiza kurroa*, *Carum carvi*, *Angelica glauca*, and *Nardostachys grandiflora* are still cultivated on the fringes of the forest (encroached land) and agricultural land by people in some studied villages in Govind Pashu Vihar Wildlife Sanctuary, Central Himalaya, India (Negi et al. 2010).

5.9 Contribution to the Rural Economy

Medicinal plants are an important subject for livelihood-focused projects in India, which supports the traditional agricultural system to give farmers an alternate means of income (Uniyal et al. 2000). In addition to being a significant source of resources for the herbal and traditional medicine industries, medicinal plants also gave an important means of livelihood and health security to the population region (Olsen and Larsen 2003). In 2017–18, exports of value-added extracts from medicinal plants and herbal products totalled USD 456.12 million, marking a 12.23% growth compared to the previous year. There are only 242 species that witness high-volume trade in the Himalayan region (>100 MT) annually (Goraya and Ved 2017). Medicinal plants can support small-scale farmers by enhancing their livelihoods through both income generations from their trade and access to healthcare services, particularly in rural and remote areas. The estimated value of medicinal plants from both cultivation and forest sources in Uttarakhand was 70 billion USD in 2013 (Kuniyal et al. 2015) and growing between 12% and 15% a year.

The majority of medicinal plants that are naturally grown are found in India. Approximately 15,000 species are utilised in various healthcare systems across Asia, with around 7000 found in China and 8000 in India (Negi et al. 2010 The global market for herbal medicines is valued at nearly 2700 crore rupees annually, whereas in India, the trade amounts to about 350 crore rupees per year and it is increasing at the rate of 7% per year (Kala et al. 2006). Just 14% of all land is under cultivation in Uttarakhand, but agriculture provides a living for roughly 65% of the people. The state has a lot of potential for growing medicinal plants, and this could be one of the key ways for the hilly areas to have a sustainable way of life (Joshi and Joshi 2014). Thousands of medicinal species are primarily found in the mid-altitudes and highlands. The State Forest Department of Uttarakhand identifies around 175 species that are commercially harvested and traded. However, the state has not been able to fully utilise the potential of these medicinal plants. Annual earnings from this place are estimated at about 100 million in Indian currency (Sati 2013). In the Gulmarg Himalaya Region of Jammu & Kashmir, people use medicinal plants like *Podophyllum hexandrum*, *Bergenia ligulata*, *Trillium govanianum* etc., and while grazing they collect those plants and sell them to the nearby villages (Najar and Mansoor 2020). Phondani et al. (2016) reported that 132 farmers adopted the cultivation of

MPs under 14 hectares of land in the Champawat district of Uttarakhand. The study reported the highest net profit to the villagers about 1469.69 USD per hectare per year for *Rosmarinus officinalis* and 523.24 USD per hectare per year for *Matricaria chamomilla* after 3 years of cultivation.

5.10 Value Chain and Conservation Initiatives

Conservation of threatened medicinal plants is a complicated geopolitical process involving regional and national-local components for the creating effervescent environment protection rights and preparing a taskforce to develop a practical framework for collaborative action. The major challenges in creating the international value chain for medicinal plants are finding synergy between environmental sustainability targets, developmental agendas of countries, cultural and social divergence, and convergence between institutions and organisations (Volenzo and Odiyo 2020). There is increased necessity for complementary action among the numerous factors to integrate the different aspects of sustainability. In developing nations, benefits to growers and producers remain low as well as corporate and intermediaries or agents, despite the increasing trade in medicinal plants in the international market. The lack of organisation and networking by the poor collectors of medicinal plants from the wild is the central region behind this financial dispute. Quality certification is becoming a non-tariff barrier for the majority of developing countries during entry and competition in the global value chain market.

European Union requires the supply of comprehensive physicochemical documentation, microbiological, pharmacological, and toxicological analysis and widespread clinical trials to validate the safety, and quality of plant-based medication products. Thus, a successful value chain requires complete coverage to avail a product from its grower to consumers that satisfies their needs and preferences. Among the various issues in the medicinal plant sector, the paucity of collection centres at different regional locations and low-price fixation needs to be addressed. There is a shortage of reliable information about prices and the size of the demand for various species. Currently, marketing of medicinal plant produce is operated through Mandis, wholesalers, and generally intermediaries in a few states of the Indian Himalayan region (IHR).

In general, traders through middlemen, exercise substantial control over the collectors/growers because they later depend on them for loans. This limits the direct access of growers to the market; thus, the maximum benefit still goes to traders and middlemen. Whether the raw material is grown or gathered from the wild, the intermediaries and traders purchase it illegally. Thus, at the regional level, cooperatives for medicinal plant cultivation and purchase might be established. Given the scope and complexity of the medicinal plant trade and collection in the area, it appears likely that neither wild collection nor trade regulations have been put in place. Local commerce and export of 29 medicinal plants in their raw or plant forms continue despite the Indian government's prohibition on their gathering from the environments (Ved et al. 2003); the ban has however been reported to increase illegal trade (Chauhan et al. 2013). Due to habitat loss and overharvesting, 21% of medicinal plant species worldwide are currently threatened (Schippmann et al. 2006; Chen et al. 2016).

5.11 Trade and Market Potential

The continuous availability of raw materials as a result of an expanding market is a vital requirement in MPs marketing. The supply chain, appropriate market locations, and post-harvesting techniques are all interdependent on the availability of raw resources and agronomic practices. The marketing system in MPs sector is largely unregulated and inequitable. Currently, over 90% of plant species utilised in the herbal industry are harvested from the wild, mostly occurring in the Himalayan sub-alpine and alpine regions (Singh and Dey 2005). These include aspirin, atropine, artemisinin, colchicine, digoxin, ephedrine, morphine, physostigmine, pilocarpine, quinine, quinidine, reserpine, taxol, tubocurarine, vincristine and vinblastine. The majority of these plant-derived medications were initially found through the study of traditional remedies and indigenous peoples' folk knowledge, and some of these could not be replaced despite the enormous developments in synthetic chemistry (Kumar et al. 2011).

The worldwide herbal market Products are constantly growing, and by 2050, they are predicted to reach 5 trillion USD, compared with $62 billion in 2004. Although a major trader, its share of India in the global trade of medicinal and aromatic plants is mainly 0.5%, and in the countries like China export raw drugs and plants, therapeutics, with additional medicinal and aromatic plants valued at Rs 18,000 crores per year. The medicinal plants of 162 species are expected to increase by about 16–18% from 2010 to 2015 to fulfil the demand of both domestic and export markets.

As per the estimate of the National Medicinal Plant Board, the herbal drug industries of India consumed more than 1178 medicinal plant species (NMPB 2024). Among these, 242 medicinal plant species are utilised annually in more than 100 MT in India. Among these, 11 threatened medicinal plants including, *Abies spectabilis*, *Aconitum heterophyllum*, *Aegle marmelos*, *Anacyclus pyrethrum*, *Coptis teeta*, *Ephedra gerardiana*, *Nardostachys jatamansi*, *Picrorhiza kurroa*, *Saussurea costus*, *Taxus wallichiana*, and *Trillidium govanianum* have annual trade of 100 MT in India and collected manly from the wild population (Table 5.1).

Natural product markets are expanding, particularly for food, nutrition, and medications. The industry for medicinal and aromatic plant products alone is estimated to be worth over USD 80 billion. The agricultural economy of Himachal Pradesh has significantly benefited from using aromatic and medicinal plants. This state is home to 24 of the 100 most significant species of medicinal plants traded nationwide. The state exports 2500 tons of medicinal plants and their valuable portions. At current market rates, the state's yearly legal trade in medicinal plants is valued at approximately 10 crores. The state government receives approximately

Table 5.1 Threatened medicinal plants in the Himalaya that have more than 100 metric tons annual trade in India

SL.No.	Botanical name	Trade name	Estimated annual trade (in Metric tons)
1	*Abies spectabilis*	Talispatra, Talisa	100–200
2	*Aconitum heterophyllum*	Atis, Ativisa	100–200
3	*Aegle marmelos*	Bael, Belgiri, Bilva	2000–5000
4	*Anacyclus pyrethrum*	Akarkara, Akarkarabha	200–500
5	*Coptis teeta*	Mamira, Rohini	100–200
6	*Ephedra gerardiana*	Somalatha	100–200
7	*Nardostachys jatamansi*	Balchad, Jatamansi	500–1000
8	*Picrorhiza kurroa*	Kutki, Katuka	1000–2000
9	*Saussurea costus*	Kuth, Uplet, Kustha	100–200
10	*Taxuswallichiana*	Talispatra, Sthauneya	100–200
11	*Trillidium govanianum*	Nag Chhatri, Satva	200–500

Source: NMPB (2024)

40 lakhs annually from the export of medicinal plants. Disorganised marketing was the main obstacle to producing several species, including Safed Musli, Aloe vera, Stevia, and Lemon grass, in Himachal Pradesh. Thus, medicinal plants in controlled market places and location identification are necessary (Guleria et al. 2014). Restriction on gathering wild harvesting of *Saussurea lappa* and *Picrorhiza kurroa,* which are important optional crops for cultivation at higher elevations (>2400 m a.s.1.) (Kuniyal et al. 2013).

5.12 Conclusions

Medicinal plants (MPs) serve as a vital bridge between traditional knowledge and modern healthcare. In the Himalayan region, their significance goes far beyond pharmacological use, they are embedded in the cultural, economic, and ecological fabric of the region. These plants provide essential primary healthcare services to local communities, contribute to household incomes, and represent a living repository of centuries-old ethnobotanical wisdom. However, the unchecked and indiscriminate harvesting of these plants from the wild, driven by growing demand and lack of sustainable alternatives, has led to the depletion of several high-value and ecologically sensitive species. Ineffective conservation strategies, limited awareness, and insufficient involvement of local institutions have compounded the threat to these valuable resources.

While a few Indian Himalayan states, notably Uttarakhand and Himachal Pradesh, have initiated large-scale cultivation of medicinal and aromatic plants, much of the region still relies heavily on cheaper, wild-sourced materials, an approach that is ecologically damaging in the long run. To reverse this trend, there is an urgent need to develop and implement region-specific agro-techniques, provide financial and institutional support for cultivation, and establish robust conservation frameworks. Additionally, many threatened species identified by the IUCN are geographically restricted and require habitat-specific protection. Species that have not yet been assessed for conservation status must be prioritised for evaluation and monitoring.

A critical yet often underappreciated aspect of medicinal plant conservation and commercialisation involves legal protections and benefit-sharing mechanisms, particularly for traditional knowledge (TK) associated with these resources. In this context, the Nagoya Protocol to the Convention on Biological Diversity (CBD) plays a pivotal role. Adopted in 2010, the Nagoya Protocol is an international agreement designed to ensure fair and equitable sharing of benefits arising from the utilisation of genetic resources and traditional knowledge. It empowers resource-providing countries and indigenous communities by requiring users, such as researchers, pharmaceutical companies, or exporters—to obtain Prior Informed Consent (PIC) and establish Mutually Agreed Terms (MAT) before accessing or using biological resources. The Protocol safeguards the rights of local communities and ensures they are appropriately compensated for their knowledge and resources. In India, these provisions are enforced through the Biological Diversity Act, 2002, and regulated by the National Biodiversity Authority (NBA), which oversees compliance with access and benefit-sharing (ABS) rules.

To ensure the long-term sustainability of the medicinal plant sector, a multi-pronged approach is essential, one that integrates conservation, ethical resource use, scientific research, and equitable economic development. Legal harmonisation of policies related to biodiversity, trade, and intellectual property rights must be strengthened. Collection permits granted to harvesters need stringent monitoring, and high-risk wild populations must be periodically reassessed at regional and national levels. Moreover, fostering awareness and collaboration among local communities, policymakers, researchers, and industry stakeholders is crucial to safeguarding the future of medicinal plant biodiversity. With the right combination of scientific innovation, policy enforcement, and respect for traditional knowledge systems, the Himalayan region can become a global model for sustainable and equitable utilisation of medicinal plant resources.

Acknowledgements We thank the Director of GBPNIHE for providing facilities and encouragement. We also thank colleagues in the Sikkim Regional Centre for their support and help. Partial financial support received from the In-House Project no 04 (CBCM) project funded by MoEF&CC, Govt of India is deeply acknowledged.

References

Adla K, Dejan K, Neira D, Dragana S (2022) Degradation of ecosystems and loss of ecosystem services. In: Prata JC, Ribeiro AI, Rocha-Santos T (eds) One health. Academic, Cambridge, MA, USA, pp 281–327

Ashraf MV, Khan S, Misri S, Gaira KS, Rawat S, Rawat B, Khan MH, Shah AA, Asgher M, Ahmad S (2024) High-altitude medicinal plants as promising source of phytochemical antioxidants to combat lifestyle-associated oxidative stress-induced disorders. Pharmaceuticals 17:975

Bahar M, Deng Y, Fletcher JN, Kinghorn AD (2007) Plant-derived natural products in drug discovery and development: an overview. In: Selected topics in the chemistry of natural products. World Scientific, Singapore, pp 11–48

Bahulikar RA, Stanculescu D, Preston CA, Baldwin IT (2004) ISSR and AFLP analysis of the temporal and spatial population structure of the post-fire annual, *Nicotiana Attenuata*, in SW Utah. BMC Ecol 4:1–13

Baricevic D, Mathe A, Bartol T (2015) Conservation of wild crafted medicinal and aromatic plants and their habitats. In: Mathe A (ed) Medicinal and aromatic plants of the world. Springer, pp 131–144

Bhatt ID, Dauthal P, Rawat S, Gaira KS, Jugran A, Rawal RS, Dhar U (2012) Characterization of essential oil composition, phenolic content, and antioxidant properties in wild and planted individuals of *Valeriana jatamansi* Jones. Sci Hort 136:61–68

Boakye YD, Osafo N, Danquah CA, Adu F, Agyare C (2019) Antimicrobial agents: antibacterial agents, anti-biofilm agents, antibacterial natural compounds, and antibacterial chemicals. In: Kırmusaoğlu S (ed) Antimicrobials, antibiotic resistance, antibioflm strategies and activity methods. IntechOpen, London, pp 75–98

Bodare S, Tsuda Y, Ravikanth G, Uma Shaanker R, Lascoux M (2013) Genetic structure and demographic history of the endangered tree species *Dysoxylum malabaricum* (Meliaceae) in Western Ghats, India: implications for conservation in a biodiversity hotspot. Ecol Evol 3:3233–3248

Bohlin L, Bruhn JG (eds) (1999) Bioassay methods in natural product research and drug development, vol 43. Springer Science and Business Media, The Netherlands

Bradshaw CJ, Brook BW (2014) Human population reduction is not a quick fix for environmental problems. Proc Natl Acad Sci 111:16610–16615

Chauhan RS, Nautiyal BP, Nautiyal MC (2013) Trade of threatened Himalayan medicinal and aromatic plants-socio-economy. Manage Conserv Issues Garhwal Himalaya India 13:1–10

Chen SL, Yu H, Luo HM, Wu Q, Li CF, Steinmetz A (2016) Conservation and sustainable use of medicinal plants: problems, progress, and prospects. Chinese Med 11:1–10

Connor LH, Samuel G (eds) (2001) Healing powers and modernity: traditional medicine, shamanism, and science in Asian societies. Bloomsbury Publishing USA

Ganie AH, Tali BA, Khuroo AA, Reshi ZA, Nawchoo IA (2019) Impact assessment of anthropogenic threats to high-valued medicinal plants of Kashmir Himalaya, India. J Nat Conserv 50: 125715

Gautam K, Raina R, Dikshit N (2023) Current knowledge on sustainability and conservation of endangered Himalayan medicinal herb A. glauca Edgew.-a review. J Herb Med 42:100764

Glade RS, Richter GT, James CA, Suen JY, Buckmiller LM (2010) Diagnosis and management of pediatric cervicofacial venous malformations: retrospective review from a vascular anomalies center. Laryngoscope 120:229–235

Goraya GS, Ved DK (2017) Medicinal plants in India: an assessment of their demand and supply. National Medicinal Plants Board, Ministry of AYUSH, Government of India, New Delhi, and Indian Council of Forestry Research and Education, Dehradun

Guleria C, Vaidya MK, Sharma R, Dogra D (2014) Economics of production and marketing of important medicinal and aromatic plants in mid hills of Himachal Pradesh. Econ Affairs 59:363–378

Hamilton AC (2004) Medicinal plants, conservation and liveli-hoods. Biodivers Conserv 13:1477–1517

IUCN (2022) The IUCN red list of threatened species (Ver 2021-3). https://www.iucnredlist.org

Jain AK (2014) When did India-Asia collide and make the Himalaya. Curr Sci 106:254–266

Joshi BC, Joshi RK (2014) The role of medicinal plants in livelihood improvement in Uttarakhand. Int J Herb Med 2014:55–58

Kala CP (2005) Current status of medicinal plants used by traditional Vaidyas in Uttaranchal state of India. Ethnobo Res Appl 3:267–278

Kala CP, Dhyani PP, Sajwan BS (2006) Developing the medicinal plants sector in northern India: challenges and opportunities. J Ethnobiol Ethnomed 2:1–15

Khanday ZH, Singh S, Mir F (2018) Traditional botanical drugs of Kashmir and their therapeutic value. Int J Basic Appl Biol 5:184–188

Kumar PA, Mishra S (2012) Some belief, practices and prospects of folk healers of Sikkim. Indian J Tradition Knowl 11:369–373

Kumar D, Rawat S (2022) Modeling the effect of climate change on the distribution of threatened medicinal orchid *Satyriumnepalense* D. Don in India. Environ Sci Pollut Res 29:72431–72444

Kumar S, Kumar R, Khan A (2011) Medicinal plant resources: manifestation and prospects of life-sustaining healthcare system. Cont J Biol Sci 4:19–29

Kumar A, Kumar R, Sharma M, Kumar U, Gajula MP, Singh KP (2018) Uttarakhand medicinal plants database (UMPDB): a platform for exploring genomic, chemical, and traditional knowledge. Data 3:7

Kuniyal CP, Kuniyal PC, Butola JS, Sundriyal RC (2013) Trends in the marketing of some important medicinal plants in Uttarakhand, India. Int J Biodiv Sci Ecosys Serv Manag 9:324–329

Kuniyal CP, Bisht VK, Negi JS, Bhatt VP, Bisht DS, Butola JS, Sundriyal RC, Singh SK (2015) Progress and prospect in the integrated development of medicinal and aromatic plants (MAPs) sector in Uttarakhand, Western Himalaya. Environ Dev Sustain 17:1141–1162

Lambert J, Srivastava J, Vietmeyer N (1997) Medicinal plants: rescuing a global heritage, vol 355. World Bank Publications

Meke GS, Mumba RF, Bwanali RJ, Williams VL (2017) The trade and marketing of traditional medicines in southern and Central Malawi. Int J Sustain Dev World Ecol 24:73–87

Morrone JJ (2015) Bio-geographical regionalization of the world: a reappraisal. Austr System Bot 28:81–90

Najar NA, Mansoor A (2020) Ethnobotanical use, conservation, and livelihood of important medicinal plants grown in the Gulmarg Himalaya region of Jammu and Kashmir (India). Indian J Appl Pure Bio 35:33–38

Nautiyal MC (1995) Agro-technique of some high altitude medicinal herbs: cultivation of medicinal plants and orchids in Sikkim Himalaya. In: Sundriyal RC, Sharma E (eds). Bishen Singh Mahendra Pal Singh, Dehra Dun, India, pp 53–64

Nautiyal S, Maikhuri RK, Rao KS, Saxena KG (2001) Medicinal plant resources in Nanda Devi biosphere reserve in the Central Himalayas. Int J Geogr Inf Syst 8:47–64

Negi VS, Maikhuri RK, Phondani PC, Rawat LS (2010) An inventory of indigenous knowledge and cultivation practices of medicinal plants in Govind Pashu Vihar Wildlife Sanctuary, Central Himalaya, India. Int J Biodiv Sci Ecosyst Serv Manag 6:96–105

Negi VS, Maikhuri RK, Maletha A, Phondani PC (2019) Ethnobotanical knowledge and population density of threatened medicinal plants of Nanda Devi biosphere reserve, Western Himalaya, India. Iran J Sci Tech Transact A Sci 43:63–73

Newman DJ, Cragg GM, Snader KM (2000) The influence of natural products upon drug discovery. Nat Prod Rep 17:215–234

NMPB (2024) Data and trade. https://nmpb.nic.in/content/marketing-trade. Retrieved on 28/10/2024

Olsen CS, Larsen HO (2003) Alpine medicinal plant trade and Himalayan mountain livelihood strategies. Geogr J 169:243–254

Pan SY, Zhou SF, Gao SH, Yu ZL, Zhang SF, Tang MK, Sun JN, Ma DL, Han YF, Fong WF, Ko KM (2013) New perspectives on how to discover drugs from herbal medicines: CAM′ S outstanding contribution to modern therapeutics. Evid-Based Complemen Alternat Med 2013: 6273–6275

Panda AK, Misra S (2010) Health traditions of Sikkim Himalaya. J Ayurveda Integr Med 1:183

Panmei R, Gajurel PR, Singh B (2019) Ethnobotany of medicinal plants used by the Zeliangrong ethnic group of Manipur, Northeast India. J Ethnopharmacol 235:164–182

Perme N, Choudhury SN, Choudhury R, Natung T, De B (2015) Medicinal plants in traditional use at Arunachal Pradesh, India. Int J Phytopharm 5:86–98

Phondani PC, Bhatt ID, Negi VS, Kothyari BP, Bhatt A, Maikhuri RK (2016) Promoting medicinal plants cultivation as a tool for biodiversity conservation and livelihood enhancement in Indian Himalaya. J Asia-Pac Biodivers 9:39–46

Rao KS, Saxena KG (1994) Sustainable development and rehabilitation of degraded village lands in Himalaya. Bishen Singh Mahendra Pal Singh, Dehradun, India

Raut AA, Rege NN, Tadvi FM, Solanki PV, Kene KR, Shirolkar SG, Pandey SN, Vaidya RA, Vaidya AB (2012) Exploratory study to evaluate tolerability, safety, and activity of Ashwagandha (*Withania somnifera*) in healthy volunteers. J Ayurveda Integr Med 3:111

Rawal RS, Bhatt ID, Sekar KC, Nandi SK (2013) The Himalayan biodiversity: richness, representativeness, uniqueness and life-support values. GB Pant Institute of Himalayan Environment and Development, Kosi-Katarmal, Almora, India

Rawal R, Negi VS, Bhatt ID (2021) Changing outlook on harnessing biodiversity values–a special focus on Indian Himalaya. J Graphic Era Univ 9:5–82

Rawat S, Bhatt ID, Rawal RS, Nandi SK (2017) Geographical and environmental variation in chemical constituents and antioxidant properties in Roscoea procera wall. J Food Biochem 41: e12302

Rawat S, Jugran AK, Bhatt ID, Rawal RS (2018) Hedychium spicatum: a systematic review on traditional uses, phytochemistry, pharmacology and future prospectus. J Pharm Pharmacol 70: 687–712

Samant SS, Dhar U (1997) Diversity, endemism and economic potential of wild edible plants of Indian Himalaya. Int J Sustain Dev World Ecol 4:179–191

Samant SS, Palni LMS (2000) Diversity, distribution and indigenous uses of essential oil yielding plants of Indian Himalayan Region. J Med Arom Plants Sci 22:671–687

Samant SS, Pant S (2003) Diversity, distribution pattern and traditional knowledge of sacred plants in Indian Himalayan Region. Indian J Forest 26:201–213

Samant Pant S, Singh M, Lal M, Singh A, Sharma A, Bhandari S (2007) Medicinal plants in Himachal Pradesh, north western Himalaya, India. Int J Biodiv Sci Manag 3:234–251

Samant SS, Dhar U, Palni LMS (1998) Medicinal plants of Indian Himalaya: diversity, distribution, potential values. Gyanodaya Prakashan, Nainital

Sangma DB, Manohara TN (2018) The role of Garo tribes of Meghalaya (India) in the conservation and management of medicinal plants diversity used in treating livestock diseases. Plant Sci Today 5:155–162

Sati VP (2013) Cultivation of medicinal plants and its contribution to livelihood enhancement in the Indian Central Himalayan Region. Adv Med Plant Res 1:17–23

Schippmann UWE, Leaman DJ, Cunningham AB (2002) Impact of cultivation and gathering of medicinal plants on biodiversity: global trends and issues. In: Biodiversity and the ecosystem approach in agriculture, Forestry and Fisheries. Food and Agriculture Organization, Rome, pp 142–167

Schippmann UWE, Leaman D, Cunningham AB (2006) A comparison of cultivation and wild collection of medicinal and aromatic plants under sustainability aspects. In: Bogers RJ, Craker LE, Lange D (eds) Medicinal and aromatic plants: agricultural, commercial, ecological, legal, pharmacological and social aspects. Springer, Dordrecht, pp 75–95

Sen S, Chakraborty R, De B (2011) Challenges and opportunities in the advancement of herbal medicine: India's position and role in a global context. J Herb Med 1:67–75

Shaanker RU, Ganeshaiah KN, Krishnan S, Ramya R, Meera C, Aravind NA, Kumar A, Rao D, Vanaraj G, Ramachandra J, Gauthier R (2004) Livelihood gains and ecological costs of non-timber forest product dependence: assessing the roles of dependence, ecological knowledge and market structure in three contrasting human and ecological settings in South India. Environ Conserv 31:242–253

Sherpa MT, Mathur A, Das S (2015) Medicinal plants and traditional medicine system of Sikkim: a review. World J Pharm Pharmaceut Sci 4:161–184

Shivali V, Namrata S (2022) Threat status of three important medicinal Himalayan plant species and conservation implications. Nature Conservation Research 7:27–41
Shrestha AB, Agrawal NK, Alfthan B, Bajracharya SR, Maréchal J, Oort BV (2015) The Himalayan climate and water atlas: impact of climate change on water resources in five of Asia's major river basins. International Centre for Integrated Mountain Development (ICIMOD), Nepal, Kathmandu, Nepal
Singh MP, Dey S (2005) Indian medicinal plants. Satish Serial Publishing House, Delhi, India
Singh DK, Hajra PK (1996) Floristic diversity. In: Gujral GS, Sharma V (eds) Changing perspectives of biodiversity status in the Himalaya. British Council, New Delhi, pp 23–38
Singh DK, Pusalkar PK (2020) Floristic diversity of the Indian Himalaya. In: Dar GH, Khuroo AA (eds) Biodiversity of the Himalaya: Jammu and Kashmir state, vol 18. Springer, Singapore, pp 93–126
Singh L, Bhatt ID, Negi VS, Nandi SK, Rawal RS, Bisht AK (2021) Population status, threats, and conservation options of the orchids in Indian Western Himalaya. Reg Environl Change 21:1–16
Tali BA, Khuroo AA, Ganie AH, Nawchoo IA (2019) Diversity, distribution and traditional uses of medicinal plants in Jammu and Kashmir state of Indian Himalayas. J Herb Med 17:100280
Thakur D, Chawla A (2019) Functional diversity along elevational gradients in the high altitude vegetation of the western Himalaya. Biodivers Conserv 28:1977–1996
Timmermans K (2003) Intellectual property rights and traditional medicine: policy dilemmas at the interface. Social Sci Med 57:745–756
Uniyal RC, Uniyal MR, Jain P (2000) Cultivation of medicinal plants in India: a reference book. TRAFFIC, New Delhi
Ved DK, Kinhal GA, Ravikumar K, Prabhakaran V, Ghate U, Vijayshankar R, Indresha JH (2003) Conservation Assessment and Management prioRitization (CAMP) for medicinal plants of Jammu and Kashmir, Himachal Pradesh and Uttaranchal. India. Foundation for Revitalization of local health and tradition (FRLHT), Bangalore
Volenzo T, Odiyo J (2020) Integrating endemic medicinal plants into the global value chains: the ecological degradation challenges and opportunities. Heliyon 6:e04970
WHO (2002) WHO (World Health Organization). Traditional medicine strategy 2002–2005. World Health Organization, Rome
Woodruff DS (2001) Declines of biomes and biotas and the future of evolution. Proc Natl Acad Sci 98:5471–5476
Zhang S, Zhang L, Zou H, Qiu L, Zheng Y, Yang D, Wang Y (2021) Effects of light on secondary metabolite biosynthesis in medicinal plants. Front Plant Sci 12:781236

Chapter 6
Wetland Flora of the Himalayan Hotspot

Anupam Kundu, Shouvik Pramanik, Sourav Roy, Basabi Barua Hazra, and Amitava Nayek

Abstract The Himalayan region is a globally significant biodiversity hotspot, home to a rich variety of native and endemic species. Himalayan wetlands, including seepages, lakes, marshes, riverine floodplains, and high-elevation peats, are vital for maintaining ecological balance and serve as critical habitats for various plant and animal species. These wetlands provide habitats for diverse plants and contribute to vital processes like water purification, nutrient cycling, and carbon sequestration. However, knowledge about the region's variety and distribution of wetland plants is still limited. Human activities, such as urban expansion, farming, tourism, dam construction, and industrial pollution, along with climate change and invasive species, are major threats to this region's flora and ecosystems. This chapter explores the unique wetlands flora of the Himalayas, highlights their ecological, medicinal and cultural importance and discusses the challenges they face. It also calls for focused conservation efforts and research to protect the wetland flora for future generations.

Keywords Himalaya · Aquatic Flora · Biodiversity hotspot · Ecosystem conservation

6.1 Introduction

The Himalayan region, a global biodiversity hotspot, is renowned for its abundance of rare, indigenous, and unique species (Chaps. 1, Sect. 1.3, and 2, Sect. 2.1). Still, despite the ubiquitous presence of wetlands throughout the globe, specific and current information is scarce regarding management, ecological health, and even

A. Kundu · S. Pramanik · S. Roy
Applied Phycology Laboratory, Department of Botany, Visva-Bharati, Santiniketan, West Bengal, India

B. Barua Hazra · A. Nayek (✉)
Department of Botany, Triveni Devi Bhalotia College, Raniganj, West Bengal, India

D. Das et al. (eds.), *Plant Diversity and Conservation Strategies of the Himalayan Biodiversity Hotspot*, Topics in Biodiversity and Conservation 21,
https://doi.org/10.1007/978-3-032-21697-7_6

classification in tropical regions. Wetlands under the Himalayas consist of lakes and marshes, supporting unique ecosystems that harbour a large number of endemic plant species. Such shallow aquatic settings are important as ecotones or transition areas between the permanent aquatic ecosystem and dry terrestrial landscapes (Mitsch and Gosselink 2015). According to the Ramsar Convention, wetlands comprise diverse aquatic habitats like rivers, lakes, marshes, and mangroves and are crucial for maintaining ecological balance by linking water and land. These are spread over many different climates and serve essential functions as natural biofilters to absorb nutrients from the enriched water bodies through what is termed processes such as the water cycle as well as nutrient cycling (Halls 1997). This is crucial in matters of conservation and mitigation of climate change. For instance, it is this role that has also earned the name "Biological Supermarket" and the "Kidney of the Landscape" because of the high species variety and ecological equilibrium it can sustain. It becomes important in conservation, thus maintaining biodiversity and ecosystem health (Allen-Diaz et al. 2004; Adhikari and Babu 2008).

In the Indian Himalayas, wetlands maintain indigenous ways of life by supplying food, medicine, and energy, in addition to supporting a variety of flora and animals. Important biological, physical, and socioeconomic advantages are provided by these extremely productive ecosystems, including as fisheries, fresh water, wood, flood control, and habitat for endangered species (Rawal et al. 2013; Soni 2020). However, they are devastatingly threatened by habitat destruction, over-exploitation, urbanisation, and invasive aquatic weeds badly jeopardizing their ecological balance (Kingsford et al. 2016). Himalayan wetlands in India are vital biodiversity hotspots, home to numerous flora and fauna, including endangered species. They support local livelihoods and play a crucial role in natural cycles like water, carbon, nitrogen, phosphorus, and sulphur sustaining complex food chains (Reddy et al. 2022). Wetlands serve as crucial carbon sinks, boosting ecosystem productivity and aiding in climate change mitigation (Adhikari and Babu 2008). Their conservation supports biodiversity and climate regulation. However, they face threats like illegal hunting, deforestation, overgrazing, chemical contamination, unmanaged tourism, and invasive aquatic weeds. While some weeds degrade wetlands, other plants protect against erosion and nutrient loss (Dar et al. 2021). Wetland plants, vital for ecological balance, face threats like habitat destruction and over-exploitation in Himalayan wetlands, endangering species (Alam et al. 2011). Initiatives like "Saving Wetlands Sky High" aim to address these issues through community and partner collaboration. As defined by the Ramsar Convention, wetlands include marshes, ferns, and peatlands, supporting fisheries, flood control, and wildlife habitats (Chatterjee et al. 2010). With wetland services under growing pressure, their value is increasing, highlighting the need for quantitative valuation to guide conservation. The Ministry of Environment and Forests, India, has designated 21 nationally important sites, including Pong Dam, Renuka, and Chandertal in Himachal Pradesh, emphasising the urgency of conservation efforts (Kohli 2015; Pathania and Kumar 2017).

This chapter aims to provide a comprehensive overview of the wetland flora within the Himalayan hotspot, emphasising the importance of these ecosystems, the unique adaptations of wetland plants, their challenges, and the ongoing conservation efforts to protect and manage these vital natural resources.

6.2 Types of Himalayan Wetlands

The Himalayas are traditionally divided into four parallel altitudinal ranges as previously discussed in Chap. 3 (Sect. 3.1), each with distinct physiographic features originating from geological events. The Tethys (Trans or Tibetan) Himalayan Range, the Great Himalayan Range (Inner or Greater Himalaya), the smaller (Lower) Himalayan Range, and the Outer (sub-Himalayan or Shiwalik) Range are the names given to these mountains from north to south (Singh and Pusalkar 2020). With their deep, carved terrain and interconnected network of valleys between various elevational bands, the Himalayas include a complicated geological structure. Additionally, there are steep, jagged snowy peaks, deep gorges, and massive glaciers that combine to produce a unique biogeographical unit that supports a variety of wildlife. However, the flanks of the Himalayas are very different from one another because of their geographical position, trends, and climate (Singh and Pusalkar 2020). Types of wetlands and their location in the Himalayas were mentioned in Table 6.1.

Table 6.1 Types of wetlands and their location in the Himalayas

Region	Type	States	Features	Examples
Trans-Himalaya-Tibetan Plateau	Himalaya–Ladakh Mountains	Jammu & Kashmir	Cold water bodies (sub-zero temperatures, snow-fed)	Pangong Lake, Tsomoriri Lake
	Himalaya–Tibetan Plateau	Jammu & Kashmir, Himachal Pradesh (Lahaul-Spiti, Kinnaur), Uttarakhand (Uttarkashi, Chamoli, Pithoragarh districts)	Cold water bodies (high-altitude, snow-fed)	Tso Kar Lake, Dal Lake
	Himalaya–Sikkim	Sikkim (Northern Sikkim)	Cold, snow-fed water bodies	Gurudongmar Lake
Himalaya	Himalaya–Western	Jammu & Kashmir, Himachal Pradesh	Temperate, cold-water lakes	Dal Lake, Renuka Lake
	Himalaya–Central	Uttarakhand	Cold, moderate water bodies	Naini Lake, Bhimtal Lake
	Himalaya–Eastern	Sikkim-Darjeeling Himalaya (Sikkim and adjoining parts of West Bengal: Kalimpong, Darjeeling, Kurseong, Mirik subdivisions of Darjeeling district, and the northern fringe of Jalpaiguri district)	Cold and temperate, often perennial	Tsomgo Lake, Ranjit Sagar Lake

6.3 Biodiversity-Rich Himalayan Hotspot

The Himalayas are a vast area, hosting tropical forests and alpine meadows; the wetlands that are critically important are also involved. The lush diversity and uniqueness of the flora sheltered in these wetlands play a crucial role in holding the ecological balance in these environments (Tse-Ring et al. 2010). Many species contribute to the richness of the wetland flora in the region due to altitudinal gradients at higher altitudes. The remarkable biodiversity of Himalayan wetlands provides a fragility that makes it an area that seriously warrants preservation as this ecosystem supports rare species and millions of people's livelihoods in the region (Iqbal and Shang 2020).

6.3.1 Rich Diversity of Plant Species Due to Varied Altitude and Climatic Conditions

The Himalayan region is a mosaic of diverse habitats, including tropical forests, alpine meadows, and wetlands, contributing to its remarkable biodiversity. Due to its unique climate, varied topography, and altitudinal gradients, the region is a critical area for conservation (Ning et al. 2014). The Himalayas are home to approximately 10,000 plant species, many of which are endemic to the region. Wetland plants in the Himalayas, like species of *Cyperus, Potentilla*, and *Carex,* thrive in these ecosystems and play crucial roles in maintaining the ecological balance of these habitats. The wetlands are particularly rich in species diversity, hosting plants that are adapted to the unique hydrological and climatic conditions of the Himalayas (Butt et al. 2021). *Potamogeton* spp., commonly found in Himalayan lakes, is essential for maintaining water quality and providing habitat for aquatic life (Ganie et al. 2015). Similarly, *Nymphaea* spp. (water lilies) is significant in the region's wetland ecosystems, contributing to both these water bodies' aesthetic and ecological value. The altitudinal variation across the Himalayas, ranging from around 100 m.a.s.l. in the foothills to over 8000 m at the peaks, creates distinct vegetation zones (Pant et al. 2018). Wetlands at lower altitudes often feature dense growths of broadleaf evergreen trees, while higher elevations are characterised by species adapted to colder, harsher conditions. The wetlands in the sub-Alpine and Alpine zones (3000–4500 m) support hardy genera such as *Eleocharis* and *Juncus* adapted to the cold, nutrient-poor environments (Rawat 2017). In the context of the Himalayan biodiversity hotspot, the degree of endemism is a very critical feature. A large number of wetland plants developed ecological adaptations specific to certain conditions and are confined only to this area. A few species of *Rhododendron* and *Saxifraga* within these high-altitude wetlands exist exclusively in this mountain range and are therefore considered at the top of the list for conservation (Manikandan et al. 2023).

Given the high levels of biodiversity and endemism, the Himalayas are recognised as one of the world's 36 biodiversity hotspots, characterised by significant species richness and high levels of human-induced threats. These unique ecosystems are threatened by various factors, including climate change, habitat degradation, and the spread of invasive species (Banerjee et al. 2019). Conservation efforts of the unique biodiversity of this wetland region and its associated crucial ecological services for millions of people dwelling in and around the Himalayan region would be a matter of key interest in this regard (Bhatta et al. 2018; Kattel 2022).

6.4 Vegetation Types in the Himalayan Wetland

The vegetation in the Himalayan wetlands is predominantly composed of herbaceous and shrub species, which account for 58% and 30% of the total species, respectively. In contrast, trees and lianas are much less common, representing only 7% and 5% of the species. The high prevalence of shrubs and herbs is a clear indicator of the characteristic vegetation structure of wetland ecosystems (Tenzin and Tendar 2021).

6.4.1 Bryophyta

The Himalayan wetlands, mainly East Sikkim, Garhwal Himalaya, and Northern Sikkim, are a rich reservoir of bryophyte diversity as also mentioned in Chap. 3 (Sect. 3.3.1), particularly mosses, which are crucial for ecological balance, soil formation, water retention, and nutrient cycling. These wetlands harbour a remarkable array of Bryophyta, which play a vital role in maintaining the ecological balance of these unique ecosystems and contribute to water retention, soil formation, and nutrient cycling, making them integral to the health and sustainability of the Himalayan wetlands. In East Sikkim, 49 bryophyte taxa across 18 families, including 45 moss and four liverwort species, with 12 new records for Sikkim were documented (Dhyani et al. 2023). *Fissidens* (47 species), *Brachythecium* (34 species), and *Bryum* (27 species) are major genera, and other mosses are also found, like *Buxbaumia* sp., *Cyathophorella* sp., and *Ganguleea* sp. Few other interesting genera within the order include *Leptobryum* sp., *Campyliadelphus* sp., *Conostomium* sp., and *Cyathothecium* sp. Few endemic species, such as *Habrodon kashmiriensis* and *Hygroamblystegium gangulianum*, show regional features, and genera such as *Octogonella* sp., and *Orthotheciadelphus* sp. represent the richness of the region in the wetland (Bahuguna et al. 2016; Dar and Khuroo 2020; Yadav et al. 2022).

6.4.2 Pteridophyta

Wetland flora of the Himalayas, particularly in the Indian Himalayan Region (IHR), contains many species of unique pteridophytes that can be largely endemic. However, overall, endemism within this group is relatively less rich than in other plant groups (Singh and Pusalkar 2020). Some notable endemic species include *Selaginella adunca* and *Athyrium kumaonicum* in the western Himalaya, and *Thelypteris kingii*, *Selaginella pentagona*, *Arthromeris notholanoides* and *Osmunda cinnamomea* in the central and eastern Himalayas. The region also hosts critically endangered taxa such as *Psilotum nudum*, *Anogramma leptophylla*, *Athyrium duthiei*, *Christella kaumanica*, *C. clarkei*, *Christopteris tricuspis* and *Cyclogramma squamaestipes*. Other significant species of the wetland flora comprise *Metathelypteris decipiens*, *Dryopteris gamblei* and *Stenogramma himalaica*. Despite these so varied floras, many species remain on a very vulnerable conservation level. Many species face extinction threats, and recommended conservation strategies are of immense importance in such wetland ecosystems (Dar and Khuroo 2020).

6.4.3 Gymnosperm

The wetland species of gymnosperms found in the Himalayas include notable examples like *Ephedra gerardiana* and *Ephedra nebrodensis*, which are important due to their medicinal and ecological significance. *Ephedra gerardiana* is known for its high alkaloid content, particularly ephedrine, extensively used in traditional medicine and pharmaceuticals. Similarly, *Ephedra nebrodensis* is alkaloid-rich and is an important local component of the ecosystem. Other gymnosperm species included are *Gnetum montanum* which is endemic to Eastern Himalayas and is part of the exclusive biodiversity of this region. With many more species, these are essential components of the wetland ecosystems of the Himalayas, sustaining a variety of ecological functions and very valuable for local communities in terms of resource utilisation (Dar and Khuroo 2020).

6.4.4 Angiosperm

The wetland species of angiosperms found in the Himalayas include a range of unique and fascinating plants adapted to the region's diverse and challenging environments. Angiosperm plants can be categorised into the following categories (Fig. 6.1).

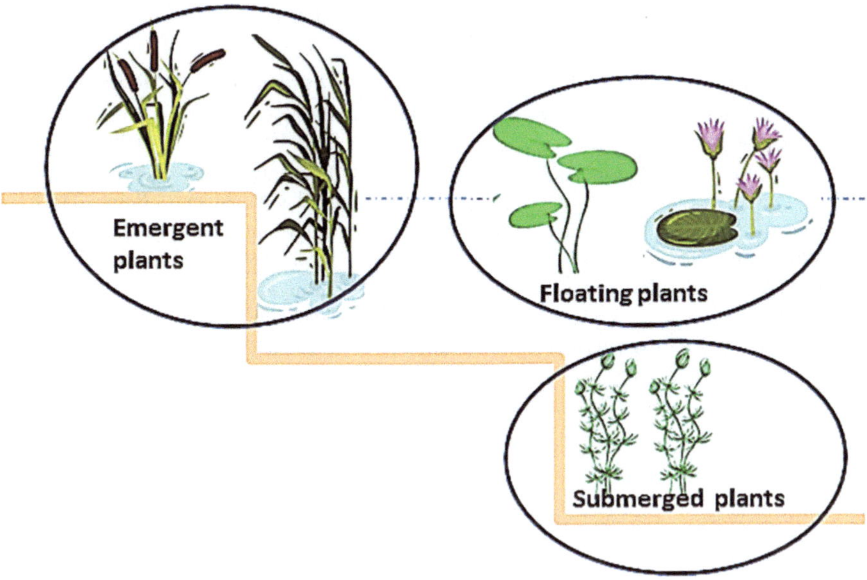

Fig. 6.1 Types of aquatic plants (Emergent, Floating and Submerged)

6.4.4.1 Submerged Plants

Submerged or aquatic plants of Himalayan wetlands exhibit diverse adaptations to their environment. Submerged wetland flora thrives in fresh and saline conditions, with some species maintaining underwater vegetation while others develop extensive stems to reach the water surface (Guntenspergen et al. 2020). These submerged plants, eelgrass, and seagrasses are an important food source for many animals in wetlands. They also help in filtering water. All these different types of vegetation are useful and sustain the balance and sound health of Himalayan wetlands. Common submerged plants in the Himalayan wetland are mentioned in Table 6.2.

6.4.4.2 Emergent Plants

The feature of emergent plants in Himalayan wetlands is the most striking in this regard, and they strongly dominate the region's ecosystems, which makes this region unique and diverse. These plants are characterised by their ability to grow under conditions that are waterlogged, with their roots submerged and stems and leaves extending above the water surface are important for soil stabilisation and provision of habitat for a wide range of wildlife.

These plants play a crucial role in nutrient cycling, water filtration, and carbon sequestration, making them vital components of the wetland's ecological functions. They supply shelter and breeding habitats for aquatic and terrestrial species, which

Table 6.2 Common submerged plants in the Himalayan wetland

Sl. No.	Scientific name	Family	Common name	Distribution	References
1.	*Ceratophyllum demersum*	*Ceratophyllaceae*	Coontail	Baanganga Wetland, Uttarakhand	Adhikari and Babu (2008)
2.	*Chara sp.*	*Characeae*		Baanganga Wetland, Uttarakhand	Adhikari and Babu (2008)
3.	*Cyperus bulbosus*	*Cyperaceae*	Bulbous Flat sedge	Baanganga Wetland, Uttarakhand	Adhikari and Babu (2008)
4.	*Eleocharis palustris*	*Cyperaceae*	Creeping Spikerush	Baanganga Wetland, Uttarakhand	Adhikari and Babu (2008)
5.	*Enydra fluctuans*	*Asteraceae*	Buffalo Spinach	Baanganga Wetland, Uttarakhand	Adhikari and Babu (2008)
6.	*Hydrilla verticillata*	*Hydrocharitaceae*	Hydrilla	Ladakh Union Territory, Trans Himalaya, India	Bhojak et al. (2024)
7.	*Hydrobryum griffithii*	*Podostemaceae*	–	Rahgit River, Bijanbari, Darjeeling	Chowdhury and Chowdhury (2022)
8.	*Lemna perpusilla*	*Lemnaceae*	Dwarf Duckweed	Baanganga Wetland, Uttarakhand	Adhikari and Babu (2008)
9.	*Najas minor*	*Najadaceae*	Lesser Naiad	Baanganga Wetland, Uttarakhand	Adhikari and Babu (2008)
10.	*Nymphoides cristata*	*Menyanthaceae*	Crested Floating Heart	Baanganga Wetland, Uttarakhand	Adhikari and Babu (2008)
11.	*Potamogeton gramineus*	*Potamogetonaceae*	Grass-leaved pondweed	Himalayan and sub-himalayan streams	Chowdhury et al. (2015)
12.	*Rorippa nasturtium-aquaticum*	*Brassicaceae*	Watercress.	Baanganga Wetland, Uttarakhand	Adhikari and Babu (2008)
13.	*Rotala densiflora*	*Lythraceae*	Dense Rotala	Baanganga Wetland, Uttarakhand	Adhikari and Babu (2008)
14.	*Rungia pectinata*	*Acanthaceae*	Small Flowered Rungia	Baanganga Wetland, Uttarakhand	Adhikari and Babu (2008)
15.	*Sagittaria sagittifolia*	*Alismataceae*	Arrow head	Baanganga Wetland, Uttarakhand	Adhikari and Babu (2008)
16.	*Vallisneria natans*	*Hydrocharitaceae*	Eelgrass	Ladakh Union Territory, Trans Himalaya, India	Bhojak et al. (2024)

increases the levels of biodiversity. The emergent plants actually provide support to the balance of the Himalayan wetlands' ecological system and make these ecosystems more resilient to environmental change. Common Emergent plants in the Himalayan wetland are mentioned in Table 6.3.

Table 6.3 Common emergent plants in the Himalayan wetland

Sl. No.	Scientific name	Family	Common name	Distribution	References
1.	*Acorus calamus*	*Acoraceae*	Sweet Flag	Throughout the Himalayas	Srivastava et al. (2023)
2.	*Anaphalis contorta*	*Asteraceae*	Himalayan Pearly Everlasting	Ladakh Union Territory, Trans Himalaya, India	Bhojak et al. (2024)
3.	*Artemisia biennis*	*Asteraceae*	Biennial Wormwood	Ladakh Union Territory, Trans Himalaya, India	Bhojak et al. (2024)
4.	*Carum carvi*	*Apiaceae*	Caraway	Baanganga Wetland, Uttarakhand	Adhikari and Babu (2008)
5.	*Ceratophyllum demersum*	*Ceratophyllaceae*	Hornwort	Shallabugh Wetland, Ganderbal, Jammu, and Kashmir, India	Javid et al. (2024)
6.	*Cirsium arvense.*	*Asteraceae*	Canada Thistle	Ladakh Union Territory, Trans Himalaya, India	Bhojak et al. (2024)
7.	*Corydalis meifolia*	*Papaveraceae*	Thread leaf Corydalis	Ladakh Union Territory, Trans Himalaya, India	Bhojak et al. (2024)
8.	*Dalbergia sissoo*	*Fabaceae*	Indian Rosewood	Baanganga Wetland, Uttarakhand	Adhikari and Babu (2008)
9.	*Desmodium triflorum*	*Fabaceae*	Three-flowered Desmodium	Baanganga Wetland, Uttarakhand	Adhikari and Babu (2008)
10.	*Euphrasia himalayica*	*Orobanchaceae*	Himalayan Eyebright	Ladakh Union Territory, Trans Himalaya, India	Bhojak et al. (2024)
11.	*Ficus racemosa*	*Moraceae*	Cluster Fig	Uttarakhand	Adhikari and Babu (2008)
12.	*Gnaphalium pensylvanicum*	*Asteraceae*	Pennsylvania cudweed	Baanganga Wetland, Uttarakhand	Adhikari and Babu (2008)
13.	*Gnaphalium polycaulon*	*Asteraceae*	Spreading cudweed	Baanganga Wetland, Uttarakhand	Adhikari and Babu (2008)

(continued)

Table 6.3 (continued)

Sl. No.	Scientific name	Family	Common name	Distribution	References
14.	*Grangea maderaspatana*	*Asteraceae*	Madras Carpetweed	Ladakh Union Territory, Trans Himalaya, India	Bhojak et al. (2024)
15.	*Grubovia dasyphylla*	*Amaranthaceae*	–	Baanganga Wetland, Uttarakhand	Adhikari and Babu (2008)
16.	*Heracleum lallii*	*Apiaceae*	Lall's Hogweed	Gosaikunda Wetlands, Nepal higher altitude	Kumar and Lamsal (2016)
17.	*Hygrophila polysperma*	*Acanthaceae*	Indian Swampweed	Baanganga Wetland, Uttarakhand	Adhikari and Babu (2008)
18.	*Ipomoea carnea*	*Convolvulaceae*	Pink Morning Glory	Baanganga Wetland, Uttarakhand	Adhikari and Babu (2008)
19.	*Juncus effusus*	*Juncaceae*	Wet meadows	Sub-Himalayan regions	Jocou and Gandullo (2022)
20.	*Juncus membranaceus*	*Juncaceae*	Membrane-leaved Rush	Ladakh Union Territory, Trans Himalaya, India	Bhojak et al. (2024)
21.	*Kobresia gandakiensis*	*Cyperaceae*	Sedge	Gokyo Wetlands, Nepal higher altitude	Kumar and Lamsal (2016)
22.	*Lactuca tatarica*	*Asteraceae*	Blue Lettuce or Tartarian Lettuce	Ladakh Union Territory, Trans Himalaya, India	Bhojak et al. (2024)
23.	*Lancea tibetica*	*Phrymaceae*	Tibetan Lancea	Ladakh Union Territory, Trans Himalaya, India	Bhojak et al. (2024)
24.	*Martynia annua*	*Martyniaceae*	Martynia	Baanganga Wetland, Uttarakhand	Adhikari and Babu (2008)
25.	*Myriophyllum aquaticum*	*Haloragaceae*	Parrot's Feather	Shallabugh Wetland, Ganderbal, Jammu, and Kashmir, India	Javid et al. (2024)
26.	*Oxalis corniculata*	*Oxalidaceae*	Creeping Wood Sorrel	Baanganga Wetland, Uttarakhand	Adhikari and Babu (2008)
27.	*Pedicularis cheilanthifolia*	*Orobanchaceae*	Fernleaf Lousewort	Ladakh Union Territory, Trans Himalaya, India	Bhojak et al. (2024)
28.	*Pedicularis poluninii*	*Orobanchaceae*	Lousewort	Gokyo Wetlands, Nepal higher altitude	Kumar and Lamsal (2016)

(continued)

Table 6.3 (continued)

Sl. No.	Scientific name	Family	Common name	Distribution	References
29.	*Pedicularis pseudoregelina*	*Orobanchaceae*	Lousewort	Gokyo Wetlands, Nepal higher altitude	Kumar and Lamsal (2016)
30.	*Phragmites australis*	*Poaceae*	Common Reed	Throughout the Himalayas	Srivastava et al. (2023)
31.	*Pithecellobium dulce*	*Mimosaceae*	Manila Tamarind	Baanganga Wetland, Uttarakhand	Adhikari and Babu (2008)
32.	*Polygonum amphibium*	*Swamp Smartweed*	Water Knotweed	Central and Eastern Himalayas	Shah and Reshi (2014)
33.	*Polypogon fugax*	*Poaceae*	Annual Beard Grass	Baanganga Wetland, Uttarakhand	Adhikari and Babu (2008)
34.	*Rhododendron cowanianum*	*Ericaceae*	Cowan's *Rhododendron*	Gosaikunda Wetlands, Nepal higher altitude	Kumar and Lamsal (2016)
35.	*Saussurea graminifolia*	*Asteraceae*	Grass-leaved Snow Lotus	Ladakh Union Territory, Trans Himalaya, India	Bhojak et al. (2024)
36.	*Scirpus lacustris*	*Cyperaceae*	Common Bulrush	Throughout the Himalayas	Davidson (1993)
37.	*Stellaria media*	*Caryophyllaceae*	Common chickweed	Baanganga Wetland, Uttarakhand	Adhikari and Babu (2008)
38.	*Syzygium cumini*	*Myrtaceae*	Java Plum	Baanganga Wetland, Uttarakhand	Adhikari and Babu (2008)
39.	*Trigonella foenumgraecum*	*Fabaceae*	Fenugreek	Baanganga Wetland, Uttarakhand	Adhikari and Babu (2008)
40.	*Typha angustifolia*	*Typhaceae*	Narrow-leaf Cattail	Sub-Himalayan regions	Chettri et al. (2010)

6.4.4.3 Floating Plants

Floating plants in Himalayan wetlands exhibit unique adaptations that allow them to thrive on the water's surface. They are categorised into three types: submerged floating, free-floating and trailing floating plants. As the name suggests, free-floating plants freely float on the water's surface, often forming dense mats. Submerged floating plants, on the other hand, anchor themselves to the bottom of the water with roots while their foliage remains buoyant. Trailing floating plants are rooted along the banks of water bodies, exhibiting a creeping growth habit that extends over the water surface. These floating plants are very important in the wetland ecosystem because they provide habitat and food for many aquatic organisms besides playing a role in enhancing water quality by shading and limiting algal blooms. Common floating plants in Himalayan wetlands are mentioned in Table 6.4.

Table 6.4 Common floating plants in Himalayan wetlands

Sl No.	Scientific name	Family	Common name	Distribution	References
1.	*Askellia flexuosa*	*Asteraceae*	Slender hawksbeard	Gokyo Wetlands, Nepal higher altitude	Kumar and Lamsal (2016)
2.	*Carex diandra*	*Cyperaceae*	Lesser Pan-icled Sedge	Eastern Himalayan Highlands of Gasa, Bhutan	Tendar and Sridith (2021)
3.	*Dilophia salsa*	*Asteraceae*	–	Gokyo Wetlands, Nepal higher altitude	Kumar and Lamsal (2016)
4.	*Eichhornia crassipes*	*Pontederiaceae*	Water Hyacinth	Lower Himalayan regions	Abba and Sankarannair (2024)
5.	*Enkianthus deflexus*	*Ericaceae*	Drooping Enkianthus	Eastern Himalayan Highlands of Gasa, Bhutan	Tendar and Sridith (2021)
6.	*Nymphaea abhayana*	*Nymphaeaceae*	Dhisa	Sub-himalayan water bodies in various conservatory	Chowdhury and Chowdhury (2016)
7.	*Nymphaea nouchali*	*Nymphaeaceae*	Blue Water Lily	Eastern Himalayas	Dwivedi et al. (2023)
8.	*Nymphaea odorata*	*Nymphaeaceae*	American White-Water Lily	Shallabugh Wetland, Ganderbal, Jammu, and Kashmir, India	Javid et al. (2024)
9.	*Parrya nudicaulis*	*Brassicaceae*	Naked-stemmed parrya	Gokyo Wetlands, Nepal higher altitude	Kumar and Lamsal (2016)
10.	*Pedicularis poluninii*	*Orobanchaceae*	Lousewort	Gokyo Wetlands, Nepal higher altitude	Kumar and Lamsal (2016)
11.	*Pistia stratiotes*	*Araceae*	Water Lettuce	Lower Himalayan regions	Ghosh et al. (2021)
12.	*Schoenoplectus mucronatus*	*Cyperaceae*	Bog Bulrush	Eastern Himalayan Highlands of Gasa, Bhutan	Tendar and Sridith (2021)
13.	*Spirodela polyrhiza*	*Lemnaceae*	Giant Duckweed	Sub-Himalayan regions	Adhikari and Babu (2008)
14.	*Trapa natans*	*Trapaceae*	Water Chestnut	Eastern Himalayas	Rajput and Singh (2023)
15.	*Zizania aquatica*	*Poaceae*	Wild Rice	Shallabugh Wetland, Ganderbal, Jammu, and Kashmir, India	Javid et al. (2024)

Himalayan wetlands harbour many endemic and rare plants that are of great value to the region's ecology, culture and biodiversity (Palni and Rawal 2010). Endemic plants like the Himalayan poppy (*Meconopsis* sp.) highlight the vulnerability of Himalayan wetlands to habitat loss and climate change (Majid et al. 2015). Similarly, species such as *Nymphoides peltata, Potamogeton pectinatus, Aconitum heterophyllum*, and *Primula florindae* are crucial to the health of Kashmir Himalaya (Shah and Reshi 2014). The loss of endemic plants in Himalayan wetlands threatens biodiversity, disrupts ecosystems, and endangers their cultural, medicinal, and ecological value, underscoring the need for targeted conservation (Palni and Rawal 2010; Sharma and Singh 2021). *Saussurea obvallata* (Brahma Kamal) and *S. gossipiphora* have anti-inflammatory properties (Thapliyal et al. 2024). *Euphorbia sikkimensis* is used in treatment of poisoning and jaundice (Dahal et al. 2017). *Meconopsis betonicifolia* (Blue Poppy) is added with ornamental and medicinal value. *Gentiana kurroo*, known as Kashmir Gentian, has fevers and ailments related to digestion. *Rheum nobile* a natural Sunscreen and Purgative *Primula denticulate* aid in respiratory problems; *Nardostac hysjatamansi* used for perfumes and sedative. *Swertia chirayita* (Chirayita) is valued for its antimalarial properties, while *Aconitum ferox* (Indian Aconite) treats neuralgia in controlled doses. *Corydalis govaniana* (Govan's Corydalis) has analgesic effects, and *Dactylorhiza hatagirea* (Marsh Orchis) is a tonic. *Aster diplostephioides* (Himalayan Aster) may help with respiratory conditions, and *Roscoea alpina* (Alpine Roscoea) and *Pinguicula alpina* (Alpine Butterwort) have unique adaptations. *Saxifraga brunonis* and *Cypripedium himalaicum* (Himalayan Lady's Slipper Orchid) are of ornamental and medicinal importance. *Utricularia wallichiana* is used in controlling mosquito populations, with *Epipactis gigantean* and *Polygonum polystachyum* aiding in achieving ecosystem balance. *Pedicularis siphonantha* has medicinal uses. *Phragmites karka*is are used for crafting. *Impatiens glandulifera* provides nectar to insects. *Carex nivalis*, *Scirpusl ittoralis*, and *Arisaema costatum* (Cobra Lily) have cultural and medical importance. *Schoenoplectu slacustris* and *Blyxaau bertii* (Water Spike-moss) have significant roles in aquatic ecosystems, while *Nymphoides indica* (Floating Heart) add ornamental and ecological dimensions (Bhojak et al. 2024; Dey et al. 2021; Kumar and Lamsal 2016; Tendar and Sridith 2021). Images of a few common plants in the Himalaya wetland (Darjeeling, Assam, South Sikkim) are given in Fig. 6.2.

6.5 Important Wetland Flora

6.5.1 Medicinal Plants Used as Local and Traditional Medicine

In the wetlands of the Himalayas, medicinal plants hold significant value in local and traditional medicine, which reflects the deep connection between these ecosystems and indigenous health practices. These plants are often used to treat various ailments,

Fig. 6.2 A few common plants in Himalaya wetland (Darjeeling, Assam, South Sikkim) (**1**) *Nymphaea alba* (**2**) *Lemna minor* (**3**) *Typha latifolia* (**4**) *Eichhornia crassipes* (**5**) *Centella asiatica* (**6**) *Equisetum diffusum* (**7**) *Hydrocotyle ehimalaica* (**8**) *Acorus calamus* (**9**) *Colocasia esculenta* (**10**) *Hygrophila auriculata* (**11**) *Marsilea minuta* (**12**) *Sagittaria latifolia*

drawing on centuries of traditional knowledge. For instance, plants such as *Swertia chirayita* and *Rheum australe* are prized for their therapeutic properties, including anti-inflammatory, anti-microbial, and digestive benefits (Samant et al. 2007). Wetland medicinal plants are often used in local communities for basic respiratory, gastrointestinal, and skin diseases. Saving these ecosystems means saving traditional medicine and providing future sources of natural remedies. As these plants are applied to local healthcare, conservation augments well-being and, at the same time, accentuates the importance of saving wetlands.

In Khecheopalri Lake, West Sikkim, the following plants are found (Fig. 6.3). Among these, the *Acorus calamus* is well-known for its use in treating digestive disorders. *Rubus paniculatus* and *Rubus ellipticus* (Himalayan Raspberry) are both used in traditional medicine for their antioxidant properties and inflammatory conditions. *Houttuynia cordata* (Chameleon Plant) is valued for its anti-Inflammatory and antiviral properties.

Fig. 6.3 Medicinal plants collected from Khecheopalri Lake, West Sikkim (**1**) *Hydrocotyl ehimalaica* (**2**) *Cynoglossum zeylanicum* (**3**) *Swertia bimaculate* (**4**) *Drymaria cordata* (**5**) *Houttuynia cordata* (**6**) *Phytolacca acinose* (**7**) *Paederia foetida* (**8**) *Rubus paniculatus* (**9**) *Hydrangea febrifuga*

Colocasia esculenta (Taro) is used for skin and digestive issues, while *Hydrangea febrifuga* (Chinese Quinine) is valued for its antimalarial properties. *Drymaria cordata* treats sleep disorders and convulsions. *Arisaema tortuosum*, *Cynoglossum zeylanicum*, and *Paederia foetida* are used in traditional medicine for arthritis, diarrhoea, and diabetes. *Begonia hatacoa* addresses dysentery and gastric issues, and *Phytolacca acinosa* is known for its anti-inflammatory and urinary benefits. *Bulbophyllum leopardinum* treats burns, and *Rubus paniculatus* is used for scabies. *Swertia bimaculata* helps with dyspepsia, while *Astilberi vulgaris* treats peptic ulcers. *Dactylicapnos scandens* is used for inflammation and hypertension. Other notable Himalayan medicinal plants include *Hydrocotyle himalaica*, *Thalictrum foliolosum*, *Podophyllum hexandrum*, *Phalaenopsis difformis*, *Hoya linearis*, and *Ophiopogon intermedius*. Furthermore, *Berberis aristata* (Indian Barberry) is renowned for its antibacterial and anti-inflammatory effects and is commonly used for skin diseases in North Sikkim. *Spiranthes sinensis* (Chinese Spiranthes) is noted for its diuretic and anti-inflammatory properties. *Rosa brunonii* (Himalayan Musk Rose) is used for skin ailments and as a mild laxative. *Holboellia latifolia* treats digestive and respiratory issues, while *Enkianthus deflexus* is valued for gastrointestinal problems and as a general tonic (Tenzin and Tendar 2021).

The Western Himalayas host diverse medicinal plants used in traditional medicine, which include *Caltha palustris* var. *alba* and *Cardamine hirsuta*, which are valued for their healing properties. *Cyperus alternifolius* is also recognised for its medicinal uses. *Persicaria nepalensis*, *Geranium nepalense*, *Rorippa islandica*, *Veronica anagallis-aquatica*, and *Veronica beccabunga* offer various therapeutic benefits. *Adiantum capillus-veneris* aids respiratory conditions, while *Polygonum aviculare* is known for its diuretic and anti-inflammatory effects. *Nelumbo nucifera* (sacred lotus) is revered for its broad therapeutic effects, and *Plantago lanceolata* supports wound healing. *Cyperus glomeratus* is also valued for its medicinal properties (Butt et al. 2021).

6.5.2 Plants Used for Other Purposes

Plants in the Himalayan wetlands serve several roles (Fig. 6.4), besides playing ecological functions. For instance, *Nelumbo nucifera* and *Nymphaea* sp. provide nutritious seeds and tubers important to local diets. At the same time, *Echinochloa crus-galli* (barnyard grass) offers valuable fodder for livestock, supporting agricultural practices. *Saussurea obvallata* (Brahma Kamal) is used ornamentally in religious ceremonies due to their striking white flowers, and *Rhododendron campanulatum* adds vibrant floral decoration. Wetland plants are also utilised in traditional weaving; *Phragmites australis* (common reed) and *Carex* sp. are crafted into mats and baskets, while *Typha angustifolia* (cattail) is used for various

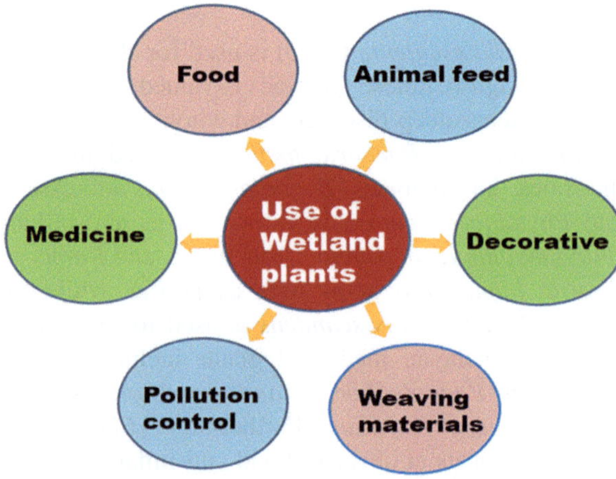

Fig. 6.4 Uses of wetland plants for different purposes

household items. They further contribute to the removal of pollutants. Both *Typha latifolia* and *Phragmites australis* are crucial in the decrease of pollutants, absorbing heavy metals as well as reducing nutrients that pose problems in constructed wetlands; these are efforts toward bettering the quality of water and environmental health (Wang et al. 2012).

6.6 Ecological Importance of Wetland Flora

6.6.1 Role in Water Purification and Nutrient Cycling

Aquatic plants play an important role in freshwater ecosystems and represent an ideal resource for cleaning water and nutrient cycling. The beneficial availability of aquatic plants, primarily in environmentally sensitive regions like the Himalayas, is crucial to maintaining freshwater quality and biodiversity (Bhowmik 2022; Thomaz 2023). Species like *Potamogeton* sp., *Carex* sp. and *Myriophyllum* sp. absorb these nutrients through their roots and leaves, reducing their concentrations in the water and preventing eutrophication, especially in nutrient-rich waters (Hilt and Gross 2008; Wang et al. 2022). In the Himalayan region, where nutrient loads can vary significantly due to agricultural runoff, sewage discharge, and glacial melt, the role of aquatic plants in nutrient regulation is even more critical. Wetlands and aquatic vegetation in Tso Moriri Lake in Ladakh effectively reduce nutrient levels, maintaining the lake's oligotrophic (nutrient-poor) status (Ahmad et al. 2020). This nutrient regulation is essential for sustaining the lake's unique biodiversity, including several endemic flora and fauna species. Without aquatic plants, these ecosystems would be at greater risk of nutrient overload, leading to algal blooms and hypoxic conditions that could devastate local wildlife and disrupt the ecological balance (Burkholder and Glibert 2013; Zhang et al. 2018). Aquatic plants also play a significant role in stabilising sediments and controlling turbidity of water bodies in the Himalayan region, where glacial-fed rivers and lakes often carry high sediment loads (Krishna 2011; Clarke 2012). The resuspension of re-deposited sediments can, in turn, replenish the nutrients and contaminants into the water furthering the worst-case water quality (Davies-Colley and Smith 2007; Bloesch 2009). This ground is a solid basis for aquatic plants to settle, thus supporting various aquatic organisms whose life cycle depends on clear water for feeding and breeding purposes (Peters and Lodge 2009). In addition, the stabilisation of sediment by aquatic plants will hold together the structural framework of wetland ecosystems otherwise at risk of erosion and degradation from changing water levels and human-induced activities (Barznji 2014; O'Hare et al. 2018).

Aquatic plants also provide surfaces for colonising microbial communities that play essential roles in biogeochemical cycling, particularly in the transformation and immobilisation of nutrients (Srivastava et al. 2017). These microbes, including bacteria, fungi, and algae, form biofilms on the rhizosphere, where they carry out processes such as nitrification, denitrification, and phosphorus cycling (Shahid et al.

2020; Srivastava et al. 2017). In the rhizosphere, plant roots exude organic compounds to stimulate microbial activity and break down organic matter, contributing toward nutrient cycling. For example, nitrifying bacteria convert ammonium (NH_4^+) to nitrate (NO_3^-), which the plants can uptake or further process by denitrifying bacteria into nitrogen gas (N_2), thus removing the excess nitrogen from the water. This microbial-mediated nutrient cycling is crucial in the nutrient-rich wetlands of the Himalayas, where it helps to mitigate the impacts of nutrient loading from external sources (McNear Jr. 2013; Kuvaja 2018; De Mandal et al. 2020; Ma et al. 2022). For instance, in the Himalayan wetland, for example, in the Kashmir Valley, plant-microbe interaction is influencing nutrient cycling. At the same time, aquatic vegetation supports diverse microbial communities that enhance nutrient cycling and reduce eutrophication, hence maintaining the health of the ecosystem in the aquatic vegetation of *Typha* sp. and *Phragmites* sp. (Gao and Shi 2018; Dar et al. 2021). The following case studies from the Himalayas and elsewhere highlight the significance of aquatic macrophytes in water purification and nutrient cycling. For instance, Dal Lake in Kashmir has been an ecotourism and cultural centre for centuries. However, the lake faces serious problems with agrochemical inputs related to agriculture runoff, sewage, and other forms of human activity, leading to eutrophication, algal blooms, and quality decline (Parveen et al. 2013; Kumar et al. 2022). Measures have been put in place to re-equilibrate the ecological balance by conserving and reintroducing native aquatic plants. This includes introducing species, such as *Nymphaea alba*, that will consume the excess nutrients found in the lake. Such species, though limited, have effectively enabled curtailment of nutrient levels, algal growth, and other factors afflicting the lake's ecosystem (Dar et al. 2021). The other case study is that of Tso Moriri Lake, a high-altitude wetland in Ladakh. This lake is among the largest in the Trans-Himalayan region, and the plant and animal species are unique. Aquatic vegetation, such as *Potamogeton* sp. and *Ranunculus* sp., plays a decisive role in nutrient cycling and maintenance of oligotrophic status at the lake level. Nutrients are absorbed from glacial meltwater and sediments by these plants, thereby preventing nutrient accumulation in the lake, which supports clear waters and high biodiversity (Mishra 2015; Panigrahy et al. 2012).

Aquatic plants play an important role in water purification and nutrient cycling, which is crucial for supporting clean water availability, fisheries, and biodiversity (Carey and Migliaccio 2009; Dugan et al. 2006). These Himalayan plants are also vulnerable to climatic changes, pollution, and human activities such as agricultural development, urban development, and tourism. Increased temperature and changed precipitation patterns can alter plant growth and increase nutrient loads. For successful conservation purposes, wetland restoration, regulation of nutrient inputs, and preservation of natural water flows shall be implemented (Upadhyay et al. 2022; Kattel 2022). Aquatic plants are important for maintaining water quality and ecosystem stability. Case studies on Dal Lake and Tso Moriri illustrate their importance in sustaining sensitive environments and maintaining the ecological balance (Siddiqui et al. 2023).

6.6.2 Carbon Sequestration

Wetland flora is crucial in carbon sequestration, particularly in cold habitats like the Himalayan region. High-altitude wetlands in areas such as Ladakh, the Kashmir Valley, and the Ruoergai Plateau capture and store atmospheric CO_2 through the accumulation of organic matter and peat formation (Kayranli et al. 2010; Lolu et al. 2020). Plants in wetlands, such as sedges (*Carex* sp.), grasses (*Eriophorum* sp.), and mosses (*Sphagnum* sp.), capture CO_2 from the atmosphere during photosynthesis, incorporating it into their biomass (Archibold 1995; Batzer and Baldwin 2019). Peatlands in the Himalayas and other cold habitats can store large amounts of carbon for thousands of years; making them significant carbon sinks (Biancalani and Avagyan 2014). In the Himalayan region, high-altitude wetlands, such as those in Ladakh and the Kashmir Valley, are critical in carbon sequestration (Medhi and Saikia 2020; Rawat et al. 2021). Globally, cold habitats such as the Arctic tundra and boreal peatlands act as significant carbon sinks due to slow decomposition rates. Therefore, preserving and restoring wetland ecosystems are crucial for maintaining their carbon sequestration capabilities and supporting global climate goals (Lolu et al. 2020; Villa and Bernal 2018). The preservation and restoration of these ecosystems are vital for maintaining their carbon sequestration potential and supporting global climate goals. Protecting Himalayan wetlands from degradation is essential to ensure their continued role in carbon storage and climate regulation.

6.7 Climate Change Impact on Wetland Flora

Climate change significantly threatens species and ecosystems globally, impacting wetlands through hydrology, temperature, and land use changes. Projected extreme climate events, such as altered hydrology, increased heat stress, and expanded ranges of pests and diseases, further threaten wetland ecosystems. Flooding and landslides disturb wetlands, resulting in soil erosion, reducing water quality, risking fire, and affecting invasive species. Climate change will affect the regimes and hydro-periods of wetlands, forcing a change in these sensitive ecosystems. The effects require tailored recovery strategies, which include the development of climate models into plans for conservation. Extreme droughts can rapidly alter the vegetation, ranges and migration of species. The case studies focus on the urgency for management action for species ranges and their migrations affected by climate change (Erwin 2009; Kattel 2022; Lolu et al. 2020).

6.8 Invasive Species Outcompeting Native Flora

Invasive species like water hyacinth (*Eichhornia crassipes*) and *Lantana camara* pose a significant threat to the native flora of the Himalayan wetlands by outcompeting indigenous plants and disrupting the ecological balance. Water hyacinth, introduced for ornamental purposes, rapidly forms dense mats on water surfaces, blocking sunlight, depleting oxygen, and leading to the decline of native aquatic plants like *Potamogeton* sp. and *Myriophyllums* sp. This aggressive growth not only displaces native flora but also disrupts habitats for aquatic fauna, undermining the wetlands' ecological functions, such as water purification, flood control, and biodiversity support, ultimately threatening the sustainability of these critical ecosystems (Ghosh et al. 2021; Chowdhury and Chowdhury 2022).

6.8.1 The Management of Invasive Alien Plant Species

Managing invasive alien plant species, a priority highlighted in IUCN conferences, is essential but costly and labour-intensive. Effective eradication is most successful when invasive species are targeted early, though removing established invasive species can sometimes lead to new issues with other species, requiring continuous monitoring. Organisations like Centre for Agriculture and Biosciences International (CABI), Invasive Species Specialist Group (ISSG), and Natural Resources Conservation Service (NRCS) focus on prevention, early detection, eradication, and management of invasive species. In regions like the Indian Himalayas, techniques like the cut rootstock method and restoration with native grass species have been effective. Biological control using natural enemies is a safe, cost-effective method, though underutilised in India. Systematic ecological studies and long-term monitoring are crucial for sustainable invasive species management (Edgerton 2014; Chowdhury and Chowdhury 2022).

6.8.2 Biocontrol Options for the Invasive Species

Biological control alternatives for invasive species involve their suppression using natural adversaries like predators, parasitoids, or pathogens. Aside from offering a restoration of ecological balance, they provide a low-cost solution when compared to the use of chemicals (Gutierrez et al. 2013). Successful biocontrol needs sustained efforts for research to guarantee that the agent will not negatively affect native species and will only target the invasive species. Several traditional biocontrol initiatives have successfully decreased the use of pesticides while enhancing native biodiversity, soil quality, and agricultural productivity. To control invasive insects and weeds, these efforts have brought in thousands of insect predators, parasitoids, herbivores, and microbial agents (Myers and Cory 2017). Each programme typically releases two to three species, and over half of them achieve notable decreases in their target populations. For example, 38 weed biocontrol projects worldwide have employed 28 fungal species with only minor, anticipated non-target effects. The need for careful ecosystem-level planning in biocontrol efforts is highlighted by ecological shifts, such as the emergence of reed canary grass following the decline of purple loosestrife or the rise of cheatgrass following diffuse knapweed control. Classical biocontrol using exotic fungi has demonstrated significant success, exemplified by the effective control of skeleton weed in Australia through *Puccinia chondrillina*, highlighting its potential as a sustainable and targeted weed management strategy (Hasan and Ayres 1990). Notable fungal genera explored as potential bioherbicides include *Colletotrichum*, *Phoma*, and *Sclerotinia*, while *Xanthomonas* and *Pseudomonas* are the leading bacterial candidates in this field (Harding and Raizada 2015).

6.9 Conservation Efforts

6.9.1 Protected Areas and Ramsar Sites

Wetlands are among the Earth's most productive and biologically different ecosystems, making up around 6% of the planet's land area. Wetlands, which include shallow seawater areas, are defined by the Ramsar Convention as areas of marsh, fens, and peatlands, whether natural or constructed, with fresh, brackish, or saline water (Hails 1996). However, disjointed conservation measures require a thorough, integrated, and inclusive strategy for effective management. The Ramsar Convention, established in 1971, emphasises regional collaboration while promoting the conservation and sustainable use of wetlands. Launched in 2009 by the Ramsar Secretariat, Wetlands International, International Centre for Integrated Mountain Development (ICIMOD), and World-Wide Fund for Nature (WWF), the Himalayan Wetlands Initiative (HWI) promotes interdisciplinary research, builds stakeholder capacity, and addresses transboundary challenges in high-altitude wetlands (HAWs) (Chandan et al. 2007). HAWs are crucial for protecting biodiversity, particularly in the arid Trans-Himalaya, in addition to their hydrological significance at the basin level, wetlands, and ecosystems. In addition to serving as vital habitats and breeding grounds for significant local and migratory species of waterfowl and other wildlife, HAWs are stores of rich aquatic biodiversity (Chatterjee et al. 2010). Started in 1987, India's wetland conservation initiatives mainly relied on biological approaches rather than technological fixes. Since then, an integrated conservation plan has been supported by initiating a nationwide wetland mapping effort. In some areas, cooperation between local people, NGOs, and government agencies has aided in achieving the objectives of the Ramsar Convention (Erwin 2009; Medhi and Saikia 2020). Institutions, non-governmental organisations, and professional groups must coordinate for effective conservation. Wetland status assessment, aquatic species inventories, and creating a Red Data Book for wetland biodiversity are among the top goals. Famous wetlands in India include Tsomoriri (Ladakh), Pong Dam Lake (Himachal Pradesh), Renuka Wetland (Himachal Pradesh), Hokera Wetland (Kashmir), Chandertal Wetland (Himachal Pradesh), and Asan Conservation Reserve (Uttarakhand) need immediate conservation.

6.9.2 Community-Based Conservation Programmes

Community-based conservation programmes play a vital role in preserving wetlands' ecological integrity and biodiversity. These initiatives should focus on key actions such as identifying critical habitats, rehabilitating degraded areas, and protecting rare or endangered species. By matching their interventions with local community expectations and offering socioeconomic advantages to society, community-based conservation (CBC) makes long-term conservation goals more attainable (Brooks et al. 2013). The Ramsar Convention emphasises wetland conservation and its sustainable and wise use. To ensure effective wetland management, engaging local communities and stakeholders is essential, fostering inclusive and participatory governance. This approach enhances collective decision-making and increases the success of conservation efforts (Sharma and Singh 2021).

The Convention also underscores the importance of legal accountability for those responsible for wetland degradation and the need for performance evaluation systems to ensure effective governance. Strengthening wetland conservation laws and enforcement at national and international levels is crucial for improving wetland health (Jiang et al. 2024). Capacity building is another key aspect involving the development of reliable training programmes for wetland managers and policymakers through both online and offline platforms. Interdisciplinary research and collaboration among stakeholders are critical to mitigating the adverse impacts on wetlands. Education and awareness are foundational to the success of conservation initiatives (Wood et al. 2013). Managing invasive species, such as *Lantana camara*, in the Himalayan wetlands is crucial to safeguarding native biodiversity. Additionally, sustainable development requires comprehensive land-use planning, including eco-zoning and robust environmental impact assessments (EIAs) (Kunwar et al. 2022). Wetland legislation must be regularly reviewed and updated to align with international frameworks, such as the Ramsar Convention. Strengthening institutions through training, resource allocation, and awareness programmes is essential. Examples from the Hindukush Himalayan region highlight the effectiveness of community-based approaches in wetland conservation (Ritika et al. 2024). These initiatives demonstrate the value of engaging local populations in sustainable wetland management, ecotourism, and the preservation of indigenous practices, ensuring long-term ecosystem stewardship and sustainability (Papastergiadou et al. 2008; Wu 2017).

6.9.3 Research and Monitoring Initiatives for Wetland Flora of Himalaya

The Himalayan wetlands are ecologically vital, supporting a rich and unique diversity of flora, including numerous rare and endemic species. With their distinctive and varied vegetation, these wetlands play a significant ecological role. Understanding the effects of industrialisation, human activities, and climate change on these fragile ecosystems relies heavily on sustained research and monitoring initiatives (Thapliyal et al. 2024). These efforts encompass comprehensive biodiversity surveys, the development of a Red Data Book for Himalayan wetland plants, and ecosystem monitoring using advanced technologies such as remote sensing and GIS mapping. Significant progress in wetland mapping has been achieved through the efforts of the Indian Space Research Organisation's (ISRO) Space Application Center (SAC) and state-level remote sensing agencies; however, smaller wetlands still demand finer spatial resolution for accurate representation (Khinchi and Yadav 2024; Prasad et al. 2002).

An integral role in this process is played by the Wetlands Division of the Ministry of Environment, Forest and Climate Change (MoEF&CC), which oversees the implementation of wetland conservation and management initiatives across India. The Division coordinates the National Plan for Conservation of Aquatic Ecosystems (NPCA), works with state governments to identify and designate wetlands of national and international importance, and provides technical and financial support for site-specific conservation actions. The Division also manages the indianwetlands.in portal, which serves as a central repository for wetland-related data, policies, and management strategies.

In collaboration with the National Wetland Committee and other scientific bodies, MoEF&CC has recognised 22 wetlands as sites of national and international significance, and, through the Standing Committee on Bioresources of the National Natural Resource Management System, is spearheading a national inventory of wetlands. While SAC's 1:250,000 scale wetland mapping represents an important foundational step (Prasad et al. 2002), future conservation planning must focus on high-resolution mapping and long-term monitoring. Combining these scientific advancements with local knowledge through active community engagement enhances the effectiveness of conservation strategies (Lichvar et al. 2016). To ensure the long-term protection of Himalayan wetland biodiversity and their ecological integrity, it will be critical to update existing wetland legislation, reinforce institutional mechanisms, and promote integrated conservation approaches driven by both community participation and technological innovation.

6.10 Challenges, Gaps, and Solutions in Wetland Management

Wetland management has great challenges; already 50% of wetlands in the world have been destroyed. Between the 1980s and early 2000s, it was reported that 95 km^2 were lost every year due to human activity. The main factors attributed to the destruction of wetlands are road construction, agriculture, deforestation, pollution, overgrazing, river canalisation, over-harvesting, and invasive species from both developed and developing countries (Zedler and Kercher 2005). These threats result in habitat loss, species decline, especially among migratory birds and reduced biodiversity due to pollution, overexploitation, and habitat modification. Climate change and natural disasters cause significant threats. A major constraint is the lack of awareness about wetlands' economic and ecological value, where wetland development often leads to biodiversity loss (Zebardast et al. 2021). Despite various comprehensive policies and programmes for wetland conservation, several problems persist, including:

1. Lack of a national policy and action plan for wetland conservation and management.
2. Poor integration of environmental factors in resource management, reflecting low awareness among government officials and the public about the environmental implications of development proposals. A lack of technical expertise, funding, workforce, and resource information compounds this.
3. A rigid sectoral approach with minimal intersectoral planning and coordination.
4. Conflicting priorities among different resource management agencies, which do not encourage sustainable resource management practices.
5. Lack of a clear mandate, leading to overlapping activities among various agencies.
6. Absence of comprehensive wetland management legislation and enforcement mechanisms (Fois et al. 2021; Warner and Asada 2006).

Effective conservation of the Himalayan wetland requires incorporating local perspectives through feedback mechanisms, ensuring that communities are informed and actively involved in decision-making. This fosters genuine participation and prevents residents from feeling exploited. Continuous engagement is crucial, as interruptions or long breaks in conservation programmes can damage credibility and reduce effectiveness (Dar et al. 2024). Economic incentives, such as addressing livelihood needs and employing local labour, can further enhance community support. To avoid ineffective top-down solutions, conservation initiatives must adopt flexible, context-specific strategies considering the community's socioeconomic and cultural realities (Agbeti 2023). Various organisations contribute significantly to global wetland conservation. Wetlands International focuses on restoring and protecting wetland ecosystems through partnerships with governments, NGOs, and local communities. The Nature Conservancy emphasises sustainable land and water management in wetland habitats while engaging local populations in conservation. WWF works to safeguard wetland habitats, promote sustainable water management, and alleviate pressures on wetland species. Ducks Unlimited is dedicated to preserving waterfowl habitats by restoring wetlands and advocating for sustainable farming practices. Environmental NGOs like the Sierra Club and Greenpeace also play key roles through advocacy, direct action, and raising public awareness about the importance and sustainability of wetlands (Khinchi and Yadav 2024; Sharma and Naik 2024).

6.11 Conclusions

The Himalayas are an important hotspot for biodiversity, and wetlands are key for maintaining rare species and carrying out fundamental ecological tasks. However, invasive species, habitat degradation, biodiversity loss, and climate change pose major risks to these wetlands. Gaps in long-term climate data and wetland-specific preservation techniques frequently hinder conservation efforts, making many wetlands susceptible to neglect. This study highlights the necessity of thorough research, ongoing observation, and management techniques based on scientific principles to address these issues. Inadequate scientific evidence should not limit conservation efforts; rather, the emphasis should be on filling such knowledge gaps. Strengthening the scientific basis for wetland conservation requires increased cooperation with NGOs, academic institutions, and national and international organisations. For sustainable wetland management, it is essential to model future inflows, predict changes in wetland water levels, and understand their ecological impacts. Addressing these gaps will provide valuable insights for communities, conservationists, and water resource managers, ensuring that Himalayan wetlands and their biodiversity are preserved for future generations.

Acknowledgements We extended our sincere gratitude to Mr. Kingsuk Sarkar (JRF at the Regional Ayurveda Research Institute, Gangtok) for providing valuable photographs and details of Ayurvedic flora from Khecheopalri Lake, Sikkim. We also thank Dr. Animesh Sen (Research Assistant in Botany at the Regional Ayurveda Research Institute, Gangtok, Sikkim) for sharing images of wetland flora from Darjeeling, Assam, and Sikkim.

References

Abba A, Sankarannair S (2024) Global impact of water hyacinth (Eichhornia Crassipes) on rural communities and mitigation strategies: a systematic review. Environ Sci Pollut Res Int 31(31): 43616–43632

Adhikari BS, Babu MM (2008) Floral diversity of Baanganga wetland, Uttarakhand, India. Check List 4(3):279–290. https://doi.org/10.15560/4.3.279

Agbeti JA (2023) Sustainable conservation of Sakumo wetlands for social and environmental benefits. PhD Thesis

Ahmad A, Jehangir A, Yousuf AR, Shah WA, Tanveer A (2020) Major ion chemistry of surface sediments of brackish endorheic lake Tso Moriri-a high altitude Ramsar site in Western Himalaya. J Himal Ecol Sustain Dev 15:41–57

Alam A, Rashid SM, Bhat MS, Sheikh AH (2011) Impact of land use/land cover dynamics on Himalayan wetland ecosystem. J Exp Sci 2(3):60–64

Allen-Diaz B, Jackson RD, Bartolome JW, Tate KW, Oates LG (2004) Long-term grazing study in spring-fed wetlands reveals management tradeoffs. Calif Agric 58(3):144–148

Archibold OW (1995) Terrestrial wetlands. In: Ecology of world vegetation. Springer Netherlands, Dordrecht, pp 319–353

Bahuguna YM, Gairola S, Uniyal PL, Bhatt AB (2016) Moss Flora of Kedarnath Wildlife Sanctuary (KWLS), Garhwal Himalaya, India. Proc Natl Acad Sci India Sect B Biol Sci 86: 931–943

Banerjee A, Chakraborty P, Bandopadhyay R (2019) Urgent conservation needs in the Sikkim Himalaya biodiversity hotspot. Biodiversity 20(2–3):88–97. https://doi.org/10.1080/14888386.2019.1656547

Barznji DAM (2014) Role of aquatic plants in improving water quality. UJPBS 2(05):12–16

Batzer DP, Baldwin AH (2019) 9. Northern peatlands. In: Wetland Habitats of North America. University of California Press, pp 119–134

Bhatta LD, Ning W, Udas E, Agrawal NK, Ranabhat S, Basnet D (2018) Wetlands in the Himalaya: securing services for livelihoods. International Centre for Integrated Mountain Development (ICIMOD)

Bhojak P, Dey D, Sekar KC, Thapliyal N, Danu H (2024) Floristic diversity in high-altitude wetlands of Ladakh Union Territory, trans Himalaya, India. Check List 20(4):915–939

Bhowmik S (2022) Ecological and economic importance of wetlands and their vulnerability: a review. In: Research Anthology on Ecosystem Conservation and Preserving Biodiversity. University of Calcutta, pp 11–27

Biancalani R, Avagyan A (2014) Towards climate-responsible peatlands management. Food and Agiculture Organization of the United Nations (FAO)

Bloesch J (2009) Sediments of aquatic ecosystems. In: Likens GE (ed) Encyclopedia of inland waters. Academic, Oxford, pp 479–490

Brooks J, Waylen KA, Mulder MB (2013) Assessing community-based conservation projects: a systematic review and multilevel analysis of attitudinal, behavioral, ecological, and economic outcomes. Environ Evid 2(1):2. https://doi.org/10.1186/2047-2382-2-2

Burkholder JM, Glibert PM (2013) Eutrophication and oligotrophication. In: Levin SA (ed) Encyclopedia of biodiversity, 2nd edn. Academic, Waltham, pp 347–371

Butt MA, Zafar M, Ahmed M, Shaheen S, Sultana S (2021) Types of wetland and wetland plants. In: Wetland plants. Springer, Cham, pp 35–54

Carey RO, Migliaccio KW (2009) Contribution of wastewater treatment plant effluents to nutrient dynamics in aquatic systems: a review. Environ Manag 44:205–217

Chandan P, Chatterjee A, Gautam P (2007) Management planning of Himalayan high altitude wetlands: a case study of Tsomoriri and Tsokar wetlands in Ladakh, India. In: Proceedings of Taal, pp 1446–1452

Chatterjee A, Blom E, Gujja B, Jacimovic R, Beevers L, O'Keeffe J, Beland M, Biggs T (2010) WWF initiatives to study the impact of climate change on Himalayan high-altitude wetlands (HAWs). Mt Res Dev 30(1):42–52

Chettri N, Sharma E, Shakya B, Thapa R, Bajracharya B, Uddin K, Oli K, Choudhury D (2010) Biodiversity in the Eastern Himalayas: status, trends and vulnerability to climate change. In: Climate change impact and vulnerability in the Eastern Himalayas—Technical report 2. ICIMOD

Chowdhury A, Chowdhury M (2016) *Nymphaea abhayana* sp. nov. of Nymphaeaceae from Duars of West Bengal, India. Int J Plant Sci 5(4):57–60

Chowdhury M, Chowdhury A (2022) Wetland flora of West Bengal. Bluerose Publishers, New Delhi

Chowdhury A, Chowdhury M, Das AP (2015) *Potamogeton gramineus* L. (Potamogetonaceae): a new record for Indian sub-continent with a complete morphological description. Int J Curr Res 7(2):12362–12364

Clarke SJ (2012) Aquatic plants. In: Bengtsson L, Herschy RW, Fairbridge RW (eds) Encyclopedia of lakes and reservoirs. Springer Netherlands, Dordrecht, pp 39–42

Dahal S, Sharma TP, Borthakur SK (2017) Database on medicinal plants of Tamze medicinal plants conservation area (MPCA) of Sikkim Himalaya, India. NeBIO–Int J Env Biodivers 8:45–56

Dar GH, Khuroo AA (2020) Biodiversity of the Himalaya: Jammu and Kashmir State. Springer Nature

Dar SA, Bhat SU, Rashid I (2021) The status of current knowledge, distribution, and conservation challenges of wetland ecosystems in Kashmir Himalaya, India. In: Sharma S, Singh P (eds) Wetlands conservation, 1st edn. Wiley, pp 175–200

Dar SA, Rashid I, Bhat SU, Dar JA (2024) Evaluation of ecosystem services vis-à-vis perceptions and attitudes of local communities towards wetland conservation in Kashmir Himalaya. Environ Dev 51:101041

Davidson R (1993) The vegetation of Lough Neagh wetlands. In: Lough Neagh: the ecology of a multipurpose water resource. Springer, pp 473–490

Davies-Colley R, Smith D (2007) Turbidity, suspended sediment, and water clarity: a review. JAWRA J Am Water Resour Assoc 37:1085–1101. https://doi.org/10.1111/j.1752-1688.2001.tb03624.x

De Mandal S, Laskar F, Panda AK, Mishra R (2020) Microbial diversity and functional potential in wetland ecosystems. In: Recent advancements in microbial diversity. Elsevier, pp 289–314

Dey D, Bhojak P, Sekar KC, Arya D (2021) An annotated checklist of vascular plants in and around two major high-altitude wetlands of Lahaul-Spiti, Himachal Pradesh, India. Check List 17(6): 1715–1730

Dhyani A, Shantanu K, Kholia BS, Uniyal PL (2023) A preliminary assessment of bryophyte diversity in some alpine areas of East Sikkim, India. Indian J For 46(1):43–49

Dugan P, Dey MM, Sugunan VV (2006) Fisheries and water productivity in tropical river basins: enhancing food security and livelihoods by managing water for fish. Agric Water Manag 80(1–3):262–275

Dwivedi DH, Shrivastava U, others (2023) Wetland horticulture: new dimensions. Progress Hortic 55(2):79–94

Edgerton EA (2014) Prevention and management of aquatic invasive plants in Texas. PhD Thesis

Erwin KL (2009) Wetlands and global climate change: the role of wetland restoration in a changing world. Wetl Ecol Manag 17(1):71–84

Fois M, Cuena-Lombraña A, Bacchetta G (2021) Knowledge gaps and challenges for conservation of Mediterranean wetlands: evidence from a comprehensive inventory and literature analysis for Sardinia. Aquat Conserv Mar Freshwat Ecosyst 31(9):2621–2631. https://doi.org/10.1002/aqc.3659

Ganie AH, Reshi ZA, Wafai BA, Puijalon S (2015) Phenotypic plasticity: cause of the successful spread of the genus Potamogeton in the Kashmir Himalaya. Aquat Bot 120:283–289

Gao T, Shi X-Y (2018) Taxonomic structure and function of seed-inhabiting bacterial microbiota from common reed (Phragmites australis) and narrowleaf cattail (Typha angustifolia L.). Arch Microbiol 200(6):869–876. https://doi.org/10.1007/s00203-018-1493-3

Ghosh R, Özaslan C, Ray P (2021) Invasive alien freshwater hydrophytes: co-facilitating factors with emphasis on Indian scenario. ISWS 53(3):216–229

Guntenspergen GR, Stearns F, Kadlec JA (2020) Wetland vegetation. In: Constructed wetlands for wastewater treatment. CRC Press, pp 73–88

Gutierrez AP, Messenger PS, van den Bosch R (2013) An introduction to biological control. Springer Science & Business Media

Hails AJ (ed) (1996) Wetlands, biodiversity and the Ramsar convention: the role of the convention on wetlands in the conservation and wise use of biodiversity. Ramsar Convention Bureau, Gland

Halls AJE (1997) Wetlands, biodiversity and the Ramsar convention: the role of the convention on wetlands in the conservation and wise use of biodiversity. Ramsar Convention Bureau, Gland

Harding DP, Raizada MN (2015) Controlling weeds with fungi, bacteria and viruses: a review. Front Plant Sci 6:659. https://doi.org/10.3389/fpls.2015.00659

Hasan S, Ayres PG (1990) The control of weeds through fungi; principles and prospects. New Phytol 115(2):201–222. https://doi.org/10.1111/j.1469-8137.1990.tb00447

Hilt S, Gross EM (2008) Can allelopathically active submerged macrophytes stabilise clear-water states in shallow lakes? Basic Appl Ecol 9(4):422–432

Iqbal A, Shang Z (2020) Wetlands as a carbon sink: insight into the Himalayan region. In: Shang Z, Degen AA, Rafiq MK, Squires VR (eds) Carbon management for promoting local livelihood in the Hindu Kush Himalayan (HKH) region. Springer, Cham, pp 125–144

Javid M, Ahmad K, Rahmani AR, Ahmad F, Nigam P (2024) Species diversity and composition of macrophytes in Shallabugh Wetland, Ganderbal, Jammu, and Kashmir, India. Int J Plant Soil Sci 36(6):436–443

Jiang W, Zhang Z, Ling Z, Deng Y (2024) Experience and future research trends of wetland protection and restoration in China. J Geogr Sci 34(2):229–251

Jocou A, Gandullo R (2022) Juncus in Northern Patagonian Wetlands. Soc Wetl Sci Bull 40:16–25

Kattel GR (2022) Climate warming in the Himalayas threatens biodiversity, ecosystem functioning and ecosystem services in the 21st century: is there a better solution? Biodivers Conserv 31(8–9):2017–2044. https://doi.org/10.1007/s10531-022-02417-6

Kayranli B, Scholz M, Mustafa A, Hedmark Å (2010) Carbon storage and fluxes within freshwater wetlands: a critical review. Wetlands 30:111–124

Khinchi SS, Yadav S (2024) Fundamental concepts of aerial photography, remote sensing. In: GPS & Geographical Information System (GIS). Academic Guru Publishing House

Kingsford RT, Basset A, Jackson L (2016) Wetlands: conservation's poor cousins. Aquat Conserv Mar Freshw Ecosyst 26(5):892–916. https://doi.org/10.1002/aqc.2709

Kohli M (2015) Wetlands of international importance in Himachal Pradesh. Int J Econ Plants 2(1): 023–027

Krishna AP (2011) Characteristics of snow and glacier fed rivers in mountainous regions with special reference to Himalayan Basins. In: Singh VP, Singh P, Haritashya UK (eds) Encyclopedia of snow, ice and glaciers. Springer Netherlands, Dordrecht, pp 128–133

Kumar L, Lamsal P (2016) High altitude wetlands of Nepal. In: The wetland book. Springer, Dordrecht

Kumar R, Parvaze S, Huda MB, Allaie SP (2022) The changing water quality of lakes—a case study of Dal Lake, Kashmir Valley. Environ Monit Assess 194(3):228

Kunwar BB, Adhikari BR, Muensit N, Techato K, Gyawali S (2022) Role of vegetation for the protection of Phewa Watershed, Kaski, Nepal. Environ Ecol Res 10(2):161–173

Kuvaja V (2018) Nitrogen removal from industrial wastewaters. Master's Thesis

Lichvar RW, Banks DL, Kirchner WN, Melvin NC (2016) The national wetland plant list: 2016 wetland ratings. Phytoneuron 30:1–17

Lolu AJ, Ahluwalia AS, Sidhu MC, Reshi ZA, Mandotra SK (2020) Carbon sequestration and storage by wetlands: implications in the climate change scenario. In: Upadhyay AK, Singh R, Singh DP (eds) Restoration of wetland ecosystem: a trajectory towards a sustainable environment. Springer Singapore, Singapore pp.45–58

Ma W, Tang S, Dengzeng Z, Zhang D, Zhang T, Ma X (2022) Root exudates contribute to belowground ecosystem hotspots: a review. Front Microbiol 13:937940. https://doi.org/10.3389/fmicb.2022.937940

Majid A, Ahmad H, Saqib Z, Ali H, Alam J (2015) Conservation status assessment of Meconopsis aculeata Royle: a threatened endemic of Pakistan and Kashmir. Pak J Bot 47:1–5

Manikandan R, Nithya SP, Devi RM (2023) Checklist flora of Sunderdhunga Valley, Western Himalaya, with emphasis on ethno-medicinal plants. In: Arunachalam K, Yang X, Puthanpura Sasidharan S (eds) Bioprospecting of tropical medicinal plants. Springer Nature Switzerland, Cham, pp 159–225

McNear DH Jr (2013) The rhizosphere-roots, soil and everything in between. Nat Educ Knowl 4:1

Medhi B, Saikia S (2020) High altitude wetlands in Arunachal Pradesh: a review on its importance and way forward for conservation and management. J Crit Rev 7(3):618–622

Mishra PK (2015) Late Quaternary climate variability in the Indian monsoon domain. PhD Thesis

Mitsch WJ, Gosselink JG (2015) Wetlands, 5th edn. Wiley

Myers JH, Cory JS (2017) Biological control agents: invasive species or valuable solutions? In: Vilà M, Hulme PE (eds) Impact of biological invasions on ecosystem services. Springer, Cham, pp 191–202

Ning W, Rawat GS, Sharma E (2014) High-altitude ecosystem interfaces in the Hindu Kush Himalayan region. In: High-altitude rangelands their interfaces Hindu Kush Himalayas, p 3

O'Hare MT, Aguiar FC, Asaeda T, Bakker ES, Chambers PA, Clayton JS, Elger A, Ferreira TM, Gross EM, Gunn ID (2018) Plants in aquatic ecosystems: current trends and future directions. Hydrobiologia 812:1–11

Palni LMS, Rawal RS (2010) Conservation of Himalayan bioresources: an ecological, economical and evolutionary perspective. In: Sharma VP (ed) Nature at work: ongoing saga of evolution. Springer India, New Delhi, pp 369–402

Panigrahy S, Patel JG, Parihar JS (2012) National Wetland Atlas: high altitude lakes of India. Space Applications Centre, ISRO, Ahmedabad India

Pant GB, Pradeep Kumar P, Revadekar JV, Singh N, Pant GB, Kumar PP, Revadekar JV, Singh N (2018) The Himalaya. Springer

Papastergiadou ES, Retalis A, Apostolakis A, Georgiadis T (2008) Environmental monitoring of spatio-temporal changes using remote sensing and GIS in a Mediterranean wetland of Northern Greece. Water Resour Manag 22:579–594

Parveen M, Shafi Bhat M, Haq S (2013) Studies on physico-chemical characteristics of Dal Lake, Srinagar Kashmir. Int J Curr Res 5(6):1352–1354

Pathania MS, Kumar GA (2017) Economic valuation and causes of degradation of Pong Dam wetland of Himachal Pradesh. Int J Soc Sci 6(1):21–29

Peters JA, Lodge DM (2009) Littoral zone. In: Likens GE (ed) Encyclopedia of inland waters. Academic, Oxford, pp 79–87

Prasad SN, Ramachandra TV, Ahalya N, Sengupta T, Kumar A, Tiwari AK, Vijayan VS, Vijayan L (2002) Conservation of wetlands of India-a review. Trop Ecol 43(1):173–186

Rajput JD, Singh SP (2023) Water chestnut (*Trapa natans* L.): functional characteristics, nutritional properties and applications in food industry: a review. J Phytopharm 12(2):119–126

Rawal RS, Bhatt ID, Sekar KC, Nandi SK (2013) The himalayan biodiversity: richness, representativeness, uniqueness and life-support values. GB Pant Institute of Himalayan Environment and Development (GBPIHED), Kosi-Katarmal, Almora, p 84

Rawat GS (2017) The Himalayan vegetation along horizontal and vertical gradients. In: Prins HT, Namgail T (eds) Bird migration in the Himalaya. Cambridge University Press, pp 189–204

Rawat D, Sati SP, Khanduri VP, Riyal M, Mishra G (2021) Carbon sequestration potential of different land use sectors of Western Himalaya. In: Pant D, Kumar Nadda A, Pant KK, Agarwal AK (eds) Advances in carbon capture and utilization. Springer Singapore, Singapore, pp 273–294

Reddy KR, DeLaune RD, Inglett PW (2022) Biogeochemistry of wetlands: science and applications. CRC press. https://doi.org/10.1201/9780429155833

Ritika KC, Rai U, Ghimire M, Khadka UR (2024) Wetlands in the Hindu Kush Himalayan region: eco-economic function and conservation strategies. In: Shang Z, Degen A, Gauchan D, Koirala M, Rafiq MK (eds) Sustainable ecological restoration and conservation in the Hindu Kush Himalayan region. CABI, GB, pp 234–250

Samant, Pant S, Singh M, Lal M, Singh A, Sharma A, Bhandari S (2007) Medicinal plants in Himachal Pradesh, north western Himalaya, India. Int J Biodivers Sci Manag 3(4):234–251. https://doi.org/10.1080/17451590709618177

Shah MA, Reshi ZA (2014) Characterization of alien aquatic flora of Kashmir Himalaya: implications for invasion management. Trop Ecol 55(2):143–157

Shahid MJ, Al-Surhanee AA, Kouadri F, Ali S, Nawaz N, Afzal M, Rizwan M, Ali B, Soliman MH (2020) Role of microorganisms in the remediation of wastewater in floating treatment wetlands: a review. Sustainability 12(14):5559

Sharma LKN, Naik R (2024) Conservation of saline wetland ecosystems: an initiative towards UN decade of ecological restoration. Springer Nature

Sharma S, Singh P (2021) Wetlands conservation: current challenges and future strategies. Wiley

Siddiqui AJ, Jahan S, Adnan M, Ashraf SA, Singh R (2023) Macrophytes and their role in wetland ecosystems. In: Kumar S, Bauddh K, Singh R, Kumar N, Kumar R (eds) Aquatic macrophytes: ecology, functions and services. Springer Nature Singapore, pp 119–138

Singh DK, Pusalkar PK (2020) Floristic diversity of the Indian Himalaya. In: Dar GH, Khuroo AA (eds) Biodiversity of the Himalaya: Jammu and Kashmir State. Springer, Singapore, pp 93–126

Soni HB (2020) Biodiversity of wetlands and forests: a Nature Trail. Google Book Publishers

Srivastava JK, Chandra H, Kalra SJS, Mishra P, Khan H, Yadav P (2017) Plant–microbe interaction in aquatic system and their role in the management of water quality: a review. Appl Water Sci 7(3):1079–1090. https://doi.org/10.1007/s13201-016-0415-2

Srivastava N, Negi S, Kaushal S (2023) Xenobiotics interaction: consequences and remediation

Tendar P, Sridith K (2021) Vegetation structure of wetlands in Eastern Himalayan highlands of Gasa, Bhutan. Science Asia 47(1):78–85

Tenzin S, Tendar P (2021) Wetland medicinal plants of eastern Himalayan highlands of Gasa District, Bhutan. Asian Plant Res J 7(2):1–13

Thapliyal N, Bhojak P, Mehta P, Sekar KC (2024) Threats and conservation of plant diversity in the Himalayan biodiversity hotspot. In: Biodiversity hotspot of the Himalaya. Apple Academic Press, pp 459–511

Thomaz SM (2023) Ecosystem services provided by freshwater macrophytes. Hydrobiologia 850(12):2757–2777

Tse-Ring K, Sharma E, Chettri N, Shrestha AB (2010) Climate change vulnerability of mountain ecosystems in the Eastern Himalayas. International Centre for Integrated Mountain Development (ICIMOD)

Upadhyay KD, Mainali J, Ghimire NP (2022) Diversity of aquatic plants and macroinvertebrates and their spatial patterns in a Himalayan Watershed, Central Nepal. Aquat Bot 180:103529

Villa JA, Bernal B (2018) Carbon sequestration in wetlands, from science to practice: an overview of the biogeochemical process, measurement methods, and policy framework. Ecol Eng 114: 115–128

Wang X, Yu J, Zhou D, Dong H, Li Y, Lin Q, Guan B, Wang Y (2012) Vegetative ecological characteristics of restored reed (Phragmites australis) wetlands in the Yellow River Delta, China. Environ Manage 49(2):325–333. https://doi.org/10.1007/s00267-011-9757-6

Wang L, Wu X, Song H, An J, Dong B, Wu Y, Wang Y, Li B, Liu Q, Yu W (2022) Influence of Potamogeton crispus harvesting on phosphorus composition of Lake Yimeng. Sci Rep 12(1): 17616. https://doi.org/10.1038/s41598-022-22484-7

Warner BG, Asada T (2006) Knowledge gaps and challenges in wetlands under climate change in Canada. Clim Change Manag Ecosyst:355–372. https://doi.org/10.1201/9780849330971.ch18

Wood AP, Dixon A, McCartney MP (2013) Wetland management and sustainable livelihoods in Africa. Routledge/Earthscan from Routledge, Oxford

Wu Q (2017) GIS and remote sensing applications in wetland mapping and monitoring. Earth Syst Environ 2:140–157

Yadav H, Dhyani A, Uniyal PL (2022) Contribution to the moss flora of Northern Sikkim, India. J Threat Taxa 14(10):22008–22015

Zebardast L, Akbarpour S, Jafari HR, Bagherzadeh Karimi M (2021) Sustainable wetland management through bridging the communication gap between conservation projects and local communities. Environ Dev Sustain 23:11098–11119

Zedler JB, Kercher S (2005) Wetland resources: status, trends, ecosystem services, and restorability. Annu Rev Environ Resour 30(1):39–74. https://doi.org/10.1146/annurev.energy.30.050504.144248

Zhang H, Shang Y, lyu T, Chen J, Pan G (2018) Switching harmful algal blooms to submerged macrophytes in shallow waters using geo-engineering methods: evidence from a 15N tracing study. Environ Sci Technol 52(20):11778–11785. https://doi.org/10.1021/acs.est.8b04153

Chapter 7
Diversity and Impact of Alien Plants in the Himalayan Hotspot and Their Management

Saini Sultana and Chandra Ghosh

Abstract The introduction and spread of alien plant species in the Himalayan range pose a growing threat to the region's ecological balance, leading to severe consequences such as biodiversity loss, species extinction, agricultural destruction, and significant alterations to ecosystem functions. These non-native plants disrupt the natural environment by changing key factors like soil structure, nutrient cycling, and hydrological patterns, often outcompeting native flora. The varied topography, microclimates, and altitudes in the region create favorable conditions for the successful establishment and rapid spread of alien species, making the Himalayan landscapes particularly vulnerable to plant invasions. Contributing factors such as climate change, human activity, and trade further accelerate this process, exacerbating the ecological challenges already faced by the region. This study aims to provide a comprehensive review of the current research on alien plant species in the Himalayan range, focusing on their ecological and socio-economic impacts, document the processes and causes driving plant invasions, and assess the potential consequences of these invasions on native biodiversity, agriculture, and local ecosystems. Additionally, the study will propose effective strategies for managing the spread of alien species, offering recommendations for conservation efforts and sustainable ecosystem management. The findings aim to inform policy and guide practical solutions to preserve the ecological integrity of the Himalayan region.

Keywords Himalayan ecosystem · Species extinction · Soil structure alteration · Habitat degradation · Landscape heterogeneity

S. Sultana · C. Ghosh (✉)
Tea Taxonomy and Ecology Laboratory, Department of Tea Science, University of North Bengal, Darjeeling, West Bengal, India
e-mail: chandraghosh19@nbu.ac.in

D. Das et al. (eds.), *Plant Diversity and Conservation Strategies of the Himalayan Biodiversity Hotspot*, Topics in Biodiversity and Conservation 21,
https://doi.org/10.1007/978-3-032-21697-7_7

7.1 Introduction

Alien species, defined as organisms introduced outside their natural range due to human activities, have been spreading across the globe for centuries, with an accelerated pace in the last 50 years (Richardson et al. 2000; Hulme et al. 2009). The prevalence of alien plants varies regionally, with some areas heavily impacted by their proliferation, while others remain less affected (Guo et al. 2021; Pyšek et al. 2017). Climate plays a pivotal role in the successful establishment of alien species, as regions with similar climatic conditions to their native habitats facilitate invasion (Sax et al. 2002). Presently, approximately one-sixth of the Earth's land is at high risk of invasion, including major biodiversity hotspots and regions with developing economies (Early et al. 2016). While some alien species have enhanced agricultural productivity, others, particularly invasive weeds, compete for vital resources like space, light, moisture, and nutrients, leading to reduced crop yields and economic loss (Sujay et al. 2010).

The Himalayas, a global biodiversity hotspot, face mounting threats from biological invasions (Wani et al. 2022). The Himalayan range covers about 750,000 km^2, divided into two regions: the Eastern Himalaya, including parts of Nepal, northeastern India (West Bengal, Sikkim, Assam, Arunachal Pradesh), Bhutan, southeastern Tibet, and northern Myanmar, and the Western Himalaya, which includes Kumaon-Garhwal, northwestern Kashmir, and northern Pakistan. Most of the range lies within India, spanning from Jammu & Kashmir to Arunachal Pradesh, and covering approximately 16.2% of the country's territory (Khuroo et al. 2012). Stretching across South Asia, this region is ecologically fragile, containing over 8000 species of angiosperms alone (Singh et al. 2000). Despite its ecological richness, the Indian Himalayan Region (IHR) which constitutes a major portion of the Himalayas remains poorly documented in terms of its alien flora. This gap in documentation mirrors concerns highlighted in Chap. 6 (Sect. 6.1), where limited understanding of the floral composition of Himalayan wetlands is identified as a key challenge to effective ecological management. The rise of invasive species is largely driven by human activities, economic development, and changing land-use patterns, leading to land degradation, reduced biodiversity, and agricultural damage (Weber and Li 2008; Nunez and Pauchard 2010; Everard 2020). Despite the region's rich floral diversity and susceptibility to environmental changes, there has been little effort to inventory, predict, and map the alien flora. Studies by Pathak et al. (2019) stress the importance of such investigations for effective management. Significant research has been conducted in areas such as Kashmir Himalaya (Khuroo et al. 2008; Haq et al. 2021), Himachal Pradesh (Jaryan et al. 2013), Arunachal Pradesh (Kosaka et al. 2010), Assam's hilly regions (Barua et al. 2013), West Bengal (Maiti and Guha-Bakshi 1981), and Tripura (Debnath et al. 2017). Comprehensive inventories and predictive mapping of alien plant species are crucial to develop effective management strategies and preserve the ecological integrity of this fragile mountain ecosystem.

7.2 Alien Plants of Himalayan Hotspot

The study by Khuroo et al. (2012) identified 771 alien plant species in the Indian Himalayan Region, accounting for approximately 48% of India's total alien flora. A total of 297 naturalised alien plant species from 65 families have been recorded in the region, collectively representing about 84% of India's naturalised flora (Inderjit et al. 2018). Some of the common alien species are recorded in Table 7.1.

Among the 297 naturalised alien species in the Indian Himalayan Region, the highest numbers were found in Himachal Pradesh (232 species; 78.1%), followed by Jammu & Kashmir (192 species; 64.6%), and Uttarakhand (181 species; 60.9%). Notable invasive species (Figs. 7.1 and 7.2) documented in several states of the region include *Lantana camara*, *Ageratina adenophora*, *Parthenium hysterophorus*, and *Ageratum conyzoides*, which have proliferated extensively (Pathak et al. 2019).

In the Sikkim-Darjeeling Himalaya, 334 out of 4920 identified angiosperm species (6.78%) are non-native (Singh and Pusalkar 2020). Arunachal Pradesh, despite its larger geographical size, remains comparatively under-researched, with fewer documented native species than neighboring Bhutan (Rana et al. 2019). This uneven research focus reflects the broader floristic and climatic gradients that extend from the Eastern to the Western Himalayas, as discussed in Chap. 1 (Sect. 1.3.1.3), which significantly influence vegetation diversity and patterns of endemism.

Bhutan's exotic flora, including cultivated taxa, consists of 139 families, 545 genera, and 964 species. Of these, 626 species occur only in cultivation, while the remaining 338 species are found in the wild (spontaneous), with 131 (39%) classified as casuals, 103 (31%) as naturalised, and 101 (30%) as invasive (Dorjee et al. 2020).

In Nepal, five major alien plant species *Ageratum conyzoides*, *Ageratina adenophora*, *Lantana camara*, *Parthenium hysterophorus*, and *Erigeron karvinskianus*, are present in both agricultural lands and wild areas (Shrestha and Shrestha 2021). *Ageratum conyzoides*, *Lantana camara*, and *Parthenium hysterophorus* are well-established invasive species that have spread and impacted biodiversity across the Western to Eastern Himalayan regions of India. Other species, such as *Argemone mexicana*, *Ageratina adenophora*, *Senna tora*, and *Sapium sebiferum*, continue to proliferate in the Western Himalayas. Thus, more attention should be paid to the management of flourishing alien species, as well as invasive species that have not yet spread widely, to prevent their potential invasiveness in new areas.

Some species are confined to specific regions. For example, *Anthemis cotula* is exclusively found in the Kashmir Valley (Khuroo et al. 2008), while *Sapium sebiferum* has been observed to proliferate only in the Himachal Pradesh (Jaryan et al. 2013). *Ageratina adenophora* predominantly thrives in moist environments, such as roadsides and forests in the Garhwal Himalayas and other parts of the Eastern Himalayas. This species negatively impacts plant diversity and agricultural land by enhancing its competitive advantage through allelopathic effects, while also altering soil microbial populations (Tripathi and Yadav 1982). Similarly, *Ageratum*

Table 7.1 Common Alien plants present in agricultural land and in wild of Himalayan hotspot

Plant species	Habitat	Ecological impact	References
Ageratina adenophora	*AL*, *W*	Affect phytodiversity by negative allelopathy	Tripathi and Yadav (1982)
Ageratum conyzoides	*AL*, *W*	Reduces yield by competing with crops	Batish et al. (2011)
Argemone mexicana	*AL*, *W*	Affect germination, and chlorophyll content of other plant	Khuroo et al. (2008)
Bidens pilosa	*AL*, *W*	Produces allelochemicals and affects different agroecosystems	Ndam et al. (2025)
Cannabis sativa	*AL*, *W*	Causes soil erosion	Zheng et al. (2021)
Chenopodium album	*AL*, *W*	Threat to native biota	Singh et al. (2020)
Chromolaena odorata	*AL*, *W*	Changes soil microbial characteristics	Wei et al. (2017)
Crotalaria pallida	*AL*, *W*	Has high potential to invade a new land	Fonseca et al. (2006)
Cuscuta reflexa	*AL*, *W*	Parasitic exotic climber that derives water and nutrients from host plants	Singh et al. (2020)
Eclipta prostrata	*AL*, *W*	Alternative host of *Meloidogyne* spp. in tea plantation	Setyawati et al. (2015)
Lantana camara	*AL*, *W*	Suppress native tree seedlings growth	Shrestha and Shrestha (2021)
Mikania micrantha	*AL*, *W*	Renowned as one of the worst 100 invasive species of the world	Vijay (2015)
Parthenium hysterophorus	*AL*, *W*	Named famine weed as it causes terrible failure of crop production	Shrestha and Shrestha (2021)
Pontederia crassipes	*W*	An aquatic invasive plant that causes huge ecological problem	Datta et al. (2021)
Solanum torvum	*AL*, *W*	Invades variety of ecosystems	Cuda et al. (2002)
Synedrella nodiflora	*AL*, *W*	Causes negative allelopathic effect on seed germination of other plants	Ghayal et al. (2013)
Tithonia diversifolia	*AL*, *W*	Affect different physiological activity of different plants by producing allelochemicals	Ndam et al. (2025)
Urena lobata	*AL*, *W*	Very common invasive alien species	Singh et al. (2020)

Abbreviations: *AL* Agricultural Land, *W* Wild

conyzoides, another invasive species, has been shown to harm agricultural fields, reducing yields of important crops like tea by affecting seed germination, altering soil properties, and decreasing species richness in areas it infests (Batish et al. 2011).

Fig. 7.1 Common alien plants of Himalayan hotspot (**a**) *Parthenium hysterophorus;* (**b**) *Lantana camara;* (**c**) *Argemone mexicana*; (**d**) *Ageratum conyzoides;* (**e**) *Mikania micrantha;* (**f**) *Synedrella nodiflora*

Fig. 7.2 Common alien plants of Himalayan hotspot (**a**) *Solanum torvum;* (**b**) *Tithonia diversifolia;* (**c**) *Urena lobata*; (**d**) *Cuscuta reflexa;* (**e**) *Crotalaria pallida;* (**f**) *Chenopodium album*

7.3 Path of Introduction of Exotic Plants

The introduction of exotic plant species into the Himalayan region is closely linked to a range of anthropogenic pressures, including population growth, agricultural expansion, urbanisation, and tourism. This pattern of human-driven ecological disruption is echoed in Chap. 2 (Sect. 2.1), which highlights population pressure and unregulated development as critical threats to the region's biodiversity. These activities have created opportunities for the establishment and spread of alien species, often leading to significant ecological changes. According to Pathak et al. (2019), the region's rapid population growth and development initiatives have fostered the spread of non-native plants, exacerbating challenges associated with biodiversity conservation.

Additionally, the Himalayan region's historical context plays a crucial role in the introduction of alien species. Not only is the region susceptible to plant invasions due to its elevated vulnerability, but its colonial past and status as a key trade route across the Central Asia have also facilitated the introduction of foreign flora. The interaction between traders, colonial powers, and local communities led to the unintentional introduction of plant species, which were often transported as part of goods, food, or medicinal products. This historical trade network, combined with later colonisation efforts, provided a conduit for the exchange of plant species across vast distances (Walther et al. 2009).

The establishment of transportation corridors in more recent times, such as roads, railways, and highways, has further exacerbated the spread of alien plant species. These corridors act as highways for the dispersal of non-native species, enabling them to colonise new habitats and outcompete native vegetation. The movement of plant species along these routes is often inadvertent, with seeds hitching rides on vehicles, machinery, and even human clothing, further promoting the spread of invasive species across the region (Walther et al. 2009).

Biogeographically, most alien plant species in the Himalayan region are believed to originate from the Americas particularly North and South America home to some of the world's most notorious invaders, such as *Lantana camara* and *Ageratina adenophora* (Khuroo et al. 2012). These floristic introductions stand in sharp contrast to the native plant distributions shaped by the Indian-Eurasian plate collision and the ensuing waves of floral migration, as outlined in Chap. 1 (section "Establishment of Land Connections and Biogeographic Heterogeneity"). These species have adapted well to the varied climates and ecosystems of the Himalayas, making them formidable invaders. This trend of the American-origin species dominating the alien flora of the region is in contrast to the relatively higher concentration of native species from regions such as Africa, Europe, and Australia (Khuroo et al. 2012).

It is important to note, however, that the relatively low presence of alien species from Africa and Australia in the Himalayan region might reflect a broader issue in the classification and documentation of plant species. There are indications that some species from these regions may have been misclassified as native species, leading to a skewed understanding of the region's floristic composition. This misclassification can result in inaccurate records and complicate efforts to track the spread and impact of alien species in the region (Boivin et al. 2016).

The movement of species across different regions whether through trade, transportation, or historical processes has undeniably influenced the flora of the Himalayan region. Understanding the origins and pathways of introduction of alien species is critical for developing management strategies to mitigate their ecological impact and preserve native biodiversity. Therefore, more accurate documentation and research are needed to assess the full extent of alien plant introductions, as well as to distinguish between native and non-native species in the region.

7.4 Process of Naturalisation

The naturalisation process refers to alien species establishing self-sustaining populations in a new area, while invasive species are those that rapidly proliferate and spread, often disrupting ecosystems (Richardson et al. 2000). Alien plants employ diverse reproductive strategies, including sexual and asexual reproduction, to adapt to environmental variability in their new habitats, enhancing their fitness in fluctuating conditions (Goodwillie et al. 2005).

Invasive Alien Plant Species (IAPS) possess traits such as high growth rates, environmental adaptability, short generation times, frequent reproduction, small seeds, and extensive dispersal ability, which aid their rapid colonisation and spread (Whitney and Gabler 2008). Additionally, they often exhibit greater phenotypic plasticity than native species, allowing them to perform better under diverse conditions and outcompete native plants in disturbed areas (Walther et al. 2009). These factors contribute to the success of alien species in invading and dominating new environments.

7.5 Climate Change and Alien Plants

Climate change is driving the proliferation of invasive alien species, which often replace slower-growing native plants that require stable conditions (Sax et al. 2002). Alien plants typically exhibit rapid growth, high reproductive rates, and a greater responsiveness to elevated CO_2 levels, enabling them to migrate to higher altitudes faster than native species (Bradley 2013). As temperatures rise, species from lower elevations are expected to move upwards, increasing the potential for invasions in high-elevation regions (Simberloff 2000).

Studies in the Western Himalayas, such as by Dar and Reshi (2015), show invasive species are homogenising ecosystems, while Priyanka and Joshi (2013) predict that rising temperatures will make areas of the Western Himalayas more conducive to the spread of *Lantana camara* (Aryal et al. 2022). While high-elevation areas are less susceptible to invasions due to energy limits and low disturbance, climate change and human activity will likely lead to increased incursions of lowland species into these regions (Priyanka and Joshi 2013).

The Indian National Action Plan on Climate Change (NAPCC) recognises the Himalayan region's ecological importance but highlights its vulnerability to climate change and biological invasions (Adhikari et al. 2015; Negi et al. 2019). As temperatures rise, invasive species are expected to expand their range, altering biodiversity and ecosystem dynamics in the region.

7.6 Harmful Effects of Alien Plants

Biological invasion by non-native species is widely recognised as the second most significant threat to global biodiversity, after direct habitat degradation (McNeely 2001). Economic globalisation has facilitated the spread of species across borders, a process often referred to as "biotic globalisation" or "biological pollution." This phenomenon has led to severe ecological and economic impacts, as invasive species disrupt native ecosystems, compete with local species, and alter ecosystem functions (Gurevitch and Padilla 2004).

When plants migrate from their native habitats to new ecosystems, they are often exposed to different environmental conditions. In such environments, these species may outcompete native flora due to their superior adaptability, reproductive strategies, or lack of natural predators and competitors. This process, driven by land-use changes such as urbanisation, agriculture, and transportation corridors, results in the introduction and establishment of invasive plant species (Negi et al. 2019).

Invasive plants can have profound ecological effects, including the displacement of native species, alteration of soil and water properties, and changes in habitat structure, leading to a loss of biodiversity. Economically, invasive species can reduce crop yields, increase management costs, and harm industries reliant on natural resources, such as forestry and tourism. The broader ecological and economic consequences of biological invasions emphasise the need for effective management strategies to mitigate their impacts.

7.7 Problem to Agronomic Crop

The invasion of alien plant species has been closely linked to global economic liberalisation, beginning in the mid-twentieth century (Lowry et al. 2013). Nitrogen-fixing alien plants, in particular, can disrupt local ecosystems by releasing compounds that hinder nitrogen fixation in other species, altering nutrient availability, and influencing soil erosion and fire regimes. These species also affect hydrological cycles by modifying evapotranspiration rates, evaporation timing, water runoff, and groundwater levels. The most significant impacts occur when invasive species differ from native species in traits such as transpiration rates, foliar area index, photosynthetic biomass, rooting depth, and phenology (Charles and Dukes 2007).

The unchecked spread of invasive alien plants poses a severe threat to agriculture, potentially causing financial losses. Invasive species can reduce crop yields by up to 50–70% due to their competition with crops for nutrients, water, and light, as well as by altering soil and water conditions (Deka and Barua 2015). The failure to control the spread of these species can lead to long-term detriment to agricultural productivity and economic stability (Rai 2022).

7.8 Threat to Endemic Flora

Invasive alien plant species, influenced by both naturalisation processes and climate change, have been increasingly migrating into less disturbed areas, such as deep forests, posing significant risks to native plant species and ecosystems. These species often establish self-sustaining populations that can outcompete native plants for resources like nutrients, light, and water. Their rapid proliferation alters the structure and function of ecosystems, leading to a reduction in biodiversity. Moreover, invasive species contribute to climate change by modifying ecological processes such as carbon storage, soil composition, and water cycles, further exacerbating environmental degradation.

The Himalayan region, known for its rich biodiversity and unique ecological features, is particularly susceptible to the threat posed by invasive species. The region's vulnerability is compounded by the influx of international tourists, along with increased human activity such as agriculture, urbanisation, and infrastructure development, all of which facilitate the spread of non-native plant species. Many of these invasive species are more adaptable to varying climatic conditions, giving them a competitive advantage over native species. As a result, the native flora, particularly endemic species that are already limited in distribution, are increasingly at risk.

Despite the growing threat of invasive alien species, comprehensive research on their spread, ecological impact, and classification within the Himalayan ecosystem is still limited. While efforts have been made to document and categorise naturalised and invasive plants in the region, much of the work remains incomplete. For instance, studies by Tripathi et al. (2006) and Moktan and Das (2014) have provided some insights, but the full extent of the problem remains inadequately understood. This lack of in-depth research hampers effective management and conservation strategies, as the true impact of invasive species on native ecosystems and biodiversity is not yet fully recognised.

Many exotic plants are associated with a wide variety of pests. Presence of alien plants like *Chromolaena odorata* and *Lantana camara* are harmful for native plants, as they provide an ideal hiding spot for sucking pest like *Helopeltis theivora*. Invasive alien plant like *Ageratum conyzoides* acts as a host plant for red spider mite *Oligonychus coffeae*, which may jump onto different indigenous plants (Roy et al. 2014).

The spread of invasive species poses significant challenges because they often lack natural predators and competitors in their new environment, which allows them to proliferate rapidly. This can lead to the displacement of indigenous species, many of which may already be vulnerable due to their specialised habitat requirements.

The long-term ecological consequences of these invasions include altered habitat conditions that prevent the recovery of native species, further disrupting the delicate balance of ecosystems.

To mitigate the threat posed by invasive alien species in the Himalayas, detailed studies are urgently needed to assess their distribution, impact, and potential for further spread. Such research is crucial for developing targeted management and conservation strategies that protect the region's endemic flora and preserve its unique biodiversity in the face of growing environmental pressures. Without such efforts, the Himalayan region's biodiversity may face irreversible harm, with long-term consequences for both local ecosystems and the broader global environment.

7.9 Management

The proliferation of invasive species has become a significant global concern, especially in mountainous regions where steep terrain and limited accessibility make management challenging (McDougall et al. 2011). Once alien species establish themselves, they often proliferate rapidly, outcompeting native flora for essential resources and disrupting local ecosystems. The absence of natural predators and competitors in these areas further accelerates their spread, making control efforts more complex, especially in remote, rugged environments where climate change may also facilitate their expansion.

Effective management in mountainous regions requires early detection, continuous monitoring, and a combination of mechanical, biological, and chemical control measures. Due to the complex interactions between invasive species, native plants, and environmental factors, a comprehensive, landscape-level approach is necessary. Collaboration among local communities, researchers, and policymakers is crucial to developing strategies that mitigate the impacts of invasive species, ensuring the protection of biodiversity and the preservation of essential ecosystem services (McDougall et al. 2011).

7.9.1 *Chemical Control*

Herbicides play a significant role in managing invasive plant species, with their impact on soil microflora varying based on the chemical composition, formulation, dosage, application timing, and soil type (Barman and Varshney 2008). Specific herbicides like 2,4-D, diuron, and simazine are effective in controlling dicotyledonous weeds, while paraquat can eradicate broadleaf weeds, including species like *Ipomoea* spp., *Borreria hispida*, *Mikania micrantha*, *Eupatorium* spp., and *Polygonum* spp. Additionally, herbicides such as dalapon, paraquat and glyphosate are commonly used to control monocotyledonous weeds. It is recommended to manually remove resistant species before they flower and produce seeds to prevent further spread and resistance development.

Despite their effectiveness, herbicides remain the primary method of weed management in many regions. However, concerns about their environmental and health impacts have led to a growing interest in non-chemical control methods. These alternative approaches are increasingly popular as they offer a more sustainable solution to managing invasive species while minimising adverse effects on the environment and human health (Barman and Varshney 2008).

7.9.2 Biological Control and Allelopathy

Biological control and allelopathy are effective methods for managing invasive plant species and improving agricultural productivity. Allelopathy, where plants release chemicals to inhibit the growth of competing species, has been shown to play a role in agricultural weed management. For example, *Urochloa brizantha* inhibits the growth of *Panicum repens* in tea cultivation (Prematilake 2003). Additionally, certain rhizobacteria and fungi, such as *Fusarium* spp., can suppress seed germination of parasitic plants like *Striga hermonthica. Trichoderma* spp., which is a low-cost soil amendment, has been used effectively for organic weed control in paddy and tea fields.

Various biological agents have demonstrated potential for controlling invasive species. *Zygogramma bicolorata* and *Crocidosema lantana* are known to suppress *Parthenium hysterophorus* and *Lantana camara*, respectively. Furthermore, commercial mycoherbicides, such as *Phytophthora palmivora*, have been used to control *Araujia odorata* in grapevines. Organic herbicides such as corn gluten meal, a natural bio-herbicide, effectively affect the growth of *Solanum nigrum* (Tiwari and Kanissery 2022). Additionally, the use of fresh leaves and twigs of *Schima wallichii* at 10 t/ha has been shown to enhance rice productivity while reducing weed growth (Behera et al. 2020).

Further studies have demonstrated that natural substances can be effective in weed management. Acetic acid, clove, cinnamon, lemongrass, and organic herbicidal soaps like ammonium nonanoate and pelargonic acid can serve as natural alternatives to synthetic herbicides. Maize gluten meal (MGM) at 1–2 g/dm^2 has been found effective in controlling a variety of weeds, including *Medicago lupulina*, *Plantago lanceolata*, *Agrostis palustris*, *Portulaca oleracea*, and *Amaranthus retroflexus* (Tiwari and Kanissery 2022). Furthermore, mulching with sorghum in maize and rice fields has been reported to reduce weed growth by 26–37% and 50%, respectively (Marshall 1992). These biological and allelopathic strategies represent an environmentally friendly approach for managing invasive plant species and improving agricultural practices.

7.9.3 *Manual Control*

Manual weeding remains a safe and cost-effective method for controlling weeds, particularly in small-scale farming. Techniques such as hand and slash weeding, where tools like knives or hoes are used to remove unwanted vegetation, are commonly employed (Prematilake 2003). Additionally, modern agricultural technologies, including smart sensors, remote sensing, aerial vehicles, satellites, and the Internet of Things (IoT), have been increasingly integrated into farming to enhance crop yields, reduce waste, and lower operational costs (Monteiro and Santos 2022). These technologies help farmers detect weed infestations early and apply targeted interventions, improving efficiency and reducing the need for extensive manual labor.

Commercial mechanical weeders, such as Robocrop and Robovator, are also used for intra-row weed management, complementing manual methods by pulling or cutting weeds between crop rows. Mulching is another widely used, low-cost, and highly effective technique for weed suppression. It is especially favoured by small farmers and, when combined with other weed management strategies, can yield significant economic benefits. Innovations in physical weeding techniques, such as electroporation (Fogelberg 2000) and Carbon Dioxide laser technology (Heisel et al. 2001), show promise, but their high costs make them less feasible for widespread use in agriculture. The push-pull technique, which uses repelling "push" plants and attractive "pull" plants to control pests, offers a sustainable approach for pest and weed management (Rana et al. 2017). Furthermore, a participatory management strategy, which involves local stakeholders in weed control and ecosystem monitoring, can foster collaborative efforts to manage invasive species effectively. Educating visitors in tourism destinations about regulations and conservation efforts can also contribute to the broader management of invasive species in natural areas.

7.10 Conclusions

Rapid urbanisation, road development, and increased tourism in the Himalayan region have accelerated the spread of alien species, especially in high-altitude ecosystems. The region, known for its unique biodiversity, faces significant threats from invasive plants, which are encroaching into higher altitudes, threatening native species and agroecosystems. The extensive proliferation of these species poses a growing challenge to biodiversity conservation. Effective management strategies must be developed to address plant invasions, with an emphasis on early detection and control. Recent advancements in GIS and remote sensing technologies offer valuable tools for studying and managing invasive species in the Indian Himalayan region.

Acknowledgments The first author is thankful to the University of North Bengal for providing University Fellowship.

References

Adhikari D, Tiwary R, Barik SK (2015) Modelling hotspots for invasive alien plants in India. PLoS One 10(7):e0134665

Aryal PC, Aryal C, Bhusal K, Chapagain D, Dhamala MK, Maharjan SR, Chhetri PK (2022) Forest structure and anthropogenic disturbances regulate plant invasion in urban forests. Urban Ecosyst 25(2):367–377

Barman KK, Varshney JG (2008) Impact of herbicides on soil environment. Indian J Weed Sci 40(1&2):10–17

Barua IC, Deka J, Devi M (2013) Invasive weeds and vegetation dynamics in Assam. In: Paper presented at the weed science society conference, Bandung, Indonesia, pp 22–25

Batish DR, Kohli R, Singh H (2011) Invasive plants in the Indian subcontinent. In: Biological invasions: economic and environmental costs of alien plant, animal, and microbe species. CRC Press, Boca Raton, pp 245–257

Behera B, Nanda SS, Gulati JML, Sahoo BK, Behura AK, Mohapatra AK, Behera SD (2020) Overview of agronomic research for greening rice (*Oryza sativa*)-fallows during winter in India. Ind J Agro 65(3):243–263

Boivin NL, Zeder MA, Fuller DQ, Crowther A, Larson G, Erlandson JM et al (2016) Ecological consequences of human niche construction: Examining long-term anthropogenic shaping of global species distributions. Proc Natl Acad Sci USA 113(23):6388–6396

Bradley BA (2013) Distribution models of invasive plants over-estimate potential impact. Biol Invas 15:1417–1429

Charles H, Dukes JS (2007) Impacts of invasive species on ecosystem services. In: Biological invasions. Springer, Berlin, pp 217–237

Cuda JP, Parker PE, Coon BR (2002) Evaluation of exotic *Solanum spp*. (Solanales: Solanaceae) in Florida as host plants for the leaf beetles *Leptinotarsa defecta* & L. Texan (Coleoptera: Chrysomelidae). Flor Entomol 85(4):599–610

Dar PA, Reshi ZA (2015) Do alien plant invasions cause biotic homogenization of terrestrial ecosystems in the Kashmir Valley, India. Trop Ecol 56(1):111–123

Datta A, Maharaj S, Prabhu GN, Bhowmik D, Marino A, Akbari V, Rupavatharam S, Sujeetha JARP, Anantrao GG, Poduvattil VK, Kumar S, Kleczkowski A (2021) Monitoring the spread of water hyacinth (*Pontederia crassipes*): challenges and future developments. Front Ecol Evol 9: 631338. https://doi.org/10.3389/fevo.2021.631338

Debnath A, Paul C, Debnath B (2017) Two invasive alien giant sensitive weeds-new additions to the flora of Tripura, India. Biosci Discov 8(4):720–725

Deka J, Barua IC (2015) Weeds of tea fields and their control. In: Paper presented at National seminar on Plant Protection in Tea: Recent Advances organized by Tea Research Association, TOCKLAI, Jorhat in association with Tea Improvement Consortium, National working group in Plant Protection in Tea

Dorjee, Johnson SB, Buckmaster AJ, Downey PO (2020) Weeds in the land of Gross National Happiness: knowing what to manage by creating a baseline alien plant inventory for Bhutan. Biol Invasions 22(10):2899–2914. https://doi.org/10.1007/s10530-020-02306-5

Early R, Bradley BA, Dukes JS, Lawler JJ, Olden JD, Blumenthal DM, Gonzalez P, Grosholz ED, Ibanez I, Miler LP, Sorte CJB, Tatem AJ (2016) Global threats from invasive alien species in the twenty-first century and national response capacities. Nat Commun 7(1):12485

Everard M (2020) Can management of 'thirsty'alien trees improve water security in semi-arid India? Sci Total Environ 704:135451

Fogelberg F (2000) Research on pest control and pesticide reduction in Sweden, Denmark and The Netherlands

Fonseca R, Guimarães P, Morbiolo S, Scachetti-Pereira R, Peterson A (2006) Predicting invasive potential of smooth crotalaria (*Crotalaria pallida*) in Brazilian national parks based on African records. Weed Sci 54(3):458–463. https://doi.org/10.1614/WS-05-059R4.1

Ghayal N, Dhumal K, Deshpande N, Ruikar A, Phalgune U (2013) Phytotoxic effects of leaf leachates of an invasive weed *Synedrella nodiflora* and characterization of its allelochemical. Int J Pharm Sci Rev Res 19(1):79–86

Goodwillie C, Kalisz S, Eckert CG (2005) The evolutionary enigma of mixed mating systems in plants: occurrence, theoretical explanations, and empirical evidence. Annu Rev Ecol Evol Syst 36:47–79

Guo Q, Cade BS, Dawson W, Essl F, Kreft H, Pergl J, Kleunen MV, Weigelt P, Pyšek P (2021) Latitudinal patterns of alien plant invasions. J Biogeogr 48(2):253–262

Gurevitch J, Padilla DK (2004) Are invasive species a major cause of extinctions? Trend Ecol Evol 19(9):470–474

Haq SM, Shah AA, Yaqoob U, Hassan M (2021) Floristic quality assessment index of the Dagwan stream in Dachigam National Park of Kashmir Himalaya. Proc Natl Acad Sci India Sect B Biol Sci 91(3):657–664

Heisel T, Schou J, Christensen S, Andreasen C (2001) Cutting weeds with a CO2 laser. Weed Res 41(1):19–29

Hulme PE, Roy DB, Cunha T, Larsson TB (2009) A pan-European inventory of alien species: rationale, implementation and implications for managing biological invasions. In: Handbook of alien species in Europe, pp 1–14

Inderjit, Pergl J, van Kleunen M, Hejda M, Babu CR, Majumdar S et al (2018) Naturalized alien flora of the Indian states: biogeographic patterns, taxonomic structure and drivers of species richness. Biol Invasions 20:1625–1638

Jaryan V, Uniyal SK, Gupta RC, Singh RD (2013) Alien flora of Indian Himalayan state of Himachal Pradesh. Environ Monitor Assess 185:6129–6153

Khuroo AA, Reshi Z, Rashid I, Dar GH, Khan ZS (2008) Operational characterization of alien invasive flora and its management implications. Biodivers Conserv 17:3181–3194

Khuroo AA, Ahmad R, Mehraj G, Rashid I, Malik AH, Dar GH (2012) Diversity and distribution of alien flora in the Indian Himalayan region. Plant Divers Himal Hotspot Reg 22:567–611

Kosaka Y, Saikia B, Mingki T, Tag H, Riba T, Ando K (2010) Roadside distribution patterns of invasive alien plants along an altitudinal gradient in Arunachal Himalaya, India. Mount Res Develop 30(3):252–258

Lowry E, Rollinson EJ, Laybourn AJ, Scott TE, AielloLammens ME (2013) Biological invasions: a field synopsis, systematic review, and database of the literature. Ecol Evol:182–196. https://doi.org/10.1002/ece3.431

Maiti GG, Guha Bakshi DN (1981) Invasion of exotic weeds in West Bengal since 1903: dicotyledons and monocotyledons. J Econ Taxon Bot 2:1–21

Marshall T (1992) Weed control in organic farming systems. First Int Weed Control Congress 72(1):311–314. https://doi.org/10.2478/contagri-2023-0007

McDougall KL, Khuroo AA, Loope LL, Parks CG, Pauchard A, Reshi ZA et al (2011) Plant invasions in mountains: global lessons for better management. Mount Res Develop 31(4): 380–387

McNeely JA (2001) An introduction to human dimensions of invasive alien species. In: The great reshuffling: human dimensions of invasive alien species, pp 5–20

Moktan S, Das AP (2014) Plant species richness and phytosociological attributes of the vegetation in the cold temperate zone in Darjiling Himalaya, India. Int Res J Environ Sci 3(10):14–19

Monteiro A, Santos S (2022) Sustainable approach to weed management: the role of precision weed management. Agronom 12(1):118

Ndam LM, Toumguem FO, Juru VN, Agbor DT, Njilar RM, Fonge BA (2025) Allelopathic potential of selected invasive alien weed species and mathematical modelling of rhizospheric

soil impact of *Ageratum conyzoides* on *Phaseolus vulgaris* L. Agric Sci 16(2). https://doi.org/10.4236/as.2025.162019
Negi VS, Pathak R, Rawal RS, Bhatt ID, Sharma S (2019) Long-term ecological monitoring on forest ecosystems in Indian Himalayan region: criteria and indicator approach. Ecol Indic 102: 374–381. https://doi.org/10.1016/j.ecolind.2019.02.035
Nunez MA, Pauchard A (2010) Biological invasions in developing and developed countries: does one model fit all? Biol Invasions 12:707–714
Pathak R, Negi VS, Rawal RS, Bhatt ID (2019) Alien plant invasion in the Indian Himalayan region: state of knowledge and research priorities. Biodivers Conserv 28(12):3073–3102
Prematilake KG (2003) Weed management in tea-recent developments. Trop Agri Res Exten 6:98–107
Priyanka N, Joshi PK (2013) A review of *Lantana camara* studies in India. Int J Sci Res Publ 3(10): 1–11
Pyšek P, Pergl J, Essl F, Lenzner B, Dawson W, Kreft H et al (2017) Naturalized alien flora of the world: species diversity, taxonomic and phylogenetic patterns, geographic distribution and global hotspots of plant invasion. Preslia 89(3):203–274
Rai PK (2022) Environmental degradation by invasive alien plants in the anthropocene: challenges and prospects for sustainable restoration. Anthr Sci 1(1):5–28
Rana SS, Sharma N, Badiyala D, Kumar R (2017) Weed management in turmeric. Ind J Weed Sci 49(1):51–57. https://doi.org/10.5958/0974-8164.2017.00013.2
Rana SK, Gross K, Price TD (2019) Drivers of elevational richness peaks, evaluated for trees in the east Himalaya. Ecol 100(1):e02548
Richardson DM, Pyšek P, Rejmanek M, Barbour MG, Panetta FD, West CJ (2000) Naturalization and invasion of alien plants: concepts and definitions. Divers Distrib 6(2):93–107
Roy S, Muraleedharan N, Mukhopadhyay A (2014) The red spider mite, Oligonychus coffeae (Acari: Tetranychidae): its status, biology, ecology and management in tea plantations. Exp Appl Acarol 63:431–463
Sax DF, Gaines SD, Brown JH (2002) Species invasions exceed extinctions on islands worldwide: a comparative study of plants and birds. Am Nat 160(6):766–783
Setyawati T, Narulita S, Bahri IP, Raharjo GT (2015) A Guide Book to Invasive Alien Plant Species in indonesia. BLI KLHK, Bogor. [Indonesian]
Shrestha BB, Shrestha KK (2021) Invasions of alien plant species in Nepal: patterns and process. Invasive Alien Species 23(2):168–183
Simberloff D (2000) Global climate change and introduced species in United States forests. Sci Total Environ 262(3):253–261
Singh DK, Pusalkar PK (2020) Floristic diversity of the Indian Himalaya. In: Biodiversity of the Himalaya: Jammu and Kashmir State, pp 93–126
Singh NP, Vohra JN, Hajra PK, Singh DK (2000) Flora of India. Introductory Volume Part II. Botanical survey of India, Calcutta
Singh S, Afshan G, Rehman F, Khan SJ (2020) An initial survey of invasive alien angiosperms of district Rampur of Uttar Pradesh, India. Trend Biosci 13(16):1259–1266
Sujay YH, Sattagi HN, Patil RK (2010) Invasive alien insects and their impact on agroecosystem. Karnatka J Agri Sci 23(1):26–34
Tiwari R, Kanissery R (2022) Weed management strategies for tomato plasticulture production in Florida. Plants 11(23):3292
Tripathi RS, Yadav AS (1982) Population regulation of *Eupatorium adenophorum* Spreng. and E. *riparium* Regel: effect of population density, soil nitrogen and light intensity. Plant Soil 65: 35–49
Tripathi OP, Pandey HN, Tripathi RS (2006) Plant diversity and fragmentation of major forest types of Meghalaya, North-East India. In: Ecology, diversity and conservation of plants and ecosystems in India, pp 414–436
Vijay B (2015) Invasive plant Mikania Micrantha a major threat to biological diversity-a review. IJARSE 8354(4):5

Walther GR, Roques A, Hulme PE, Sykes MT, Pysek P, Kuhn I, Zobel M (2009) Alien species in a warmer world: risks and opportunities. Trends Ecol Evol 24(12):686–693

Wani SA, Ahmad R, Gulzar R, Rashid I, Malik AH, Khuroo AA (2022) Diversity, distribution and drivers of alien flora in the Indian Himalayan region. Global Ecol Conserv 38:e02246

Weber E, Li B (2008) Invasive alien plants in China: diversity and ecological insights. Biol Invas 10:1411–1429

Wei H, Xu J, Quan G, Zhang J, Qin Z (2017) Invasion effects of *Chromolaena odorata* on soil carbon and nitrogen fractions in a tropical savanna. Ecosphere 8(5):e01831. https://doi.org/10.1002/ecs2.1831

Whitney KD, Gabler CA (2008) Rapid evolution in introduced species,'invasive traits' and recipient communities: challenges for predicting invasive potential. Divers Distrib 14(4): 569–580

Zheng Z, Fiddes K, Yang L (2021) A narrative review on environmental impacts of cannabis cultivation. J Cannabis Res 3(1):35. https://doi.org/10.1186/s42238-021-00090-0

Chapter 8
The Knotweed Conservation Challenge: Protecting *Polygonaceae* in the Himalayan Range

Payel Paul, Sourish Dasgupta, Dipayan Das, and Monoranjan Chowdhury

Abstract The Himalayan region, renowned for its biodiversity, is home to numerous knotweed species (family *Polygonaceae*), including those with ecological and medicinal significance. However, the region faces a dual challenge: while certain native knotweed species are threatened by habitat loss, climate change, and overharvesting, invasive species are rapidly spreading and disrupting local ecosystems. Protecting knotweed species in the Himalayas demands a balanced approach that addresses both the conservation of endangered species and the management of invasive ones. This study examines the ecological role of native knotweeds, their contributions to soil stabilisation and local economies, and the growing threat posed by invasive species. Knotweeds are critical in preventing erosion in the region's rugged landscapes, yet deforestation and unsustainable harvesting practices have led to declining populations. Additionally, invasive species are outcompeting native varieties, further jeopardizing the region's biodiversity. Effective conservation efforts must integrate sustainable harvesting, ecological research, and stronger policy frameworks to safeguard the rich biodiversity, ecological balance, and cultural heritage associated with the *Polygonaceae* family in the Himalayas.

Keywords Invasive species · IUCN · Biodiversity · Habitat · Ecosystem

---

Authors Sourish Dasgupta and Dipayan Das contributed equally to this chapter.

P. Paul (✉)
Department of Botany, North Bengal St. Xavier's College, Jalpaiguri, West Bengal, India
e-mail: payel_bot@nbxc.org

S. Dasgupta
Department of Botany, North Bengal St. Xavier's College, Rajganj, West Bengal, India

D. Das
Department of Microbiology, Royal School of Biosciences, The Assam Royal Global University, Guwahati, Assam, India

M. Chowdhury
Department of Botany, University of North Bengal, Siliguri, West Bengal, India

D. Das et al. (eds.), *Plant Diversity and Conservation Strategies of the Himalayan Biodiversity Hotspot*, Topics in Biodiversity and Conservation 21,
https://doi.org/10.1007/978-3-032-21697-7_8

8.1 Introduction

The *Polygonaceae* family, commonly known as the knotweed family, encompasses over 50 genera and around 1200 species, exhibiting a wide range of morphological diversity (Hao et al. 2015). This family is predominantly composed of herbaceous plants, though it also includes shrubs, vines, and small trees. Key characteristics of *Polygonaceae* species include simple, alternate leaves with sheathing stipules (ocreae), swollen nodes, and small, often inconspicuous flowers that can be unisexual or bisexual. The fruits are typically three-angled achenes, which aid in seed dispersal (Simpson 2010).

Polygonaceae species are distributed globally, thriving in diverse habitats ranging from wetlands and coastal areas to alpine and arid region (Paul and Chowdhury 2019). This adaptability has allowed them to become significant ecological players in many ecosystems. Notable genera within the family include *Polygonum*, *Rheum*, *Rumex*, and *Fagopyrum*. Each genus contributes uniquely to both natural ecosystems and human use, particularly in the context of the Himalayan Belt.

The Himalayan Belt, stretching across parts of Nepal, India, Bhutan, China and Pakistan, is a biodiversity hotspot and a critical region for the conservation of many plant species, including those within the *Polygonaceae* family. The region's distinctive blend of climatic zones from subtropical to alpine gives rise to a rich mosaic of habitats that sustain an extraordinary diversity of flora and fauna. This distribution closely mirrors the elevational gradients in floristic diversity outlined in Chap. 3 (Sect. 3.3.2), where altitude-related shifts in climate and topography are shown to drive vegetation patterns across the Himalayan Belt.

Polygonaceae species play vital ecological roles in the Himalayan Belt. They are often pioneers in disturbed soils, contributing to soil stabilisation and preventing erosion, which is particularly important in the rugged, erosion-prone terrains of the Himalayas (Shaheen et al. 2023). Their deep roots systems help bind the soil, maintaining soil structure and fertility. Additionally, these plants support various wildlife species. The flowers of many *Polygonaceae* attract pollinators such as bees and butterflies, while the leaves and stems serve as food for herbivores, thus supporting the region's biodiversity. For instance, *Rheum nobile*, known as the noble rhubarb, is a distinctive high-altitude plant in the Himalayas. It has adapted to extreme conditions and plays a role in supporting other alpine species by providing microhabitats and stabilising the soil. Similarly, species of the genus *Fagopyrum*, including buckwheat, are cultivated both for their nutritional value and their role in supporting agricultural biodiversity (Nalinkumar and Singh 2020).

Polygonaceae species hold significant medicinal and economic value. Many species have been used traditionally in local medicine for centuries. *Rheum emodi* (Himalayan rhubarb) is known for its laxative properties and is used to treat various ailments, while *Rumex nepalensis* (Nepal dock) is utilised in traditional remedies for skin diseases and digestive issues.

The economic importance of *Polygonaceae* extends to agriculture. Buckwheat (*Fagopyrum esculentum*), for instance, is a staple crop in the Himalayan region. It provides a vital source of nutrition and income for local communities, supporting food security and economic stability (Majzoobi et al. 2023). The cultivation of buckwheat also promotes agricultural biodiversity, offering habitat and food for pollinators and other beneficial insects (Vyas et al. 2012).

The *Polygonaceae* family is deeply embedded in the cultural practices of Himalayan communities. These plants are often integral to traditional rituals, festivals, and daily practices. The knowledge of their uses and benefits is passed down through generations, contributing to the rich cultural heritage of the region. For example, various *Polygonaceae* species are used in religious ceremonies, traditional medicine, and as food sources, highlighting their importance beyond just ecological and economic roles (Uddin et al. 2014).

Despite their ecological, economic, and cultural significance, *Polygonaceae* species in the Himalayan Belt are increasingly threatened (Kandel et al. 2019). Habitat destruction driven by deforestation, agricultural expansion, and infrastructure development remains a primary concern, leading to the fragmentation and degradation of natural ecosystems. As a result, suitable habitats for these species are rapidly diminishing, disrupting the region's ecological equilibrium. While members of the *Polygonaceae* family display remarkable ecological adaptability ranging from wet meadows to rocky alpine slopes these habitats are becoming progressively more vulnerable to human-induced pressures such as road construction, tourism, and unregulated land-use practices (Paul 2020). These concerns align with those highlighted in Chap. 2 (Sect. 2.1), which identifies such anthropogenic factors as major threats to the Himalayan Biodiversity.

Climate change poses another significant threat. The Himalayan region is particularly vulnerable to the impacts of climate change, including rising temperatures, changing precipitation patterns, and increased frequency of extreme weather events (Sharma et al. 2009). These changes affect the distribution and phenology of *Polygonaceae* species, forcing them to migrate to higher altitudes and disrupting their relationships with pollinators and other species.

Overexploitation for medicinal and economic purposes also threatens *Polygonaceae* populations. Unsustainable harvesting practices, driven by high demand for traditional medicines and economic products, deplete natural populations and reduce the plants' regenerative capacity (Saravanan and Das 2024). Invasive species further exacerbate these threats by outcompeting native *Polygonaceae* for resources and altering their habitats.

The conservation of *Polygonaceae* in the Himalayan Belt is imperative for preserving the region's unique biodiversity, ecological health, and cultural heritage (Haq et al. 2024). By implementing a combination of in-situ and ex-situ conservation strategies, promoting sustainable use practices, conducting research and monitoring, and strengthening policies and legislation, the long-term survival of these vital plants can be ensured. The present review deals with the status and conservation of *Polygonaceae* in the Himalayan belt.

8.2 The *Polygonaceae* Family

The *Polygonaceae* family is classified under the order Caryophyllales. It includes several well-known genera such as *Polygonum, Rheum, Rumex*, and *Fagopyrum*. Members of this family are distinguished by their jointed stems, sheathing stipules known as ocreae, and small, often inconspicuous flowers (Berry 1935).

The leaves of *Polygonaceae* plants are usually simple and alternate, with a distinct sheathing base. The flowers are typically arranged in clusters and can be bisexual or unisexual. The fruit is generally a three-angled achene, which aids in seed dispersal (Zuo-Cheng et al. 2000).

The Himalayan region hosts a rich diversity of *Polygonaceae* species. Notable genera include *Rheum*, known for its rhubarb, and *Fagopyrum*, known for buckwheat. Other significant genera include *Polygonum* and *Rumex*, which comprise numerous species with diverse ecological roles (Ayodele and Olowokudejo 2006; Paul and Chowdhury 2016).

Some species, such as *Rheum nobile* (noble rhubarb), are adapted to high-altitude environments and are iconic plants of the Himalayan alpine zone. These species contribute to the unique flora of the region and have adapted to the challenging climatic conditions of the high Himalayas.

Polygonaceae species play vital roles in the ecosystems of the Himalayan Belt. They contribute to soil stabilisation, particularly in alpine and subalpine regions, where they prevent soil erosion (Srivastava 2014). Their flowers provide nectar and pollen for a variety of pollinators, including bees and butterflies.

Moreover, many *Polygonaceae* species serve as food sources for herbivores. Their presence in diverse habitats supports a range of wildlife, contributing to the overall biodiversity and ecological resilience of the region.

8.3 The Himalayan Belt: A Biodiversity Hotspot

The Himalayan Belt spans approximately 2400 km across five countries and includes some of the highest and most rugged mountains in the world. The region's geography is characterised by deep valleys, towering peaks, and diverse climatic zones, ranging from tropical at the base to permanent ice and snow at the highest elevations (Paul and Chowdhury 2019).

These climatic gradients create a range of habitats, from lush forests and alpine meadows to arid plateaus and glacial landscapes. The diversity of habitats supports a wide array of plant and animal species, making the Himalayas one of the most biodiverse regions on the planet.

The Himalayas are renowned for their high levels of biodiversity and endemism. The region is home to approximately 10,000 species of plants, of which nearly 4000 are endemic. This remarkable biodiversity is due to the region's varied climates, altitudes, and geological history (Kandel et al. 2019).

Endemic species, those found nowhere else on the Earth, are particularly significant as they contribute to the unique biological heritage of the Himalayas. The high levels of endemism highlight the importance of conserving this region to preserve its unique flora and fauna.

The ecosystems of the Himalayan Belt provide a wide range of services that are crucial for both nature and human communities. These services include:

- **Provisioning Services**: The region supplies resources such as water, food, medicinal plants, and raw materials.
- **Regulating Services**: Himalayan ecosystems regulate climate, water cycles, and air quality. They also play a critical role in carbon sequestration and soil stabilisation.
- **Cultural Services**: The Himalayas hold cultural, spiritual, and recreational significance for many communities. They are a source of inspiration, traditional knowledge, and ecotourism.

8.4 Significance of *Polygonaceae* in the Himalayas

Polygonaceae species are integral to the ecological health of the Himalayan Belt. They contribute to soil stabilisation, particularly in areas prone to erosion and landslides. Their deep roots systems help bind the soil, preventing erosion and maintaining soil structure (Table 8.1).

In addition to soil stabilisation, *Polygonaceae* plants support a variety of wildlife. Their flowers attract pollinators, while their leaves and stems provide food for herbivores. By supporting pollinators and herbivores, *Polygonaceae* species contribute to the overall biodiversity and ecological balance of the region (Figs. 8.1, 8.2, and 8.3). Many *Polygonaceae* species have significant medicinal properties and are used in traditional medicine.

The economic value of *Polygonaceae* species extends beyond their medicinal uses. Buckwheat (*Fagopyrum esculentum*) is an important crop in the Himalayan region, providing a source of food and income for local communities (Babu et al. 2019). The cultivation of buckwheat also supports biodiversity by providing habitats for pollinators and other wildlife.

Polygonaceae species hold cultural significance for many communities in the Himalayan Belt. They are used in traditional rituals, festivals, and medicinal practices. The knowledge of these plants and their uses is often passed down through generations, contributing to the cultural heritage of the region.

Ethnobotanical studies highlight the importance of *Polygonaceae* species in the daily lives of Himalayan communities. By conserving these plants, we also preserve the cultural practices and traditional knowledge associated with them.

Table 8.1 Significance of *Polygonaceae* in the Himalayas

Aspect	Significance	Examples	References
Ecological Role	Soil stabilisation and prevention of erosion	*Rheum nobile* (Noble rhubarb), *Rumex hastatus* (Himalayan dock)	Wangchuk (2015)
	Supporting pollinators and herbivores	*Polygonum affine* (Himalayan knotweed), *Fagopyrum esculentum* (Buckwheat)	Campbell et al. (2016)
Medicinal Value	Traditional medicine for various ailments	*Rheum emodi* (Himalayan rhubarb), *Rumex nepalensis* (Nepal dock)	Kumar et al. (2018)
Economic Value	Source of food and income through agriculture	*Fagopyrum esculentum* (Buckwheat), *Polygonum multiflorum* (Fo-Ti)	Zhang et al. (2023)
	Supporting biodiversity in agricultural systems	*Fagopyrum dibotrys* (Perennial buckwheat), *Rheum australe* (Himalayan rhubarb)	Pandith et al. (2018)
Cultural Significance	Integral to traditional rituals, festivals, and daily practices	*Polygonum* spp. used in religious ceremonies, traditional remedies involving various species	Namsa et al. (2011)
Adaptation to Extreme Conditions	Thriving in high-altitude and alpine environments, supporting other alpine species	*Rheum nobile* (Noble rhubarb), *Polygonum affine* (Himalayan knotweed)	Shaheen et al. (2023)
Genetic Diversity	Contribution to the genetic pool, important for plant breeding and resilience	Diverse species like *Fagopyrum tataricum* (Tartary buckwheat) and *Rumex acetosa* (Sorrel)	Mondal et al. (2021)
Environmental Indicators	Indicators of ecosystem health and climate change impact	Species distributions and phenology changes in *Rheum nobile* and *Polygonum bistorta* (Bistort)	Srivastava (2014)
Invasive Control	Some species help manage invasive plant populations	*Persicaria perfoliata* (Knotgrass) used in habitat restoration and control of invasive species	Kim et al. (2021)

8.5 Threats to *Polygonaceae* in the Himalayan Belt

8.5.1 Habitat Destruction

Habitat destruction is one of the most significant threats to *Polygonaceae* species in the Himalayan Belt. Deforestation, driven by agricultural expansion, logging, and infrastructure development, leads to the loss of natural habitats. This destruction reduces the availability of suitable habitats for *Polygonaceae* species and disrupts ecological balance (Table 8.2).

Urbanisation and the construction of roads, dams, and other infrastructure projects further fragment habitats, making it difficult for species to thrive. Habitat fragmentation also isolates populations, reducing genetic diversity and increasing the risk of local extinctions.

Fig. 8.1 Different taxa of *Polygonaceae* in their natural habitat of Himalayan belt (**a**) *Persicaria orientalis*, (**b**) *Rumex dentatus*, (**c**) *Polygonum viscosum*, (**d**) *Persicaria campanulata*, (**e**) *Persicaria nepalensis*, (**f**) *Persicaria tenella*, (**g**) *Rheum nobile*, (**h**) *Antigonon leptopus*

Fig. 8.2 Different taxa of *Polygonaceae* in their natural habitat of Himalayan belt (**a**) *Persicaria lapathifolia* var. *lanata*, (**b**) *Homalocladium platyclada*, (**c**) *Persicaria hydropiper*, (**d**) *Persicaria chinensis* var. *chinensis*, (**e**) *Persicaria chinensis* var. *hispida*, (**f**) *Persicaria capitata*, (**g**) *Fagopyrum dibotrys*

Fig. 8.3 Different taxa of *Polygonaceae* in their natural habitat of Himalayan belt (**a**) *Polygonum plebeium*, (**b**) *Rumex maritimus*, (**c**) *Rumex nepalensis*, (**d**) *Polygonum molle*, (**e**) *Persicaria runcinatum*, (**f**) *Persicaria glabra*

Table 8.2 *Polygonaceae* species found in the Himalayan Belt, along with their habitats, distribution, and threat levels

Species	Habitat	Distribution	Threat level	Comment	References
Aconogonon alpinum	Alpine meadows, rocky slopes	Nepal, Northern India, Bhutan	Least Concern	Thrives in high-altitude meadows and rocky slopes, not under significant threat.	Khan et al. (2020)
Bistorta affinis	Moist meadows, streambanks	Nepal, Northern India, Bhutan	Least Concern	Found in moist meadows and streambanks, generally not at risk.	Sultana et al. (2022)
Fagopyrum dibotrys	Forest edges, disturbed areas	Nepal, Northern India, Bhutan	Near Threatened	Near threatened due to habitat loss and overharvesting.	Zhang et al. (2021)
Fagopyrum esculentum	Cultivated fields, disturbed areas	Nepal, Northern India, Bhutan	Least Concern (Cultivated)	Widely cultivated for its seeds (buckwheat), not threatened due to agricultural propagation.	Campbell et al. (1997)
Persicaria amplexicaulis	Moist slopes, riverbanks	Nepal, Northern India, Bhutan	Least Concern	Grows in moist slopes and riverbanks, abundant and not currently threatened.	Srivastava (2014)
Persicaria capitata	Roadsides, waste grounds, moist areas	Nepal, Northern India, Bhutan	Least Concern	Grows in moist, disturbed areas and is not threatened.	Singhurst et al. (2013)
Persicaria hydropiperoides	Wetlands, marshes, riverbanks	Nepal, Northern India, Bhutan	Least Concern	Common in wetlands and riverbanks, with no immediate threats.	Keskin and Severoğlu (2021)
Polygonum aviculare	Roadsides, waste areas, grasslands	Widespread in Himalayas	Least Concern	A versatile species common in disturbed areas and roadsides, not currently threatened.	Choudhury et al. (2011)
Polygonum cuspidatum	Roadsides, forest edges, disturbed areas	Nepal, Northern India, Bhutan Tibet	Invasive	An invasive species that thrives in disturbed areas, turbed areas, posing a threat	Xiao et al. (2002)

(continued)

Table 8.2 (continued)

Species	Habitat	Distribution	Threat level	Comment	References
				to native biodiversity.	
Polygonum hydropiper	Wetlands, marshes, moist areas	Nepal, Northern India, Bhutan, Tibet	Least Concern	Prefers wetlands and marshes, with a stable population.	Paul and Chowdhury (2019)
Polygonum plebeium	Moist, disturbed soils	Nepal, Northern India, Bhutan	Least Concern	Prefers moist, disturbed soils, with a stable population.	Nayak et al. (2024)
Polygonum viviparum	High-altitude grasslands, alpine meadows	Nepal, Northern India, Bhutan	Least Concern	Commonly found in alpine meadows, this species is resilient and widespread, classified as least concern	Brandt et al. (2013)
Rheum emodi	Alpine and sub-alpine zones, rocky areas	Nepal, Northern India, Bhutan	Vulnerable	Found in rocky alpine zones, this species faces threats from overharvesting for medicinal use.	Singh and Chaturvedi (2018)
Rheum nobile	Alpine meadows, rocky slopes	Nepal, Sikkim (India), Bhutan	Vulnerable	It is found in high-altitude areas and is considered vulnerable due to habitat loss and climate change	Pandith et al. (2018)
Rumex acetosella	Moist meadows, forest clearings	Nepal, Northern India, Bhutan	Least Concern	Found in moist meadows and clearings, not currently at risk.	Korpelainen and Pietiläinen (2020)
Rumex nepalensis	Moist meadows, forest clearings	Nepal, India, Bhutan, Tibet	Least Concern	Thrives in moist meadows and forest clearings; widely distributed and of least concern.	Kumar et al. (2018)

8.5.2 Climate Change

Climate change poses a major threat to the biodiversity of the Himalayan Belt, including *Polygonaceae* species. Rising temperatures, changing precipitation patterns, and the increased frequency of extreme weather events impact the distribution and phenology of plants (Pérez-Ramos et al. 2020).

As the climate warms, some species may be forced to migrate to higher altitudes in search of suitable conditions. However, the availability of suitable habitats at higher elevations is limited, leading to increased competition and potential declines in populations. Climate change also affects the timing of flowering and fruiting, disrupting the relationships between plants and their pollinators (Geissler et al. 2023).

8.5.3 Overexploitation

The overexploitation of *Polygonaceae* species for medicinal, economic, and other purposes threatens their survival. Unsustainable harvesting practices, driven by high demand for traditional medicines and economic products, deplete natural populations and reduce the regenerative capacity of these plants.

The lack of regulation and enforcement further exacerbates the problem, allowing overharvesting to continue unchecked. Overexploitation not only threatens the survival of *Polygonaceae* species but also undermines the livelihoods of local communities that depend on these plants.

8.5.4 Invasive Species

Invasive species pose a significant threat to the native biodiversity of the Himalayan Belt. Non-native plants and animals can outcompete native species for resources, alter habitats, and disrupt ecological processes. Invasive plants, such as *Lantana camara* and *Ageratina adenophora*, compete with native *Polygonaceae* species for light, water, and nutrients (Chaudhary et al. 2021). These invasions can lead to declines in native plant populations and reduce the overall biodiversity of the region. Another particularly aggressive invader is *Reynoutria japonica* (Japanese knotweed), a member of the *Polygonaceae* family, which has been increasingly reported in disturbed and riparian habitats across parts of the Himalayan foothills. Known for its rapid clonal spread and dense monocultural stands, *R. japonica* significantly alters soil chemistry and outcompetes native vegetation, thereby diminishing floral diversity and negatively impacting associated fauna. Its robust rhizome network makes eradication extremely challenging, compounding the ecological threats already posed by other invasive species in the region (Sharma and Thapa 2018).

8.6 Conservation Strategies and Efforts

8.6.1 In-Situ Conservation

In-situ conservation involves protecting species in their natural habitats. This approach is essential for maintaining the ecological interactions and evolutionary processes that support biodiversity (Mahanayak 2024).

Protected areas, such as national parks and wildlife reserves, play a crucial role in in-situ conservation. Effective management of these areas, including habitat restoration and anti-poaching measures, helps safeguard *Polygonaceae* species and their habitats (Table 8.3).

Table 8.3 *Polygonaceae* species found in the Himalayan Belt along threat level, reasons for threat, and conservation strategies

Species	Threat level	Reason for threat	Conservation strategy	References
Fagopyrum esculentum (Buckwheat)	Moderate	Habitat loss due to agricultural expansion	Sustainable agriculture, in-situ conservation, promotion of traditional farming methods	Luitel et al. (2021)
Persicaria amplexicaulis (Red Bistort)	Moderate	Habitat loss, overharvesting for ornamental use	Sustainable harvesting, habitat protection, in-situ conservation	Srivastava (2014)
Polygonum affine (Himalayan Knotweed)	High	Overharvesting for ornamental use, habitat loss	Ex-situ conservation, cultivation in botanical gardens, habitat protection	Negi et al. (2017)
Polygonum aviculare (Common Knotgrass)	Moderate	Habitat disturbance, competition with invasive species	Habitat restoration, control of invasive species, in-situ conservation	Costea and Tardif (2005)
Polygonum cuspidatum (Japanese Knotweed)	High (invasive)	Habitat invasion, competition with native species; Overharvesting for medicine	Invasive species control, habitat restoration	Ke et al. (2023)
Polygonum hydropiper (Water Pepper)	Moderate	Habitat degradation, pollution	Wetland protection, pollution control, in-situ conservation	Mihaylova et al. (2019)
Polygonum viviparum (Alpine Bistort)	Moderate	Habitat loss, climate change	In-situ conservation, monitoring climate impacts, habitat restoration	Mundra et al. (2016)
Rheum emodi	High	Overharvesting for medicinal use	In-situ conservation, sustainable harvesting practices, cultivation programmes	Zargar et al. (2011)

(continued)

Table 8.3 (continued)

Species	Threat level	Reason for threat	Conservation strategy	References
Rheum nobile (Noble Rhubarb)	High	Climate change, overharvesting	In-situ conservation, ex-situ conservation, monitoring climate impacts	Feng et al. (2023)
Rumex nepalensis	Low	Overharvesting for traditional medicine	In-situ conservation, sustainable harvesting practices	Kumar et al. (2018)

Community-based conservation initiatives also contribute to in-situ conservation (Esmail et al. 2023). Engaging local communities in conservation efforts, providing them with incentives, and incorporating traditional knowledge into management practices are key to the success of these initiatives.

8.6.2 Ex-Situ Conservation

Ex-situ conservation involves preserving species outside their natural habitats. This approach is complementary to in-situ conservation and provides a safety net for species at risk of extinction (Lee 2023). Botanical gardens, seed banks, and tissue culture laboratories are important ex-situ conservation tools. They provide facilities for propagating and preserving *Polygonaceae* species, ensuring the availability of genetic material for future restoration efforts. Ex-situ conservation also supports research and education, raising awareness about the importance of *Polygonaceae* species and the need for their conservation.

8.6.3 Restoration Ecology

Restoration ecology focuses on rehabilitating degraded ecosystems and restoring their ecological functions. Techniques such as reforestation, soil stabilisation, and the reintroduction of native species are employed to restore habitats and support biodiversity (Liu et al. 2024).

Successful restoration projects in the Himalayan Belt demonstrate the potential for recovering degraded landscapes. Case studies of restored habitats highlight the importance of collaboration between scientists, local communities, and policymakers in achieving conservation goals.

8.6.4 Policy and Legislation

Effective conservation policies and legislation are essential for protecting biodiversity and ensuring sustainable use of natural resources. National and international frameworks, such as the Convention on Biological Diversity (CBD) and national biodiversity action plans, provide guidelines for conservation efforts. Government agencies, non-governmental organisations, and local communities play critical roles in implementing and enforcing conservation policies. Strengthening legal frameworks, improving enforcement mechanisms, and promoting cross-border cooperation are key to addressing the threats facing *Polygonaceae* species in the Himalayan Belt.

8.6.5 Research and Monitoring

Research is fundamental to understanding the biology, ecology, and conservation needs of *Polygonaceae* species. Scientific studies provide insights into species distribution, population dynamics, genetic diversity, and ecological interactions. Identifying knowledge gaps and prioritising research efforts are essential for effective conservation planning. Research also supports the development of sustainable management practices and informs policy decisions.

Long-term ecological monitoring is crucial for assessing the health of ecosystems and the effectiveness of conservation measures. Monitoring programmes track changes in species populations, habitat conditions, and environmental factors over time. The use of technology, such as remote sensing, geographic information systems (GIS), and DNA barcoding, enhances monitoring efforts. These tools provide valuable data for conservation planning and decision-making.

Citizen science involves the participation of local communities and volunteers in data collection and monitoring activities. Engaging citizens in conservation efforts not only increases the scope of monitoring programmes but also raises awareness and fosters a sense of stewardship. Successful citizen science projects in the Himalayan Belt demonstrate the potential for community involvement in biodiversity conservation. These projects empower local communities and contribute to the collection of valuable data for conservation research.

8.7 Sustainable Utilisation

8.7.1 Sustainable Harvesting Practices

Promoting sustainable harvesting practices is essential for ensuring the long-term survival of *Polygonaceae* species and the livelihoods of local communities. Guidelines and techniques for sustainable use include:

- Rotational harvesting to allow for plant regeneration.
- Harvesting only a portion of the plant to minimise impact.
- Cultivating *Polygonaceae* species to reduce pressure on wild populations.
- Training programmes and capacity-building initiatives support the adoption of sustainable practices by local communities.

8.7.2 Alternative Livelihoods

Providing alternative sources of income reduces the reliance on harvesting wild *Polygonaceae* species. Ecotourism, agroforestry, and the cultivation of medicinal plants are potential alternative livelihoods that support conservation goals.

Promoting ecotourism, in particular, offers economic benefits while raising awareness about the importance of biodiversity conservation. Sustainable tourism initiatives showcase the natural beauty and cultural heritage of the Himalayan Belt, attracting visitors and generating revenue for local communities.

8.7.3 Education and Awareness

Raising awareness about the importance of *Polygonaceae* species and the threats they face is crucial for garnering support for conservation efforts. Educational programmes, outreach initiatives, and public awareness campaigns play vital roles in promoting conservation.

Schools, universities, and community organisations are key partners in education and awareness efforts. By integrating conservation topics into curricula and organising community events, these institutions help build a culture of conservation and stewardship.

8.8 Future Directions

Future research and conservation efforts should focus on filling knowledge gaps, enhancing monitoring programmes, and promoting sustainable use practices. International cooperation and collaboration between scientists, policymakers, and local communities are essential for addressing the complex challenges facing the Himalayan Belt.

8.9 Call to Action

Conserving the *Polygonaceae* family in the Himalayan Belt requires collective effort and commitment from all stakeholders. Governments, non-governmental organisations, researchers, and local communities must work together to implement effective conservation strategies and ensure the long-term survival of these important plants.

By taking action now, we can safeguard the biodiversity and ecological health of the Himalayan Belt for future generations. Let us recognise the value of *Polygonaceae* species and commit to their conservation, ensuring that these vital plants continue to thrive in the unique and diverse landscapes of the Himalayas.

8.10 Conclusions

The conservation of *Polygonaceae* in the Himalayan Belt is essential for maintaining the region's ecological balance, preserving biodiversity, and supporting the livelihoods of local communities. The diverse roles of *Polygonaceae* species in ecosystem functioning, traditional medicine, and cultural practices highlight their significance.

The Himalayan Belt faces numerous threats, including habitat destruction, climate change, overexploitation, and invasive species. Addressing these threats requires a multifaceted approach that includes in-situ and ex-situ conservation, habitat restoration, effective policies, and community engagement.

Acknowledgements P.P. and S.D.G would like to thank North Bengal St. Xavier's College, D.D. would like to thank The Assam Royal Global University, and M.C. would like to thank University of North Bengal for continuous support and motivation and for providing the necessary infrastructure to conduct the present review work.

References

Ayodele AE, Olowokudejo JD (2006) The family Polygonaceae in West Africa: taxonomic significance of leaf epidermal characters. South Afr J Bot 72:442–459. https://doi.org/10.1016/j.sajb.2005.12.009

Babu S, Yadav G, Singh R, Avasthe R, Das A, Mohapatra K, Tahashildar M, Kumar K, Thoithoi M, Devi MT, Rana D, Pande P, Prakash N (2019) Production technology and multifarious uses of buckwheat *(Fagopyrum spp.)*: a review. Indian J Agron 63:415–427

Berry EW (1935) Families of Flowering Plants. II. Monocotyledons. By J. Hutchinson. xiii + 243 pp. 107 figs. Macmillan and Company. London. $6.00. Science 82:196–197. https://doi.org/10.1126/science.82.2122.196

Brandt JS, Haynes MA, Kuemmerle T, Waller DM, Radeloff VC (2013) Regime shift on the roof of the world: Alpine meadows converting to shrublands in the southern Himalayas. Biol Conserv 158:116–127. https://doi.org/10.1016/j.biocon.2012.07.026

Campbell CG, Heller J, Engels J (1997) Buckwheat. *Fagopyrum esculentum* Moench. IPK and IPGRI

Campbell J, Irvin A, Irvin H, Stanley-Stahr C, Ellis J (2016) Insect visitors to flowering Buckwheat, *Fagopyrum esculentum* (Polygonales: Polygonaceae), in North-Central Florida. Fla Entomol 99:264–268. https://doi.org/10.1653/024.099.0216

Chaudhary A, Sarkar MS, Adhikari BS, Rawat GS (2021) *Ageratina adenophora* and *Lantana camara* in Kailash Sacred Landscape, India: current distribution and future climatic scenarios through modeling. PLoS One 16:e0239690. https://doi.org/10.1371/journal.pone.0239690

Choudhary RK, Oh S, Lee J (2011) An ethnomedicinal inventory of knotweeds of Indian Himalaya. J Med Plant Res 5:2095–2103

Costea M, Tardif F (2005) The biology of Canadian weeds. 131. *Polygonum aviculare* L. Can J Plant Sci 85:481–506. https://doi.org/10.4141/P03-187

Esmail N, McPherson JM, Abulu L, Amend T, Amit R, Bhatia S, Bikaba D, Brichieri-Colombi TA, Brown J, Buschman V, Fabinyi M, Farhadinia M, Ghayoumi R, Hay-Edie T, Horigue V, Jungblut V, Jupiter S, Keane A, Macdonald DW, Mahajan SL, McVey A, Moehrenschlager A, Nelson F, Noshirwani MM, Ntiamoa-Baidu Y, Postigo JL, Rakotondrazafy V, Rao M, Roe D, Sierra Huelsz JA, Stolton S, Tawake A, Wintle B (2023) What's on the horizon for community-based conservation? Emerging threats and opportunities. Trends Ecol Evol 38:666–680. https://doi.org/10.1016/j.tree.2023.02.008

Feng T, Pucker B, Kuang T, Song B, Yang Y, Lin N, Zhang H, Moore MJ, Brockington SF, Wang Q, Deng T, Wang H, Sun H (2023) The genome of the glasshouse plant noble rhubarb (*Rheum nobile*) provides a window into alpine adaptation. Commun Biol 6:706. https://doi.org/10.1038/s42003-023-05044-1

Geissler C, Davidson A, Niesenbaum RA (2023) The influence of climate warming on flowering phenology in relation to historical annual and seasonal temperatures and plant functional traits. PeerJ 11:e15188. https://doi.org/10.7717/peerj.15188

Hao DC, Gu X-J, Xiao PG (2015) 12 - Phytochemical and biological research of Polygoneae medicinal resources. In: Hao DC, Gu X-J, Xiao PG (eds) Medicinal plants. Woodhead Publishing, pp 465–529. https://doi.org/10.1016/B978-0-08-100085-4.00012-8

Haq SM, Khoja AA, Waheed M, Pieroni A, Siddiqui MH, Bussmann RW (2024) Plant cultural indicators of forest resources from the Himalayan high mountains: implications for improving agricultural resilience, subsistence, and forest restoration. J Ethnobiol Ethnomed 20:44. https://doi.org/10.1186/s13002-024-00685-w

Kandel P, Chettri N, Chaudhary RP, Badola HK, Gaira KS, Wangchuk S, Bidha N, Uprety Y, Sharma E (2019) Plant diversity of the Kangchenjunga Landscape, Eastern Himalayas. Plant Divers 41:153–165. https://doi.org/10.1016/j.pld.2019.04.006

Ke J, Li M-T, Xu S, Ma J, Liu M-Y, Han Y (2023) Advances for pharmacological activities of *Polygonum cuspidatum* - a review. Pharm Biol 61:177–188. https://doi.org/10.1080/13880209.2022.2158349

Keskin M, Severoğlu Z (2021) A new species as a member of the Flora of Turkey: *Persicaria hydropiperoides* (Polygonaceae). Front Life Sci Relat Technol 2:85–91. https://doi.org/10.51753/flsrt.951412

Khan M, Shabbir M, Saqib Z, Gilani S, Jogezai N-U, Kiyani M, Arshad M (2020) Investigation of polyphenol profile, antioxidant activity and hepatoprotective potential of *Aconogonon alpinum* (All.) Schur roots. Open Chem 18:516–536. https://doi.org/10.1515/chem-2020-0062

Kim J, Huebner C, Park Y-L (2021) Plant species composition and interactions within communities invaded by *Persicaria perfoliata* (Polygonaceae). Northeast Nat 28:340–356. https://doi.org/10.1656/045.028.0308

Korpelainen H, Pietiläinen M (2020) Sorrel (*Rumex acetosa* L.): not only a weed but a promising vegetable and medicinal plant. Bot Rev 86:234–236. https://doi.org/10.1007/s12229-020-09225-z

Kumar V, Shaikh S, Barve P, Shriram V (2018) A critical review on Nepal Dock (*Rumex nepalensis*): a tropical herb with immense medicinal importance. Asian Pac J Trop Med 11: 405. https://doi.org/10.4103/1995-7645.237184

Lee H (2023) Ex situ conservation efforts for plant diversity protection with a focus on seeds. The Open Agri J 17:1–8. https://doi.org/10.2174/18743315-v17-230822-2023-15
Liu J, Dou Y, Chen H (2024) Stepwise ecological restoration: a framework for improving restoration outcomes. Geogr Sustain 5:160–166. https://doi.org/10.1016/j.geosus.2024.02.003
Luitel D, Siwakoti M, Joshi M, Rangaswami M, Jha P (2021) Potential suitable habitat of Buckwheat (*Fagopyrum* spp) under the climate change scenarios in Nepal. J Crop Sci Biotech 24:401–410. https://doi.org/10.1007/s12892-021-00089-2
Mahanayak B (2024) Ex-situ and in-situ conservation of wild life. World J Biol Pharm Health Sci 18:277–282. https://doi.org/10.30574/wjbphs.2024.18.3.0371
Majzoobi M, Jafarzadeh S, Teimouri S, Ghasemlou M, Hadidi M, Brennan CS (2023) The role of ancient grains in alleviating hunger and malnutrition. Foods 12:2213. https://doi.org/10.3390/foods12112213
Mihaylova D, Vrancheva R, Ivanov I, Popova A (2019) Phytochemical profile and antioxidant activity of water pepper (*Polygonum hydropiper* L.). J Microbiol Biotech Food Sci 8:1205–1208. https://doi.org/10.15414/jmbfs.2019.8.5.1205-1208
Mondal DS, Bhar K, Sahoo SK, Seru G, Ashfaquddin M, Pradhan N, Anjum M, Molla S (2021) Silver hull buckwheat (*Fagopyrum esculentum* Moench) is a part of nature that offers best health and honour. J Complement Altern Med Res 15:22–52. https://doi.org/10.9734/JOCAMR/2021/v15i230262
Mundra S, Bahram M, Eidesen P (2016) Alpine bistort (*Bistorta vivipara*) in edge habitat associates with fewer but distinct ectomycorrhizal fungal species: a comparative study of three contrasting soil environments in Svalbard. Mycorrhiza 26:809–818. https://doi.org/10.1007/s00572-016-0716-1
Nalinkumar A, Singh P (2020) An overview of buckwheat (*Fagopyrum* spp)-an underutilized crop in India-nutritional value and health benefits. Int J Med Res Health Sci 9:39–44
Namsa ND, Mandal M, Tangjang S, Mandal SC (2011) Ethnobotany of the Monpa ethnic group at Arunachal Pradesh, India. J Ethnobiol Ethnomed 7:31. https://doi.org/10.1186/1746-4269-7-31
Nayak S, Kar DM, Choudhury SK, Dalai MK (2024) An update on medicinal importance of the plant: *Polygonum plebeium* R. Br. Int J Pharm Investig 14:273–288. https://doi.org/10.5530/ijpi.14.2.35
Negi V, Maikhuri R, Maletha A (2017) *Polygonum polystachyum*: peril to biodiversity of the alpine ecosystem, Western Himalaya, India. Curr Sci 113:2249–2251
Pandith SA, Dar RA, Lattoo SK, Shah MA, Reshi ZA (2018) *Rheum australe*, an endangered high-value medicinal herb of North Western Himalayas: a review of its botany, ethnomedical uses, phytochemistry and pharmacology. Phytochem Rev 17:573–609. https://doi.org/10.1007/s11101-018-9551-7
Paul P (2020) Studies on systematics of some members of Polygonaceae A. Jussieu in West Bengal, India, Ph.D Thesis, University of North Bengal, India
Paul P, Chowdhury M (2016) *Persicaria chinensis* (Linnaeus) H. Gross var. hispida (Hooker f.) Kantachot (Polygonaceae): a new distributional record for West Bengal, India. Asian J Biol Life Sci 5:260–262
Paul P, Chowdhury M (2019) Diversity of members of polygonaceae from West Bengal, India. Plant Arch 19:157–164
Pérez-Ramos IM, Cambrollé J, Hidalgo-Galvez MD, Matías L, Montero-Ramírez A, Santolaya S, Godoy Ó (2020) Phenological responses to climate change in communities of plants species with contrasting functional strategies. Environ Exp Bot 170:103852. https://doi.org/10.1016/j.envexpbot.2019.103852
Saravanan R, Das M (2024) Medicinal plants industry in India: challenges, opportunities and sustainability. Med Plants Int J Phytomedicines Relat Ind 16:1–14. https://doi.org/10.5958/0975-6892.2024.00001.7
Shaheen H, Aziz S, Nasar S, Waheed M, Manzoor M, Siddiqui MH, Alamri S, Haq SM, Bussmann RW (2023) Distribution patterns of alpine flora for long-term monitoring of global change along a wide elevational gradient in the Western Himalayas. Glob Ecol Conserv 48:e02702. https://doi.org/10.1016/j.gecco.2023.e02702

Sharma P, Thapa S (2018) Emergence and ecological impact of Reynoutria japonica in the central Himalayan region: a rising concern. J Mt Ecol 13:45–52

Sharma E, Chettri N, Eriksson M, Mool PK, Jing F, Shrestha AB, Tse-ring K (2009) Climate change impacts and vulnerability in the Eastern Himalayas. International Centre for Integrated Mountain Development (ICIMOD), Kathmandu, Nepal. https://doi.org/10.53055/ICIMOD.497

Simpson MG (2010) Diversity and classification of flowering plants: Eudicots. In: Simpson MG (ed) Plant systematics, 2nd edn. Academic, San Diego, pp 275–448. https://doi.org/10.1016/B978-0-12-374380-0.50008-7

Singh R, Chaturvedi P (2018) *Rheum emodi* Wall ex. meissn (Indian Rhubarb): highly endangered medicinal herb. J Med Plants Stud 5:13–16

Singhurst JR, Mink JN, Holmes WC (2013) *Persicaria Capitata* (Polygonaceae) naturalized in texas. Phytoneuron 51:1–3

Srivastava R (2014) Family Polygonaceae in India. Int J Plant Sci 3:2319–3824

Sultana K, Ahmad B, Ahmad Z, Rauf A, Al-Awthan Y, Bahattab O, Bawazeer S, Khattak A, Hassanien M (2022) Effect of *Bistorta affinis* (D. Don) and *Malcolmia cabulica* (Boiss) essential oils on human and foodborne pathogens, insects and termites. J Pure Appl Microbiol 16:2747–2755. https://doi.org/10.22207/JPAM.16.4.45

Uddin K, Rahman AHM, Islam A (2014) Taxonomy and Traditional Medicine Practices of Polygonaceae (Smartweed) Family at Rajshahi, Bangladesh. Int J Adv Res 2:459–469

Vyas GK, Kumar V, Sharma R, Sharma RA, Sharma S, Singh JP, Kumar S (2012) Chemical and genetic diversity among some wild stands of *Calligonum polygonoides* (Polygonaceae) from the Thar Desert of Rajasthan. Rev Biol Trop 60:1097–1108

Wangchuk K (2015) Himalayan dock (*Rumex nepalensis*): the flip side of obnoxious weed. J Anim Sci Technol 57:34. https://doi.org/10.1186/s40781-015-0067-z

Xiao K, Xuan L, Xu Y, Bai D, Zhong D (2002) Constituents from *Polygonum cuspidatum*. Chem Pharm Bull (Tokyo) 50:605–608. https://doi.org/10.1248/cpb.50.605

Zargar BA, Masoodi MH, Ahmed B, Ganie SA (2011) Phytoconstituents and therapeutic uses of *Rheum emodi* wall. ex Meissn. Food Chem 128:585–589. https://doi.org/10.1016/j.foodchem.2011.03.083

Zhang L, He Y, Sheng F, Hu Y-F, Song Y, Li W, Chen J, Zhang J, Zou L (2021) Towards a better understanding of *Fagopyrum dibotrys*: a systematic review. Chin Med 16:89. https://doi.org/10.1186/s13020-021-00498-z

Zhang Z, Fan S, Duncan GJ, Morris A, Henderson D, Morrice P, Russell WR, Duncan SH, Neacsu M (2023) Buckwheat (*Fagopyrum esculentum*) hulls are a rich source of fermentable dietary fibre and bioactive phytochemicals. Int J Mol Sci 24:16310. https://doi.org/10.3390/ijms242216310

Zuo-Cheng Z, Ming-De Z, Ding-Ze LUO, Guo-Kun S, Xin HOU (2000) Morphological characteristics of fruits of *Fagopyrum* (Polygonaceae) from China. J Syst Evol 38:486

Chapter 9
Status of Endemic Flora of the Himalayan Hotspot

Shouvik Pramanik, Kingsuk Sarkar, Anupam Kundu, Biswajit Paul, Soumitra Halder, and Nikhilesh Joardar

Abstract The Himalayan region is a vital biodiversity hotspot renowned for its diverse ecosystems and endemic flora. This chapter examines the current status of indigenous plant species in this critical area, highlighting their ecological significance and the severe threats they face. With habitats ranging from subtropical woods to alpine meadows, the Himalayas support unique species like *Trillium govanianum* and *Taxus wallichiana*, crucial for ecological balance and human needs. However, these plants are increasingly endangered due to rapid urbanisation, infrastructure expansion, unsustainable agriculture, and climate change. The chapter discusses how habitat loss, over-exploitation, invasive species, and illegal trade affect plant diversity. It also reviews local, national, and international conservation strategies emphasising the urgent need to sustain these valuable species properly. The chapter underlines the critical need for effective, comprehensive action in preserving the Himalayan flora by looking deeper into the challenges as well as the ongoing measures of conservation.

Keywords Endemic Flora · Himalayan hotspot · Conservation · Medicinal plant · Diversity

---

Supplementary Information The online version contains supplementary material available at https://doi.org/10.1007/978-3-032-21697-7_9.

S. Pramanik · A. Kundu · S. Halder
Department of Botany, Visva-Bharati, Santiniketan, West Bengal, India

K. Sarkar
Regional Ayurveda Research Institute, Gangtok, India

B. Paul
Botany Section, Central Ayurveda Research Institute, CCRAS, Ministry of AYUSH, Guwahati, Assam, India

N. Joardar (✉)
Department of Microbiology and Immunology, The University of North Carolina at Chapel Hill, School of Medicine, Chapel Hill, NC, USA

Department of Bioengineering, School of Engineering and Technology, Dr. Vishwanath Karad MIT World Peace University, Pune, Maharashtra, India

D. Das et al. (eds.), *Plant Diversity and Conservation Strategies of the Himalayan Biodiversity Hotspot*, Topics in Biodiversity and Conservation 21,
https://doi.org/10.1007/978-3-032-21697-7_9

9.1 Introduction

The Himalayan region, a vast and awe-inspiring mountain range stretching around 2400 km across eight countries, is the world's most vital biodiversity hotspots (Chap. 1, Sect. 1.1). Renowned for its striking landscapes and towering peaks like Mount Everest, it is home to an array of extraordinary flora and fauna (Rana and Rawat 2017). The region is home to a vast array of ecosystems, ranging from subtropical woods to alpine meadows, supporting an incredibly rich assortment of endemic species that have evolved in isolation over millennia (Lepcha 2010; Morgan 2019). This region, recognised as one of the world's 36 biodiversity hotspots by Conservation International, is home to numerous endemic species. However, this unique biodiversity is severely threatened, risking the region's ecological treasures (Chettri et al. 2008). The ecological stratification and altitudinal diversity discussed here are well aligned with foundational insights presented in Chap. 1 (section "Altitudinal and Climatic Variations") on altitudinal and climatic variations. The flora of the Himalayan region, in particular, plays an important role in maintaining the ecological balance, providing essential services such as food, medicine, and shelter to local communities (Joshi and Joshi 2019; Negi 2022). Research on plant species richness in the Himalayas highlights varying patterns influenced by elevation, endemism, and environmental factors. A study of 2781 angiosperms in the Eastern Himalayas shows a hump-shaped richness pattern, with endemics peaking at higher elevations. In the Indian Himalayan Region (IHR), 1076 endemic plant taxa are concentrated in mid-elevation zones, with 31% of endemic-rich areas unprotected. A comprehensive flora analysis identifies 10,503 species, but over 1000 remain taxonomically unresolved (Khan et al. 2021; Manish et al. 2017; Rana and Rawat 2017). Endemic plant species in the Himalayas, including *Trillium govanianum*, *Rheum tibeticum*, *Arnebia euchroma*, *Amentotaxus assamica*, *Gymnocladus assamicus*, *Ilex venulosa*, and *Pittosporum eriocarpum*, represent the region's unique botanical wealth (Shivali and Namrata 2022; Tiwari et al. 2024).

Endemic species are vital for biodiversity conservation, significantly contributing to the unique biodiversity of a region and playing key roles in maintaining ecosystem balance. They are integral to essential ecological processes, such as nutrient cycling, pollination, and the food webs within their native habitats. Furthermore, endemic species are of extreme importance for studying the stages of evolutionary processes and with their attunement to specific environmental conditions, they often develop specific traits. They are very good indicators of the health of an ecosystem and changes in the environment because they are sensitive to changes in their habitat. Many of these endemic species also hold cultural importance for the indigenous people, and economic benefits are derived from tourism and other sustainable farming and medicinal practices. They are also very valuable subjects in scientific research, providing insights into the mechanisms of adaptation, speciation, and conservation of species in their habitats (Fig. 9.1). Despite its ecological and socio-economic importance, the endemic flora of the Himalayan hotspot is facing unprecedented challenges. A combination of human activities and natural factors

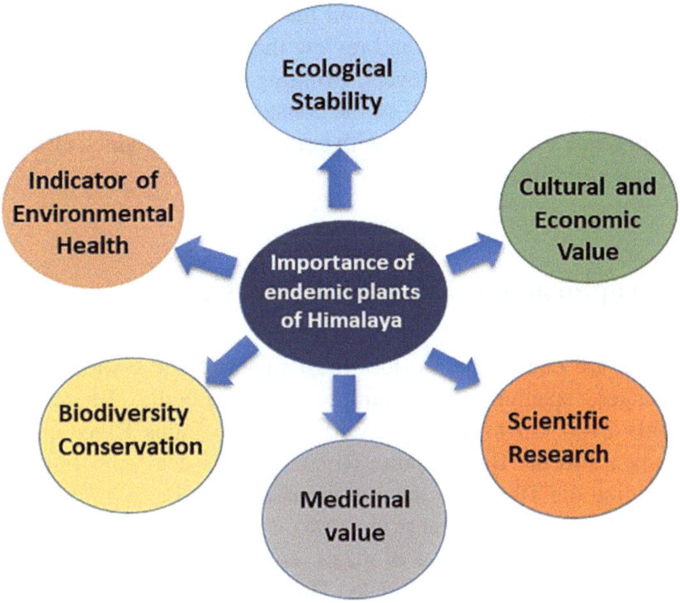

Fig. 9.1 Importance of endemic plants of the Himalayas

contribute to these invaluable species' decline. Rapid urbanisation, expanding infrastructure development, unsustainable agricultural practices, and the extraction of natural resources have resulted in habitat destruction and fragmentation. This disrupts ecosystems, reducing the natural habitats available for endemic species to thrive (Dhyani 2023; Premlata et al. 2024).

In many areas, traditional farming methods and the reliance on forest resources have given way to more intensive land use practices, often at the expense of biodiversity (Negi et al. 2012). Moreover, changing climate is exacerbating the situation by altering temperature and precipitation patterns across the region (Tse-ring et al. 2010). These changes are affecting the phenology and distribution of many plants. Some endemic species, adapted to narrow climatic conditions, may struggle to survive as their habitats shift to higher altitudes or disappear altogether (Dolezal et al. 2016; Tripathi et al. 2022; Patni et al. 2022). For instance, alpine species that thrive in colder conditions are particularly vulnerable to rising temperatures, as they have limited scope to migrate to new areas (Singh et al. 2010). In addition to habitat loss and climate change, over-exploitation of certain plant species for medicinal, ornamental, or commercial purposes poses significant risks to the survival of Himalayan endemic flora (Upreti and Bajpai 2022). Plants like the *Taxus wallichiana*, prized for its cancer-fighting properties, have been over-harvested, pushing them to extinction (Dhanjal et al. 2021; Patel et al. 2021). Similarly, the illegal trade of plants and the introduction of invasive species (Dhyani 2023). The present status of indigenous flora at this Himalayan hotspot will be discussed in this chapter, with a focus on threats they pose and present conservation efforts. It will concentrate on key species threatened and the implications of such pressures from human and environmental factors and review the present conservation strategies at local, national, and international levels. Bringing out the challenges to these plants and the associated ecosystems shall be the considerations in this chapter, to forward the ultimate urgency of conservation with complete and sustained efforts to salvage the fading beauty of Himalayan flora.

9.2 Geographical and Ecological Overview

The Himalayas are one of the world's most majestic and ecologically significant mountain ranges, spanning five countries: Nepal, India, Bhutan, China, and Pakistan. This 2400 km long range, starting at the Indus River in Pakistan and reaching eastward to the Brahmaputra River in India, acts as a formidable natural border, separating the Tibetan Plateau to the north and the lush plains of the Indian subcontinent to the south (Sharma et al. 2008). The Himalayas are more than simply a physical barrier; they are also a major environmental and climatic division, significantly impacting regional climate. The Himalayas act as a barrier to chilly winds from the Central Asia, helping to temper the climate of the northern Indian plains and considerably impact the monsoon patterns, which are vital for agriculture and water supplies in South Asia (Ramisch et al. 2016; Shrestha et al. 2015). These topographical and climatic distinctions closely mirror the regional classifications and biodiversity patterns emphasised in Chap. 2 (Sect. 2.3.1). The Himalayas' wide altitudinal and latitudinal reach results in various climatic zones, each sustaining unique biological systems, making this range one of the world's most significant and fragile ecological areas (Tse-ring et al. 2010). The Himalayan range, extending from west to east, can be broadly classified into the Western, Central, and Eastern Himalayas, each with unique climatic and vegetative patterns, further divided by altitude into the Great Himalaya, Lesser Himalaya, and Shiwaliks (Rawat 2017). In the Western Himalayas (Jammu & Kashmir, Himachal Pradesh), the Great Himalayas encounter cold alpine conditions with permanent snow and glaciers over 4500 m, where only patchy mosses and lichens survive (Rawat 2007). The Lesser Himalaya (1500–4500 m) is colder but more hospitable, with oak woods, deodar, and conifers such as pine and cedar. The Shiwaliks (600–1500 m) are warmer, with subtropical flora, including extensive Sal woods in the lower parts (Timothy and Nyaupane 2022). Winters in the Western Himalayas are colder and drier, while summers have less severe monsoons compared to the eastern sections (Dimri et al. 2015). The Central Himalaya (Uttarakhand, Nepal) has a more varied climate and vegetation because of its mid-latitude position (Sharma 2012). The Great Himalaya still has permanent snow and alpine tundra flora, while the Lesser Himalayas are rich in mixed forests of oak, *Rhododendrons*, and conifers, particularly in the upper parts (Singh and Pusalkar 2020). The Shiwaliks are characterised by tropical and subtropical forests, with more rainfall during the summer monsoon than the Western Himalayas (Lodhiyal et al. 2016; Bhatia et al. 2022). The Eastern

Himalayas (Sikkim, Bhutan, and Arunachal Pradesh) experience significant monsoon rains and a more humid climate (Chettri et al. 2010; Tse-ring et al. 2010). In the Great Himalayas, the snowline and glaciers start at lower altitudes, but despite cold temperatures, a more diverse array of alpine vegetation, including flowering plants, thrives (Suyal et al. 2022). The Lesser Himalaya is covered with dense temperate forests of bamboo, oak, magnolia, and *Rhododendron*, but the Shiwaliks are warmer and wetter, supporting deep subtropical rainforests and thick bamboo groves. This region is one of the wettest, with extended monsoons and tremendous biodiversity (Subedi 2004).

9.3 Diversity of Endemic Flora

The study of endemic vascular plants in the IHR highlights the region's rich biodiversity, with 1076 endemic plant taxa identified, predominated mainly by angiosperms. Endemics in the highest proportion were found in Arunachal Pradesh, Meghalaya, and Sikkim. *Orchidaceae*, *Asteraceae*, and *Poaceae* were the major families represented. More than 81% of these endemics exist in the mid-elevation range (1001–1500 m) (Tiwari et al. 2024). Endemic and Threatened species like *Gastrodia sikkimensis*, *Gastrodia indica, Nepenthes khasiana*, *Amentotaxus assamica* and *Gymnocladus assamicus* need the urgency of protection. A total of 395 species were documented from various regions of the Himalayas (Supplementary Table 9.1), with images of important endemic plants presented in Fig. 9.2.

9.4 Threats to Endemic Flora of the Himalayas

Although endemic taxa completely depend on national policies and actions, they play a critical role in determining national conservation priorities and guiding successful conservation programmes (Orsenigo et al. 2018). The endemic flora of the Indian Himalaya Hotspot, which includes species classified as Near Threatened (NT), Vulnerable (VU), Endangered (EN), and Critically Endangered (CR) on the IUCN Red List, is facing significant threats. This region, a crucial reservoir of biodiversity, is home to many critically endangered and threatened plant species (Tamang et al. 2017). Species like Indian Rosewood, Sandalwood Trees, and orchids underscore the ecological importance of the Himalayan region. The IUCN Red List includes the Himalayan medicinal plants such as *Sterculia khasiana* and *Ilex khasiana*, which are now classified as Extinct and Critically Endangered due to habitat loss (Mehta et al. 2020). Severely endangered species like *Meizotropis Pellita* (Patwa) and *Fritillaria cirrhosa* (Himalayan Fritillary), as well as the high-altitude orchid *Dactylorhiza hatagirea*, further emphasise the urgent need for targeted conservation efforts (Barik et al. 2018). Protecting unique and vulnerable species is crucial for the Himalayas' biodiversity and ecological integrity.

Fig. 9.2 Important endemic plants in the Himalayan region: (**a**) *Gastrodia indica*; (**b**) *Gastrodia sikkimensis*; (**c**) *Impatiens rajibgogoii*; (**d**) *Nepenthes khasiana*; (**e**) *Saussurea obvallata*; (**f**) *Impatiens scabrida*; (**g**) *Geranium wallichianum*; (**h**) *Rhododendron formosum*; (**i**) *Paphiopedilum villosum*. (Photos taken by Kingsuk Sarkar, Arka Mondal, Soumitra Halder)

Dipterocarpus macrocarpus dominates well-drained areas, while *Shorea assamica* thrives in drier, gravelly soil in Namdapha. The area is home to endangered wild relatives of cultivated plants, rare species like *Camellia caudata, Coffea benghalensis*, and *Mangifera sylvatica*, and extremely rare species like *Gnetum ula* and *Pentasacme wallichii* (Arunachalam et al. 2004). The most critically endangered species in the Indian Himalaya Hotspot include *Lilium polyphyllum, Nardostachys jatamansi, Coptis teeta, Aconitum heterophyllum, Angelica glauca*, and *Gentiana kurroo*, making their conservation a priority. Additionally, species such as *Saussurea costus, Illicium griffithii, Piper pedicellatum, Gymnocladus assamicus*, and *Angelica glauca*, which are endemic to the region and heavily trafficked, require special attention in conservation planning (Mehta et al. 2020). The vulnerability of these species and their cultural and economic importance to Himalayan communities have also been extensively addressed in Chap. 5, especially in Sect. 5.6 on threatened medicinal plants. Human activities, climate change, and livestock grazing pose significant threats to endemic plant species in the Himalayas, necessitating robust national conservation efforts to protect their unique biodiversity. Major threats to endemic flora of the Himalayas are discussed as follows:

9.4.1 Livestock Grazing

Livestock grazing in the trans-Himalayas and other regions exerts significant pressure on endemic plant species, leading to several detrimental impacts on the ecosystem. One of the primary consequences of continuous grazing is soil compaction, which causes soil to become densely packed, reducing its porosity and permeability (Chandrasekhar et al. 2007). Overgrazing leads to stunt plant growth and hinders the regeneration of endemic species. Livestock, mainly goats and sheep, feed on young shoots and plants, preventing them from maturing and reproducing, further reducing plant populations over time (García et al. 2012). The Blue Sheep (Bharal) and domestic yak in the higher Himalayas often graze in alpine meadows where rare and endemic plant species like the *Saussurea obvallata* (Brahma Kamal) are found (Adhikari et al. 2021). Overgrazing in the Himalayan region disrupts young plant regeneration, threatening endemic species and disrupting ecosystem balance. Fieldwork and herbarium consultations have assessed the threat status of three valuable medicinal species *Trillium govanianum, Rheum tibeticum*, and *Arnebia euchroma*; in the Trans-Himalayan Ladakh region and Kashmir Himalaya (Shivali and Namrata 2022). The endemic *Primula* species in the Himalayas, adapted to moist, shaded environments, can be severely impacted by grazing, leading to a decline in their populations and the potential loss of these unique plants (Khan et al. 2021). Overgrazing creates a feedback loop that hinders plant growth and further exacerbates the decline of endemic flora. Over time, this process can transform alpine meadows into barren landscapes, making it even harder for endemic species to survive.

9.4.2 Logging for Construction and Fuelwood

High demand for construction materials and fuelwood in the Himalayan region drives deforestation, threatening endemic plant species. Heavy reliance on firewood by local communities, trekkers, and herders exacerbates forest pressure. Traditional logging practices and inadequate monitoring worsen forest degradation, highlighting the urgent need for sustainable resource management and modern logging techniques to protect these fragile ecosystems (Baral et al. 2012). Logging removes critical habitats, such as the cool, shaded understory that the Himalayan yew (*Taxus wallichiana*) relies on, leaving these species exposed to harsh environmental conditions (Gaire et al. 2023). Removing trees also disrupts the microclimate, making the area drier and warmer, adversely affecting species like the endemic *Rhododendrons* that thrive in moist, excellent conditions. Additionally, logging destabilises the soil, leading to erosion and loss of nutrient-rich topsoil, which is crucial for the growth of plants like the threatened endemic *Meconopsis* sp., confined to the Himalayan ranges and Kashmir (Majid et al. 2015). The altered conditions from deforestation can also promote the invasion of non-native species like *Lantana camara*, which outcompete and further endanger the native flora.

9.4.3 Farmland Expansion

The expansion of agricultural land in the Himalayan region is threatening the natural habitats of endemic plant species, leading to habitat fragmentation, reduced biodiversity, and environmental degradation. Around 60% of the land conversion occurs in community forests, 35% in protected forests, and 5% in reserve forests, particularly at elevations of 1800–2600 m and on slopes of 10–30°. Traditional crops like buckwheat and millet have been replaced mainly by cash crops, especially potatoes. Agricultural zones vary by elevation: rice and finger millet are prevalent at low elevations (1120–1850 m), potatoes and amaranth at mid-elevations (1850–2400 m), and potatoes dominate high elevations (2400–2600 m) during the rainy season. Winter crops like barley, lentil, and mustard are grown at lower elevations, while only mustard is cultivated at higher altitudes. This shift in farming disrupts traditional practices and fragments habitats, isolating endemic plant populations (Thapliyal et al. 2024). In the Eastern Himalayas, large areas of forest have been cleared for crops like cardamom, maize, and rice, fragmenting the habitat of the endemic *Rhododendron* sp. (Singh et al. 2009). This isolation reduces genetic diversity and increases vulnerability to environmental changes. Furthermore, the expansion of agriculture reduces overall biodiversity as forests, home to various endemic species, are replaced with monocultures. The intensive use of chemical fertilizers and pesticides in the Himalayan region degrades soil health, impacting genera such as *Abies*, *Picea*, *Cedrus*, and *Taxus*. These high-altitude plants are particularly affected by changes in soil composition resulting from these farming practices (Kumar et al. 2023). Also, the commercialisation of agriculture and forestry and practices like jhum cultivation pose significant threats to endemic orchids in the Himalayan region. Furthermore, the use of pesticides harms non-target plant species and pollinators, which are essential for the reproduction of these orchids, leading to a decline in their populations (Kumar et al. 2022).

9.4.4 Anthropogenic Pressures

The increasing human population and activities around protected areas in the Himalayas pose significant threats to the sustainability of endemic flora. Similar to other tropical regions, the Himalayas face pressures from tourism, urbanisation, and infrastructure development, leading to habitat degradation and pollution (Kala 2005a). In Sikkim and Darjeeling, the rapid growth of tourism has brought visitors to sensitive ecological zones like Khangchendzonga National Park, a UNESCO World Heritage site (Choudhury et al. 2023). The increased foot traffic and development of tourist infrastructure have disturbed fragile habitats supporting endemic species like *Rhododendron hodgsonii*, leading to soil erosion, compaction, and the introduction of non-native species that outcompete native flora (Singh et al. 2009). Urbanisation is also a significant challenge, as expanding towns and cities like Shimla and Gangtok encroach on natural habitats, converting them into residential and commercial spaces (Anees et al. 2022). This expansion reduces available habitats and fragments what remains, isolating plant populations and decreasing genetic diversity. For instance, the construction in Shimla has encroached upon forests home to the already threatened *Taxus baccata* and *Taxus wallichiana* (Himalayan yew) (Lanker et al. 2010). Infrastructure development, such as dams and hydroelectric projects, further exacerbates the problem. In Uttarakhand, large dams like the Tehri Dam have submerged forested areas home to unique endemic species, altering river flows and local microclimates, making survival difficult for some species (Pandit and Grumbine 2012). Pollution from urbanisation and tourism also degrades habitats. In destinations like Manali and Leh, waste accumulation has polluted soil and water sources, harming ecosystems that support endemic flora. For example, the alpine meadows near Rohtang Pass, home to the rare *Pedicularis kansuensis*, have been impacted by litter and pollution, degrading soil quality and water sources essential for these plants' survival (Xi et al. 2024).

9.4.5 Climate Change

Climate change threatens Himalayan flora, especially endemic species like *Rhododendron anthopogon*, by forcing them to migrate to higher elevations in search of cooler habitats. However, the harsh conditions at higher altitudes, such as steep slopes, reduced oxygen, and increased UV radiation, limit available habitats. This upward shift increases the vulnerability of these species, like *Rhododendron anthopogon* in Sikkim, to extinction due to shrinking habitable areas (Singh et al. 2009). Moreover, as species migrate, they encounter new competitors from lower elevations that are better adapted to broader environmental conditions. This competition threatens endemic species like the *Meconopsis paniculata* (Himalayan blue poppy) and the alpine plant species *Aconitum gymnandrum*. These species now face rivals in their traditional alpine meadow habitat, reducing their chances of survival in the Tibeto-Himalayan and Arunachal Himalayan region (Kanwal and Lodhi 2018; Singh et al. 2009). Additionally, the slow adaptation rate of many endemic species to rapid environmental changes exacerbates these threats. Projections suggest that 2050, around 17% of Himalayan endemic species will lose their habitats, rising to 18% by 2070 (Malcolm et al. 2006; Manish et al. 2016). The upward migration of plant species also disrupts the overall ecosystem dynamics in the Himalayas. As plants move to higher altitudes, the delicate balance between flora and fauna is disturbed, leading to cascading effects throughout the ecosystem (Yadav et al. 2021).

9.4.6 Invasive Alien Species

Invasive species like *Eucalyptus camaldulensis*, *Eucalyptus globulus*, *Casuarina equisetifolia*, and *Grevillea robusta* pose significant threats to the endemic flora of the Indian Himalayan Region by competing with native plants for resources and disrupting ecosystem dynamics (Ayalew and Asfaw 2020). The IHR is home to 297 naturalised non-indigenous species across 65 families, with the highest numbers reported in Himachal Pradesh (232 species), followed by Jammu & Kashmir (192 species) and Uttarakhand (181 species) (Pathak et al. 2019). Among these, *Lantana camara*, *Ageratina adenophora*, *Parthenium hysterophorus*, and *Ageratum conyzoides* are particularly widespread and invasive. Projections indicate that by 2070, *Parthenium hysterophorus* and *Ageratum conyzoides* will lose suitable habitat, while *Chromolaena odorata*, *Ageratina adenophora*, and *Lantana camara* will expand into new regions, maintaining most of their current habitats, especially in the Himalayan foothills (Pathak et al. 2019). With 204 threatened plant species, Sikkim faces significant biodiversity loss due to deforestation, overgrazing, and unsustainable harvesting. The suitability of habitats decreases with increasing elevation, further exacerbating the issue. Recognising this, the management of invasive plant species was prioritised at the IUCN World Park Congress (2003), World Conservation Congress (2012), and World Parks Congress (2014) (Spenceley 2017). In the Darjeeling Himalayas, the invasive Maling bamboo (*Yushania maling*) poses a significant threat to endemic species and ecosystem health. By forming dense thickets, it outcompetes native plants, reducing biodiversity and disrupting the ecological balance. This invasion threatens the survival of indigenous flora and fauna, endangering the region's unique vegetation and ecological stability (Roy et al. 2016). Managing invasive species is complicated by high costs and labour intensity. Effective control relies on targeting invasive early in their propagation, but removing well-established species can sometimes allow previously suppressed non-invasive species to emerge. Continuous monitoring is essential to prevent further invasions and maintain ecological balance.

In addition to the direct ecological impacts, invasive species can also introduce or harbour pests and diseases that further threaten native biodiversity. For instance, *Lantana camara* is known to harbour injurious insect pests, including malarial mosquitoes, which can exacerbate public health concerns in affected areas. Moreover, the spread of invasive species can alter fire regimes in forests, as seen with *Lantana camara*, which can increase the risk of fires spreading to the canopy, leading to the loss of large areas of natural ecosystems (Pathak et al. 2019). Furthermore, the introduction of invasive species can lead to the spread of diseases that native species have not evolved to resist. This can result in significant declines or even extinctions of native populations. Therefore, addressing the threat of invasive species requires a comprehensive approach that includes monitoring, early detection, and management strategies to mitigate their impact on native ecosystems and public health.

9.4.7 Illegal Trade of Rare Plants

Illegal business activities, such as poaching, unregulated harvesting, and the trade of rare plants, pose significant threats to the endemic plant species of the Himalayas (Mozer and Prost 2023). These activities are driven by the high demand for medicinal herbs, ornamental plants, and other natural resources found in the region, often leading to the overexploitation of these species and pushing them toward extinction. The illegal collection and trade of *Nardostachys jatamansi*, a critically endangered plant used in traditional medicine, has severely depleted its populations in the wild (Shah 2023). Despite legal protections, poachers continue to harvest this plant from its natural habitats in the Himalayan highlands, often uprooting entire plants, which prevents regeneration and causes long-term ecological damage. Similarly, *Saussurea costus*, another endangered medicinal plant endemic to the Himalayas, is threatened due to illegal extraction (Singh et al. 2021). The root of this plant is highly valued in traditional medicine, leading to unsustainable harvesting practices. As a result, the species' natural populations have dramatically declined, and its habitat has been severely fragmented. In addition to medicinal plants, the illegal trade of ornamental plants such as *Cymbidium* orchids and *Rhododendron* species is also a significant concern (De et al. 2016). These plants are often collected illegally and sold in domestic and international markets. Removing these species from their natural habitats reduces their populations and disrupts the ecosystems they are part of, as these plants often play key roles in maintaining ecological balance.

9.4.8 Tourism

Tourism in the Himalayas has profoundly impacted the region's endemic plants, often jeopardizing their survival. The increased foot traffic in popular trekking routes like the Everest Base Camp has led to soil erosion and habitat degradation, threatening the delicate growth of endemic species such as *Rhodiola quadrifida* and the snow lotus (*Saussurea involucrata*) (Zhao et al. 2018). Ecotourism in the Himalayan ecosystem shows mixed impacts on forest conservation, with outcomes varying by region. In China, ecotourism hubs had lower forest loss rates, while in India, forest loss was higher in these areas compared to non-ecotourism zones. In Nepal and Bhutan, differences were negligible, suggesting limited conservation benefits. These findings highlight the context-dependent nature of ecotourism's environmental effects in the Himalayas (Brandt et al. 2019). In the Western Himalayas, eleven narrow endemic species, including *Aconitum heterophyllum*, *Aquilegia nivalis*, and *Atropa acuminata*, face critical extinction risks due to environmental pressures and human activities. *Aconitum heterophyllum* is threatened by habitat loss and overharvesting, *Aquilegia nivalis* by climate change-induced habitat shifts, and *Atropa acuminata* by habitat destruction and illegal collection. These species are susceptible to deforestation, land conversion, and climate fluctuations, impacting biodiversity and traditional medicinal practices. Immediate conservation efforts are crucial to protect these species and their fragile ecosystems (Ganie et al. 2019). Tourist trampling and plant collection disrupt natural habitats and hinder the reproduction and survival of vegetation. In-depth fieldwork and herbarium consultations in the Trans-Himalayan Ladakh region and the Kashmir Himalayas assessed the threat status of three significant medicinal plant species: *Trillium govanianum*, *Rheum tibeticum*, and *Arnebia euchroma* (Shivali and Namrata 2022). Tourism development, including infrastructure like hotels and trails, encroaches on the habitats of unique Himalayan plants, increasing their vulnerability. Additionally, the influx of tourists can introduce non-native species and diseases, further threatening local flora. Overall, tourism-related environmental pressures pose a significant threat to the conservation of endemic plants and their delicate ecosystems. Tourism-related activities can inadvertently introduce pests and diseases to the Himalayan region. The movement of tourists and goods can facilitate the spread of invasive insect pests and plant pathogens. For instance, the introduction of non-native species through tourism can lead to the establishment of new pests and diseases, which may not have natural predators in the new environment, allowing them to proliferate unchecked and cause significant harm to native plant species. Additionally, the disturbance of natural habitats by tourists can stress native plants, making them more susceptible to pest infestations and disease outbreaks.

9.5 Conservation Status and Efforts

The conservation of endemic flora in the Himalayas presents a multifaceted challenge shaped by the region's exceptional biodiversity, escalating human pressures, climate change, and habitat degradation (Chaudhary et al. 2024). Home to approximately 10,000 plant species including around 3160 endemics such as medicinal herbs, alpine flowers, and ancient forest trees the Himalayas are under significant ecological threat, with many species now classified as endangered or critically endangered (Singh and Pusalkar 2020).

Key drivers of ecosystem degradation include habitat destruction caused by agricultural expansion, urban development, and infrastructure projects, compounded by climate-induced altitudinal shifts that disrupt sensitive species distributions (Upadhyay 2020). In response, a network of protected areas and national parks spanning Nepal, India, Bhutan, and Pakistan play a crucial role in safeguarding these unique ecosystems (Dhyani 2023; Table 9.1). In India, conservation sites such as the Great Himalayan National Park, Valley of Flowers National Park, Nanda Devi National Park, and Khangchendzonga National Parks provide vital protection across diverse ecological zones (Rawat 2005; Sathyakumar and Bashir 2010). Similarly, Nepal's Sagarmatha National Park and Annapurna Conservation Area secure essential alpine and subalpine habitats (Shrestha and Bajracharya 2018). Bhutan's Jigme Dorji, Thrumshingla, and Royal Manas National Parks safeguard endemic biodiversity across elevational gradients, while Pakistan's Deosai and Khunjerab National Parks conserve high-altitude floral and faunal communities (Banerjee and Bandopadhyay 2016; Khan and Baig 2020; Tshewang et al. 2021; Baig et al. 2022). Transboundary collaborations, including the Sacred Himalayan Landscape and the Kailash Sacred Landscape initiatives, further strengthen ecological connectivity and promote coordinated regional conservation (Pandey et al. 2016; Thomaz 2023).

Community-based conservation approaches have emerged as critical components of effective biodiversity management. In India's Great Himalayan National Park, traditional ecological knowledge is leveraged through Biodiversity Management Committees. Nepal's Community Forest Programme and the Annapurna Conservation Area Project engage local residents in sustainable tourism and forest stewardship (Bajracharya et al. 2007; Chettri et al. 2012). Bhutan employs a Community-Based Natural Resource Management model, while Pakistan's Community Conservation Programme in Deosai emphasises local participation in ecosystem protection (Chaudhury et al. 2022). The Sacred Himalayan Landscape project exemplifies cross-border cooperation by involving communities in habitat restoration and sustainable land-use practices (Pandey et al. 2016). These initiatives are complemented by national policies and international frameworks such as the Convention on Biological Diversity (CBD), CITES, and UNESCO World Heritage designations, which collectively support the protection of Himalayan biodiversity (Khan et al. 2013; Sharma and Pradhan 2022). As highlighted in Chap. 5 (Sects. 5.1

Table 9.1 Key national parks in the Himalayas are recognized as UNESCO World Heritage Sites

Country	Protected area	Size (approx.)	Conservation importance	References
India	Valley of Flowers National Park	87.5 km^2	Diverse sub-alpine habitats and is home to over 600 plant species, including rare ones like the blue Himalayan poppy, making it a Centre of Plant Diversity with 45 medicinal plants used by locals	Kala (2005b), Pananjay et al. (2011)
	Nanda Devi National Park	630.3 km^2	This park is home to 312 floral species, including rare fir, birch, *Rhododendron*, and juniper, with limited vegetation in the inner sanctuary aside from notable mosses, lichens, and alpine species	Uniyal (2001), Rana et al. (2013)
	Great Himalayan National Park	1171 km^2	Diverse plant life, from pines and chestnuts to alpine herbs, with a rich subalpine zone and largely undisturbed forests, though some areas have been affected by grazing	Singh (2002), Singh and Singh (2009), Das et al. (2017)
	Khangchendzonga National Park	1784 km^2	Temperate broadleaf and mixed forests of oaks, birch, fir, willow, maple, and with alpine grasses, shrubs, and medicinal plants	O'Neill et al. (2017), O'Neill (2019)
Nepal	Sagarmatha National Park	1148 km^2	The subalpine forests of the park are dominated by fir, Himalayan birch, and *Rhododendron*, with juniper and *Rhododendron* prevailing at 4000–5000 m. Mosses and lichens thrive above 5000 m, and over 1000 floral species have been recorded in the park	Buffa et al. (1998), Bhuju et al. (2007)

and 5.3), conservation planning must also consider the economic dependence of indigenous communities on medicinal plant resources. Likewise, the ecoregional zoning frameworks presented in Chaps. 1 (Sect. 1.2) and 2 (Sect. 2.3) offer a critical foundation for tailoring conservation strategies to specific ecological and cultural contexts. Despite the breadth of these efforts, persistent threats underscore the need for adaptive, inclusive, and science-informed management. Ensuring the long-term survival of the Himalayas' rich endemic flora will require integrated conservation models that balance ecological protection with community well-being.

9.6 Sustainable Practices and Restoration

The Himalayan region is considered a biodiversity hotspot; however, human intervention, climate change, deforestation, and over-harvesting have negatively impacted indigenous plants, exacerbating current threats (Raj et al. 2019). Various long-term measures and restorative activities have been implemented to protect and restore the area's remarkable native vegetation. Community-based conservation intertwines ancient and new approaches, fostering intensified efforts, such as Uttarakhand's agroforestry project, which promotes the sages by propagating self-cultivated *Nardostachys jatamansi*, thereby preventing wild collection (Purohit et al. 2012; Rao et al. 2015). Ex-situ conservation of *Aconitum* species involves cultivating and preserving plants in controlled environments such as botanical gardens, seed banks, and research centres. One such initiative is in the village of Ghesh in Chamoli District, where local conservation efforts are being supported by the Department of High-Altitude Plant Physiology Research Center, Hemvati Nandan Bahuguna Garhwal University Srinagar, Uttarakhand, Garhwal, India. This strategy helps safeguard genetic diversity, reduce the risk of extinction, and support future reintroduction into their natural habitats (Chauhan et al. 2021). The Kailash Sacred Landscape rehabilitation project is an example of restoring plant habitats by reintroducing local species such as *Rhododendron campanulatum* and *Juniperus indica* (Kala et al. 2022; Maiti et al. 2022). Mainly, Nanda Devi Biosphere Reserve protects endemic species within its precincts while balancing the requirements of the local population and conservation (Maikhuri et al. 2001; Silori 2001). Nonetheless, climate change, potentially invasive species, guided ecotourism, and overuse continue to pose severe threats to vegetation loss in this region, highlighting the need for creativity and various forms of cooperation to promote survival (Ashok et al. 2022).

9.7 Research and Monitoring of Endemic Flora of Himalaya

The knowledge base on this unique biodiversity characteristic of the Himalayas as a resource crucial to maintaining ecological balance and supporting local communities through cultural, medicinal, and economic resources must be developed through research and monitoring of endemic flora (Tang and Gavin 2016). It is, however seriously threatened through soil erosion, pollution, tourism, and unsustainable harvesting practices. Most susceptible are the endemic species with low distribution levels and population densities and high utilisation value. Additionally, overexploitation, either as a medicinal product or cultural article, or as an economic resource, continues to further narrow the species' distribution toward narrow endemism. Traditional Ecological Knowledge (TEK) includes the collective environmental knowledge, resource management techniques, associated social structures, and worldviews of local and indigenous people. Because it may give locally relevant environmental data, direct adaptive management, and influence human-environment interactions (Gómez-Baggethun et al. 2013). Although there has been some progress, quite several areas with endemic species remain unconserved; noncharismatic but ecologically vital plants are overlooked. 69% of the rich areas in endemics are still protected while the remaining 31% require intense collaborative efforts. Herbariums and botanical gardens play a crucial role in the study, monitoring, and conservation of Himalayan endemic flora, these are sustainable management centres (Das and Lungphi 2019). The Govind Ballabh Pant National Institute of Himalayan Environment, Almora, Uttarakhand, India is fulfilling its considerable responsibility in developing research and management schemes on Himalayan resources (Rana et al. 2022). However, conservation of endemic species through location-specific population assessments, in-situ and ex-situ conservation practices, and reproductive biology studies deal with the problems of germination and survival (Tiwari et al. 2024).

Only through the involvement of local communities, conservation policy, as well as its implementation, would be ensured. National Action Plan on Climate Change (NAPCC) has introduced the National Mission for Sustaining the Himalayan Ecosystem in a quest to understand this relationship between the Himalayan ecosystem and the factors of climate (Deshpande et al. 2023). The mission is aimed at eliciting frank insight into the promotion of sustainable development within the Himalayas, and hence safeguards the fragile ecosystem. Climate change, habitat degradation, and human activities have to be attacked to preserve the region's ecological balance and biodiversity. Organisations like WWF continue to play a significant role by protecting, restoring, and reconnecting natural landscapes in the Eastern Himalayas. All these efforts have a holistic impact on sustaining the region's ecosystems through research, conservation, and community engagement that ensure long-term environmental and socio-economic health of the Himalayas (Anand et al. 2012). More often than not, policymakers have neglected the views of such communities and opted for their urban-centric models of conservations. Such a narrow focus of conservation might forfeit better ecological health for economic growth (Dhiman et al. 2022). Traditional knowledge assimilation, community participation enhancement, and education awareness programmes regarding sustainable uses of plant genetic resources should be prioritised to implement effective, inclusive, and adaptive conservation strategies. In such a scenario, integrated efforts involving the researcher community, policymakers, local communities, and conservation organisations become pivotal in safeguarding the unique biodiversity of the Himalayas through the protection of not only flagship but also lesser-known endemic species, addressing anthropogenic threats, and ensuring sustainable livelihoods for the local population while promoting biodiversity conservation (Ecosystem Profile 2005; Gómez-Baggethun et al. 2013; Dhyani 2023; Sahana et al. 2023).

9.8 Conclusions

The conservation of endemic plants in the Himalayan hotspot is crucial for preserving the region's unique biodiversity. To prevent the loss of these plant species, location-specific population assessments, in-situ and ex-situ conservation efforts, and studies on climate impacts and reproductive biology are essential. Adaptive strategies integrating scientific research and community involvement are key, with climate-resilient conservation plans and predictive models helping to protect habitats and allow species migration. Expanding seed banks, botanical gardens, and tissue culture facilities offer hope for restoring endangered species, while biotechnology, such as genome editing, can aid in population recovery. Promoting ecotourism, sustainable harvesting, and agro forestry can reduce pressure on wild populations and raise environmental awareness. Education, international cooperation, and policy reforms to increase legal protections and regulate resource harvesting are also critical. A comprehensive approach combining science, community action, and policy reform will be vital for safeguarding the Himalayan endemic flora for future generations.

Acknowledgements We extend our sincere gratitude to Dr. Arka Mondal, Department of Botany, Visva-Bharati, Santiniketan, West Bengal, for providing valuable photographs.

References

Adhikari BS, Rawat GS, Garbyal U, Srikar R (2021) Status of low and high altitude grazingland and issues of livestock (migratory pastoralism) in Kailash sacred landscape - India

Anand A, Chandan P, Singh RB (2012) Homestays at Korzok: supplementing rural livelihoods and supporting green tourism in the Indian Himalayas. Mt Res Dev 32(2):126–136

Anees MM, Sharma R, Joshi PK (2022) Urbanization in Himalaya—An interregional perspective to land use and urban growth dynamics. In: Schickhoff U, Singh RB, Mal S (eds) Mountain landscapes in transition. Springer, Cham, pp 517–538

Arunachalam A, Sarmah R, Adhikari D, Majumder M, Khan ML (2004) Anthropogenic threats and biodiversity conservation in Namdapha nature reserve in the Indian Eastern Himalayas. Curr Sci 87(4):447–454

Ashok S, Behera MD, Tewari HR, Jana C (2022) Developing ecotourism sustainability maximization (ESM) model: a safe minimum standard for climate change mitigation in the Indian Himalayas. Environ Monit Assess 194(12):914. https://doi.org/10.1007/s10661-022-10548-0

Ayalew A, Asfaw Z (2020) Allelophatic effects of *Gravellia Robusta, Eucalyptus camaldulensis* and *Casuarina equisetifolia* on germination and root length of maize and wheat. Inte J Rese Stud Agric Sci (IJRSAS) 6:15–20

Baig SU, Malik AJ, Khan H (2022) Wildlife habitat-suitability analysis for high mountain national parks in the Hindu Kush, Karakoram and Himalayan region of northern Pakistan. Model Earth Syst Environ 8(3):3941–3956

Bajracharya SB, Gurung GB, Basnet K (2007) Learning from community participation in conservation area management. J Fores Live 6(2):54–66

Banerjee A, Bandopadhyay R (2016) Biodiversity hotspot of Bhutan and its sustainability. Curr Sci 110:521–527

Baral NR, Acharya DP, Rana CJ (2012) Study on drivers of deforestation and degradation of forests in high mountain regions of Nepal. REDD Cell, Ministry of Forest and Soil Conservation, Nepal
Barik SK, Tiwari ON, Adhikari D, Singh PP, Tiwary R, Barua S (2018) Geographic distribution pattern of threatened plants of India and steps taken for their conservation. Curr Sci 114:470–503
Bhatia H, Srivastava G, Adhikari P, Tao S, Utescher T, Paudayal KN, Mehrotra RC (2022) Asian monsoon and vegetation shift: evidence from the Siwalik succession of India. Geo Mag 159(8): 1397–1414
Bhuju UR, Shakya PR, Basnet TB, Shrestha S (2007) Nepal biodiversity resource book: protected areas, Ramsar sites, and world heritage sites
Brandt JS, Radeloff V, Allendorf T, Butsic V, Roopsind A (2019) Effects of ecotourism on forest loss in the Himalayan biodiversity hotspot based on counterfactual analyses. Conserv Biol 33(6):1318–1328. https://doi.org/10.1111/cobi.13341
Buffa G, Ferrari C, Lovari S (1998) The upper subalpine vegetation of Sagarmatha National Park (Khumbu Himal area, Nepal) and its relationship with Himalayan tahr, musk deer and domestic yak. In: An outline Top of the World environmental research: Mount Everest–Himalayan ecosystem, pp 167–175
Chandrasekhar K, Rao KS, Maikhuri RK, Saxena KG (2007) Ecological implications of traditional livestock husbandry and associated land use practices: a case study from the trans-Himalaya, India. J Arid Environ 69(2):299–314
Chaudhary RP, Joshi SP, Rimal SK, Belbase N, Subedi CK (2024) Plant diversity conservation in the Nepal Himalaya: status, policies, and legislative frameworks. In: Rokaya MB, Sigdel SR (eds) Flora and Vegetation of Nepal. Springer, Cham, pp 383–447
Chaudhry S, Chettri N, Uddin K (2022) Implementation of the convention on biological diversity in the Hindu Kush Himalayan countries: a retrospective analysis of Aichi targets
Chauhan J, Ranjan R, Purohit VK, Prasad P (2021) Ex-situ conservation-a best method for restoring aconitum species in Indian Himalayan region. Envis Bull Himal Ecol 29:74
Chettri N, Shakya B, Thapa R, Sharma E (2008) Status of a protected area system in the Hindu Kush-Himalayas: an analysis of PA coverage. Int J Biodiver Sci Manag 4(3):164–178. https://doi.org/10.3843/Biodiv.4.3:4
Chettri N, Sharma E, Shakya B, Thapa R, Bajracharya B, Uddin K, Choudhury D, Oli KP (2010) Biodiversity in the eastern Himalayas: status, trends and vulnerability to climate change. International Centre for Integrated Mountain Development (ICIMOD)
Chettri N, Sherchan U, Chaudhary S, Shakya B (2012) Mountain biodiversity conservation and management: selected examples of good practices and lessons learned from the Hindu Kush Himalayan region
Choudhury G, Choudhury D, Sharma P (2023) Sustainable development and livelihood opportunities in the Himalayan region of eastern India. In: Leal Filho W, Ng TF, Iyer-Raniga U, Ng A, Sharifi A (eds) SDGs in the Asia and Pacific region. Springer, Cham, pp 1–33
Das AP, Lungphi P (2019) Floristic studies vis-à-vis medicinal plants in the eastern Himalaya–past and the present status. In: Medicinal plants of India: conservation and sustainable use. Today & Tomorrow's Printers and Publishers, New Delhi, pp 39–56
Das DS, Rawat DS, Shrivastava N, Ambrish K, Sinha BK, Singh P, Dash SS (2017) A contribution to the flora of great Himalayan National Park, Himachal Pradesh, India. Nelumbo 59(1):33–43
De LC, Rao AN, Singh DR (2016) Endangered orchids and their conservation in north East India. In: Purkayastha J (ed) Bioprospecting of indigenous bioresources of north-east India. Springer Singapore, Singapore, pp 61–75
Deshpande T, Mukherji R, Sastry M (2023) Policy styles and India's national action plan on climate change (NAPCC). Policy Stud 46(3):1–18. https://doi.org/10.1080/01442872.2023.2261388
Dhanjal DS, Chopra C, Singh R (2021) Himalayan herbal medicine for cancer. Modern cancer therapies and traditional medicine: an integrative approach to combat cancers 268:5839–5842

Dhiman MR, Muthanarasimha GP, Dhiman MR, Muthanarasimha GP (2022) Biodiversity conservation of Western Himalayas: a pluralistic approach. In: Floristic diversity - biology and conservation. IntechOpen

Dhyani S (2023) Are Himalayan ecosystems facing hidden collapse? Assessing the drivers and impacts of change to aid conservation, restoration and conflict resolution challenges. Biodivers Conserv 32(12):3731–3764

Dimri AP, Niyogi D, Barros AP, Ridley J, Mohanty UC, Yasunari T, Sikka DR (2015) Western disturbances: a review. Rev Geophy 53(2):225–246. https://doi.org/10.1002/2014RG000460

Dolezal J, Dvorsky M, Kopecky M, Liancourt P, Hiiesalu I, Macek M, Altman J, Chlumska Z, Rehakova K, Capkova K (2016) Vegetation dynamics at the upper elevational limit of vascular plants in Himalaya. Sci Rep 6(1):24881

Ecosystem Profile (2005) Eastern Himalayas region. WWF-US, Asia Program

Gaire D, Jiang L, Adhikari B, Panthi S (2023) Predicting habitat suitability and threats assessment of himalayan yew (*Taxus Wallichiana* Zucc.) in Nepal. Appl Ecol Environ Res 21(3): 2417–2439

Ganie AH, Tali BA, Khuroo AA, Reshi ZA, Nawchoo IA (2019) Impact assessment of anthropogenic threats to high-valued medicinal plants of Kashmir Himalaya, India. J Nat Conserv 50: 125715. https://doi.org/10.1016/j.jnc.2019.125715

García RR, Celaya R, García U, Osoro K (2012) Goat grazing, its interactions with other herbivores and biodiversity conservation issues. Small Rumin Res 107(2–3):49–64

Gómez-Baggethun E, Corbera E, Reyes-García V (2013) Traditional ecological knowledge and global environmental change: research findings and policy implications. Ecol Soc 18(4):72

Joshi AK, Joshi PK (2019) Forest ecosystem services in the Central Himalaya: local benefits and global relevance. Proc Natl Acad Sci India Sect B Biol Sci 89(3):785–792. https://doi.org/10.1007/s40011-018-0969-x

Kala CP (2005a) Indigenous uses, population density, and conservation of threatened medicinal plants in protected areas of the Indian Himalayas. Conserv Bio 19(2):368–378. https://doi.org/10.1111/j.1523-1739.2005.00602.x

Kala CP (2005b) A multifaceted review on the biodiversity conservation of the valley of flowers National Park, India. The Int J Biodiver Sci Manag 1(1):25–32

Kala A, Chaudhary A, Adhikari BS, Rawat GS (2022) Local communities' perception and attitude towards dynamic ecosystem in a part of Kailash sacred landscape-India. Environ Challen 9: 100603

Kanwal KS, Lodhi MS (2018) Climate change impact on plant biodiversity of Arunachal Himalaya: a review. Bull Arunachal For Res 33(2):15–26

Khan H, Baig SU (2020) Biodiversity conservation in the Hindu Kush-Karakoram-Himalayan mountain region of northern Pakistan: overview of big mammal protection. J Mount Sci 17(6): 1360–1373

Khan SM, Page SE, Ahmad H, Harper DM (2013) Sustainable utilization and conservation of plant biodiversity in montane ecosystems: the western Himalayas as a case study. Ann Bot 112(3): 479–501

Khan S, Shaheen H, Aziz S, Nasar S (2021) Diversity and distribution of Genus Primula in Kashmir region: an indicator genus of the western Himalayan mountain wetlands and glacial forelands. Biodivers Conserv 30(6):1673–1688

Kumar S, Devi RS, Choudhury R, Mahapatra M, Biswal SK, Kaur N, Tudu J, Rath SK (2022) Orchid diversity, conservation, and sustainability in northeastern India. In: Furze JN, Eslamian S, Raafat SM, Swing K (eds) Earth systems protection and sustainability. Springer, Cham, pp 111–139

Kumar A, De Jong W, Kumar M, Pandey R (2023) Climate change in the Himalayas: vulnerability and resilience of biodiversity and forest ecosystems. Elsevier

Lanker U, Malik AR, Gupta NK, Butola JS (2010) Natural regeneration status of the endangered medicinal plant, *Taxus baccata* Hook. F. syn. *T. wallichiana*, in northwest Himalaya. Int J

Biodivers Sci Ecosys Serv Manag 6(1–2):20–27. https://doi.org/10.1080/21513732.2010.527302
Lepcha SR (2010) Studies of the angiospermic flora of alpine East Sikkim with special reference to Pangolakha wildlife sanctuary. PhD Thesis, University of North Bengal
Lodhiyal LS, Lodhiyal N, Pathak GC (2016) Plant diversity and vegetation composition of Shiwalik forests along an altitudinal gradient in the Kumaun Himalaya, India. In: Chand R, Leimgruber W (eds) Globalization and marginalization in mountain regions. Springer, Cham, pp 205–227
Maikhuri RK, Nautiyal S, Rao KS, Saxena KG (2001) Conservation policy–people conflicts: a case study from Nanda Devi biosphere reserve (a world heritage site), India. Forest Policy Econ 2(3–4):355–365
Maiti P, Kuniyal JC, Sekar KC, Satish KV, Singh D, Bisht N, Kumar A, Arya SC, Nand M, Sundriyal RC (2022) Landscape level ecological assessment and eco-restoration strategies for alpine and sub-alpine regions of the central Himalaya. Ecol Eng 180:106674
Majid A, Ahmad H, Saqib Z, Ali H, Alam J (2015) Conservation status assessment of Meconopsis aculeata Royle: a threatened endemic of Pakistan and Kashmir. Pak J Bot 47:1–5
Malcolm JR, Liu C, Neilson RP, Hansen L, Hannah L (2006) Global warming and extinctions of endemic species from biodiversity hotspots. Conserv Biol 20(2):538–548. https://doi.org/10.1111/j.1523-1739.2006.00364.x
Manish K, Telwala Y, Nautiyal DC, Pandit MK (2016) Modelling the impacts of future climate change on plant communities in the Himalaya: a case study from eastern Himalaya, India. Model Earth Syst Environ 2(2):92. https://doi.org/10.1007/s40808-016-0163-1
Manish K, Pandit MK, Telwala Y, Nautiyal DC, Koh LP, Tiwari S (2017) Elevational plant species richness patterns and their drivers across non-endemics, endemics and growth forms in the eastern Himalaya. J Plant Res 130(5):829–844. https://doi.org/10.1007/s10265-017-0946-0
Mehta P, Sekar KC, Bhatt D, Tewari A, Bisht K, Upadhyay S, Negi VS, Soragi B (2020) Conservation and prioritization of threatened plants in Indian Himalayan region. Biodivers Conserv 29(6):1723–1745. https://doi.org/10.1007/s10531-020-01959-x
Morgan N (2019) In the mountains: the health and wellbeing benefits of spending time at altitude. Hachette UK
Mozer A, Prost S (2023) An introduction to illegal wildlife trade and its effects on biodiversity and society. Forensic Sci Int: Anim Environ 3:100064
Negi GCS (2022) Trees, forests and people: the central Himalayan case of forest ecosystem services. TFP 8:100222
Negi VS, Maikhuri RK, Rawat LS (2012) Paradigm and ecological implication of changing agricultural land-use: a case study from Govind wildlife sanctuary, central Himalaya, India. J Mt Sci 9(4):547–557. https://doi.org/10.1007/s11629-012-2216-x
O'Neill AR (2019) Evaluating high-altitude Ramsar wetlands in the eastern Himalayas. Global Ecol Conserv 20:e00715
O'Neill AR, Badola HK, Dhyani PP, Rana SK (2017) Integrating ethnobiological knowledge into biodiversity conservation in the eastern Himalayas. J Ethnobiol Ethnomed 13(1):21. https://doi.org/10.1186/s13002-017-0148-9
Orsenigo S, Montagnani C, Fenu G, Gargano D, Peruzzi L, Abeli T, Alessandrini A, Bacchetta G, Bartolucci F, Bovio M, Brullo C, Brullo S, Carta A, Castello M, Cogoni D, Conti F, Domina G, Foggi B, Gennai M, Gigante D, Iberite M, Lasen C, Magrini S, Perrino EV, Prosser F, Santangelo A, Selvaggi A, Stinca A, Vagge I, Villani M, Wagensommer RP, Wilhalm T, Tartaglini N, Duprè E, Blasi C, Rossi G (2018) Red listing plants under full national responsibility: extinction risk and threats in the vascular flora endemic to Italy. Biol Conserv 224:213–222. https://doi.org/10.1016/j.biocon.2018.05.030
Pananjay GBG, Tiwari K, Shashi GBG, Tiwari K, Tiwari SC (2011) An overview of potential ecotourism resources and their prospects in valley of flowers National Park, Uttaranchal, India. Caspian J Environ Sci 9(1):105–110

Pandey A, Kotru R, Pradhan N (2016) Kailash sacred landscape: bridging cultural heritage, conservation and development through a transboundary landscape approach. In: Asian Sacred natural sites. Routledge, pp 167–180

Pandit MK, Grumbine RE (2012) Potential effects of ongoing and proposed hydropower development on terrestrial biological diversity in the Indian Himalaya. Conserv Biol 26(6):1061–1071. https://doi.org/10.1111/j.1523-1739.2012.01918.x

Patel BP, Upadhaya A, Kumar S (2021) Conservation, cultivation and sustainable utilization of lesser known plants in Nepal. In: Lesser known plants. Walnutpublication.com, p 193

Pathak R, Negi VS, Rawal RS, Bhatt ID (2019) Alien plant invasion in the Indian Himalayan region: state of knowledge and research priorities. Biodivers Conserv 28(12):3073–3102

Patni B, Bhattacharyya M, Kumari A, Purohit VK (2022) Alarming influence of climate change and compromising quality of medicinal plants. Plant Physiol Rep 27(1):1–10. https://doi.org/10.1007/s40502-021-00616-x

Premlata, Kumar R, Hajam YA, Giri A (2024) Threats, challenges, and conservation strategies of Himalayan faunal biodiversity. In: Sobti RC (ed) Role of science and technology for sustainable future. Springer Nature Singapore, Singapore, pp 321–344

Purohit V, Chauhan R, Andola H, Pratti P, Nautiyal MC, Nautiyal A (2012) Nardostachys jatamansi DC: conservation, multiplication and policy issues. Med Plants 4:162. https://doi.org/10.5958/j.0975-4261.4.3.019

Raj A, Jhariya MK, Yadav DK, Banerjee A (2019) Conservation issues, challenges, and management of medicinal plant resources: a new dimension toward sustainable natural resource management. In: Agroforestry and climate change. Apple Academic Press, pp 123–153

Ramisch A, Lockot G, Haberzettl T, Hartmann K, Kuhn G, Lehmkuhl F, Schimpf S, Schulte P, Stauch G, Wang R (2016) A persistent northern boundary of Indian summer monsoon precipitation over Central Asia during the Holocene. Sci Rep 6(1):25791

Rana SK, Rawat GS (2017) Database of Himalayan plants based on published floras during a century. Data 2(4):36

Rana CS, Tiwari JK, Dangwal LR, Gairola S (2013) Faith herbal healer knowledge document of Nanda Devi biosphere reserve, Uttarakhand, India

Rana SK, Dangwal B, Negi VS, Bhatt ID (2022) Scientific research in the Himalaya: current state of knowledge, funding paradigm and policy implications. Environ Sci Pol 136:685–695

Rao KS, Saxena KG, Tiwari BK (2015) Biodiversity, climate change and socio-economic development in the Indian Himalaya

Rawat GS (2005) Terrestrial vegetation and ecosystem coverage within Indian protected areas. Nat Acad Sci lett 28(7/8):241

Rawat GS (2007) Alpine vegetation of the western Himalaya: species diversity, community structure, dynamics and aspects of conservation. PhD Thesis, D. Sc Thesis, Kumaun University, Nainital. p 239

Rawat GS (2017) The Himalayan vegetation along horizontal and vertical gradients. In: Bird migration across the Himalayas: wetland functioning amidst mountains and glaciers. Cambridge University Press, p 189

Roy A, Bhattacharya S, Ramprakash M, Senthil Kumar A (2016) Modelling critical patches of connectivity for invasive Maling bamboo (Yushania maling) in Darjeeling Himalayas using graph theoretic approach. Ecol Model 329:77–85. https://doi.org/10.1016/j.ecolmodel.2016.02.016

Sahana M, Areendran G, Sivadas A, Raj K, Sharma D, Sultan MS, Ghoshal A, Parameswaran S (2023) Assessing high conservation value areas for rare, endemic and threatened (RET) species: a study in high altitude Changthang landscape of India. J Nat Conserv 73:126406

Sathyakumar S, Bashir T (2010) Wildlife of the Himalaya: conservation issues and the way forward. In: Mountains, Ecosystem and Man. Soil Conservation Society of India, pp 322–343

Shah NC (2023) Historical profile of Nardostachys jatamansi: an ancient incense & aromatic medicinal herb from Kumaon, Uttarakhand. Indian J Hist Sci 58(2):111–118. https://doi.org/10.1007/s43539-023-00081-5

Sharma S (2012) Catastrophic hydrological event of 18 and 19 September 2010 in Uttarakhand, Indian Central Himalaya–an analysis of rainfall and slope failure. Curr Sci 102:327–332
Sharma G, Pradhan BK (2022) Traditional knowledge and access and benefit sharing in the context of Himalayan states. In: Oommen OV, Laladhas KP, Nelliyat P, Pisupati B (eds) Biodiversity conservation through access and benefit sharing (ABS). Springer, Cham, pp 127–162
Sharma BR, Amarasinghe UA, Sikka A (2008) Indo-Gangetic river basins: summary situation analysis. International Water Management Institute, New Delhi, India, p 14
Shivali V, Namrata S (2022) Threat status of three important medicinal Himalayan plant species and conservation implications. Nat Conserv Res 7(1):27–41
Shrestha KK, Bajracharya SB (2018) Biodiversity in Nepal. In: Global biodiversity, vol 1. Apple Academic Press
Shrestha AB, Agrawal NK, Alfthan B et al. (2015) The Himalayan climate and water atlas: impact of climate change on water resources in five of Asia's major river basins
Silori CS (2001) Biosphere reserve management in theory and practice: case of Nanda Devi biosphere reserve, Western Himalaya, India. J Int Wildlife Law Policy 4(3):205–219. https://doi.org/10.1080/13880290109353987
Singh SP (2002) Western Himalayan ecoregional biodiversity strategy and action plan. Prepared under the National Biodiversity Strategy and Action Plan, India
Singh DK, Pusalkar PK (2020) Floristic diversity of the Indian Himalaya. In: Dar GH, Khuroo AA (eds) Biodiversity of the Himalaya: Jammu and Kashmir state. Springer Singapore, Singapore, pp 93–126
Singh SK, Singh DK (2009) Hepaticae and anthocerotae of great Himalayan National Park and its environs (HP), India. Botanical Survey of India, Dehradun
Singh KK, Rai LK, Gurung B (2009) Conservation of rhododendrons in Sikkim Himalaya: an overview. World J Agri Sci 5(3):284–296
Singh SP, Singh V, Skutsch M (2010) Rapid warming in the Himalayas: ecosystem responses and development options. Clim Dev 2(3):221–232. https://doi.org/10.3763/cdev.2010.0048
Singh M, Raina RK, ul Hassan W, Malik IA (2021) Threat status of Saussurea lappa-a study in Bhaderwah area of Jammu & Kashmir India. Int Res J Ayur Yoga 4(5):97–109
Spenceley A (2017) Tourism and protected areas: comparing the 2003 and 2014 IUCN world parks congress. Tour Hospit Res 17(1):8–23. https://doi.org/10.1177/1467358415612515
Subedi BP (2004) Linking plant-based enterprises and local communities to biodiversity conservation in Nepal Himalaya. Unpublished doctoral Dissertation, Kumaun University, India
Suyal DC, Joshi D, Kumar S, Bhatt P, Narayan A, Giri K, Singh M, Soni R, Kumar R, Yadav A, Devi R, Kaur T, Kour D, Yadav AN (2022) Himalayan microbiomes for agro-environmental sustainability: current perspectives and future challenges. Microb Ecol 84(3):643–675. https://doi.org/10.1007/s00248-021-01849-x
Tamang M, Pal K, Kumar Rai S, Kalam A, Rehan Ahmad S (2017) Ethnobotanical survey of threatened medicinal plants of West Sikkim. Int J Bot Stud 2(6):116–125
Tang R, Gavin MC (2016) A classification of threats to traditional ecological knowledge and conservation responses. Conserv Soc 14(1):57–70
Thapliyal N, Bhojak P, Mehta P, Sekar KC (2024) Threats and conservation of plant diversity in the Himalayan biodiversity hotspot. In: Biodiversity hotspot of the Himalaya. Apple Academic Press, pp 459–511
Thomaz SM (2023) Ecosystem services provided by freshwater macrophytes. Hydrobiologia 850(12):2757–2777
Timothy DJ, Nyaupane GP (2022) The Himalaya: physiographic and cultural perspectives. In: Tourism and development in the Himalaya. Routledge, pp 18–36
Tiwari H, Sekar KC, Pandey A, Tiwari A, Mehta P, Kanwal KS, Arya D (2024) Diversity, distribution and need of urgent conservation of endemic plants in Himalaya. Biodivers Conserv 33(8–9):2285–2303. https://doi.org/10.1007/s10531-024-02815-y

Tripathi AK, Pandey PC, Sharma JK, Triantakonstantis D, Srivastava PK (2022) Climate change and its impact on forest of Indian Himalayan region: a review. In: Rani S, Kumar R (eds) Climate change. Springer, Cham, pp 207–222

Tse-ring K, Sharma E, Chettri N, Shrestha AB (2010) Climate change vulnerability of mountain ecosystems in the eastern Himalayas. International Centre for integrated mountain development (ICIMOD)

Tshewang U, Tobias MC, Morrison JG (2021) Bhutan: conservation and environmental protection in the Himalayas. Springer, Cham

Uniyal VP (2001) Nanda Devi expedition-2001. Wildlife Institute of India, Dehradun

Upadhyay RK (2020) Markers for global climate change and its impact on social, biological and ecological systems: a review. Amer J Clim Change 9(03):159

Upreti DK, Bajpai R (2022) Status, issues and challenges of biodiversity: lower plants (non-vascular). In: Kaur S, Batish DR, Singh HP, Kohli R (eds) Biodiversity in India: status, issues and challenges. Springer Nature Singapore, Singapore, pp 15–24

Xi S, Hu T, Mou X, Kou X, Wang X, Yu Y (2024) Reduced plant species diversity and soil carbon and nitrogen contents driven by vegetation patchiness in alpine meadows. J Veg Sci 35(1): e13238. https://doi.org/10.1111/jvs.13238

Yadav RR, Negi PS, Singh J (2021) Climate change and plant biodiversity in Himalaya, India. Proc Indian Natl Sci Acad 87(2):234–259. https://doi.org/10.1007/s43538-021-00034-5

Zhao R, Zhang H, An L (2018) Anthropogenic disturbances affect population size and biomass allocation of two alpine species from the headwater area of the Urumqi river, China. Pak J Bot 50(1):199–209

Part II
Threat to Himalayan Hotspot

Chapter 10
Himalayan Ecotourism and Plant Diversity: Perspective and Prospects

Deep Shekhar Das, Dinesh Singh, and Debabrata Maity

Abstract The Himalayan Mountains, famous for their intricate and diverse ecosystems, are home to an extraordinary array of plant species. These mountains serve as a vital global resource, providing essential water, food, and other resources for countries like Nepal, India, Bhutan, China (Tibet Autonomous Region), Pakistan, Afghanistan, Bangladesh, and Myanmar. The Indian Himalayan Region alone accounts for more than 50% of India's total flowering plants, with about 13% being endemic. The region also boasts a rich cultural heritage, with ancient civilisations, archaeological sites, and a wealth of historical landmarks, making it a cradle for vibrant indigenous cultures and knowledge systems. This unique blend has transformed the Indian Himalayan Region into a popular destination for saints, seers, and spiritual seekers in search of peace, enlightenment, touristy (including adventure tourism, trekking, rafting, mountaineering), expedition, and reflection. However, the rapid development of infrastructure, including roads, hotels, lodges, and tourism services, has led to a significant surge in tourism activities, resulting in the loss of wildlife, depletion of plant species, and degradation of habitats. In response, ecotourism has emerged as a sustainable alternative to mass tourism. Ecotourism offers opportunities for wilderness experiences and cultural immersion, benefiting economically marginalised communities in remote areas while supporting the conservation of biodiversity and landscapes. In ecologically vulnerable regions, tourism has a dual impact, generating employment opportunities while contributing to environmental degradation, making local community involvement crucial for sustainable conservation efforts.

Keywords Biodiversity · Biogeography · Indian Himalayan region · Mass tourism · Socio cultural value · Socio economic development · Sustainable development

---

D. S. Das · D. Maity (✉)
Taxonomy and Biosystematics Laboratory, Department of Botany, University of Calcutta, Kolkata, West Bengal, India

D. Singh
Department of Botany, Pt. L. M. S. Campus, Sri Dev Suman Uttarakhand University, Rishikesh, Uttarakhand, India

D. Das et al. (eds.), *Plant Diversity and Conservation Strategies of the Himalayan Biodiversity Hotspot*, Topics in Biodiversity and Conservation 21,
https://doi.org/10.1007/978-3-032-21697-7_10

10.1 Introduction

The majestic Himalayas, one of the youngest and longest mountain ranges on the Earth, are renowned for their complex and diverse ecosystems. As a distinct geographical and ecological entity, the region plays a crucial role in the planet's major biophysical landscapes (Timothy and Nyaupane 2022). Not only does the Himalaya create its own unique climate, but it also significantly influences the weather patterns across much of Asia (Chauhan et al. 2023). The region's varied topography, spanning latitudinal, longitudinal, and altitudinal gradients, gives rise to diverse climate and habitat conditions, fostering an extraordinary richness and uniqueness in its biodiversity, from genes to ecosystems (Loewen et al. 2023). This altitudinal and climatic stratification and its influence on vegetation patterns and species richness is thoroughly explained in Chap. 1 (Sects. 1.2 and 1.3.1), which outlines the geological orogeny and spatial divisions of the Himalayas. With over 400 peaks towering above 7000 m and 14 reaching over 8000 m, the Himalayas include the Tibetan Plateau, the largest and highest plateau in the world (Douglas 2021). The range is the source of 10 major rivers, whose basins cover more than 4.2 million km^2 (Wester et al. 2019). Often referred to as the 'Third Pole', it also holds the largest area of permanent ice (water tower of Asia) outside of the Arctic and Antarctic, playing a pivotal role in global hydrological cycles (Yao et al. 2012). The region's topography and climate create an ideal environment for diverse flora and fauna, making the Himalayas one of the most biologically rich areas on the planet (Chettri et al. 2008; Wester et al. 2019).

The Himalayas are not only a natural wonder but also a vital lifeline for more than half a billion people across countries such as Nepal, India, Bhutan, Tibet, China, Pakistan, Afghanistan, Bangladesh, Myanmar. An additional 1.9 billion people in downstream basins depend on the region for water and other resources, while three billion people rely on food produced in these basins, making the Himalayas essential to the livelihoods of nearly half the world's population (Molden et al. 2022). Thus, the Himalayas are a critical global asset with immense economic, sociocultural and environmental significance. Their diverse physiography, climates, and ecosystems, along with rich cultural heritage, make the region a prime destination for tourism. It is also home to a rich cultural mosaic, with ancient civilisations, archaeological sites, and a wealth of historical assets, all contributing to the region's vibrant indigenous cultures and knowledge systems (Ives and Messerli 2003). Despite these attractions, tourism in the Himalayas remains underdeveloped, with its economic contribution still below the global average due to challenges related to geography, infrastructure, socio-political issues, and poverty (Timothy and Nyaupane 2022).

As one of the planet's youngest mountain ranges, surrounded by densely populated areas facing extreme poverty, the Himalayan region is also highly vulnerable to a wide range of environmental and social changes both natural and human-induced, occurring at local, national, regional, and global levels (Matthew 2012). These challenges include natural disasters, population growth, climate change, and political instability (Matthew 2012). The region's unique geophysical processes, cultural diversity, and remoteness further distinguish the Himalayas from other mountain ranges, with fragility, marginality, and diversity emerging as key characteristics (Nyaupane and Chhetri 2009). This distinctiveness has profound implications for socioeconomic development and tourism, underscoring the need for careful attention to sustainable development and responsible tourism practices in the region.

10.2 The Indian Himalayan Region

The Indian Himalayan Region, spanning over 537,000 km^2 (0.537 million km^2), represents a significant portion of the Himalayan biodiversity hotspot. Located between latitudes 26°20′ N and 35°40′ N and longitudes 74°50′ E and 95°40′ E, the Indian Himalayan Region covers approximately 16.2% of the India's total geographical area (Singh et al. 2019). Administratively, it encompasses all of five states Jammu & Kashmir, Himachal Pradesh, Uttarakhand, Sikkim and Arunachal Pradesh; while partially covering West Bengal (hill districts); and one union territory Ladakh. The Indian Himalayan Region occupies a strategically crucial position along India's northern boundary, stretching from the northwest to the northeast of the country.

In addition to its national security significance, the Indian Himalayan Region is vital for its rich forest cover. Over 41.5% of the region is forested, accounting for one-third of India's total forest area and nearly half (47%) of the country's 'very good' forest cover (Pandit 2017; G-SHE 2018). The critical role of this forest cover in supporting endemic, threatened, and medicinally valuable plant species is extensively documented in Chap. 5 (Sect. 5.1), which highlights how these forests are a livelihood source for Himalayan communities. These forests provide a range of vital ecosystem services, including acting as a carbon sink, absorbing large amounts of carbon dioxide. The region also features extensive areas of permanent snow and glaciers about 17% of the Indian Himalayan Region and seasonal snow cover that blankets 30–40% of the area, creating a unique water reservoir (Molden et al. 2022). This snow and glacial melt water feed several major perennial rivers, which are crucial sources of water for drinking, irrigation, and hydropower generation. Each year, approximately 1.2×10^{12} m^3 of water flow from the rivers originating in the Himalayas (Molden et al. 2022). The Indian Himalayan Region shelters nearly 4% of the country's population and also provides livelihoods for its residents, offering both natural resources and ecosystem services that support local communities and the broader economy (G-SHE 2018; Halder et al. 2025).

10.3 The Plant Diversity of Indian Himalayan Region

The temporal and spatial variations driven by diverse geological orogeny have resulted in significant differences in climate and physiography across the Indian Himalayan Region (Thiede and Ehlers 2013). These variations have greatly contributed to the richness and representativeness of the region's biodiversity at all levels (Singh et al. 2019). The region's unique spatial position and the heterogeneous distribution of its biodiversity elements have led to complex biogeographic patterns (Wambulwa et al. 2021). Despite the fact that our understanding of the plant diversity in the Indian Himalayan Region is still incomplete, some regional and local floras provide sporadic information. It is estimated that the Eastern Himalaya harbours over 8000 species of flowering plants, while the Western Himalaya supports more than 5000 species, with some overlap in species between the two regions (Hajra and Mudgal 1997; Singh et al. 2019). This east-to-west biodiversity gradient and the dominance of endemic species in the Eastern Himalayas is supported by Chap. 4 (Sect. 4.2) and Chap. 9 (Sect. 9.1), which discusses endemic taxa patterns and conservation challenges.

The Indian Himalayan Region accounts for more than 50% of India's total recorded flowering plants (Singh et al. 2019), with around 13% of these species being endemic to the region (Singh et al. 2015). This high endemicity aligns with observations in Chap. 1 (Sect. 1.3) and Chap. 9 (Sect. 9.1), which emphasise the Himalayan region's position as a global biodiversity hotspot with significant conservation value. The region is home to over 1125 tree species, 675 edible species, and nearly 1743 species with medicinal value (Samanta et al. 2007; Dash and Singh 2011). Additionally, the Indian Himalayan Region supports some of the richest alpine flora in the world, with a high level of endemism (Tsering et al. 2010).

A recent study by Singh et al. (2019) has reported the occurrence of 11,157 taxa of flowering plants in the Indian Himalayan Region, belonging to 2359 genera and 241 families (Table 10.1). Out of the total taxa, dicotyledons comprise 8184 taxa from 1805 genera across 200 families, while monocotyledons consist of 2973 taxa from 554 genera across 41 families. Among the dominant families (Table 10.2), *Poaceae* is the most represented, with 912 taxa, followed by *Asteraceae* (820 taxa), *Orchidaceae* (819 taxa), and *Fabaceae* (534 taxa). The top 10 families contribute nearly 45% of the total flowering plant species found in the Indian Himalayas. Among the most dominant genera (Table 10.3), *Carex* (183 species) is the largest, followed by *Impatiens* (122), *Rhododendron* (112), *Primula* (99), *Pedicularis* (92), *Bulbophyllum* (88) and *Saxifraga* (86 species). The top 10 genera alone contribute 9.06% of the total species found in the region.

Table 10.1 Conspectus of plant diversity in the Indian Himalayan Region

Plant group	No. of taxa	No. of genera	No. of families
Angiosperms	11,157	2359	241
Dicotyledons	8184	1805	200
Monocotyledons	2973	554	41

Table 10.2 Dominant plant families of the Indian Himalayan Region

Sl. No.	Name of the family	No. of species	No. of genera
1.	*Poaceae*	912	204
2.	*Asteraceae*	820	185
3.	*Orchidaceae*	819	140
4.	*Fabaceae*	534	105
5.	*Rosaceae*	393	37
6.	*Cyperaceae*	380	29
7.	*Scrophulariaceae*	300	53
8.	*Rubiaceae*	277	70
9.	*Lamiaceae*	276	71
10.	*Ericaceae*	234	12
Total		**4945**	**906**

Table 10.3 Dominant plant genera of the Indian Himalayan Region

Sl. No	Name of the genera (family)	Number of species
1.	*Carex (Cyperaceae)*	183
2.	*Impatiens (Balsaminaceae)*	122
3.	*Rhododendron (Ericaceae)*	112
4.	*Primula (Primulaceae)*	99
5.	*Pedicularis (Scrophulariaceae)*	92
6.	*Bulbophyllum (Orchidaceae)*	88
7.	*Saxifraga (Saxifragaceae)*	86
8.	*Taraxacum (Asteraceae)*	82
9.	*Corydalis (Fumariaceae)*	78
10.	*Astragalus (Fabaceae)*	69
Total		**1011**

Table 10.4 Distribution of angiosperms in the Himalayan states of India

Attributes	Arunachal Pradesh	Himachal Pradesh	Jammu & Kashmir (including Ladakh)	Sikkim	Uttarakhand	West Bengal (Darjeeling & Kalimpong)
No. of taxa	4523	3264	4252	5931	4781	1052
No. of genera	1295	1046	1220	1596	1391	550
No. of families	192	178	189	215	211	127

Table 10.4 provides a comparative breakdown of the number of species found in various states within the Indian Himalayan Region, including Arunachal Pradesh, Himachal Pradesh, Jammu & Kashmir, Sikkim and Uttarakhand. Sikkim (5931 taxa)

is the most diverse among the Himalayan states followed by Uttarakhand (4781 taxa) and Arunachal Pradesh (4523 taxa).

10.4 Mass Tourism in the Indian Himalayan Region

Mass tourism refers to a type of tourism where large groups of people visit the same destination, often during peak seasons, for leisure or recreational activities (Naumov and Green 2015). In the Indian Himalayan Region, mass tourism is far from a recent phenomenon. For centuries, the region has been a destination for saints, seers, and spiritual seekers in search of peace, enlightenment, and sacred reflection (Bleie 2003). This is evident in the numerous major pilgrimage sites scattered across the high Himalayas, including revered Hindu, Buddhist, and Sikh shrines such as Badrinath, Kedarnath, Gangotri, Yamunotri, and Hemkund in Uttarakhand, as well as Vaishno Devi and the Amarnath Cave in Jammu & Kashmir. These sacred destinations have historically been located in remote areas, often difficult to access, limiting the number of pilgrims who could visit. However, with the development of an extensive network of roads, hotels, lodges, and tourism services in recent decades, the number of pilgrims has increased significantly (Kala 2014). In addition to religious pilgrimage, the Himalayas have long served as a corridor for trade, with people crossing high mountain passes for centuries (Wu et al. 2022). Regional and domestic pilgrimage remains a dominant form of tourism in the region today (Bleie 2003). Beyond its spiritual significance, the mountains and their passes offer stunning natural beauty and serene environments, drawing tourists from around the world. During British colonial rule, hill stations like Shillong, Darjeeling, Nainital, Almora, Ranikhet, Dehra Dun, Mussoorie, Chakrata, Shimla, Solan, and Srinagar have developed as recreational retreats, and these areas continue to attract both domestic and international visitors.

Historically, mass tourism in the Indian Himalayan Region began in the mid-twentieth century, driven by significant global and regional changes, including the end of World War II in 1945, India's independence in 1947, and the opening of new roads and airports in the mountains (Sezgin and Yolal 2012). Over the last few years, tourism in this region has grown exponentially. For instance, Sikkim welcomed 720,768 tourists in 2010, whereas in 2019, the number increased to 15,55,211. Similarly, Uttarakhand attracted 3,03,33,288 tourists in 2010 and welcomed 3,77,38,193 visitors from abroad and other Indian states in 2019. The tourism trends in the Himalayan states of India from 2010 to 2019 are presented in Table 10.5.

Table 10.5 Tourism trends in the Himalayan states of India

State	No. of tourists in 2010			No. of tourists in 2019		
	Domestic	Overseas	Total	Domestic	Overseas	Total
Arunachal Pradesh	227,857	3395	231,252	555,639	7825	563,464
Himachal Pradesh	12,811,986	453,616	13,265,602	16,829,231	382,876	17,212,107
Jammu & Kashmir (including Ladakh)	9,973,189	480,99	10,021,288	16,163,330	57,920	16,221,250
Sikkim	700,011	20,757	720,768	1,421,823	133,388	1,555,211
Uttarakhand	30,206,030	127,258	30,333,288	37,585,920	152,273	37,738,193
West Bengal (Darjeeling)	372,741	27,218	399,959	750,000 (approx.)	42,000 (approx.)	792,000 (approx.)

Source: Indian tourist statistics at a glance and State Tourism Departments

10.5 Impact of Mass Tourism on Indian Himalayan Region

Mass tourism in the Indian Himalayas has had both positive and negative impacts. It contributes to economic growth in positive means by generating income, creating employment opportunities, and promoting infrastructure development (Sati and Rawat 2024; Wani et al. 2025). However, it also takes a toll on the environment, leading to the extinction of wildlife, depletion of plant species, and degradation of habitats (Apollo et al. 2021). Besides, mass tourism affects local culture, customs, and traditions, with both positive and negative consequences (Rai et al. 2024; Sati and Rawat 2024).

10.5.1 Economic Impacts

Tourism in the Himalayas is a complex and dynamic sector that plays a crucial role in the region's economy (Mieczkowski 1995). It is a major source of income, with states like Uttarakhand and Himachal Pradesh reportedly generating 10–20% of their revenue from taxes and levies on pilgrims annually (Rangan 2000). Tourism has played a pivotal role in shaping and influencing the development of the Himalayan region particularly from the mid-twentieth century (Sati and Rawat 2024). Presently, the economy of the Himalayan states is largely dependent on tourism. For instance, nearly 6 crore tourists and pilgrims visited Uttarakhand in 2023 contributing significantly to both revenue generation and employment creation (UTDB 2023).

10.5.2 Environmental Impacts

Tourism practices have significant negative impacts on air, water, and land resources (Chandel 2022; Voumik et al. 2024). The high levels of carbon emissions in popular tourist destinations, particularly during peak seasons, worsen environmental pollution (Voumik et al. 2024). Tourism agencies often prioritise the economic benefits of tourism over its environmental costs (Baloch et al. 2023). The increasing energy demand in major tourist destinations further intensifies environmental challenges (Baloch et al. 2023; Voumik et al. 2024).

In the Himalayas, the rising number of tourists is gradually contributing to the degradation of the natural environment (Tiwari and Rani 2024). Unplanned development, combined with increasing population pressures, is leading to water scarcity, the accumulation of solid waste, and sewage problems, all of which are diminishing the region's environmental quality (NITI Aayog 2018). The growing number of construction projects is contributing to habitat degradation. The hill towns such as Darjeeling and Kalimpong (West Bengal), Gangtok (Sikkim), Dharamsala, Simla, Solan, Kullu and Manali (Himachal Pradesh), Nainital and Mussoorie (Uttarakhand), Srinagar (Jammu & Kashmir) have become overcrowded, with buildings almost obscuring the natural views from the town centres (Fig. 10.1).

Fig. 10.1 Impact of mass in hilly towns (**a**) A hillside with buildings and tall trees, showcasing a densely built area in Darjeeling; (**b**) A panoramic view of Gangtok town showing crowded buildings nestled in green hills with snow-capped mountains; (**c**) buildings obstructing natural view in Mussoorie; (**d**) traffic congestion in Darjeeling

The growing volume of waste is also compromising the quality of river water and drinking water (NITI Aayog 2018). It has been observed that the drinking water sources in hill towns like Shimla, Solan, Manali, Dharamsala, Mussoorie and Darjeeling have been severely damaged by construction, leading to water scarcity. Furthermore, the indiscriminate disposal of waste in tourist areas is disrupting wildlife habitats, threatening endangered species and other local wildlife (Puri et al. 2020; Thakur et al. 2021).

Thus, the adverse effects of mass tourism are evident in heavy traffic congestion, worsening air quality, higher levels of waste generation, and increased pressure on water resources. The concentration of tourists in certain mountain destinations has led to serious environmental, socio-economic, and cultural challenges. As tourism continues to grow, both the environmental and social systems of the region are undergoing significant changes.

10.5.3 Cultural Impact

Tourism has a significant impact on the culture of the Himalayan region, with both positive and negative consequences. On the positive side, tourism can promote cultural exchange by showcasing local traditions, such as folk songs, dances, and regional cuisine, at popular tourist destinations (Sati and Rawat 2024). This not only highlights the region's rich cultural heritage but also provides economic opportunities for local communities. However, there are also negative aspects to this cultural interaction. The traditional way of life in remote Himalayan communities makes them particularly vulnerable to external influences (Banerji and Fareedi 2009). As tourism brings modern cultural trends such as changes in food preferences, lifestyles, and occupations, there is a risk of cultural erosion. Local populations may start to abandon their time-honoured practices in favour of adopting elements of the cultures brought by tourists and pilgrims, leading to the dilution of their unique cultural identity (Sati and Rawat 2024).

10.5.4 Role of Tourism in the Spread of Invasive Species and Ecosystem Disruption

Tourism activity in the Himalayan region can significantly contribute to the spread of invasive species, pests and diseases via various pathways. Tourists and their activities often unintentionally carry seeds, insect eggs, and pathogens on their clothing, footwear, camping gear, and vehicles, which can be introduced into vulnerable ecosystems. Activities such as trekking, off-road travel, and soil trampling disturb natural habitats, creating favourable conditions for invasive species to establish and proliferate. Additionally, infrastructure development including roads, resorts, and other facilities can introduce contaminated soil and non-native landscaping plants, while improperly managed waste disperses waterborne pathogens and encourages the growth of invasive aquatic species (Hall 2019). Moreover, interactions with wildlife, such as feeding or close contact, further increase the risk of disease transmission, while discarded food invites pests. The absence of awareness and biosecurity practices among tourists worsens these risks. In the Himalayas, vegetation of invasive species such as *Lantana camara* and *Ageratina adenophora* have become widespread as discussed in Chap. 7 (Sect. 7.2), while pests such as those responsible for pine wilt disease indicating the profound ecological threats posed by tourism (Chaudhary et al. 2020; Joshi et al. 2024). Ecotourism, though intended to support conservation, can threaten native biodiversity when sensitive areas are accessed without restrictions. Unregulated movement of tourists can lead to habitat degradation, disturb wildlife, and facilitate the spread of invasive species. In fragile ecosystems like the Indian Himalayas, such impacts can have long-lasting ecological consequences.

10.6 Ecotourism: An Alternative of Mass Tourism

The impacts of tourism on mountain ecosystems and biological resources are a significant concern due to the Himalayas' high biodiversity and environmental sensitivity. Additionally, cultural identities and diversity in mountain regions are increasingly threatened by the economic, social, and environmental forces linked to mountain tourism. In this context, community-based ecotourism has emerged as a sustainable alternative to the commercial tourism currently practiced in overdeveloped hill towns like Nainital, Mussoorie, Shimla, Kullu, Manali, Gangtok, and Darjeeling.

The concept of 'Ecotourism' emerged in the 1980s as a way to promote sustainable tourism in natural, forest, and cultural areas (Sati 2021). It focuses on nature-based recreation that is both sustainable and respectful to the environment (Cheia 2013). Ecotourism is often seen as a modern term for tourism practices that have long existed in natural settings (Wall 1997; Cheia 2013). It is also referred to as 'Responsible Tourism' due to its multifaceted approach, which includes assessing tourism's environmental impact, encouraging community participation in ecotourism development, and fostering nature conservation (Black 1996; Jeevishaa et al. 2023). Additionally, ecotourism emphasises educating people on the importance of conserving natural environments and managing ecosystems (Ballantine and Eagles 1994). It regulates tourist inflows in sensitive areas and implements policies to conserve natural habitats and biodiversity (Singh et al. 2009). Local communities support ecotourism as it generates economic opportunities while promoting the conservation of natural habitats, biodiversity, and water resources (Bookbinder et al. 1998). Thus, sustainable ecotourism integrates nature conservation and poverty alleviation by balancing environmental protection with socio-economic development, offering a holistic strategy to address global challenges while benefiting both ecosystems and local communities (Majeed et al. 2025).

Currently, tourism trends in the Himalayan states are shifting towards 'off-route' destinations, in addition to traditional pilgrimage sites. Domestic travellers, in particular, are opting for stays in peripheral areas rather than in the crowded hill towns. They are increasingly looking for authentic experiences with local communities, nature-focused activities, and cultural familiarity. This trend is driving the demand for eco-friendly accommodations, guided nature tours, and sustainable travel packages that offer distinctive and unforgettable experiences. Ecotourism, therefore, offers a viable alternative to mass tourism. The rationale behind alternative tourism is to counter the negative effects of mass tourism. Alternative tourism shifts the focus of governments from economic gains alone to the preservation of pristine environments and the consideration of local communities' needs (Sati and Rawat 2024). Alternative tourism should discourage external influences on development and emphasise local participation. Ecotourism offers several advantages, such as providing income through home stay accommodations, improving management and decision-making skills, enhancing family well-being, developing eco-friendly facilities for visitors and locals, reducing tourism revenue leakage, boosting the purchasing power of local people, preserving local traditions and customs, and strengthening host-guest relationships through interaction. Encouragingly, all six Himalayan states and the Union Territory of Ladakh are making efforts to incorporate ecotourism into their mainstream development agendas through policy initiatives (NITI Aayog 2018).

The sustainable ecotourism aligns with the sustainable development goals (SDGs), particularly SDG 8 (Decent Work and Economic Growth) by creating sustainable livelihoods, SDG 12 (Responsible Consumption and Production) through eco-friendly tourism, and SDG 15 (Life on Land) by conserving biodiversity. It also supports SDG 1 (No Poverty) and SDG 6 (Clean Water and Sanitation) by benefiting local communities and protecting water resources, making ecotourism a holistic approach to sustainable development. Chapter 2 (Sect. 2.1) supports this view by highlighting how local dependency on natural resources necessitates sustainable approaches for conservation and biodiversity preservation. In 2004, the United Nations World Tourism Organization (UNWTO) published a guidebook, outlining 13 key indicators for sustainable tourism (World Tourism Organization 2004). These include:

1. Wellbeing of host communities
2. Sustaining cultural assets
3. Community participation in tourism
4. Tourist satisfaction
5. Health and safety
6. Capturing economic benefits from tourism
7. Protection of valuable natural assets
8. Managing scarce natural resources
9. Limiting the impacts of tourism activities

10. Controlling tourist activities and numbers
11. Destination planning and control
12. Designing products and services
13. Sustainability of tourism operations and services

When comparing state policies in the Indian Himalayan Region against these sustainable tourism indicators, most states align with the principles outlined. However, West Bengal lags in certain areas, such as sustaining cultural assets, ensuring tourist satisfaction, and maintaining the sustainability of tourism operations and services.

10.7 Ecotourism Versus Mass Tourism

Mass tourism and ecotourism are both valued for their potential to drive social and economic progress, but they are criticised when their activities lead to the destruction of the natural environment. While both forms share common goals such as improving economic well-being, generating income and employment, and fostering societal advancement through knowledge they differ significantly in their approach to environmental conservation and long-term, sustainable development.

For instance, mass tourism typically attracts large numbers of visitors during peak seasons, resulting in road congestion and overcrowded parking areas. The high influx of tourists also creates a demand for substantial accommodation, often in the form of modern hotels, which leads to the construction of metal buildings (Fig. 10.1) that degrade natural habitats and contribute to biodiversity loss. Beyond ecosystem degradation, increased human activity fosters the spread of invasive species, disrupts native plant life, and accelerates the declination of biodiversity (Dhingra et al. 2024). The vulnerabilities of endemic and medicinal plant species due to anthropogenic pressures and habitat degradation are deeply addressed in Chaps. 5 (Sect. 5.2) and 9 (Sect. 9.3), where overharvesting, climate change, and invasive species are cited as core threats. Mass tourists often travel on packaged tours and tend to concentrate around popular tourist destinations (Fig. 10.2).

In contrast, eco-tourists usually travel in smaller groups and prioritise staying close to nature, requiring minimal accommodations such as homestays or eco-huts (Fig. 10.3). They typically arrange their own accommodations and are more widely dispersed across a larger area. Many eco-tourists also prefer nature-based or adventure travel experiences, often bringing their own camping gear to minimise their environmental footprint.

Fig. 10.2 Images depicting the traffic of tourists and pilgrims along the mountainous trek (**a**) A group of people walking through a grassy hillside with scattered rocks in Valley of Flowers National Park, Uttarakhand; (**b**) Hikers ascending a narrow, rocky path surrounded by greenery in Valley of Flowers National Park, Uttarakhand; (**c**) A bridge crossing a rushing stream with people walking across it, surrounded by rocky terrain by Bhyundar village, Uttarakhand; (**d**) Trekkers in raincoats walking along a snow-lined path with mountains in the background in Hemkund, Uttarakhand

10.8 Conclusions

Mass tourism can damage the environment and ecology when it surpasses the carrying capacity of a destination. Famous hill towns in the Indian Himalayan Region are becoming overcrowded during peak seasons, leading to increased carbon emissions, traffic congestion, waste, river pollution, and water scarcity. As visitors seek to avoid the overcrowding, many are shifting to "off-route" destinations. Ecotourism, therefore, presents a sustainable alternative to mass tourism in these areas. It not only provides opportunities for wilderness experiences and cultural immersion, but also benefits economically marginalised communities in remote regions. Additionally, ecotourism plays a crucial role in conserving biodiversity and landscapes. The economic impact of tourism helps local communities become stewards of their natural heritage, fostering protection of flora and fauna. Employment generated includes guides, porters, conservation workers, and small shopkeepers. Thus, ecotourism has the potential to be a valuable strategy for both nature conservation and economic development in the Indian Himalayan Region.

Fig. 10.3 Four images depicting scenic landscapes with ecotourism facilities (**a**) homestays in Chatakpur, Darjeeling, a village with colorful rooftops surrounded by lush greenery and mountains under a cloudy sky; (**b**) eco-huts in Hilley, Sikkim with a green roof, surrounded by prayer flags and trees, in a misty environment; (**c**) camping in the heart of nature on a grassy meadow, with a large rock nearby, set against a backdrop of snow-capped mountains and a clear sky at Sainj Wildlife Sanctuary, Himachal Pradesh; (**d**) forest hut in Great Himalayan National Park, Himachal Pradesh nestled among trees on a grassy hill, with a backdrop of distant mountain ranges under a clear blue sky

Acknowledgements We wish to express our heartfelt thanks to our respective institutions for providing the essential academic and infrastructural support throughout our research journey. Special appreciation is owed to the local people of the Himalayas whose indigenous knowledge and intimate relationship with nature continue to motivate conservation activities. We thank Editors and reviewers for their constructive comments.

References

Apollo M, Andreychouk V, Mostowska J, Jin Z, Maciuk K, Rettinger R, Jones TE (2021) Himalayan nature-based tourism. Potential, regional diversity, nature conservation and touristic load. Int J Conserv Sci 12(4):1527–1546

Ballantine JL, Eagles PFJ (1994) Defining Canadian eco-tourists. J Sustain Tour 2(4):210–214

Baloch QB, Shah SN, Iqbal N, Sheeraz M, Asadullah M, Mahar S, Khan AU (2023) Impact of tourism development upon environmental sustainability: a suggested framework for sustainable ecotourism. Environ Sci Pollut Res 30(3):5917–5930

Banerji G, Fareedi M (2009) Protection of cultural diversity in the Himalayas. In: A background paper for a workshop on addressing regional disparities: inclusive & culturally attuned development for the Himalayas, New Delhi
Black R (1996) Eco-tourism. What does it really mean? Range 34:4–7
Bleie T (2003) Pilgrim tourism in the Central Himalayas. Mt Res Dev 23(2):177–184
Bookbinder MP, Dinerstein E, Rijal A, Cauley H, Rajouria A (1998) Eco-tourism support's to biodiversity conservation. Conserv Biol 12(6):1399–1404
Chandel SPK (2022) Impacts of tourism on environment. Cen As J Inn Tour Manag Fin 3(10): 90–98
Chaudhary A, Sarkar MS, Adhikari BS, Rawat GS (2020) *Ageratina adenophora* and *Lantana camara* in Kailash Sacred Landscape, India: current distribution and future climatic scenarios through modeling. PLoS One 15(9):e0239690. https://doi.org/10.1371/journal.pone.0239690
Chauhan HK, Gallacher D, Bhatt A, Bisht AK (2023) The Himalayas: a climate change laboratory. Environ Dev 45:100814. https://doi.org/10.1016/j.envdev.2023.100814
Cheia G (2013) Ecotourism: definition and concepts. J Tour 15:56–60
Chettri N, Shakya B, Thapa R, Sharma E (2008) Status of a protected area system in the Hindu Kush-Himalayas: an analysis of PA coverage. Int J Biodivers Sci Manag 4(3):164–178
Dash SS, Singh P (2011) Trees of Sikkim. In: Arrawatia ML, Tambe S (eds) Biodiversity of Sikkim exploring and conserving a global hotspot. Information and Public Relations Department, Government of Sikkim, Gangtok, pp 89–124
Dhingra GK, Rawat A, Nautiyal OP, Rawat DS, Kuriyal SK, Rawat S, Khanduri P (2024) An introduction to the changing scenario of plant diversity in Himalayan habitats. In: Rawat A et al (eds) Shifting ecosystems: plant diversity and habitat changes in the Himalayas. Bishen Singh Mahendra Pal Singh, Dehradun, pp 1–20
Douglas E (2021) Himalaya: a human history. W.W. Norton, New York
G-SHE (2018) Governance for sustaining Himalayan ecosystem guidelines and best practices. Ministry of Environment and Forests, Government of India, New Delhi and G.B. Pant Institute of Himalayan Environment and Development, Kosi-Katarmal, Almora
Hajra PK, Mudgal V (1997) Plant diversity hotspots in India - an overview. Botanical Survey of India, Calcutta
Halder O, Sarkar A, Sanyal A, Mandal L (2025) Interconnection between agriculture and livelihoods in Har-ki-Dun Valley, Uttarakhand, Indian Himalaya. In: Borthakur A, Singh P (eds) People and mountain environments. Springer, Cham, pp 149–175
Hall CM (2019) Biological invasion, biosecurity, tourism, and globalisation. In: Timothy DJ (ed) Handbook of globalisation and tourism. Edward Elgar Publishing, United Kingdom.
Ives JD, Messerli B (2003) The Himalayan dilemma: reconciling development and conservation. Routledge, New York
Jeevishaa S, Krithick R, Divya M (2023) Eco tourism and conservation: legal frameworks for responsible tourism and biodiversity conservation. Legal Lock J 3:105
Joshi VC, Chandra N, Sundriyal RC, Arya D, Mishra AP, Abdo HG (2024) Ecological status and spatial extent of non-native shrubs *Lantana camara* L. and *Ageratina adenophora* Sprengel in the forest communities of western Himalaya. Trees For People 15:100494. https://doi.org/10.1016/j.tfp.2024.100494
Kala CP (2014) Deluge, disaster and development in Uttarakhand Himalayan region of India: challenges and lessons for disaster management. Int J Disaster Risk Reduct 8:143–152
Loewen CJG, Jackson DA, Gilbert B (2023) Biodiversity patterns diverge along geographic temperature gradients. Glob Chang Biol 29(3):603–617
Majeed LR, Rashid S, Sharma D, Majeed LF, Bhat MA (2025) Nature conservation and poverty alleviation through sustainable ecotourism: the case of Lolab Valley, India. Discov Environ 3(1). https://doi.org/10.1007/s44274-024-00181-8
Matthew R (2012) Environmental change, human security, and regional governance: the case of the Hindu Kush/Himalaya region. Glob Environ Polit 12(3):100–118

Mieczkowski Z (1995) Environmental issues of tourism and recreation. University Press of America, Lanham

Molden DJ, Shrestha AB, Immerzeel WW, Maharjan A, Rasul G, Wester P, Wagle N, Pradhananga S, Nepal S (2022) The great glacier and snow-dependent rivers of Asia and climate change: heading for troubled waters. In: Biswas AK, Tortajada C (eds) Water security under climate change. Water resources development and management. Springer, Singapore, pp 223–250. https://doi.org/10.1007/978-981-16-5493-0_12

Naumov N, Green D (2015) Mass tourism. In: Jafari J, Xiao H (eds) Encyclopedia of tourism. Springer, New York. https://doi.org/10.1007/978-3-319-01669-6_378-1

NITI Aayog (2018) Report of working group II: sustainable tourism in the Indian Himalayan region. Government of India, New Delhi

Nyaupane GP, Chhetri N (2009) Vulnerability to climate change of nature-based tourism in the Nepalese Himalayas. Tour Geogr 11(1):95–119

Pandit MK (2017) Life in the Himalaya: an ecosystem at risk. Harvard University Press, London and Cambridge

Puri K, Joshi R, Singh V (2020) Open garbage dumps near protected areas in Uttarakhand: an emerging threat to Asian elephants in the Shivalik Elephant Reserve. J Threat Taxa 12(11): 16571–16575

Rai D, Krishnaiah YV, Panja K, Chakma A, Mallick M, Das D, Hati M (2024) New horizons for ecotourism in Sikkim: an integrated MCDA approach. Int J Ecol Environ Sci 51(1):95–110

Rangan H (2000) Of myths and movements: rewriting Chipko in Himalayan History. Verso: 70, London

Samanta SS, Pant S, Singh M, Lal M, Singh A, Sharma A, Bhandari S (2007) Medicinal plants in Himachal Pradesh, north western Himalaya, India. Int J Biodivers Sci Manag 3(4):234–251

Sati VP (2021) Trends and potential of eco-tourism development in Uttarakhand Himalaya. J Sustain Tour 5(1):19–35

Sati VP, Rawat MS (2024) Types, potential, and implications of tourism in the central Himalaya: a review. J Acad Tour Stud 5(1):117–128

Sezgin E, Yolal M (2012) Golden age of mass tourism: its history and development. In: Murat K (ed) Visions for global tourism industry-creating and sustaining competitive strategies. Intech Open, London, UK. pp 73–90

Singh RB, Mal S, Kala CP (2009) Community responses to mountain tourism: a case in Bhyundar Valley, Indian Himalaya. J Mt Sci 6(4):394–404

Singh P, Kartigeyan K, Laxminarasigham P, Dash SS (2015) Endemic vascular plants of India. Botanical Survey of India, Kolkata

Singh P, Dash SS, Sinha BK (2019) Plants of Indian Himalayan region (an annotated checklist & pictorial guide), part I. Botanical Survey of India, Kolkata

Thakur A, Kumari S, Sinai Borker S, Prashant SP, Kumar A, Kumar R (2021) Solid waste management in Indian Himalayan region: current scenario, resource recovery, and way forward for sustainable development. Front Energy Res 9(609229):1–18. https://doi.org/10.3389/fenrg.2021.609229

Thiede RC, Ehlers TA (2013) Large spatial and temporal variations in Himalayan denudation. Earth Planet Sci Let 372:278–293

Timothy DJ, Nyaupane GP (2022) The Himalaya: physiographic and cultural perspectives. In: Nyaupane GP, Timothy DJ (eds) Tourism and development in the Himalaya: social, environmental and economic forces. Routledge, New York, pp 18–36

Tiwari C, Rani M (2024) Trouble in paradise: evaluating the effects of unorganized tourism on the Himalayan ecology. In: Borthakur A, Singh P (eds) The Himalayas in the anthropocene: environment and development. Springer, Cham, pp 165–183. https://doi.org/10.1007/978-3-031-50101-2_7

Tsering K, Sharma E, Chettri N, Shrestha A (2010) Climate change vulnerability of mountain ecosystems in the Eastern Himalayas; Climate change impact and vulnerability in the Eastern Himalayas—synthesis report. ICIMOD, Kathmandu

UTDB (2023) Annual report. Uttarakhand Tourism Development Board, Dehradun

Voumik LC, Islam MA, Nafi SM (2024) Does tourism have an impact on carbon emissions in Asia? An application of fresh panel methodology. Environ Dev Sustain 26(4):9481–9499

Wall G (1997) Is ecotourism sustainable? Environ Manag 21(4):483–491

Wambulwa MC, Milne R, Wu ZY, Spicer RA, Provan J, Luo YH, Zhu GF, Wang WT, Wang H, Gao LM, Li DZ (2021) Spatiotemporal maintenance of flora in the Himalaya biodiversity hotspot: current knowledge and future perspectives. Ecol Evol 11(16):10794–10812

Wani MD, Dada ZA, Bhat WA, Shah SA (2025) Community-driven ecotourism in the trans Himalayan region: a sustainable model for cultural and environmental preservation. Community Dev. https://doi.org/10.1080/15575330.2025.2451358

Wester P, Mishra A, Mukherji A, Shrestha AB (2019) The Hindu Kush Himalaya assessment: mountains, climate change, sustainability and people. Springer, Cham

World Tourism Organization (2004) Indicators of sustainable development for tourism destinations: a guidebook (English version). UNWTO, Madrid

Wu S, Yan J, Zhang Y, Peng T, Su K (2022) Evolutionary process and development implications of traditional trade routes in the Himalayan region. J Geogr Sci 32(9):1847–1865

Yao T, Thompson LG, Mosbrugger V, Zhang F, Ma Y, Luo T, Xu B, Yang X, Joswiak DR, Wang W, Joswiak ME, Devkota LP, Tayal S, Jilani R, Fayziev R (2012) Third pole environment (TPE). Environ Dev 3:52–64

Chapter 11
Threat Status to the Himalayan Ecosystem and Its Trending Preventive Strategies

Alok Sharma, Pardeep Kumar, Sapna Avinash Kondalkar, and Biswajit Paul

Abstract The Himalayan ecosystem, one of the most unique and fragile ecological zones on the planet, is under increasing threat from a variety of natural and human-induced factors. Stretching across five countries such as Nepal, India, Bhutan, China, and Pakistan, the Himalayas are native to countless species of flora and fauna, many of which are endemic to the region. The ecosystem is also critical for millions of people who rely on its resources for their livelihood, food, and water. However, this vital ecosystem is facing severe threats due to climate change, deforestation, pollution, overgrazing, invasive species and unsustainable developmental practices. Understanding these threats and implementing effective preventive strategies is crucial to preserving the Himalayan ecosystem for future generations. Several key initiatives, such as climate-resilient infrastructure, sustainable agriculture, conservation of forests and biodiversity, ecotourism and monitoring of glacial outbursts, are being implemented at a national scale to conserve the fragile ecosystem of the Himalayas. Furthermore, the launch of the National Mission for Sustaining the Himalayan Ecosystem (a pilot programme of the government of India) is a key step toward a better understanding of the coupling between climate factors influencing the Himalayan ecosystem and providing inputs for their sustainable development.

Keywords Conservation · Sustainable development · Climate change · Ecotourism

---

Alok Sharma and Pardeep Kumar contributed equally with all other contributors.

A. Sharma (✉) · S. A. Kondalkar
CCRAS- Regional Ayurveda Research Institute, Gwalior, Madhya Pradesh, India

P. Kumar
Department of Botany, Mata Gujri College, Fatehgarh Sahib, Punjab, India

B. Paul
Botany Section, Central Ayurveda Research Institute, CCRAS, Ministry of AYUSH, Guwahati, Assam, India

D. Das et al. (eds.), *Plant Diversity and Conservation Strategies of the Himalayan Biodiversity Hotspot*, Topics in Biodiversity and Conservation 21,
https://doi.org/10.1007/978-3-032-21697-7_11

11.1 Introduction

The Himalayas, one of the youngest mountain chains on Earth, exhibit remarkable ecological complexity and biodiversity (Pandit et al. 2014). This richness arises from their dynamic geological formation and pronounced altitudinal and climatic variations, which together shape a mosaic of habitats across the region (Chap. 1, Sects. 1.2 and 1.3.1). In the Himalayas, due to the presence of lofty peaks, steep slopes, and gradients, climatic conditions change abruptly at short distances. This unique geography and landscape type govern the hydrological processes that cause high surface runoff and soil erosion. Furthermore, the snow-covered glaciers in the Himalayas ensure a year-round water supply for a large proportion of the human population. Due to its varied climatic conditions, the Himalayas are home to a million of species, and some of them are endemic to the Himalayas. The Indian Himalayan Region (IHR), spanning over 2400 km (Chap. 1, Sect. 1.1) and encompassing 13 states and union territories, is of global significance (Rana and Rawat 2017). It is commonly known as the water tower of India because several key rivers originate from the Himalayas. Also, IHR is home to several ethnic communities and a rich variety of flora and fauna (Kumar et al. 2021). Additionally, IHRs are home to numerous religious sites and tourist destinations like Kedarnath, Badrinath, and Valley of Flowers, which sustain millions of people living in the hills. Furthermore, the Himalayas comprise a large population of medicinal and aromatic plants (MAPs), which provide strong economic and health benefits. A total of 1748 plant species, including majorly angiosperms (1685), followed by pteridophytes (51) and gymnosperms (12) are known to be found in the Indian Himalayan region (IHR) (Samant et al. 1998; Singh et al. 2017; Negi et al. 2018a; Singh et al. 2021). However, despite being global biodiversity hotspots, the Himalayas are among the most fragile and unstable mountains in the world (Negi et al. 2019). Due to the increasing population and subsequent anthropogenic pressure, these mountains are facing unprecedented ecological threats, mainly in the form of urbanisation, deforestation, land use planning, climate change, habitat loss and biological invasion (Reddy et al. 2016). This chapter aims to summarise the prevailing threats to the Himalayan ecosystem and their potential influence on living organisms. Additionally, the present study also consolidates the various preventive strategies employed at both local and national scales to protect the fragile Himalayan ecosystem. Some of the major threats to the Himalayan ecosystem are discussed below.

11.2 Deforestation and Habitat Loss

The Himalayas have witnessed significant deforestation due to logging, expansion of agricultural lands, and infrastructure development, such as roads and hydroelectric projects. For example, a study conducted by Brandt et al. (2017) in the Himalayan region (Nepal, India, Bhutan, China, Myanmar) found that deforestation rates in these countries vary between 0.5% to 1.7% and only 13% of forests are protected. Deforestation leads to soil erosion, landslides, and loss of habitat for numerous species, including endangered ones like the snow leopard and red panda. The study of Pandit et al. (2007) showed that in the Indian Himalayan region (IHR), if the current rate of deforestation prevails, then by the year 2100, only 10% of the land area will be under dense forest cover leading to the extinction of more than 300 vascular plant species. The depletion of forest cover disrupts local communities' access to essential resources such as firewood, fodder, and medicinal plants, intensifying socio-economic vulnerability as mentioned in Chap. 2 (Sect. 2.1). Deforestation and forest degradation together constitute a total of 15% of all greenhouse gas emissions, causing the rise in global temperatures and climatic alterations (Kumar et al. 2022).

11.3 Climate Change

Climate change is arguably the most pressing threat to the Himalayan ecosystem. Rising global temperatures are causing the glaciers in the Himalayas to melt at an unprecedented rate (Shrestha et al. 2012; Adler et al. 2022). These glaciers are known to be a source of several major rivers in Asia, including the Ganges, Brahmaputra, and Indus, which provide water to nearly 1.5 billion people. The accelerated glacial melt not only threatens freshwater availability but also increases the risk of glacial lake outburst floods (GLOFs), which can devastate downstream communities (Talukder et al. 2021; Wester et al. 2019). Additionally, fluctuating weather conditions have caused the irregularity of rainfall, prolonged droughts, and the occurrence of increased frequency of extreme weather events, such as landslides and flash floods. Studies have demonstrated that high-altitude plants are particularly vulnerable to atmospheric warming, largely due to increased exposure to photosynthetically active radiation at elevated elevations (Telwala et al. 2013). This phenomenon has triggered significant altitudinal range shifts in climate-sensitive species, including *Betula utilis* (Himalayan birch) and *Meconopsis* spp., with some populations already facing the risk of local extinction (Hamid et al. 2018). These observations align with broader patterns of species vulnerability and ecological disruption previously outlined in Chap. 9 (Sect. 9.1), highlighting how climate change is reshaping biodiversity, agriculture, and water systems across the Himalayan landscape ultimately threatening both environmental stability and human livelihoods (Table 11.1).

Table 11.1 An overview of various threats to the Himalayan ecosystem

SL. No.	Key threats	Activities	Direct/indirect impacts
1	Anthropogenic	Deforestation	Habitat loss Carbon emission Biodiversity loss Increased rate of atmospheric pollutants
		Agriculture expansion	
		Overgrazing	
		Overexploitation of resources	
		Urbanisation	
		Infrastructure development	
		Mining	Loss of fertile soil layer Pollution Biodiversity loss
		Unsustainable tourism	Biodiversity loss Pollution Habitat degradation
		Hydroelectric dams	Habitat destruction Sediment deposition Loss of aquatic life Human displacement Displaced river flows
		Poaching	Loss of keystone species like tigers, rhinos Ecological imbalance Illegal trade
2	Natural	Weak geology	Land degradation Loss of topsoil Carbon emission
		Landslides	
		Storms and avalanches	
3	Climate change	Glacial meltdown	Sea-level rise
		Atmospheric warming	Species shifting Increased soil respiration
		Extreme weather events	Floods, droughts
		Disease and pest outbreak	Economic loss Increased hunger and poverty
4	Invasive species	Loss of native biota	Community homogenisation Biodiversity loss Altered ecosystem services

Adapted from various sources: Pandit et al. (2007), Kattel (2022), Dhyani (2023)

11.4 Overgrazing and Unsustainable Agricultural Practices

Overgrazing by livestock is a significant problem in many parts of the Himalayas. The subalpine and alpine pastures located at high altitudes provide an ideal grazing habitat. These grasslands remain snow-covered during winters and available for grazing during summers only. The Gujjars (pastoralists or nomadic communities in the Himalayas) are mainly dependent on these pastures for animal rearing

Fig. 11.1 A grazing herd of buffaloes at alpine communities in the Western Himalaya. (Photo credit: Pardeep Kumar)

(Fig. 11.1). These alpine communities are facing serious ecological threats in Kashmir, Himachal and Uttarakhand Himalayas in India and elsewhere in neighbouring countries (Nautiyal et al. 2004; Cao et al. 2013). The overgrazing and foot trampling in these unique ecosystems lead to soil degradation, reduced vegetation cover, and increased vulnerability to erosion and landslides (Ahmad et al. 2018). Unsustainable agricultural practices, such as monocropping and excessive use of chemical fertilizers, further exacerbate these problems, leading to soil fertility decline and water pollution. Traditional shifting cultivation, widely practiced in India's Eastern Himalaya, continues to pose a significant threat to the region's ecological equilibrium (Hazarika et al. 2024). This age-old agricultural practice has contributed to extensive habitat degradation and a marked decline in biodiversity an issue also examined in Chap. 5 (Sect. 5.2), where the ecological consequences of such land-use patterns are explored in greater depth.

11.5 Unsustainable Development

The accelerated pace of infrastructural development in the Himalayas encompassing roads, dams, and expanding urban settlements is exerting mounting pressure on the region's fragile ecosystems. Although infrastructure is vital for economic advancement, such growth is often pursued through unsustainable means that overlook critical environmental thresholds (Singh 2013). Large-scale deforestation, habitat fragmentation, and the disruption of natural river systems especially linked to road construction and hydropower projects have significantly undermined ecological stability and biodiversity. These trends echo concerns raised in Chap. 9 (Sect. 9.3), which outlines how anthropogenic interventions are driving habitat degradation and amplifying conservation challenges across the Himalayan landscape (Khanday et al. 2024).

11.6 Tourism

The tourism sector is considered a promising aspect to stimulate the local economy. It provides job security, service provision and exchange of foreign currencies (Geneletti and Dawa 2009). Environmental degradation is one of the shortcomings commonly associated with tourism, especially in fragile ecosystems such as the Himalayas as extensively discussed in Chap. 10 (Sect. 10.4). Rapid infrastructure development, such as roads, hotels, and transportation beyond the carrying capacity of places, poses a severe burden on environmental health. For example, degradation of biodiversity due to infrastructure development, pollution of water bodies, generation of waste and human-wildlife conflicts (Kumar et al. 2023). Furthermore, the establishment of tourist camps and trespassing through natural virgin forests by trekkers may harm local biodiversity (Fig. 11.2). Pollution, particularly plastic waste and chemical pollutants, poses a growing threat to the Himalayan environment. The accumulation of non-biodegradable waste in remote mountain areas is increasing, exacerbated by tourism and poor waste management practices. This pollution affects not only the pristine landscapes but also the health of wildlife and local communities. Additionally, the presence of pollutants in rivers and streams can have far-reaching consequences for ecosystems downstream. There is a direct link between tourism and biodiversity loss. A 10% increase in international tourism will lead to a 2–4% reduction in biodiversity (Habibullah et al. 2016). Therefore, there is a need to establish harmony between the tourism sector and biodiversity conservation.

11.7 Invasive Species

The spread of invasive species across national boundaries poses a significant threat to native ecosystems and biodiversity. Recognised as one of the primary drivers of global environmental change, invasive species contribute to biodiversity loss and the degradation of ecosystem services by altering native community composition, structure, and ecological functions (Millennium Ecosystem Assessment 2005; Pathak et al. 2019). Although mountain ecosystems are generally considered less susceptible to biological invasions due to natural elevational barriers, once an invasive species becomes established, it often proliferates rapidly, outcompeting native

Fig. 11.2 Establishment of tourist camps in a Western Himalayan region. (Photo credit: Pardeep Kumar)

flora. In the Himalayan region particularly in Nepal and India severe infestations of species such as *Lantana camara*, *Ageratina adenophora*, *Parthenium hysterophorus*, and *Ageratum conyzoides* have been widely reported (Lamsal et al. 2018; Thapa et al. 2018; Pathak et al. 2019). If current trends continue, the unchecked transboundary spread of these species may lead to serious ecological consequences an issue critically examined in Chap. 7 (Sect. 7.6) and further reinforced by Chap. 9's (Sect. 9.2) focus on the challenges invasive flora pose to endemic biodiversity in the Himalayas.

11.8 Trending Preventive Strategies

Various preventive strategies are gaining momentum in response to these mounting threats. These strategies aim to balance environmental conservation with the socio-economic needs of the local population.

11.8.1 *Community-Based Conservation Programmes*

Acknowledging and incorporating traditional ecological knowledge and community practices is essential for meaningful and effective biodiversity conservation, particularly in ecologically sensitive regions such as the Himalayas. As emphasised in Chap. 2 (Sect. 2.2), such indigenous knowledge systems form the foundation of community resilience and sustainable resource use. Community-based conservation approaches actively involve local populations in the stewardship of natural resources by recognising and valuing their ecological heritage, cultural ties, and long-standing environmental practices. This approach proves especially effective in the Himalayan context, where conservation strategies must also address the socio-economic realities of mountain communities. These programmes typically identify and engage communities based on their social networks, geographic roots, and cultural lineages to foster long-term ecological sustainability (Denman 2001; Pandey et al. 2021; Dutta and Dutta 2023).

11.8.1.1 Incorporating Indigenous Knowledge

Many Himalayan communities have lived sustainably with the environment for centuries, utilising practices such as rotational grazing, traditional agriculture, and sacred grove preservation. Modern conservation programmes increasingly integrate this indigenous knowledge, ensuring that conservation efforts are both effective and culturally appropriate. Traditional ecological knowledge (TEK) also known as local knowledge or native knowledge is found to be useful in the Himalayan ecosystem's sustainability and resilience (Florence and Mishra 2024). Nearly 40% of all Indigenous tribes reside in IHR, which is home to various groups with highly diverse cultures and races. These tribes possess traditional knowledge of fauna and flora, which plays a crucial role in managing climate change and ensuring ecosystem sustainability (Florence and Mishra 2024).

11.8.1.2 Promoting Sustainable Livelihoods

Programmes like sustainable agroforestry, which combines crops with trees to create a more resilient agricultural system, provide both environmental benefits (such as reduced soil erosion and increased biodiversity) and economic opportunities for local communities to both people living in mountains and adjoining areas (Saxena et al. 2005). Similarly, sustainable grazing practices are encouraged to prevent overgrazing and land degradation. For instance, in Nepal, the Community Forestry Programme has successfully transferred the management of forests to local community groups, leading to an increase in forest cover, improved biodiversity, and enhanced local livelihoods. Other parts of the Himalayas are also replicating this model.

11.8.2 Promotion of Sustainable Tourism

Promoting green tourism that incorporates ecological safeguards and community engagement offers a promising pathway toward sustainable conservation an approach explored in depth in Chap. 10 (Sect. 10.4). Sustainable tourism seeks to mitigate the environmental and cultural impacts of tourism while enhancing its benefits for local communities and biodiversity protection. The Himalayan region, renowned for its breathtaking landscapes and ecological richness, attracts millions of visitors annually. However, when left unregulated, tourism contributes to pollution, habitat degradation, and cultural erosion, posing direct and indirect threats to wildlife and resulting in the accumulation of substantial solid and liquid waste (Singh and Mishra 2004). As such, fostering ecologically responsible tourism anchored in environmental protection, cultural sensitivity, and inclusive development emerges as a critical strategy for ensuring the long-term sustainability of the Himalayan environment (Singh and Mishra 2004).

11.8.2.1 Eco-Friendly Trekking and Travel

Sustainable tourism initiatives include promoting eco-friendly trekking routes, limiting the number of visitors in sensitive areas, and establishing eco-campsites that minimise waste and energy consumption. For example, the government of Bhutan has implemented a "high-value, low-impact" tourism policy where tourists pay a sustainable development fee that helps fund conservation and community projects.

11.8.2.2 Waste Management Campaigns

Plastic waste is a significant issue in the Himalayas, particularly in popular trekking regions. Initiatives like the "Zero Waste Himalaya" campaign promote plastic-free zones, provide tourists with reusable water bottles, and establish waste segregation and recycling facilities in remote areas. Solid waste management (SWM) is increasingly posing a significant threat to the Indian Himalayan Region (IHR), primarily due to the growing population and inadequate disposal methods in the remote and challenging mountainous regions of IHR. Therefore, the recent focus towards the proper collection, segregation, treatment, and disposal, along with approaches like waste recycling, anaerobic digestion, and composting, is playing a positive role in ecosystem sustainability (Thakur et al. 2021).

11.8.2.3 Tourism Education Programmes

Educating tourists about the environmental and cultural sensitivity of the Himalayas is key to sustainable tourism. Many local NGOs and community groups are conducting awareness programmes to encourage responsible behaviour among tourists, such as sticking to designated trails, avoiding littering, and respecting local customs (Ministry of Tourism 2022). Moreover, programmes like "polluters to pay" and "ban on single-use plastic" are also useful for highlighting ecological sensitivity.

11.8.3 Reforestation and Afforestation Initiatives

Reforestation (restoring degraded forests) and afforestation (planting new forests) are critical strategies for combating deforestation, enhancing biodiversity, and mitigating climate change impacts in the Himalayas. Some of the key initiatives involving afforestation activities are discussed below:

11.8.3.1 Large-Scale Tree Planting Campaigns

Governments and NGOs are organising large-scale tree-planting campaigns in the Himalayas. For example, an organisation named *Grow Billion Trees* is actively involved in high-density tree plantations across Uttarakhand state, India (Sharma et al. 2023). Additionally, the Government of India's *National Mission for a Green India* aims to protect, restore, and enhance India's diminishing forest cover, thereby contributing to climate change mitigation strategies (MoEF&CC 2020). Reforestation efforts now prioritise planting native tree species that are adapted to local conditions and provide habitat for wildlife. This approach helps maintain ecological balance and ensures that the reforested areas support local biodiversity (Singh and Rawat 2021; Negi et al. 2018a, b).

11.8.3.2 Agroforestry Models

Agroforestry allows the integration of trees into agricultural landscapes, which results in multiple benefits such as improving soil health, reducing erosion, and increasing crop yields (Sharma et al. 2007). Programmes promoting agroforestry in the Himalayas encourage farmers to plant fruit-bearing and timber-producing trees, which offer both economic and ecological benefits. For example, cardamom-based traditional agroforestry practices in Eastern India are highly successful, where cultivation of cardamom provides rich cash incentives to farmers and plantation of trees like *Alnus nepalensis* acts as an efficient soil binder, preventing soil erosion and land degradation (Sharma et al. 2007). Apart from this, the agroforestry systems act as a significant storehouse of carbon, offering carbon trade in the international market as well as meeting societal needs through multiple products (Pant et al. 2022).

11.8.4 Glacial Monitoring and Early Warning Systems

With the rapid melting of glaciers due to climate change, a significant increase in glacial lake outburst floods (GLOFs) threat has been observed. Monitoring glaciers and establishing early warning systems are crucial for minimising the risk of disasters (Wang et al. 2022). Glacial monitoring involves using satellite imagery, remote sensing technology, and on-the-ground data collection to assess the state of glaciers and glacial lakes. Real-time data is then used to create early warning systems that can alert communities about potential GLOF events.

Early warning systems include automated sensors placed near glacial lakes to detect changes in water levels and alert nearby communities in real-time. For example, several early warning systems have been set up in Bhutan and China for glaciers that are at risk of a GLOF (Osti et al. 2011; Wang et al. 2022). These systems significantly reduce the potential loss of life and property by providing timely evacuation alerts. In addition, countries in the Himalayan region are increasingly collaborating on glacial research and data sharing. Organisations like the International Centre for Integrated Mountain Development (ICIMOD) are working with multiple governments to build regional databases and foster cooperation on glacial monitoring.

11.8.5 Policy Reforms and International Cooperation

Policy reforms and international cooperation are essential to addressing the complex challenges facing the Himalayan ecosystem, as this often challenges cross-national borders. Governments in the Himalayan region are tightening regulations on deforestation, mining, and infrastructure development. For example, recently, the government of India introduced the Biological Diversity (Amendment) Act (2023), revising the previous act (Biological Diversity Act 2002) to align with current needs and developments. Similarly, the revised framework of National Forest Policy (2018) aimed to address climate change, human-wildlife conflict and declining tree cover in India. On a similar note, countries in the Himalayan region are increasingly coordinating their climate action plans to address shared challenges. For example, the National Mission for Sustaining the Himalayan Ecosystem (NMSHE), part of the National Action Plan on Climate Change, seeks to develop suitable policy measures and time-bound action programmes to maintain ecological resilience and ensure the ongoing provision of essential ecosystem services in the Himalayas (Dhyani et al. 2022). Several transboundary megaprojects such as Kailash Sacred Landscapes (KSL), Hindukush-Karakoram Pamir Landscapes (HKPL), and long-term ecological monitoring of alpine ecosystems are being implemented in the Eastern Himalayas (Sharma and Chettri 2005).

11.8.6 Promotion of Climate-Resilient Agricultural Practices

Climate change is a critical ecological crisis today. The agriculture sector is particularly susceptible to climate change due to its high reliance on weather conditions and the relatively lower socioeconomic status of people involved in agriculture compared to their urban counterparts (Rao et al. 2019). Climate-resilient agriculture aims to reduce vulnerability to climate change by promoting sustainable practices that improve soil health, water conservation, and crop productivity. In the context of climate change, several climate-resilient agricultural practices have been suggested for hilly regions: (1) To deal with climatic alterations, several climate-resilient crop varieties (rice, wheat, maize, black gram, pigeon pea, apple, kiwi and peach) have been introduced by the task force on Himalayan agriculture. These varieties make Himalayan farmers climate resilient by increasing income and crop diversification; (2) For soil and water conservation, an integrated farming system involving agri-horti-silvi-livestock model has been developed for Himalayan states in India for strengthening farmers; (3) For improvement of Jhum cultivation and to conserve soil and water loss in north-east India, several improved varieties of rice and maize have been developed. Apart from this, for stabilisation of sloping lands, plantation of multipurpose trees along with horti-silviculture crops has been recommended; and (4) the promotion of zero tillage crop cultivation, and construction of *Azolla* and vermicomposting tanks to increase soil fertility and crop yield are being implemented in Himalayas (ICAR 2018; Pandey and Arunachalam 2020).

11.8.6.1 Adopting Sustainable Farming Techniques

Farmers are being encouraged to adopt sustainable farming techniques such as intercropping, crop rotation, and organic farming to maintain soil fertility and reduce dependence on chemical fertilizers and pesticides. For example, Sikkim in India has declared itself an organic state, promoting the use of organic fertilizers and biopesticides to protect the fragile mountain environment (Avi and Batra 2023). Climate-smart practices, like rainwater harvesting, contour farming, and the use of drought-resistant and fast-maturing crop varieties, are being promoted to help farmers cope with changing rainfall patterns and reduced water availability. These practices increase resilience to climate variability while reducing greenhouse gas emissions (Pandey and Arunachalam 2020).

11.9 Conclusions

The Himalayan ecosystem is facing an unprecedented level of threat due to a combination of climate change, deforestation, pollution, and unsustainable developmental practices. However, there is hope in the various preventive strategies that are currently gaining popularity. By implementing community-based conservation programmes, promoting sustainable tourism, supporting reforestation efforts, and enhancing international cooperation, we can work toward preserving this unique and vital ecosystem. The success of these strategies depends on a collaborative approach involving governments, local communities, and international organisations to ensure the long-term health and sustainability of the Himalayas. The implementation of these trending preventive strategies shows promising progress in protecting the Himalayan ecosystem. While challenges remain, the integration of community-based conservation, sustainable tourism, reforestation efforts, advanced monitoring systems, policy reforms, and climate-resilient agriculture offers a holistic approach to preserving this unique and vital region. Collaboration between local communities, governments, and international organisations will be essential to ensuring the long-term health and sustainability of the Himalayas. These strategies can reduce Himalayan ecosystem threats and improve the socioeconomic well-being of millions who depend on it.

Acknowledgements The authors are very grateful to the Director General of CCRAS, Ministry of AYUSH, New Delhi, India for the encouragement and facilities provided for this work.

References

Adler C, Wester P, Bhatt I et al (2022) Crosschapter paper 5: mountains. In: Pörtner HO, Roberts DC, Tignor M et al (eds) Climate change 2022: impacts, adaptation and vulnerability. Contribution of working group II to the sixth assessment report of the intergovernmental panel on climate change. Cambridge University Press, Cambridge, pp 2273–2318. https://doi.org/10.1017/9781009325844.022

Ahmad S, Mir N, Bhat S et al (2018) High altitude pasturelands of Kashmir Himalaya: current status, issues and future strategies in a changing climatic scenario. Curr J Appl Sci Technol 27(2):1–10

Avi A, Batra V (2023) Organic farming in India: evolution, current status and policy perspectives. Space Cult India 11(2):18–34

Brandt JS, Allendorf T, Radeloff V et al (2017) Effects of national forest-management regimes on unprotected forests of the Himalaya. Conserv Biol 31(6):1271–1282

Cao J, Yeh ET, Holden NM et al (2013) The roles of overgrazing, climate change and policy as drivers of degradationof China's Grasslands. Nomadic Peoples 17:82–101. https://doi.org/10.3167/np.2013.170207

Denman R (2001) Guidelines for community-based ecotourism development. World Wide Fund for Nature

Dhyani S (2023) Are Himalayan ecosystems facing hidden collapse? Assessing the drivers and impacts of change to aid conservation, restoration and conflict resolution challenges. Biodivers Conserv 32:3731–3764. https://doi.org/10.1007/s10531-023-02692-x

Dhyani S, Sivadas D, Basu O et al (2022) Ecosystem health and risk assessments for high conservation value mountain ecosystems of South Asia: a necessity to guide conservation policies. Anthr Sci 1:211–225. https://doi.org/10.1007/s44177-022-00010-8

Dutta M, Dutta PK (2023) Community-based conservation in Eastern Himalayan biodiversity hotspot-a case study. IJTK 22(1):220–229

Florence PS, Mishra A (2024) Traditional ecological knowledge repository in the Indian Himalayas: an overview. In: Borthakur A, Singh P (eds) Addressing the climate crisis in the Indian Himalayas: can traditional ecological knowledge help? Springer, Cham, pp 293–311. https://doi.org/10.1007/978-3-031-50097-8_12

Geneletti D, Dawa D (2009) Environmental impact assessment of mountain tourism in developing regions: a study in Ladakh, Indian Himalaya. Environ Impact Assess Rev 29(4):229–242

Habibullah MS, Din BH, Chong CW et al (2016) Tourism and biodiversity loss: implications for business sustainability. Procedia Econ Finance 35:166–172

Hamid M, Khuroo AA, Charles B et al (2018) Impact of climate change on the distribution range and niche dynamics of Himalayan birch, a typical treeline species in Himalayas. Biodivers Conserv 28(8–9):2345–2370

Hazarika A, Kurmi B, Francaviglia R et al (2024) The transition from shifting cultivation to indigenous agroforestry as nature-based solution for land restoration in the Indian Eastern Himalayas. Ecol Indic 162:112031

ICAR (2018) Annual report of the Task Force on Himalayan Agriculture – 2017-18, National Mission for Sustaining the Himalayan Ecosystems (NMSHE). ICAR. Government of India

Kattel GR (2022) Climate warming in the Himalayas threatens biodiversity, ecosystem functioning and ecosystem services in the 21st century: is there a better solution? Biodivers Conserv 31: 2017–2044. https://doi.org/10.1007/s10531-022-02417-6

Khanday AA, Rather GM, Kumar MA et al (2024) Environmental sustainability and its impact on public health in Ladakh-India. In: Tripathi G, Shakya A, Kanga S et al (eds) Sustainability and health informatics, Advances in geographical and environmental sciences. Springer, Singapore. https://doi.org/10.1007/978-981-97-6706-9_9

Kumar P, Sharma PK, Kumar P et al (2021) Agricultural sustainability in Indian Himalayan region: constraints and potentials. Indian J Ecol 48(3):649–661

Kumar R, Kumar A, Saikia P (2022) Deforestation and forests degradation impacts on the environment. In: Singh VP, Yadav S, Yadav KK, Yadava RN (eds) Environmental degradation: challenges and strategies for mitigation, Water science and technology library. Springer, Cham, p 104. https://doi.org/10.1007/978-3-030-95542-7_2

Kumar P, Kumar A, Patil M et al (2023) Biodiversity resources and ecosystem services in the context of the industrialized world. In: Sobti RC (ed) Biodiversity threats and conservation, 1st edn. CRC Press, Boca Raton, pp 23–39. https://doi.org/10.1201/9781003220398-3

Lamsal P, Kumar L, Aryal A et al (2018) Invasive alien plant species dynamics in the Himalayan region under climate change. Ambio 47:697–710. https://doi.org/10.1007/s13280-018-1017-z

Millennium Ecosystem Assessment (2005) Ecosystems and human well-being: synthesis. Island Press, Washington, DC

Ministry of Environment, Forest and Climate Change. National Forest Policy (2018), https://moef.gov.in/uploads/2019/06/Draft-National-Forest-Policy-2018.pdf (2018).

Ministry of Tourism (2022) National Strategy for Ecotourism. https://tourism.gov.in/sites/default/files/202209/National%20Strategy%20for%20Ecotourism%202022.pdf

MoEFCC (2020) National mission for a green India: implementation framework. Ministry of Environment, Forest and Climate Change, Government of India

Nautiyal MC, Nautiyal BP, Prakash V (2004) Effect of grazing and climatic changes on alpine vegetation of Tungnath, Garhwal Himalaya, India. Environmentalist 24:125–134. https://doi.org/10.1007/s10669-004-4803-z

Negi GCS, Samant SS, Singh P (2018a) Forest ecosystem services in the Indian Himalaya: socioecological relevance and future challenges. Int J Ecol Environ Sci 44:215–229

Negi VS, Kewlani P, Pathak R et al (2018b) Criteria and indicators for promoting cultivation and conservation of Medicinal and Aromatic Plants in Western Himalaya, India. Ecol Indic 93:434–446

Negi VS, Pathak R, Rawal RS et al (2019) Long-term ecological monitoring on forest ecosystems in Indian Himalayan Region: criteria and indicator approach. Ecol Indic 102:374–381. https://doi.org/10.1016/j.ecolind.2019.02.035

Osti R, Egashira S, Adikari Y (2011) Prediction and assessment of multiple glacial lake outburst floods scenario in Pho Chu River basin, Bhutan. Hydrol Process 27(2):262–274

Pandey L, Arunachalam A (2020) Potential technologies for climate resilient agriculture in the Indian Himalayan region. In: Goel P, Ravindra R, Chattopadhyay S (eds) Climate change and the white world. Springer, Cham. https://doi.org/10.1007/978-3-030-21679-5_6

Pandey K, Adhiguru P, De HK et al (2021) Permaculture tomonoculture in shifting cultivation landscape of Mizoram, Northeast India: are agrobiodiversity and happiness waning? Indian J Tradit Knowl 20(2):479–448

Pandit MK, Sodhi NS, Koh LP et al (2007) Unreported yet massive deforestation driving loss of endemic biodiversity in Indian Himalaya. Biodivers Conserv 16:153–163. https://doi.org/10.1007/s10531-006-9038-5

Pandit MK, Manish K, Koh LP (2014) Dancing on the roof of the world: ecological transformation of the Himalayan landscape. Bioscience 64:980–992. https://doi.org/10.1093/biosci/biu152

Pant KS, Bishist R, Sharma PP et al (2022) Agroforestry in Himachal Pradesh: status, challenges and future perspectives. Indian J Agrofor 24(3):47–61

Pathak R, Negi VS, Rawal RS et al (2019) Alien plant invasion in the Indian Himalayan region: state of knowledge and research priorities. Biodivers Conserv 28:3073–3102. https://doi.org/10.1007/s10531-019-01829-1

Rana SK, Rawat GS (2017) Database of Himalayan plants based on published floras during a century. Data 2(4):36. https://doi.org/10.3390/data2040036

Rao CS, Kareemulla K, Krishnan P et al (2019) Agro-ecosystem based sustainability indicators for climate resilient agriculture in India: a conceptual framework. Ecol Indic 105:621–633

Reddy CS, Jha CS, Dadhwal VK et al (2016) Quantification and monitoring of deforestation in India over eight decades (1930–2013). Biodivers Conserv 25:93–116. https://doi.org/10.1007/s10531-015-1033-2

Samant SS, Dhar U, Palni LMS (1998) Medicinal plants of Himalaya: diversity, distribution potential values. Gyanodaya Prakashan, Nainital

Saxena KG, Maikhuri RK, Rao KS (2005) Changes in agricultural biodiversity: implications for sustainable livelihood in the Himalaya. J Mount Sci 2:23–31

Sharma E, Chettri N (2005) ICIMOD's transboundary biodiversity management initiative in the Hindu Kush-Himalayas. Mt Res Dev 25:278–281

Sharma R, Xu J, Sharma G (2007) Traditional agroforestry in the eastern Himalayan region: land management system supporting ecosystem services. Trop Ecol 48(2):189

Sharma R, Verma N, Singh A (2023) Community-based reforestation initiatives in Uttarakhand: opportunities and challenges. Indian Forester 149:123–131

Shrestha UB, Gautam S, Bawa KS (2012) Widespread climate change in the Himalayas and associated changes in local ecosystems. PLoS One 5:e36741. https://doi.org/10.1371/journal.pone.0036741

Singh GS (2013) Commercialization unsustainable to Himalayan environment. Ecol Quest 15(1):71–75

Singh RB, Mishra DK (2004) Green tourism in mountain regions-reducing vulnerability and promoting people and place centric development in the Himalayas. J Mt Sci 1:57–64. https://doi.org/10.1007/BF02919360

Singh SP, Rawat YS (2021) Restoration ecology in the Indian Himalaya: a review of approaches and practices. Trop Ecol 62:1–15

Singh L, Tariq M, Sekar KC et al (2017) Ecologicalniche modelling: an important tool for predicting suitable habitat and conservation of the Himalayan medicinal herbs. ENVIS Bull 27:154–155

Singh L, Bhatt ID, Negi VS et al (2021) Population status, threats, and conservation options of the orchid *Dactylorhiza hatagirea* in Indian Western Himalaya. Reg Environ Chang 21:1–16

Talukder B, Matthew R, Bunch MJ et al (2021) Melting of Himalayan glaciers and planetary health. Curr Opin Environ Sustain 50:98–108

Telwala Y, Brook BW, Manish K et al (2013) Climate-induced elevational range shifts and increase in plant species richness in a Himalayan biodiversity epicentre. PLoS One 8(2):e57103

Thakur A, Kumari S, Sinai Borker S et al (2021) Solid waste management in Indian Himalayan Region: current scenario, resource recovery, and way forward for sustainable development. Front Energy Res 9:609229. https://doi.org/10.3389/fenrg.2021.609229

Thapa S, Chitale V, Rijal SJ et al (2018) Understanding the dynamics in distribution of invasive alien plant species under predicted climate change in Western Himalaya. PLoS One 13(4): e0195752

Wang W, Zhang T, Yao T et al (2022) Monitoring and early warning system of Cirenmaco glacial lake in the central Himalayas. Int J Disaster Risk Reduct 73:102914

Wester P, Mishra A, Mukherji A et al (eds) (2019) The Hindu Kush Himalaya assessment: mountains, climate change, sustainability and people. Springer, Cham

Chapter 12
Impact of Climate Change and Vulnerability in the Himalayan Hotspot

Bhagat Suberi

Abstract The breathtaking Himalayan region, known for its stunning landscapes and delicate ecosystems, is confronting significant challenges due to climate change. This extraordinary area is not only picturesque but also highly susceptible to environmental shifts. Observations indicate that temperatures are on the rise, rainfall patterns are altering, glaciers are melting, and extreme weather events are becoming more frequent. For the local populations, these changes affect not just the environment but also their everyday lives, biodiversity, and essential water sources. Research shows that over the past century, average temperatures in the Himalayas have increased by approximately 1.5 °C, surpassing the global average. Predictions indicate that by the end of this century, temperatures may further rise by 2–4 °C. This warming trend presents significant challenges for both the ecosystem and the communities that inhabit this region.

Keywords Biodiversity · Climate change · Ecosystem · Glaciers · Himalayas

12.1 Introduction

The majestic Himalayas, a breathtaking mountain range in Asia, serve as a natural divide between the lush plains of the Indian subcontinent and the vast Tibetan Plateau. The name ‘Himalayas’ has roots in the ancient Sanskrit language, with ‘hima’ meaning ‘snow’ and ‘alayas’ translating to ‘abode’. This stunning range is the youngest in the world, stretching over 595,000 km^2 and embracing eight nations, from Afghanistan in the west to Myanmar in the east, while cradling the diverse landscapes of Nepal, India, Bhutan, China, Pakistan, and Bangladesh (Chettri and Sharma 2016). The ecosystems of the Himalayas provide essential services that directly support about 240 million people, impacting the lives of an astonishing 1.9 billion people living in the river basins below (Sharma et al. 2019).

B. Suberi (✉)
College of Natural Resources, Royal University of Bhutan, Lobesa, Bhutan

D. Das et al. (eds.), *Plant Diversity and Conservation Strategies of the Himalayan Biodiversity Hotspot*, Topics in Biodiversity and Conservation 21,
https://doi.org/10.1007/978-3-032-21697-7_12

The Himalayan region is exceptionally vulnerable to the ongoing effects of climate change, a theme echoed throughout prior chapters, particularly in geophysical complexity (Chap. 1, Sect. 1.3.1), socio-ecological vulnerability (Chap. 2, Sect. 2.1), and climate-induced ecological disruptions (Chap. 9, Sect. 9.1). The evidence of climate change in this sensitive region is alarming, with rising temperatures, shifting rainfall patterns, retreating glaciers, and an increasing frequency of extreme weather events becoming more apparent. These changes are not just numbers; they have profound impacts on the environment, biodiversity, and the livelihoods of the millions who call this region home (IPCC 2021).

The atmospheric shifts in the Himalayas are dramatic. According to the Intergovernmental Panel on Climate Change (IPCC), the average temperature in this region has risen by approximately 1.5 °C over the last century, higher than the global average. Projections suggest that by the century's end, we could see increases of 2–4 °C (Kumar and Jain 2022).

Alarming as it is, the Himalayas house some of the largest glaciers outside the Polar Regions, crucial for the freshwater supply of countless people across South Asia. However, these glaciers are vanishing at an accelerated pace, creating challenges like water shortages and disrupted water supplies. Furthermore, the shrinking glaciers raise the risk of glacial lake outburst floods, a significant danger for communities living downstream (Bajracharya and Shrestha 2020).

One of the most pressing issues facing the Himalayas is the critical loss of glaciers, which impacts the availability of water for downstream communities (Hock et al. 2019; Sun et al. 2022; You and Xu 2022). The region is grappling with several environmental challenges, such as deforestation, habitat loss, fragmentation of ecosystems, resource overexploitation, illegal hunting, and intensified agriculture, loss of genetic diversity, human-wildlife conflicts, and invasive species.

Monitoring the trends and impacts of climate change in the Himalayas is incredibly complex. While much focus has been on the melting glaciers, systematic observation is lacking, primarily due to a shortage of hydrological and meteorological stations and the rugged terrain that characterises this beautiful region.

The local communities in the Himalayas are particularly vulnerable to the impacts of climate change, as detailed in Chap. 2 (Sect. 2.1) with respect to traditional ecological dependence, and expanded upon in Chap. 11 (Sect. 11.2) in the context of adaptation challenges in mountain systems. Small-scale farmers, pastoralists, and indigenous populations are at high risk, relying on agriculture, livestock, and forest resources to survive (Wangchuk and Wangdi 2022). The effects of climate change worsen existing vulnerabilities, such as poverty, limited access to resources, and inadequate infrastructure, making it increasingly challenging for these communities to adapt to the shifting environment (Sharma and Chettri 2021). Given their sensitivity to rising temperatures and evolving rainfall patterns, the Himalayas are indeed recognised as a climate change hotspot.

The Himalayas contain some of the largest glaciers found outside of Polar Regions, and these glaciers are retreating at an alarming rate, impacting the water supply for millions residing downstream. Research indicates a significant loss of mass in many glaciers over the last few decades, with Himalayan glaciers retreating faster than previously thought, posing serious implications for water resources in

South Asia (Sun et al. 2022). The region has experienced temperature increases that exceed global averages, influencing weather patterns and consequently affecting agriculture and biodiversity.

Climate change is disrupting monsoon patterns, resulting in either excessive rainfall or droughts. Such variability can trigger landslides and flash floods, severely impacting local communities. The Himalayas have warmed by approximately 1.5 °C since the mid-twentieth century, further accelerating glacier melt (Khan et al. 2022). The unique ecosystems in the Himalayas are endangered due to these shifting climatic conditions, which lead to habitat loss and species extinction. Communities reliant on agriculture and natural resources are grappling with the repercussions of environmental instability, a challenge elaborated in Chap. 11 (Sects. 11.2 and 11.4), which emphasises the compounded vulnerabilities arising from deforestation, shifting cultivation, and climate variability particularly due to the increasing unpredictability of monsoon rainfall in the region (Sharma et al. 2023).

12.2 Impacts on Glaciers

Himalayan glaciers are experiencing accelerated retreat as a direct result of rising temperatures. Studies indicate that many glaciers have significantly lost mass over recent decades, a trend expected to continue (Khan et al. 2022). With warmer temperatures, melting during the summer months increases, initially raising river flows but eventually leading to a decrease in long-term water availability as glaciers diminish.

As glaciers melt and create glacial lakes, there is an associated risk of glacial lake outburst floods (GLOFs), which threaten downstream communities (Bhattacharya and Kaur 2023). Glaciers act as vital water reservoirs for rivers essential to millions in South Asia. While the initial rise in meltwater can cause flooding, the long-term loss of glaciers jeopardizes sustainable water supply. Changes in the timing and volume of glacial melt can disrupt traditional agricultural practices and hinder water management systems, adversely affecting aquatic ecosystems and biodiversity, resulting in potential habitat loss for various species (Kumar et al. 2023).

The ongoing consequences of climate change on Himalayan glaciers and water resources present formidable challenges for local communities and ecosystems. The combination of glacial retreat, altered hydrology, and increased risks from glacial lake outburst floods demands urgent attention from both policymakers and researchers.

12.3 Impacts on Biodiversity and Ecosystem

Climate change yields several significant impacts on biodiversity, as previously discussed in Chaps. 5 (Sect. 5.1) and 9 (Sect. 9.1), where habitat shifts, invasive species, and phenological changes were identified as major ecological stressors

(Kumar et al. 2023). As temperatures rise, many species are migrating to higher altitudes or latitudes in search of suitable habitats, which threatens those species unable to adapt or relocate quickly enough. The Himalayas are home to many endemic and endangered species, such as the snow leopard and red panda. Climate change increases their vulnerability through habitat degradation and altered food availability (Sharma et al. 2019).

Warmer temperatures may also facilitate the proliferation of invasive species, which can outcompete native flora and fauna, thereby further threatening local biodiversity. Changes in the timing of biological events, such as flowering and breeding, can disrupt ecological interactions, particularly between pollinators and plants (Olson et al. 2021).

The Himalayas are not just mountains; they play a vital role in supporting the ecosystem services needed by Asia and the world. They help maintain slope stability, ensure clean water, and sustain a diverse range of plant and animal life, all of which benefit human wellbeing. Recently, there has been a growing focus on conserving biodiversity in these mountain regions, highlighting their unique ecological importance (Pei 1995; Guangwei 2002).

The Hindu Kush-Himalayas (HKH) is celebrated for their impressive biodiversity. This region is home to diverse gene pools and a multitude of species, linking it to four of the Global Biodiversity Hotspots: the Himalayas Hotspot, the Indo-Burma Hotspot, the Mountains of South-West China Hotspot, and the Mountains of Central Asia Hotspot (Mittermeier et al. 2004). With its varied landscapes and climatic conditions, the HKH showcases a wide array of plants and animals, many of which cannot be found anywhere else in the world (Olson et al. 2001; Olson and Dinerstein 2002).

Land Use consist of approximately 39% grassland, 20% forest, 15% shrub land, and 5% agricultural land. The remaining 21% includes barren land, rocky areas, urban spaces, snow, and water bodies. The altitude in this region ranges from tropical lowlands below 500 m to the icy alpine zone above 6000 m. Vegetation here varies from tropical forests to temperate coniferous forests, and even high-altitude cold areas (Pei 1995; Chettri et al. 2008).

Unfortunately, the mountain ecosystems and the biodiversity within them are facing significant threats from human activities. This includes timber harvesting, overgrazing by livestock, and the expansion of agricultural land into forests. Climate change is perhaps the most pressing challenge, compounded by modernisation issues such as greenhouse gas emissions, air pollution, and land degradation. As a result, the landscapes and communities in the HKH are undergoing rapid environmental and socioeconomic changes.

Understanding the key ecological and socioeconomic factors is essential for managing the mountain ecosystems effectively, a point reiterated in Chap. 2 (Sect. 2.2) and Chap. 10 (Sect. 10.4). This knowledge is crucial for developing policies aimed at sustainable development and environmental management. Furthermore, the wellbeing of around 1.3 billion people living downstream is closely tied to the health of the natural resources found in the HKH region. There is an urgent need to recognise these connections for a sustainable future (Schild 2008).

The urgent challenges posed by biodiversity loss, declining ecosystem health, and diminished ecosystem services due to climate change and human activities must be tackled promptly by nations and international organisations in the Hindu Kush Himalaya region. This involves identifying vulnerable ecosystem services and developing improved sustainability strategies. A resilience framework presents effective solutions, including comprehensive and coordinated climate data monitoring, the application of multi-indicator methods and modeling, and fostering collaboration among policymakers, implementers, and analysts to address the ongoing decline in biological diversity and ecosystem services in the HKH area (Kattel 2022).

12.4 Climate Impact on Vulnerability of Communities in the Himalayas

Vulnerability refers to how likely a system is to be negatively impacted by various factors (IPCC 2021). It's not something we can easily see or measure precisely because it's influenced by numerous variables. Because of this complexity, different approaches have been developed to assess vulnerability. The method created by the Intergovernmental Panel on Climate Change (IPCC) is quite comprehensive and is widely used to evaluate vulnerability related to climate change across different sectors (Chakraborty et al. 2018; Pandey et al. 2022).

12.4.1 The IPCC Framework

The IPCC's assessment framework looks at three main aspects: exposure, sensitivity, and adaptive capacity. This framework was used in the current study to analyse how climate change affects the ecological vulnerability of forests (Fig. 12.1).

- Exposure refers to external stressors that can increase vulnerability. This includes changes in climate variability and extreme weather conditions that impact the system.
- Sensitivity is about how much a system can be influenced (either positively or negatively) by climate-related changes.
- Adaptive capacity defines how well a system can adjust to the impacts of climate change to minimise potential harm, seize new opportunities, or manage the consequences.

Several studies have utilised this indicator-based framework to examine various areas, including forests, urban informal settlements, slum communities, socio-ecological systems, and smallholder farmers (Aggarwal et al. 2023).

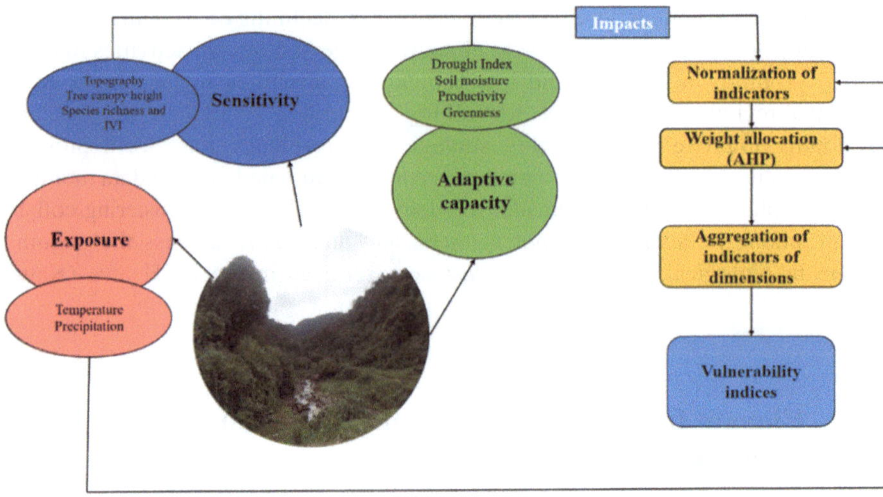

Fig. 12.1 IPCC's assessment framework (IPCC 2007)

The Himalayas, housing some of the planet's largest glaciers, are facing significant threats due to climate change. As temperatures rise, glaciers are melting at an alarming rate, which not only increases the risk of glacial lake outburst floods (GLOFs) but also leads to water shortages during dry seasons. Local communities are increasingly vulnerable to the more frequent and severe extreme weather events, like landslides and floods, which jeopardize their livelihoods and safety.

Changes in weather patterns are disrupting traditional farming, resulting in lower crop yields and a looming food insecurity crisis. The fragile ecosystems are also suffering, leading to a loss of biodiversity that many local communities depend on for survival. Rising temperatures and evolving climates can spread diseases, further impacting the health and well-being of these communities (Thayyen and Gergan 2021).

One major concern linked to climate change in this region is the rising occurrence of extreme weather events be it glacial melt, landslides, or severe flooding. The ongoing melting of glaciers is changing water availability and raising the chances of GLOFs, putting both upstream and downstream communities at risk. This environmental instability threatens agricultural practices, which serve as the backbone of many Himalayan societies (Kumar et al. 2022).

The socio-economic implications are severe as well. Many communities are tied to traditional practices intricately linked to their environment. As resources fade and weather becomes unpredictable, these communities are struggling to adapt. Vulnerable groups, particularly women and indigenous populations, face even greater challenges due to limited financial resources and a lack of information on how to navigate these changes (Singh et al. 2023).

Health concerns are also rising with the changing climate. Higher temperatures lead to a surge in vector-borne diseases, while air quality worsens due to landslides and forest fires (Rai et al. 2023). Many communities suffer from inadequate infrastructure and limited access to healthcare, making them particularly vulnerable to health risks. The psychological stress from losing livelihoods and enduring more frequent natural disasters also affects community mental health (Sharma et al. 2019).

Recent studies have highlighted that climate change is a critical threat to the vulnerability of Himalayan communities. To tackle these issues effectively, it's essential to adopt a combined approach that considers environmental, social, and economic factors. Ongoing research and local collaboration will be pivotal in developing strategies that enhance resilience in this uniquely challenging region.

12.5 Measures to Overcome Vulnerability

Vulnerability impacts various aspects of our lives, including social, economic, and environmental dimensions. It reflects the challenges that individuals and communities encounter during crises, whether from natural calamities, economic downturns, or social inequities. Addressing these vulnerabilities is crucial for helping people foster resilience and move towards a sustainable future.

Education stands out as a key tool in combating vulnerability. Access to quality education can empower individuals by equipping them with the skills and knowledge needed to improve their situations, as highlighted by the World Bank (2022). Education initiatives focused on marginalised groups can open up new avenues for resources, rights, and opportunities. Moreover, education in health equips communities with the information necessary to prevent illnesses and manage health emergencies, thereby enhancing their overall well-being (UNICEF 2023).

As climate change and natural disasters grow more common, developing resilient infrastructure is imperative. The United Nations Office for Disaster Risk Reduction (2022) stresses the importance of investing in robust structures that can withstand extreme weather conditions, including flood defenses and better drainage systems. Additionally, training communities in disaster risk management empowers them to better prepare for and recover from crises effectively (Benson and Twigg 2007).

Involving local communities in developing adaptive strategies tailored to their unique situations can further enhance their resilience. Approaches such as promoting agroecology, diversifying crops, and implementing sustainable land management can help counter some of the negative effects of climate change on agriculture (Kumar and Sharma 2020; Bhutani and Ghosh 2021).

Establishing early warning systems and improving infrastructure can reduce risks associated with extreme weather events. Efficiently managing water resources through practices like rainwater harvesting and sustainable irrigation can address water scarcity issues (Wangchuk and Wangdi 2022). Strengthening local governance and incorporating climate adaptation strategies into national policies can create a supportive framework for community resilience (Chaudhary and Sharma

2022). It's crucial to raise awareness about climate change's impacts and present effective adaptation strategies, enabling communities to respond proactively.

Finally, advocating for policy reforms to address systemic inequalities is essential. Vulnerability frequently arises from entrenched social injustices, discrimination, and limited resource access. Developing fair policies that promote social justice, combat discrimination, and increase economic opportunities is vital for creating lasting change. Legislative efforts to protect marginalised groups' rights can significantly reduce vulnerability in the long run (Amnesty International 2023).

12.6 Conclusions

The challenges posed by climate change in the Himalayan region are becoming more urgent by the day, calling for immediate action from everyone from local communities to national leaders and international organisations. We need to tackle this issue on multiple fronts. It's not just about cutting down greenhouse gas emissions; we also need to help communities adapt and build their resilience against the changes that are already happening. By joining forces, we can safeguard the delicate ecosystems of the Himalayas and pave the way for a sustainable future for the generations to come.

The rich biodiversity and unique environments of the Himalayas are under serious threat from climate change. We're seeing shifts in where species live, changes in how ecosystems function, and a rise in the presence of invasive species. This makes it clear that we need robust conservation strategies and adaptive management practices now more than ever. Taking care of these incredible landscapes isn't just about protecting nature; it's also about preserving livelihoods and cultures that have thrived here for centuries.

Acknowledgements First of all, it is a great opportunity for me to get selected to write a chapter on the topic. I would like to thank Professor Monoranjan Chowdhury for passing the information on writing the chapter. I would also like to thank the reviewer for their invaluable comments and suggestions to improve my chapter further. Lastly, I am fortunate to publish my chapter in a renowned journal, Springer.

References

Aggarwal P, Gupta A, Sharma R (2023) Integrating traditional knowledge in climate adaptation strategies in the Himalayas. Himal J Envion Clim 30(1):45–62

Amnesty International (2023) Amnesty International report 2022/23: The state of the world's human rights. Amnesty International

Bajracharya SR, Shrestha AB (2020) Climate change and glacial retreat in the Himalayas: impacts on water resources and livelihoods. Sustain Water Resour Manag 6(3):1–15. https://doi.org/10.1007/s40899-020-00409-y

Benson C, Twigg J (2007) Tools for mainstreaming disaster risk reduction: guidance notes for development organisations. ProVention Consortium

Bhattacharya P, Kaur R (2023) Adaptive strategies of Himalayan communities in response to climate change: challenges and opportunities. JEM 320:115856. https://doi.org/10.1016/j.jenvman.2022.115856

Bhutani M, Ghosh S (2021) Climate change and vulnerability in the Himalayan region: an overview. MRD 41(2):123–134. https://doi.org/10.1659/MRD-JOURNAL-D-20-00024.1

Chakraborty A, Saha SK, Sachdeva K, Joshi PK (2018) Vulnerability of forests in the Himalayan region to climate change impacts and anthropogenic disturbances: a systematic review. Reg Environ Chang 18:1783–1799. https://doi.org/10.1007/s10113-018-1309-7

Chaudhary P, Sharma A (2022) Building resilience through education in the Himalayan communities. Clim Pol 22(4):487–500. https://doi.org/10.1080/14693062.2021.1936148

Chettri N, Sharma E (2016) Reconciling the mountain biodiversity conservation and human wellbeing: drivers of biodiversity loss and new approaches in the Hindu-Kush Himalayas. Proc Indian Natl Sci Acad 82:53–73

Chettri N, Shakya B, Thapa R, Sharma E (2008) Status of a protected area system in the Hindu Kush-Himalayas: an analysis of PA coverage. Int J Biodivers Sci Manag 4:164–178

Guangwei C (2002) Biodiversity in the Eastern Himalayas: conservation through dialogue. Summary reports of the workshops on Biodiversity Conservation in the Hindu Kush-Himalayan Ecoregion. ICIMOD, Kathmandu

Hock R, Rasul G, Adler C, Caceres B, Gruber S, Hirabayashi Y, Jackson M, Kaab A, Kang S, Kutuzov S, Milner A, Molau U, Morin S, Orlove B, Steltzer H (2019) High Mountain areas. In: Special report on the ocean and cryosphere in a changing climate. Cambridge University Press, Cambridge, UK, pp 131–202

Intergovernmental Panel on Climate Change (2021) Climate change: the physical science basis. Contribution of working group I to the sixth assessment report of the IPCC. Cambridge University Press. https://www.ipcc.ch/report/ar6/wg1/

IPCC (2007) In: Solomon S, Qin D, Manning M, Chen Z, Marquis M, Averyt KB et al (eds) Climate change. The physical science basis. Contribution of Working Group I to the 92 R. Krishnan et al. Fourth assessment report of the intergovernmental panel on climate change. Cambridge University Press, Cambridge, UK

Kattel GR (2022) Climate warming in the Himalayas threatens biodiversity, ecosystem functioning, and ecosystem services in the 21st century: is there a better solution? Biodivers Conserv 31: 2017–2044. https://doi.org/10.1007/s10531-022-02417-6

Khan M, Ellahi A, Niar R, Ghoneim ME, Tag-eldin E, Rashid A (2022) Water quality assessment of alpine glacial blue water lakes and glacial-fed rivers. Geomat Nat Hazards Risk 13(1): 2597–2617. https://doi.org/10.1080/19475705.2002.1216800

Kumar R, Jain S (2022) Climate change and its impact on the Himalayan glaciers: a review. Environ Sci Pollut Res 29(15):22345–22358. https://doi.org/10.1007/s11356-021-17629-7

Kumar P, Sharma R (2020) Impact of climate change on agriculture in the Himalayan region. JEM 261:110239. https://doi.org/10.1016/j.jenvman.2020.110239

Kumar P, Singh A, Choudhury D (2022) Climate change and food security in the Himalayas. J Himal Stud 15(3):25–40

Kumar S, Hasija N, Kumar V, Sageena G (2023) Ecotourism: A Holistic Assessment of Environmental and Socioeconomic Effects towards Sustainable Development. Current World Environment 18(2):589–607

Mittermeier RA, Gils PR, Hoffman M, Pilgrim J, Brooks T, Mittermeier CG, Lamoreaux J, Fonseca GAB (2004) Hotspots revisited. Earth's biologically richest and most endangered terrestrial ecoregions. Mexico City, CEMEX, USA

Olson DM, Dinerstein E (2002) The Global 200; Priority ecoregions for global conservation. Ann Missouri Bot Gard 89:199–224

Olson DM, Dinerstein E, Wikramanayke ED et al (2001) Terrestrial ecoregions of the world: a new map of life on earth. BioSci 51(11):993–1038

Olson JR, Benjamin PH, Azman AA, Kellogg MA, Pullmann MD, Suter JC, Bruns EJ (2021) Systematic review and meta-analysis: effectiveness of wraparound care coordination for children and adolescents. J Am Acad Child Adolesc Psychiatry 60(11):1353–1366. https://doi.org/10.1016/j.jaac.2021.02.022

Pandey RJ, Sharma Singh R, Rawat M, Saklani H, Tomar PK (2022) Vegetation characteristics based climate change vulnerability assessment of temperate forests of Western Himalaya. Forests 13(6):848

Pei S (1995) Banking on biodiversity. Report on the regional consultations on biodiversity assessment in the Hindu Kush-Himalaya. ICIMOD, Kathmandu

Rai M, Shrestha A, Adhikari R (2023) Health impacts of climate change in the Himalayas: challenges and prospects. Environ Health Perspect 131(2):123–136

Schild A (2008) The case of the Hindu Kush-Himalayas: ICIMOD's position on climate change and mountain systems. MRD 28(3/4):328–331

Sharma E, Chettri N (2021) Vulnerability of mountain communities to climate change: a case study from the eastern Himalayas. MRD 41(2):1–12. https://doi.org/10.1659/MRD-JOURNAL-D-20-00001.1

Sharma E, Molden D, Rahman A, Khatiwada YR, Zhang L, Singh SP, Yao T, Wester P (2019) Climate change, sustainability and people. In: The Hindu Kush Himalaya assessment. Springer, Cham, pp 1–16

Sharma S, Hussain S, Singh NA (2023) Impacts of climate change on ecosystem services in the Himalayan region. Ecol Indic 145:109560. https://doi.org/10.1016/j.ecolind.2023.109560

Singh R, Thayyen RJ, Gergan J (2023) Glacial change and socioeconomic vulnerability in the Indian Himalayas. MRD 43(2):150–160

Sun X, Zhang Q, Zhang G, Li M, Li S, Guo J, Dong H, Zhou Y, Kang S, Wang X, Shi J (2022) Melting Himalayas and mercury export: results of continuous observations from the Rongbuk Glacier on Mt. Everest and future insights. Water Res 218:118474

Thayyen RJ, Gergan J (2021) Glacial lake outburst floods and their impact on local communities in the Himalayas. Geophys Res Lett 48(18):e2021GL095196

UNICEF (2023) Children displaced in a changing climate: Turning up the heat on displacement. UNICEF

United Nations Office for Disaster Risk Reduction (UNDRR) (2022) UNDRR annual report 2022. UNDRR

Wangchuk S, Wangdi K (2022) Climate change and its impact on agriculture in the Himalayas: a review of challenges and adaptation strategies. Agric Food Secur 11(1):1–15. https://doi.org/10.1186/s40066-022-00334-4

World Bank (2022) Safe and learning in the midst of fragility, conflict, and violence: a world bank group approach paper. World Bank

You C, Xu C (2022) Himalayan glaciers threatened by frequent wildfires. Nat Geosci 15:956–957

Chapter 13
Impact of Anthropogenic Activities on the Himalayan Biodiversity

Tej Kumar Nepal and Jeewan Kumar Bastola

Abstract With enormous biological products and ecological significance, the Himalayas are under severe pressure from augmented anthropogenic activities. Such activities, for example, deforestation, infrastructural and industrial development, pollution, and exploitation of natural resources, are gradually affecting the area's ecosystem intensely. The Himalayan region is considered one of the most biodiverse zones in the world, and numerous endemic species living in the area are especially sensitive to environmental changes resulting from human activities. This chapter focuses on specific anthropogenic factors that affect the Himalayan biota and their positive and negative impacts on species and ecosystem services. Thus, the study highlights the need for conservation measures, sustainable development activities, and community involvement to mitigate the negative impacts of these activities, as mentioned earlier, and to preserve the environmental integrity of the affected area. Additionally, it emphasises the importance of cooperation in the transboundary conservation of Himalayan ecosystems for their sustainable future.

Keywords Himalaya · Conservation · Ecosystem · Sustainable development

13.1 Introduction

13.1.1 Background on the Himalayan Region

The Himalayas are home to the most youthful and elevated mountain chain (45–55 million years old) a region marked by significant altitudinal and climatic gradients contributing to ecological diversity as outlined in Chap. 1 (Sect. 1.3.1) and are

T. K. Nepal (✉)
Independent Researcher, Brisbane, QLD, Australia

J. K. Bastola
Department of Arts and Humanities, Samtse College of Education, Royal University of Bhutan, Samtse, Bhutan

D. Das et al. (eds.), *Plant Diversity and Conservation Strategies of the Himalayan Biodiversity Hotspot*, Topics in Biodiversity and Conservation 21,
https://doi.org/10.1007/978-3-032-21697-7_13

situated on the Asian subcontinent. The development of the Himalayas was a significant geophysical occasion that prompted the vanishing of the Tethys Sea, which isolated the Eurasian Plate in the north and the Indian Plate in the south (Pandit et al. 2014). The Himalayas are located between 70°E and 105°E longitude and 26°N and 41°N latitude. The Himalayas span approximately 1700 km from north to south and 3200 km from east to west (Singh et al. 2019), extending across Nepal, India, Bhutan, China, Pakistan, and Afghanistan. The Himalayas are home to more than 10,000 seed plants, 1737 bryophytes, 600 pteridophytes, 6900 fungi, 1159 lichens, 290 fish, more than 275 amphibians and reptiles, 900 birds, and 300 mammals (Pandit et al. 2014). This extraordinary richness of flora and fauna is consistent with the biogeographic and ecological patterns described in Chap. 1 (Sect. 1.3.1), which emphasises the role of geological history and altitudinal gradients in shaping the Himalayan biodiversity. The Himalayas' glaciated region is often referred to as the Third Pole (Yao et al. 2012). Ice meltwater feeds several major water systems, including the Yangtze, the Mekong, the Salween, the Brahmaputra, the Indus, and the Ganges (Pant et al. 2018; Boral and Sen 2020; Banerjee et al. 2021).

The Himalayan region encompasses a range of ecological zones due to its diverse topography and climatic variations. These zones include the Subtropical Zone, Temperate Zone, Alpine Zone, and Nival Zone, characterised by dense forests, including tropical and subtropical moist broadleaf forests (Ashton and Zhu 2020). Temperate forests dominated by coniferous trees like pine, fir, and spruce, alpine meadows, rocky slopes, and glaciers, with specialised flora and fauna adapted to cold and harsh conditions and above the tree line, are the permanent snow and ice, respectively (Gupta et al. 2020; Ahmad et al. 2017).

The Himalayas are renowned for their high biological diversity, with a wide range of plant and animal species. This region is a biodiversity hotspot, housing numerous endemic species, including the Himalayan blue sheep, snow leopard, Himalayan monal pheasant, and various species of *Rhododendron* and orchids (Yadav et al. 2021). Besides its natural beauty, the Himalayan region is culturally rich, with diverse ethnic groups, languages, and traditions. Communities in the Himalayas have historically depended on the region's natural resources for their livelihoods, a socio-ecological dynamic discussed in Chap. 2 (Sect. 2.1) including agriculture, animal husbandry, and trade along ancient routes like the Silk Road (Sati 2014). Despite its significance, the Himalayan region faces numerous environmental challenges, including deforestation, habitat fragmentation resulting from infrastructure development, overgrazing, climate change-induced glacier retreat, and the impact of tourism on fragile ecosystems.

13.1.2 Importance of Himalayan Biodiversity

Himalayan biodiversity holds enormous importance due to several factors. The Himalayas are one of the world's biodiversity hotspots, hosting a vast array of flora and fauna, including many endemic species found nowhere else on Earth

(Rawal et al. 2013). This biodiversity richness is crucial for maintaining ecological balance and supporting various ecosystems. The biodiversity in the Himalayas provides essential ecosystem services, including clean water, air purification, soil fertility, pollination, and climate regulation. These services are critical for human well-being, agriculture, and environmental health (Joshi and Joshi 2019). Most importantly, ecosystem services by the Himalayas, like carbon sequestration, benefit local communities and the wider region, making some countries carbon neutral/ negative.

The region is rich in medicinal plants, many of which have been used in traditional medicine for centuries. The genetic diversity in Himalayan species is valuable for breeding programmes, conservation efforts, and adaptation to environmental changes, including climate change (Dahal et al. 2021). Thus, the diverse ecosystems in the Himalayas contribute to climate resilience by providing habitat for species that can adapt to changing climatic conditions, and protecting this biodiversity is crucial for mitigating the impacts of climate change. Moreover, the biodiversity in the Himalayas is intricately linked to the cultural and spiritual beliefs of the indigenous communities residing in the region. Many species have cultural significance and are part of local traditions, rituals, and folklore (Nepal 2023; Bhatia et al. 2021; Dema 2021).

Furthermore, the biodiversity of the Himalayas attracts tourists and nature enthusiasts, supporting local economies through ecotourism and recreational activities like trekking, bird-watching, and wildlife tourism (Bhardwaj 2022). However, due to the anthropogenic activities in the Himalayan regions, emerging threats are causing a huge impact on biodiversity conservation. Studying Himalayan biodiversity offers insights into evolutionary processes, ecosystem dynamics, conservation strategies, and sustainable resource management. Therefore, preserving Himalayan biodiversity requires concerted efforts in conservation, sustainable land use practices, community engagement, policy frameworks, and international cooperation to ensure the long-term survival of these unique ecosystems and their associated benefits, which this chapter will address.

Anthropogenic activities refer to human actions or processes that impact the natural environment (Bogan et al. 2015). These can include deforestation (Maurya et al. 2020; Myers 2023), urbanisation (Ababor and Zakir 2019), pollution (air, water, soil) (Artiola et al. 2019), agriculture expansion (Laurance et al. 2014), infrastructure development (Sharma et al. 2018), mining, and climate change-related activities such as greenhouse gas emissions (Odell et al. 2018).

13.2 Literature Review

13.2.1 Overview of Himalayan Biodiversity

The Himalayan region is renowned for its rich biodiversity, encompassing many ecosystems and species due to its diverse topography, climate gradients, and unique geographical features (Yadav et al. 2021). The Himalayas span a vast elevation range, from subtropical foothills to alpine and snow-covered peaks, creating diverse habitats such as forests, grasslands, wetlands, and glaciers. The region includes temperate, sub-tropical, and alpine climate zones, each supporting distinct flora and fauna adapted to specific temperatures, precipitation, and seasonal variations (Mashwani 2020).

Biodiversity faces conservation challenges in the Himalayas due to natural and human-induced factors such as habitat loss, climate change, human-wildlife conflict, poaching, and illegal trade (Dhyani 2023). There are several measures and initiatives put forward to conserve the enriching biodiversity through protected areas (national parks, wildlife sanctuaries, and biosphere reserves), community-based conservation (conservation efforts, sustainable livelihoods, and wildlife monitoring), and few international collaboration (countries sharing the Himalayan region, such as Nepal, India, Bhutan, and China) is essential for transboundary conservation efforts, research collaboration, and addressing common conservation challenges (Thomsen and Thomsen 2021).

13.2.2 Anthropogenic Activities in the Himalayan Region

The Himalayan region boasts incredible biodiversity, encompassing a vast array of flora and fauna across its varied landscapes, from alpine meadows and coniferous forests to *Rhododendron* and oak forests (Deng et al. 2023; Pathak et al. 2023). Iconic species like the snow leopard, Himalayan brown bear, red panda, and diverse bird species inhabit this region, providing vital ecosystem services such as water regulation for major rivers, carbon sequestration, and cultural significance (Sharma et al. 2022). However, this biodiversity faces threats from habitat loss due to human activities such as deforestation, infrastructure development, pollution, overexploitation of resources, and the impacts of climate change. Conservation efforts focus on protected areas, community-based initiatives, sustainable practices, and addressing climate change to safeguard the rich biodiversity of the Himalayas for future generations and global ecological balance.

13.2.2.1 Deforestation and Habitat Loss

Deforestation and habitat loss are among the most significant threats to Himalayan biodiversity, a pattern observed across Chaps. 5 (Sect. 5.2) and 11 (Sect. 11.2), which detail the consequences of land-use change on forest ecosystems and biodiversity. Deforestation, driven primarily by human activities such as logging, agriculture expansion, infrastructure development, and fuelwood collection, leads to the loss of valuable forest habitats and disrupts crucial ecological processes. This loss of habitat directly affects wildlife species (Püttker et al. 2020), including endangered ones like the snow leopard, Himalayan brown bear, and red panda, by reducing their available living space, fragmenting populations, and limiting their access to food and shelter. Moreover, deforestation contributes to soil erosion, alters local climates, reduces water retention capacity, and diminishes overall ecosystem resilience (Qu et al. 2024). These impacts endanger biodiversity and threaten the livelihoods of communities that depend on forest resources and the provision of essential ecosystem services, such as water regulation and soil conservation. Addressing deforestation and habitat loss requires a holistic approach that integrates sustainable land-use practices, community engagement, policy interventions, and conservation initiatives to ensure Himalayan ecosystems' long-term health and integrity (Dhyani et al. 2022).

13.2.2.2 Climate Change

Climate change in the Himalayan region profoundly impacts biodiversity as discussed in Chap. 9 (Sect. 9.1), through habitat loss and fragmentation, shifts in species distribution, loss of endemic species adapted to unique conditions, altered ecosystem services like water regulation and pollination, increased invasive species disrupting native ecosystems, risks to keystone species and the cascading effects on entire ecosystems, and impacts on indigenous communities reliant on biodiversity for livelihoods and cultural practices. These complexities highlight the urgent need for conservation strategies and climate change adaptation measures to mitigate these effects (Dahal et al. 2021), a concern further reinforced in Chap. 12 (Sect. 12.2) which explores the intersection of ecological stressors and community vulnerability under a changing climate.

13.2.2.3 Pollution

Pollution from human activities has a significant impact on biodiversity in the Himalayan region (Kattel 2022). Air pollution, including emissions from industries, vehicles, and biomass burning, can lead to respiratory problems in wildlife and alter vegetation composition. Water pollution from agricultural runoff, industrial waste, and sewage affects aquatic ecosystems, harming fish populations and aquatic plants (Ramanathan and Ramana 2005). Soil pollution, often caused by improper waste disposal and chemical contaminants, reduces soil fertility and harms soil-dwelling organisms crucial for nutrient cycling. Pollution also contributes to climate change, exacerbating its impacts on biodiversity (Lal 2011). Addressing pollution in the Himalayan region is crucial to safeguarding its rich biodiversity and ensuring the resilience of ecosystems and communities dependent on them.

13.2.2.4 Overexploitation of Resources

The overexploitation of natural resources in the Himalayas has severely affected its biodiversity. Key activities contributing to this issue include timber extraction, medicinal plants, and non-timber forest products (NTFPs) harvesting, hunting, poaching, and overfishing. Timber extraction has been a significant contributor to deforestation, resulting in habitat loss and fragmentation. Extensive logging activities have resulted in the depletion of forest cover, crucial for the survival of many species. For example, the Red Panda, an endangered species, depends on dense bamboo forests and specific tree species, rapidly disappearing due to logging activities (Pradhan et al. 2001). The Himalayas are rich in medicinal plants and NTFPs, which local communities have traditionally harvested for centuries. However, increasing commercial demand has led to overharvesting, threatening the survival of many plant species, such as the Himalayan Yew (*Taxus wallichiana*), valued for its anti-cancer compound taxol, leading to population declines and disrupting the ecological balance (Kunwar et al. 2006). Illegal hunting and poaching threaten wildlife, targeting species such as the Snow Leopard *(Panthera uncia*), Musk Deer (*Moschus* spp.), and various pheasants for their valuable pelts, musk, and feathers. This poaching disrupts predator-prey relationships and reduces biodiversity (McCarthy and Mallon 2016). Overfishing in the rivers and lakes has severely impacted aquatic biodiversity, with fish species, many endemic to the region, declining due to unsustainable fishing practices. This threatens fish populations and affects other species that depend on fish as a primary food source, leading to a decline in overall ecosystem health (Gurumayum and Tamang 2017).

13.2.2.5 Infrastructure Development

Infrastructure development, including road construction, hydropower projects, urbanisation, and tourism, profoundly impacts Himalayan biodiversity, a theme also explored in Chap. 9 (Sect. 9.3) and this chapter (Sect. 13.3.5) in the context of habitat fragmentation and ecological disruption. Road construction leads to habitat fragmentation, isolating wildlife populations and making them more vulnerable to genetic bottlenecks and local extinctions while increasing wildlife-vehicle collisions, threatening species such as the Himalayan Black Bear (*Ursus thibetanus*) and various ungulates (Chole et al. 2025). Hydropower projects disrupt river ecosystems, affecting aquatic life and local communities by altering the natural flow of rivers, impacting fish migration and breeding, particularly for species like the Mahseer (*Tor* spp.), which require long stretches of free-flowing rivers to complete their life cycles (Everard 2021). The expansion of urban areas and tourism infrastructure results in habitat destruction and increased pollution as urban sprawl encroaches on natural habitats, leading to biodiversity loss. Moreover, while economically beneficial, tourism often brings about waste management issues and pollution, degrading natural habitats and polluting water bodies and soil, affecting both terrestrial and aquatic species (Nepal 2003). This concern is further supported in Chap. 10 (Sect. 10.4), which discusses the ecological impacts of unregulated tourism and emphasises the need for sustainable tourism practices in the Himalayan region.

Previous studies on the impact of anthropogenic activities on Himalayan biodiversity have highlighted significant concerns related to resource overexploitation, habitat fragmentation, and environmental degradation. Extensive logging and land conversion for agriculture have led to substantial habitat loss and fragmentation, resulting in the decline of forest-dependent species and altered ecological processes (Acharya et al. 2011). The unsustainable harvesting of medicinal plants and NTFPs has been extensively documented, with overexploitation threatening the survival of high-value species and disrupting ecological balance (Kunwar et al. 2010; Singh 2007). Infrastructure development, such as road construction and hydropower projects, further exacerbates these issues by fragmenting habitats and disrupting river ecosystems, impacting species like the Mahseer (*Tor* spp.) and increasing wildlife-vehicle collisions (Khatri et al. 2024; Everard and Kataria 2011). Expanding urban areas and tourism infrastructure contributes to habitat destruction and pollution, leading to biodiversity loss and environmental degradation. Climate change, exacerbated by anthropogenic pressures, also poses an additional threat by altering temperature and precipitation patterns, affecting species distribution and ecosystem dynamics (Xu et al. 2009). These studies collectively underscore the multifaceted impacts of human activities on the fragile biodiversity of the Himalayas.

13.3 Impact of Anthropogenic Activities on Himalayan Biodiversity

The Himalayas, known for their extraordinary biodiversity and ecological significance, face profound threats from anthropogenic activities such as deforestation, habitat loss, and climate change. These activities disrupt delicate ecological balances and threaten the survival of numerous species, impacting both flora and fauna and leading to a cascade of environmental consequences (Fig. 13.1).

13.3.1 Deforestation and Habitat Loss

Deforestation in the Himalayas, driven by agricultural expansion, logging, and infrastructure development, has led to significant habitat loss and fragmentation. The clearing of vast tracts of forest for timber, agriculture, and human settlements has altered the landscape, leaving behind fragmented patches of forest that are insufficient to support diverse ecosystems. This fragmentation disrupts the habitat connectivity for species movement, breeding, and foraging. The loss of habitat

Fig. 13.1 Destruction of habitat and loss of organism's life at the site of development. (Photo taken by Tej Kumar Nepal)

directly impacts species survival by reducing the availability of food, shelter, and breeding grounds while also increasing vulnerability to predation and human-wildlife conflict (Wang et al. 2016; Pandit and Kumar 2013).

13.3.1.1 Effects on Flora and Fauna

The consequences of deforestation extend beyond habitat loss to affect the entire ecological fabric. Flora, such as high-altitude herbs and medicinal plants, suffer from overharvesting and habitat destruction. This leads to a decline in plant species and disrupts the ecosystems that depend on these plants. For example, the Himalayan Yew (*Taxus wallichiana*), a source of the anti-cancer drug taxol, is being unsustainably harvested, threatening its survival and impacting associated wildlife that depends on it for food and habitat. Fauna is also significantly affected; species like the Snow Leopard (*Panthera uncia*) and the Himalayan Tahr (*Hemitragus jemlahicus*) experience habitat reduction and food scarcity, leading to population declines and altered behaviours. The loss of key species further destabilises ecological relationships and impacts ecosystem services, such as pollination and soil fertility (Singh et al. 2021; Sharma and Sharma 2017; Parveen et al. 2013).

13.3.1.2 Loss of Biodiversity Hotspots

Loss of biodiversity and the preservation of endangered species and unique ecological systems are often central issues in today's policy choices and implementation. The Himalayas are home to several biodiversity hotspots, regions rich in endemic species, and high levels of biodiversity. The Eastern Himalayas, for instance, are recognised as a critical hotspot due to their unique assemblage of flora and fauna. The destruction of these hotspots compromises their role in global biodiversity and ecological balance. As these areas degrade, the loss of species and habitats affects broader ecological processes, including carbon sequestration and water regulation. The disappearance of biodiversity hotspots undermines the resilience of ecosystems to environmental changes and diminishes their capacity to provide essential ecosystem services (Dorji et al. 2018; Myers et al. 2000; Sala et al. 2000). This concern is also emphasised in Chap. 2 (Sect. 2.1), which discusses how the Himalayas, recognised as one of the world's biodiversity hotspots, face increasing vulnerability due to socio-ecological pressures and unsustainable human activities.

13.3.2 Melting Glaciers and Altered Water Availability

Climate change is intensifying the impacts of anthropogenic activities on Himalayan biodiversity. Rising temperatures and changing precipitation patterns are causing shifts in species distribution and altering migration patterns. Many species are migrating to higher elevations in search of suitable climatic conditions, disrupting established ecological relationships and giving rise to new interactions between species. For instance, alpine plants are migrating up the slopes, which may lead to competition with lower-altitude species and changes in plant communities. The altered migration patterns of animals, such as birds and mammals, can lead to mismatches between their life cycles and the availability of food resources, affecting their survival and reproduction (Hamid et al. 2019).

The Himalayas, often referred to as the "Third Pole" due to their extensive glaciers, are experiencing accelerated glacier melt as a result of global warming. The loss of glaciers affects the flow of rivers and streams, critical freshwater sources for millions of people and ecosystems. Glacial melt alters water availability, changing river discharge patterns and affecting aquatic habitats. Species like the Mahseer (*Tor* spp.), which depend on specific flow conditions for spawning; face declining populations as river conditions change. The alteration in water availability also impacts human communities that rely on these water resources for agriculture, drinking water, and hydroelectric power (Nie et al. 2021; Aryal et al. 2014).

13.3.3 Pollution

13.3.3.1 Air Pollution and Its Effects on Himalayan Ecosystems

Air pollution releases harmful contaminants (chemicals, toxic gases, particulates, biological molecules etc.) into the Earth's atmosphere. These contaminants are quite detrimental and, in some cases, pose serious health issues. Water pollution occurs when toxic pollutants and particulate matter are introduced into water bodies such as lakes, rivers, and seas. These contaminants are generally introduced by human activities like improper sewage treatment and oil spills. Pollution is a severe worldwide problem, resulting in the deterioration of the quality of natural biotic and abiotic factors. Water pollution is a huge problem, especially in developing countries. Water is one of the most essential urban and rural amenities and is indispensable for human activities, including drinking water, irrigation, and recreational opportunities. Pollution poses a severe risk to life, especially when water is a source of drinking and is used for domestic purposes by humans. Polluted waters are potent agents of diseases such as cholera, typhoid, and tuberculosis. The growth of the human population and industrial and agricultural practices are major causes of pollution. Motorcars and other industries have increased the number of poisonous gases like SOx, NOx, CO, and smoke in the atmosphere. This will bring unbelievable changes to the Himalayan biodiversity (Hassan et al. 2023; Sharma and Sharma 2019; Pramanik and Bhaduri 2016).

13.3.3.2 Water Pollution in Rivers and Lakes

The Global Waterkeeper Alliance and Clean Bhutan found the *E-coli* in Bhutan's rivers during the 2017 water quality test (Gyelmo 2018). A poor sewer system that connects directly to the river pollutes the water downstream. For example, the black water from the wasteland fields in Memelakha falls into the Olarongchu River, making it quite toxic. Additionally, funeral rites on the riverbank and urbanisation pose threats to water quality.

13.3.4 Overexploitation of Resources

13.3.4.1 Hunting and Poaching

Life expectancy has increased due to modern medical facilities, which has resulted in the growth of the population, thus posing a strain on natural resources and causing environmental degradation. The increasing population demands more food, clothing, and shelter and needs more space to grow food and build houses, which results in loss of biodiversity and destruction of the ecosystem (Park 2002).

The deterioration of the environment is caused by the depletion of resources (both biotic and abiotic elements) that form our surroundings. Some major environmental degradation factors in Bhutan are human (urbanisation, industrialisation, population growth, deforestation) and natural (floods, rising temperatures, and fires) causes (Fig. 13.2). This has caused serious issues in the country, such as pollution, biodiversity loss, animal extinction, deforestation, and global warming (Uddin et al. 2007).

Environmental degradation will also lead to the extinction of wild animals and species, which would indirectly disrupt the food chain and food web, resulting in indirect losses to farmers due to crop destruction and human-wildlife conflicts, including loss of human life from wild animal attacks. Hunting will also have an economic impact when excessive poaching and hunting are carried out, and the

Fig. 13.2 Wildfire burning acres of forest in Central Bhutan. (Photo taken by Tej Kumar Nepal)

secondary food sources become exhausted, resulting in tertiary animals entering cultivated lands and attacking humans as well (Manral et al. 2016).

13.3.4.2 Unsustainable Harvesting of Medicinal Plants and Herbs

The Himalayas, renowned for their rich biodiversity, are also a critical source of medicinal plants and herbs, which have been used for centuries by local communities and traditional medicine practitioners. However, the unsustainable harvesting of these plants poses a significant threat to Himalayan biodiversity. The increasing commercial demand for medicinal plants, such as the Himalayan Yew (*Taxus wallichiana*) and the *Saussurea costus*, has led to overexploitation and population declines. The Yew, valued for its anti-cancer compound taxol, is being harvested at unsustainable rates. It threatens its survival and disrupts the ecological balance, as many wildlife species rely on these plants for food and habitat (Kunwar et al. 2006). Similarly, the overharvesting of herbs like *Saussurea costus*, used in traditional medicine and for its aromatic properties, impacts its populations and disrupts the ecosystems that depend on it. The loss of these plants affects the entire ecological network, reducing biodiversity and altering ecosystem services such as nutrient cycling and habitat structure. These concerns are also addressed in Chap. 5 (Sect. 5.1), which examines the ecological significance and conservation challenges of medicinal plant species in the Himalayan region.

13.3.5 Infrastructure Development

13.3.5.1 Roads and Dams Disrupting Ecosystems

Infrastructure development, particularly roads and dams, significantly disrupts ecosystems in the Himalayas. Road construction, essential for economic development and connectivity, leads to habitat fragmentation and isolation. This fragmentation makes it challenging for wildlife to move between habitats, leading to genetic isolation, reduced biodiversity, and increased mortality due to vehicle collisions. Species such as the Himalayan Black Bear (*Ursus thibetanus*) and various ungulates face increased risks as their habitats are bisected by roads, leading to habitat loss and fragmentation (Bhandari and Bijlwan 2022).

Similarly, hydropower projects, promoted as a renewable energy source, have considerable environmental impacts. The construction of dams and reservoirs alters river ecosystems by disrupting natural water flow, sediment transport, and fish migration patterns. These changes particularly affect species like the Mahseer (*Tor* spp.), which require long stretches of free-flowing rivers for spawning. Dams can lead to the loss of critical spawning grounds and altered water conditions, which threaten the survival of aquatic species and impact the livelihoods of local communities that depend on these rivers for agriculture and drinking water (Richter et al. 2010).

13.3.5.2 Fragmentation of Habitats and Wildlife Corridors

The fragmentation of habitats and wildlife corridors due to infrastructure development further exacerbates the challenges Himalayan biodiversity faces. Wildlife corridors, which facilitate animal movement between habitats, are crucial for maintaining genetic diversity and ecosystem health. The construction of roads, settlements, and other infrastructure projects (Fig. 13.3) disrupts these corridors, making isolated populations more susceptible to inbreeding and local extinctions. For example, the fragmentation of habitats in the Eastern Himalayas has led to isolated populations of species like the Red Panda, which struggle to find suitable habitats and resources (Thapa et al. 2018). This fragmentation affects the species directly and impacts their ecological processes, including seed dispersal and predator-prey dynamics. The loss of functional wildlife corridors reduces the ability of species to adapt to environmental changes and increases their vulnerability to threats such as habitat loss and climate change.

13.4 Conservation Efforts and Mitigation Strategies

13.4.1 Policy Recommendations for Sustainable Development

Natural resources, such as forests and commercially exploitable fisheries, and environmental attributes, such as air quality, are valuable assets in that they yield flows of services to people. Public policies and the actions of individuals and firms can lead to changes in these service flows, thereby creating benefits and costs. Technological transfer and good policy can be the best combination to minimise environmental impact at different levels of development, which will neither cause major environmental effects nor impact the level of development. Education is the driving vehicle for environmental protection and sustainable development. Education, as a life-long learning process, enhances one's knowledge, skills, and positive attitudes toward preserving the environment, especially through environmental and ecosystem valuation. This makes people more conscious of rising environmental issues, leading to a lifestyle favouring the environment.

13.4.2 Community-Based Conservation Approaches

Community-based conservation in the Himalayas is vital for sustainably preserving biodiversity and managing natural resources. By involving local communities in conservation efforts, these approaches leverage traditional knowledge and practices to enhance effectiveness. Successful examples include Nepal's Community Forestry Programme (CFP), which has enhanced forest cover and biodiversity by

Fig. 13.3 Road construction in the remote part of Bhutan. (Photo taken by Tej Kumar Nepal)

empowering local communities to manage resources sustainably (Acharya 2006; Acharya et al. 2006). Similarly, Bhutan's community-managed protected areas and biological corridors help protect wildlife habitats and promote sustainable land use. Sacred groves, protected due to cultural beliefs, also serve as crucial biodiversity hotspots (Sen 2018).

13.4.3 Opportunities for International Collaboration

International collaboration addresses transboundary conservation challenges and enhances regional biodiversity protection. Transboundary conservation areas, such as the Kangchenjunga Conservation Area shared by Nepal, India, and Bhutan, facilitate joint management and conservation efforts (Shakya and Joshi 2008; Phuntsho et al. 2012). Such initiatives enhance habitat connectivity and mitigate threats such as poaching and illegal logging. Additionally, international partnerships offer opportunities for knowledge exchange, joint research, and resource mobilisation. Global organisations, such as WWF, play a key role in supporting these efforts through funding and expertise (Anyango-van Zwieten et al. 2019).

13.5 Future Research Directions

13.5.1 Gaps in Current Knowledge

Future research in Himalayan biodiversity should address several key gaps in current knowledge. First, there is a need for comprehensive baseline data on species distributions, population dynamics, and ecological interactions, which often require regular updates. This information is crucial for understanding how biodiversity is responding to various anthropogenic pressures and climate change. Second, research should focus on the impacts of lesser-studied threats, such as emerging diseases and invasive species, which may be affecting native flora and fauna. Furthermore, additional studies are necessary to investigate the socio-economic impacts of conservation measures, ensuring that strategies are both effective and equitable for local communities.

13.5.2 Emerging Threats to Himalayan Biodiversity

Emerging threats to Himalayan biodiversity include climate change, which is causing rapid alterations in habitat conditions, shifting species distributions, and affecting migration patterns. The accelerated melting of glaciers is altering river flows and impacting aquatic ecosystems. Another significant threat is the increase in human-wildlife conflicts resulting from the expansion of agricultural activities and infrastructure development. Invasive species, introduced either intentionally or accidentally, are also becoming a growing concern, as they can outcompete native species or disrupt ecosystem functions. Research should focus on understanding the interactions between these emerging threats and their cumulative impacts on biodiversity.

13.5.3 Long-Term Monitoring and Adaptive Management Strategies

Long-term monitoring and adaptive management strategies are essential to address these challenges. Monitoring programmes should be designed to track changes in biodiversity over time, using standardised methods to ensure data comparability. This includes assessing changes in species populations, habitat conditions, and ecosystem services. Adaptive management involves using monitoring data to adjust conservation strategies in response to changing conditions and new information. Research should explore the development of flexible management frameworks that can be adjusted based on real-time data and predictions of future trends. Collaborative efforts among scientists, local communities, and policymakers are crucial for implementing adaptive strategies responsive to ecological and socio-economic changes.

13.6 Conclusions

The Himalayan region, one of the global hotspots with high levels of endemicity and vulnerability, is now experiencing an alarming rate of habitat destruction through deforestation, infrastructural developments, escalating levels of pollution, over utilisation of resources, and climate change. Such pressures are becoming the leading causes of habitat destruction, species loss, and declining ecosystem services, which are critical to supporting millions of people's livelihoods, especially in water supply, food security, and income. Himalayan regions are rich with flora and fauna species, and out of them, most are endangered species, indicating that the future of these species is at the edge. To reduce these impacts and to ensure the overall conservation of the Himalayas, immediate and bio-physical integration is required. These are practices such as controlling land-use change, preserving ecosystems and minimising exploitation of natural assets. Also, it is necessary to implement climate change adaptation measures since increased temperatures and changes in precipitation patterns harm the region's wildlife even more. People should be involved, especially the locals, since they are always impacted by deforestation, possess valuable information, and are key players in environmental conservation.

Furthermore, the need to cooperate internationally is important since the Himalayan ecosystem cuts across several countries, and therefore, all countries have a role to play in ecosystem conservation. The above challenges will, therefore, require a multi-disciplinary approach to research, policy, and resource management.

Acknowledgements The authors express their sincere gratitude to the editors and reviewers for their insightful comments and constructive suggestions, which have significantly improved the quality of this work. The views and opinions presented in this chapter are solely those of the authors. Unless otherwise specified, all photographs included were taken by Tej Kumar Nepal.

References

Ababor Z, Zakir M (2019) Review on effect of climate change and urbanization on agriculture. Civil Environ Res 11(5):13

Acharya KP (2006) Linking trees on farms with biodiversity conservation in subsistence farming systems in Nepal. Biodivers Conserv 15:631–646

Acharya KP, Goutam KR, Acharya BK, Gautam G (2006) Participatory assessment of biodiversity conservation in community forestry in Nepal. Banko Janakari 16(1):46–56

Acharya BK, Chettri B, Vijayan L (2011) Distribution pattern of trees along an elevation gradient of Eastern Himalaya, India. Acta Oecol 37(4):329–336

Ahmad L, Habib Kanth R, Parvaze S, Sheraz Mahdi S, Ahmad L, Habib Kanth R, Sheraz Mahdi S (2017) Agro-climatic and agro-ecological zones of India. In: Experimental Agrometeorology: A Practical Manual. Springer, Cham. https://doi.org/10.1007/978-3-319-69185-5_15

Anyango-van Zwieten N, Lamers M, Van der Duim R (2019) Funding for nature conservation: a study of public finance networks at World Wide Fund for nature (WWF). Biodivers Conserv 28(14):3749–3766

Artiola JF, Walworth JL, Musil SA, Crimmins MA (2019) Soil and land pollution. In: Environmental and pollution science. Academic, pp 219–235. https://doi.org/10.1016/B978-0-12-814719-1.00014-8

Aryal S, Cockfield G, Maraseni TN (2014) Vulnerability of Himalayan transhumant communities to climate change. Clim Chang 125:193–208

Ashton P, Zhu H (2020) The tropical-subtropical evergreen forest transition in East Asia: an exploration. Plant Divers 42(4):255–280

Banerjee A, Chen R, Meadows ME, Sengupta D, Pathak S, Xia Z, Mal S (2021) Tracking 21st century climate dynamics of the Third Pole: an analysis of topo-climate impacts on snow cover in the central Himalaya using Google Earth Engine. Int J Appl Earth Obs Geoinf 103:102490

Bhandari BS, Bijlwan K (2022) Vulnerability of forest vegetation due to anthropogenic disturbances in Western Himalayan region of India. In: Research anthology on ecosystem conservation and preserving biodiversity. IGI Global, India. pp 1297–1312

Bhardwaj RL (2022) Role of ecotourism in sustainable development of Himalayas: a case study of Himachal Pradesh. In: Rethinking Himalaya: its scope and protection, Blue Rose Publishers, New Delhi vol 11, pp 109–129

Bhatia S, Suryawanshi K, Redpath SM, Namgail S, Mishra C (2021) Understanding people's relationship with wildlife in trans-himalayan folklore. Front Environ Sci 25(9):595169

Bogan E, Doina ST, Vărvăruc D (2015) The impact of anthropogenic activities on components of the natural environment of the Titu Plain. Georev Sci Ann Ştefan cel Mare Univ Suceava Geogr Ser 24(1):54–64

Boral S, Sen IS (2020) Tracing 'Third Pole' ice meltwater contribution to the Himalayan rivers using oxygen and hydrogen isotopes. Geochem Perspect Lett 13:48–53

Chole A, Dhadse S, Sakhare P (2025) Biodiversity loss due to road widening and a new highway. Sustainability and Biodiversity Conservation 4(2):38–52. https://doi.org/10.5281/zenodo.15623245

Dahal N, Lamichhaney S, Kumar S (2021) Climate change impacts on Himalayan biodiversity: evidence-based perception and current approaches to evaluate threats under climate change. J Indian Inst Sci 101(2):195–210

Dema T (2021) Eco-spiritual and economic perspectives in Bhutan's Haa district. In: Environmental humanities in the new Himalayas, vol 16. Routledge, pp 66–80

Deng X, Domisch S, Favre A, Jähnig S, Frandsen P, He F, Pauls S (2023) Comparative phylogeography of Himalopsyche (Trichoptera, Rhyacophilidae) in the Tibeto-Himalayan Region: an assessment of the mountain-geobiodiversity hypothesis, Authorea Preprints

Dhyani S (2023) Are Himalayan ecosystems facing hidden collapse? Assessing the drivers and impacts of change to aid conservation, restoration and conflict resolution challenges. Biodivers Conserv 32(12):3731–3764

Dhyani S, Sivadas D, Basu O, Karki M (2022) Ecosystem health and risk assessments for high conservation value mountain ecosystems of South Asia: a necessity to guide conservation policies. Anthr Sci 1(1):211–225

Dorji S, Rajaratnam R, Falconi L, Williams SE, Sinha P, Vernes K (2018) Identifying conservation priorities for threatened Eastern Himalayan mammals. Conserv Biol 32(5):1162–1173

Everard M, Kataria G (2011) Recreational angling markets to advance the conservation of a reach of the Western Ramganga River, India. Aquatic Conservation: Marine and Freshwater Ecosystms 21:101–108. https://doi.org/10.1002/aqc.1159

Everard M (2021) Ecosystem services: key issues. Routledge

Ghimire SK (2008) Sustainable harvesting and management of medicinal plants in the Nepal Himalaya: current issues, knowledge gaps and research priorities. In: Med Plants Nepal Anthol Contemp Res, Ecological Society - Kathmandu. pp 25–44

Gupta AK, Negi M, Nandy S, Kumar M, Singh V, Valente D, Pandey R (2020) Mapping socio-environmental vulnerability to climate change in different altitude zones in the Indian Himalayas. Ecol Indic 109:105787

Gurumayum SD, Tamang L (2017) Coldwater fish and fishery of Arunachal Pradesh, India. Bull Arunachal For Res 32(1&2):17–26

Gyelmo D (2018) Bhutan economy boost brings a rising waste problem. The Third Pole

Hamid M, Khuroo AA, Charles B, Ahmad R, Singh CP, Aravind NA (2019) Impact of climate change on the distribution range and niche dynamics of Himalayan birch, a typical treeline species in Himalayas. Biodivers Conserv 28:2345–2370

Hassan MA, Mehmood T, Liu J, Luo X, Li X, Tanveer M, Faheem M, Shakoor A, Dar AA, Abid M (2023) A review of particulate pollution over Himalaya region: characteristics and salient factors contributing ambient PM pollution. Atmos Environ 294:119472

Joshi AK, Joshi PK (2019) Forest ecosystem services in the central Himalaya: local benefits and global relevance. Proc Natl Acad Sci India Sect B Biol Sci 89:785–792

Kattel GR (2022) Climate warming in the Himalayas threatens biodiversity, ecosystem functioning and ecosystem services in the 21st century: is there a better solution? Biodivers Conserv 31(8):2017–2044

Khatri K, Jha BR, Gurung S, Khadka UR (2024) Freshwater fish diversity and IUCN Red List status of glacial-fed (Bheri) and spring-fed (Babai) rivers in the wake of inter-basin water transfer. J Threat Taxa 16(1):24535–24549

Kunwar RM, Nepal BK, Kshhetri HB, Rai SK, Bussmann RW (2006) Ethnomedicine in Himalaya: a case study from Dolpa, Humla, Jumla and Mustang districts of Nepal. J Ethnobiol Ethnomed 2:27

Kunwar RM, Shrestha KP, Bussmann RW (2010) Traditional herbal medicine in Far-west Nepal: a pharmacological appraisal. J Ethnobiol Ethnomed 6:1–18

Lal M (2011) Implications of climate change in sustained agricultural productivity in South Asia. Reg Environ Chang 11(Suppl 1):79–94

Laurance WF, Sayer J, Cassman KG (2014) Agricultural expansion and its impacts on tropical nature. Trends Ecol Evol 29(2):107–116
Manral U, Sengupta S, Hussain SA, Rana SA, Badola RU (2016) Human wildlife conflict in India: a review of economic implication of loss and preventive measures. Indian For 142(10):928–940
Mashwani ZUR (2020) Environment, climate change and biodiversity. In: Environment, climate, plant and vegetation growth, Springer - Cham, Switzerland. pp 473–501
Maurya PK, Ali SA, Ahmad A, Zhou Q, Castro JS, Khan E, Ali H (2020) An introduction to environmental degradation: causes, consequence, and mitigation. https://doi.org/10.26832/aesa-2020-edcrs-01
McCarthy T, Mallon D (2016) Snow leopards: biodiversity of the world: conservation from genes to landscapes. Academic. London
Myers N (2023) Tropical deforestation: rates and patterns. In: The causes of tropical deforestation, pp 27–40
Myers N, Mittermeier RA, Mittermeier CG, Da Fonseca GA, Kent J (2000) Biodiversity hotspots for conservation priorities. Nature 403(6772):853–858
Nepal SK (2003) Tourism and the environment: perspectives from the Nepal Himalaya. Sage Publications. Lalitpur Nepal
Nepal TK (2023) Preserving traditional medicinal knowledge and its transmission in Dorokha, Samtse, Bhutan. Rig Tshoel 6(1):53–74
Nie Y, Pritchard HD, Liu Q, Hennig T, Wang W, Wang X, Liu S, Nepal S, Samyn D, Hewitt K, Chen X (2021) Glacial change and hydrological implications in the Himalaya and Karakoram. Nat Rev Earth Environ 2(2):91–106
Odell SD, Bebbington A, Frey KE (2018) Mining and climate change: a review and framework for analysis. Extr Ind Soc 5(1):201–214
Pandit MK, Kumar V (2013) Land-use change and conservation challenges in the Indian Himalaya: past, present, and future. In: Conservation biology: voices from the tropics, Wiley, United Kingdom. vol 16, pp 121–133
Pandit MK, Manish K, Koh LP (2014) Dancing on the roof of the world: ecological transformation of the Himalayan landscape. BioSci 64(11):980–992
Pant NC, Ravindra R, Srivastava D, Thompson L (2018) The Himalayan cryosphere: past and present variability of the 'third pole'. Geological Soc London 462:1–6
Park C (2002) The environment: principles and applications. Routledge
Parveen S, Jan U, Kamili A (2013) Importance of Himalayan medicinal plants and their conservation strategies. Aus J Herb Med 25(2):63–67
Pathak R, Pant V, Negi VS, Bhatt ID, Belwal T (2023) Introduction to Himalayan region and wild edible diversity. In: Himalayan fruits and berries. Academic, pp 1–12. https://doi.org/10.1016/B978-0-323-85591-4.00024-6
Phuntsho K, Chettri N, Oli KP (2012) Mainstreaming community-based conservation in a transboundary mountain landscape–lessons from Kangchenjunga. International Centre for Integrated Mountain Development, Kathmandu. p 5
Pradhan S, Saha GK, Khan JA (2001) Ecology of the red panda *Ailurus fulgens* in the Singhalila National Park, Darjeeling, India. Biol Conserv 98(1):11–18
Pramanik P, Bhaduri D (2016) Impact of climate change on water resources in Indian Himalaya. In: Conservation agriculture: an approach to combat climate change in Indian Himalaya, Springer - Singapore. pp 487–507
Püttker T, Crouzeilles R, Almeida-Gomes M, Schmoeller M, Maurenza D, Alves-Pinto H et al (2020) Indirect effects of habitat loss via habitat fragmentation: a cross-taxa analysis of forest-dependent species. Biol Conserv 241:108368
Qu X, Li X, Bardgett RD, Kuzyakov Y, Revillini D, Sonne C et al (2024) Deforestation impacts soil biodiversity and ecosystem services worldwide. Proc Natl Acad Sci USA 121(13):e2318475121
Ramanathan V, Ramana MV (2005) Persistent, widespread, and strongly absorbing haze over the Himalayan foothills and the Indo-Gangetic Plains. Pure Appl Geophys 162:1609–1626
Rawal RS, Bhatt ID, Sekar KC, Nandi SK (2013) The Himalayan biodiversity: richness, representativeness, uniqueness and life-support values. In: GB Pant Institute of Himalayan Environment and Development, p 84

Richter BD, Postel S, Revenga C, Scudder T, Lehner B, Churchill A, Chow M (2010) Lost in development's shadow: the downstream human consequences of dams. Water Altern 3(2):14

Sala OE, Stuart Chapin FI, Armesto JJ, Berlow E, Bloomfield J, Dirzo R, Huber-Sanwald E, Huenneke LF, Jackson RB, Kinzig A, Leemans R (2000) Global biodiversity scenarios for the year 2100. Science 287(5459):1770–1774

Sati VP (2014) Towards sustainable livelihoods and ecosystems in mountain regions. Springer International Publishing, pp 137–150

Sen UK (2018) Assessing the social, ecological and economic impact on conservation activities within human-modified landscapes: a case study in Jhargram district of West Bengal, Cham Switzerland, India. Int J Conserv Sci 9(2):319–336

Shakya B, Joshi RM (2008) Protected areas and biodiversity conservation in the Hindu Kush-Himalayan region with special reference to the Kangchenjunga landscape. In: Biodiversity conservation in the Kangchenjunga landscape, International Centre for Integrated Mountain Development. Kathmandu. p 13

Sharma G, Sharma E (2017) Agroforestry systems as adaptation measures for sustainable livelihoods and socio-economic development in the Sikkim Himalaya. In: Agroforestry: anecdotal to modern science, pp 217–243

Sharma A, Sharma A (2019) Impact of anthropogenic activities on Himalayan ecosystem services and their management/sustainability: a review. Pharma Innov J 8(1):361–365

Sharma R, Rimal B, Stork N, Baral H, Dhakal M (2018) Spatial assessment of the potential impact of infrastructure development on biodiversity conservation in lowland Nepal. ISPRS Int J Geo-Inf 7(9):365

Sharma LK, Joshi BD, Thakur M, Chandra K (2022) Status, issues, and challenges of biodiversity: wild animals. In: Biodiversity in India: status, issues and challenges. Springer Nature Singapore, Singapore, pp 149–173

Singh SP (2007) Himalayan forest ecosystem services. Central Himalayan Environment Association, Nainital, p 53

Singh SP, Sharma S, Dhyani PP (2019) Himalayan arc and treeline: distribution, climate change responses and ecosystem properties. Biodivers Conserv 28(8):1997–2016

Singh C, Chauhan N, Upadhyay SK, Singh R, Rani A (2021) The Himalayan natural resources: challenges and conservation for sustainable development. J Pharmacogn Phytochem 10(1): 1643–1648

Thapa A, Hu Y, Wei F (2018) The endangered red panda (*Ailurus fulgens*): ecology and conservation approaches across the entire range. Biol Conserv 220:112–121

Thomsen B, Thomsen J (2021) Multispecies livelihoods: partnering for sustainable development and biodiversity conservation. In: Partnerships for the goals, Springer, Cham Switzerland. pp 758–768

Uddin SN, Taplin R, Yu X (2007) Energy, environment and development in Bhutan. Renew Sust Energ Rev 11(9):2083–2103

Wang L, Young SS, Wang W, Ren G, Xiao W, Long Y, Li J, Zhu J (2016) Conservation priorities of forest ecosystems with evaluations of connectivity and future threats: implications in the Eastern Himalaya of China. Biol Conserv 195:128–135

Xu J, Grumbine RE, Shrestha A, Eriksson M, Yang X, Wang YUN, Wilkes A (2009) The melting Himalayas: cascading effects of climate change on water, biodiversity, and livelihoods. Conserv Biol 23(3):520–530

Yadav RR, Negi PS, Singh J (2021) Climate change and plant biodiversity in Himalaya, India. Proc Indian Natl Sci Acad 87:234–259

Yao T, Thompson LG, Mosbrugger V, Zhang F, Ma Y, Luo T, Xu B, Yang X, Joswiak DR, Wang W, Joswiak ME (2012) Third pole environment (TPE). Environ Dev 3:52–64

Chapter 14
Eastern Himalayan Ecosystem Under Human Influence and Climate Crisis

Chandan Barman and Piyush Mathur

Abstract Present unsustainable human driven activities and climate crisis pose a serious threat to the Himalayan environment, one of the planet's most delicate and biologically significant areas. Deforestation, unplanned urbanisation, overgrazing, and infrastructure development are examples of human activities that have destroyed habitat, reduced biodiversity, and degraded the ecosystem of Eastern Himalayas. The stress on natural resources is also being exacerbated by pollution and irresponsible tourism. With rising temperatures speeding up glacial retreat, changing river flows, and increasing the frequency of extreme weather events like landslides, floods, and droughts, climate change is making these problems even more severe. Local residents, who rely on the natural resources of the area for their livelihoods, biodiversity, and agriculture are all significantly impacted by these changes. Millions of people who depend on river systems that originate in the Himalayas are at risk of losing their access to water due to the melting of the glaciers. The ecological equilibrium is being upset by anthropogenic pressure and climate change, which is causing species distribution changes and the extinction of endemic plants and animals. These risks have the potential to permanently harm the ecosystem and socioeconomic stability of the area if they are not addressed. Urgent actions are required to lessen these effects, such as improved disaster preparedness, afforestation initiatives, sustainable land-use planning, and responsible tourism. For the Himalayas to be preserved, international collaboration and climate adoption measures must be strengthened. To maintain the long-term resilience of this vital ecosystem, conservation initiatives must combine contemporary scientific methods with traditional ecological knowledge. Protecting the Himalayas from the combined consequences of human-caused alterations and global climate shifts requires immediate action.

C. Barman
Department of Botany, Gour Banga University, Malda, West Bengal, India

P. Mathur (✉)
Department of Botany, Institute of Science, Banaras Hindu University, Varanasi, Uttar Pradesh, India
e-mail: piyushmathur@bhu.ac.in

D. Das et al. (eds.), *Plant Diversity and Conservation Strategies of the Himalayan Biodiversity Hotspot*, Topics in Biodiversity and Conservation 21,
https://doi.org/10.1007/978-3-032-21697-7_14

Keywords Biodiversity · Elevated CO_2 · Forest fires · Mitigation · Rainfall · Urbanisation

14.1 Introduction

The Himalayan ecosystem constitutes one of the most ecologically significant yet critically fragile regions of the Earth's biosphere. Extending across Nepal, India, Bhutan, and China, this mountain range harbours exceptional biodiversity and supports diverse human communities. However, escalating anthropogenic pressures coupled with the intensifying effects of global climate change are precipitating profound ecological transformations. Key manifestations include accelerated environmental degradation, biodiversity loss, and an increased frequency of climate- and geology-related disasters.

Deforestation remains a principal driver of ecological destabilisation in the Himalayas. The extraction of timber and fuelwood, alongside the expansion of agricultural land, has resulted in extensive forest loss, diminishing carbon sequestration capacity and exacerbating soil erosion. Singh and Kumar (2022) report a marked acceleration in deforestation rates across the Indian Himalayan Region over the past three decades, primarily attributable to infrastructure expansion and agricultural intensification. The loss of forest cover destabilises steep mountain slopes, thereby increasing their susceptibility to landslides and flash floods.

Rapid urbanisation and the proliferation of large-scale infrastructure projects including highways, dams, and hydropower facilities are further disrupting ecological equilibrium in the Eastern Himalayas, as observed elsewhere in the range. The construction of high-capacity dams, such as the Tehri Dam in India, has significantly altered river hydrology and displaced substantial human populations (Kumar and Katoch 2016). Concurrently, unregulated urban expansion contributes to escalating pollution levels and unmanaged waste accumulation, compounding stress on already fragile ecosystems (Bajracharya and Sultana 2022).

Tourism while representing a vital source of economic revenue exerts considerable environmental pressure. Prominent destinations such as Everest Base Camp, Kedarnath, and the Kangchenjunga Landscape attract large numbers of visitors annually, generating significant volumes of plastic waste and increasing greenhouse gas emissions. Baloch et al. (2023) emphasise that over-tourism in such areas results in extensive habitat degradation and water resource contamination, posing severe threats to endemic biodiversity. This study examines the unique biological wealth of the Eastern Himalayas and elucidates the mounting threats arising from human activities, set against the backdrop of an accelerating global climate crisis.

14.2 Vascular and Non-vascular Plant Diversity Spectrum Eastern Himalayas

14.2.1 Geographic Extent and Ecological Significance

The Indian Himalayan Region (IHR), extending between 21°57′ – 37°5′ N latitudes and 72°40′ – 97°25′ E longitudes, covers approximately 419,873 km^2 and serves as a rich repository of wild plant diversity and associated biota, including algae, fungi, lichens, bryophytes, and pteridophytes. Vegetation patterns vary markedly with altitude, transitioning from tropical and subtropical zones at lower elevations to temperate, sub-alpine, and alpine zones at higher altitudes. This altitudinal gradient supports an extraordinary diversity of angiosperms, gymnosperms, pteridophytes, bryophytes, fungi, lichens, and algae.

From the Eastern Himalaya (Nepal and Bhutan), approximately 2302 genera of angiosperms and 20 genera of gymnosperms have been recorded (Rana and Rawat 2017), with an estimated 30% endemism (Chatterjee 1940; Samant 2021). In Assam, 994 plant species representing 529 genera and 110 families have been documented (Choudhary et al. 2012). Across the IHR, 18,940 plant species are recorded, including ~3400 endemic angiosperms, ~7 endemic gymnosperms, ~150 endemic pteridophytes, ~565 bryophytes, ~130 lichens, and ~ 1890 fungi (Singh and Hajra 1996; Samant 2021). Arunachal Pradesh alone harbours 59 endemic species across 46 genera and 27 families (Behera et al. 2002).

14.2.2 Floristic Diversity in the Eastern Himalayas

The Eastern Himalaya exhibits substantial variation in species richness across states (Naithani 2020; Samant 2021). Indian states of Sikkim and Arunachal Pradesh recorded 4458 species and 4005 species of seed plants respectively, while associated state of West Bengal and Assam also housing a significant diversity (>3000) of seed plant species. The floristic diversity of Bhutan showing very rich diversity of more than 5600 species of seed plants in various altitudinal gradation of eastern Himalayas.

Sikkim (7096 km^2; 88°00′58″ – 88°55′25″ E; 27°4′ – 28°7′48″ N) and the Darjeeling Himalaya (2417.3 km^2) form biodiversity-rich sectors of the Eastern Himalaya. Darjeeling harbours ~3000 angiosperms, 12 gymnosperms, 250 pteridophytes, and 200 bryophytes, while Sikkim hosts ~4500 angiosperms, 16 gymnosperms, 480 pteridophytes, 362 bryophytes, 506 lichens, and 346 algae taxa (Maity et al. 2019). *Orchidaceae* is especially diverse, with Coelogyne (4 spp.), Dendrobium (9 spp.), Paphiopedilum (3 spp.), and Pleione (2 spp.) as notable genera.

The Kangchenjunga Landscape (KL; 26°21′40.49″–28°7′51.25″ N; 87°30′30.67″–90°24′31.18″ E) is a transboundary biodiversity hotspot shared by

Nepal, India, and Bhutan. It supports 5198 species of seed plants belonging to 1548 genera and 216 families, including >3000 dicot species, >1000 monocots, and 23 gymnosperms (Kandel et al. 2019). KL exhibits remarkable orchid diversity (152 genera, 704 spp.) and is also rich in *Fabaceae* (105 genera, 308 spp.), *Asteraceae* (100 genera, 255 spp.), *Rubiaceae* (50 genera, 147 spp.), and *Poaceae* (96 genera, 233 spp.). Gymnosperms are dominated by *Cupressaceae* (9 spp.) and *Pinaceae* (10 spp.), with other families represented by single species. Threatened species in KL include members of *Orchidaceae* (6 spp.) and *Pinaceae* (3 spp.), while endemism is highest in *Asteraceae* (18 genera, 34 spp.) (Kandel et al. 2019; DPR 2012; ICMOD et al. 2017).

Among the seed plants, gymnosperms are chiefly represented by *Cycadaceae*, *Ephedraceae*, *Ginkgoaceae*, *Taxaceae*, *Podocarpaceae*, *Taxodiaceae*, *Cupressaceae*, and *Pinaceae* in various forest types (Rana and Rawat 2017). Dominant angiosperm families include *Brassicaceae*, *Fabaceae*, *Rosaceae*, *Apiaceae*, *Rubiaceae*, *Asteraceae*, *Scrophulariaceae*, *Acanthaceae*, *Lamiaceae*, *Polygonaceae*, *Euphorbiaceae*, *Orchidaceae*, and *Poaceae* (Samant 2015, 2021). Several families are monotypic, such as *Ochnaceae*, *Rhizophoraceae*, *Plumbaginaceae*, *Plantaginaceae*, and *Ceratophyllaceae*.

The Eastern Himalayas are considered as suitable habitats for various orchid species that are epiphytic, terrestrial and saprophytic types. *Orchidaceae* is exceptionally well represented in the Eastern Himalaya, with 165 genera and ~900 species predominantly epiphytes (~563 species), followed by terrestrial (~298 species) and mycotrophic (~39 species) taxa. Among these, 108 species are threatened and 131 are endemic to the region (Naithani 2020). Genera with high species richness include *Bulbophyllum*, *Dendrobium*, *Habenaria*, *Liparis*, *Oberonia*, and *Peristylus*.

14.2.3 Medicinal Plant and Wild Edible Resources

The IHR supports ~1748 medicinal plant species belonging to 915 genera and 223 families, dominated by *Asteraceae* (129 spp.), followed by *Fabaceae*, *Lamiaceae*, *Rubiaceae*, *Euphorbiaceae*, *Ranunculaceae*, *Rosaceae*, *Poaceae*, *Orchidaceae*, *Polygonaceae*, and *Gentianaceae* (Samant 2021). Most medicinal plants occur below 1800 m elevation, with few species found above 3800 m. This floristic richness and concentration of both orchids and medicinal plants are extensively discussed in Chap. 4 (Sect. 4.2) which highlights the taxonomic diversity and conservation value of these groups in the Himalayas.

Notable medicinal angiosperms include *Azadirachta indica*, *Berberis* spp., *Boerhavia diffusa*, *Justicia* spp., *Syzygium cumini*, *Terminalia chebula*, *T. arjuna*, *Tinospora cordifolia*, *Glycyrrhiza glabra*, *Oroxylum indicum*, *Cinnamomum tamala*, *Plantago major*, and *Rhododendron anthopogon*. Gymnosperm medicinal plants include *Ephedra gerardiana* and *Taxus wallichiana*.

Threatened medicinal plants comprise critically endangered (*Aconitum chasmanthum*, *Gentiana kurroo*, *Nardostachys jatamansi*, *Saussurea costus*, *Vatica lanceaefolia*), endangered (*A. heterophyllum*, *Angelica glauca*, *Coptis teeta*, *T. wallichiana*), and vulnerable (*Cephalotaxus mannii*, *Cayratia pedata*) species (Ved et al. 2003, 2005; Mehta et al. 2021; Samant 2021). Approximately 675 species of wild edible plants from 384 genera and 149 families are recorded in the IHR, including pteridophytes (12 spp.), fungi (7 spp.), and lichens (2 spp.). The most species-rich edible genera include *Rubus*, *Polygonum*, *Ficus*, *Allium*, *Dioscorea*, *Rhus*, *Chenopodium*, *Berberis*, *Prunus*, *Viburnum*, and *Piper*.

14.3 Human Driven Pressure on the Eastern Himalayan Ecosystem

The Eastern Himalayan ecosystem is under considerable strain due to not only natural climatic variations but also a range of human activities that have intensified environmental degradation. Key human-induced factors include deforestation, forest fires, and habitat fragmentation, each playing a role in the loss of biodiversity and the disruption of essential ecosystem functions. These activities exert pressures that have both immediate and long-term effects, leading to shifts in ecological stability and hastening the pace of degradation. This section examines these critical human-driven influences and discusses their broader implications for conservation strategies in the Himalayas.

14.3.1 Forest Shrinkage and Tree Felling

Despite numerous conservation efforts and initiatives, tree felling and forest degradation continue to pose serious threats to the natural forest ecosystems across the Himalayas (Brandt et al. 2012). The region has witnessed an accelerated rate of forest loss, and while some ecosystems may appear relatively intact, they are, in reality, experiencing a gradual decline characterised by extended recovery periods. This indicates that Himalayan ecosystems are facing a hidden degradation that could become more evident in the near future (Lindenmayer and Sato). The broader implications of forest degradation for species populations and ecological processes remain under-researched (Pandit et al. 2007). A detailed analysis of deforestation rates between 2000 and 2014 in temperate forests across Nepal, India, Bhutan, China, and Myanmar highlights that merely 13% of these forests are under formal protection, with variations in deforestation rates reaching up to 1.4% across these countries (Brandt et al. 2017). This pattern of deforestation is also elaborated in Chap. 11 (Sect. 11.2) which examines the socioeconomic and governance drivers of forest loss, as well as Chap. 13 (Sect. 13.3.2), highlighting the impacts of land-use changes and forest fragmentation in the Himalayan belt.

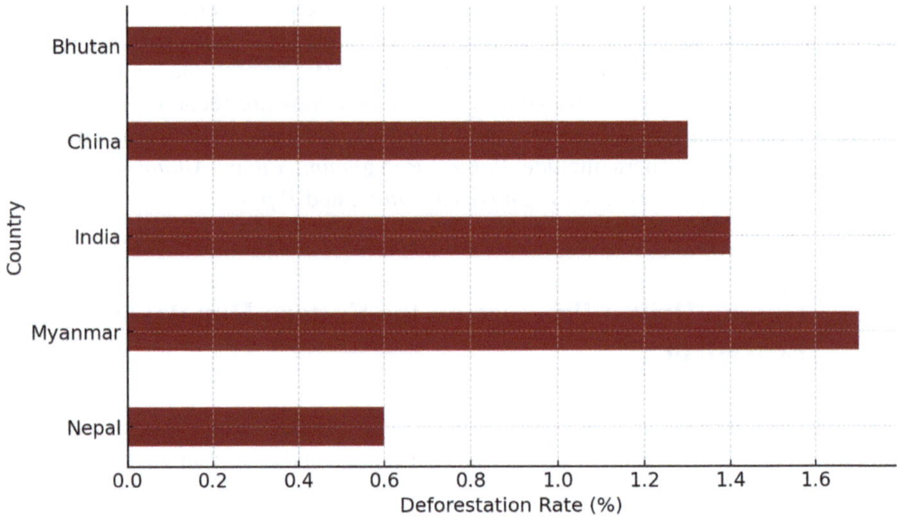

Fig. 14.1 Deforestation rates over a decade (2000–2014) in forests outside protected areas within the Himalayan temperate zone

Since the mid-1800s, forest ecosystems in the Himalayan regions have undergone substantial degradation. Bhutan has been increasingly classified as a 'crisis ecoregion' due to worsening land and forest degradation (Dorjee et al. 2020), with a loss of about 60.9% of its forest cover from 1930 to 2014, notably in the subtropical broad-leaved hill forests (64.5 km^2) and moist Sal forests (9.9 km^2) (Reddy et al. 2016). According to reports (Watershed Management Division 2017) indicated a rise in forest cover in Bhutan, the structural integrity of these forests has notably declined. Bhutan's total annual deforestation, according to statistical data, stands at 5798 hectares, while spatial analysis estimates it at 4274 hectares. In Nepal, over half of the forest cover was lost between 1960 and 1990, predominantly in the Tarai and Churia regions, with deforestation rates accelerating from 1990 to 2020 (Ives and Messerli 1989; Chaudhary et al. 2015). However, in the 1980s and 1990s, Nepal's government reevaluated forest management strategies, leading to the transfer of national forests to community forest groups under the Forestry Act of 1993. This change resulted in a notable recovery, with forest cover nearly doubling from 26% to 45% between 1992 and 2016 (Van Den Hoek et al. 2021). In contrast, deforestation rates in the Himalayan part of China remained low, staying below 0.01% between 1991 and 1999, but increasing to 0.21% from 2001 to 2010 (Fig. 14.1; Wang and Myint 2016). Forest cover in the Himalayas has decreased from 84.9% in 1970, with projections suggesting a further decline to 52.8% by 2100 (Pandit et al. 2014a). The Himalayan region is undergoing extensive deforestation, driven by a combination of factors such as unsustainable land-use practices, rapid infrastructural expansion, recurring forest fires, illegal timber extraction, agricultural encroachment, governance shortcomings, and unclear land ownership rights (Roy et al. 2015; Sheth et al. 2019). While official assessments, such as those by the Forest

Survey of India, report a net increase in forest cover in parts of the region, persistent and widespread biodiversity loss continues to be documented, highlighting deep-rooted ecological challenges (Brandt et al. 2012).

Traditional cropping has shifted to commercial agricultural intensification and has led to decrease in forest cover in many districts of Bhutan and Nepal (Fig. 14.1; Reddy et al. 2016). Between 1874 and 2010, tea cultivation in Darjeeling, India, expanded by 30.7%, leading to a 69.5% decrease in forest cover (Prokop 2018). This rapid deforestation, at a rate of 16–30 km^2 per year, is primarily attributed to the expansion of large tea plantations. As tea production areas increased by 16 km^2 annually, the overall forest cover across the Himalayas decreased by 2105.9 km^2.

Human activities in Southeast Asia have led to significant forest degradation (Fig. 14.1). In Myanmar's Hukaung Valley, opium cultivation along riverbanks has resulted in substantial forest loss (Papworth et al. 2017). In India's Tripura state, indigenous Jhum practices have been replaced by rubber plantations (Pandey et al. 2020). Selective logging has also impacted forest ecosystems. In India's Eastern Himalayan states, this practice has reduced epiphyte diversity and abundance due to altered microclimates and habitat conditions (Padmawathe et al. 2004). In Myanmar, illegal logging of valuable timber species like *Tectona grandis* (Teak) and *Xylia xylocarpa* has degraded previously undisturbed forests (Khai et al. 2016; Win et al. 2009).

14.3.2 Forest Fires

Human-induced Forest fires are a common occurrence in the Himalayan regions that are inhabited by people, especially between 800 and 2000 m above mean sea level (Singh et al. 2016). Moderate Resolution Imaging Spectroradiometer (MODIS) data analyses have highlighted India, Pakistan, and Myanmar as major hotspots for forest fire incidents (Vadrevu et al. 2019). Shifting cultivation and seasonal factors have traditionally intertwined forests and fires in the Eastern Himalayas. Anthropogenic activities, encompassing slash-and-burn agriculture, deforestation, controlled burning, and resource extraction, pose a significant threat to forest ecosystems through fire. A significant share (58%) of these fires is intentionally ignited by grazers, poachers, and non-timber forest product (NTFP) collectors, with 18 of Nepal's 75 districts identified as high-risk for forest fires (Kunwar 2006). As further detailed in Chap. 5 (Sect. 5.2), anthropogenic pressures including shifting cultivation and unsustainable harvesting practices have intensified forest fire occurrences. Chapter 12 (Sect. 12.2) also highlights how climate variability and prolonged dry spells contribute to rising forest fire events in the Himalayas.

Bhutan experienced 37 forest fires across 12 districts between 2017 and 2018, leading to the loss of 6561 hectares of natural forest (Fig. 14.2; Tariq et al. 2021). In the Western Himalayas of India, severe forest fires caused by human activities were recorded over 17 years between 1910 and 1980, with further incidents in 1995, 1999, 2010, 2012, 2016, 2018, and 2020 (Fig. 14.2; Dobriyal and Bijalwan 2017).

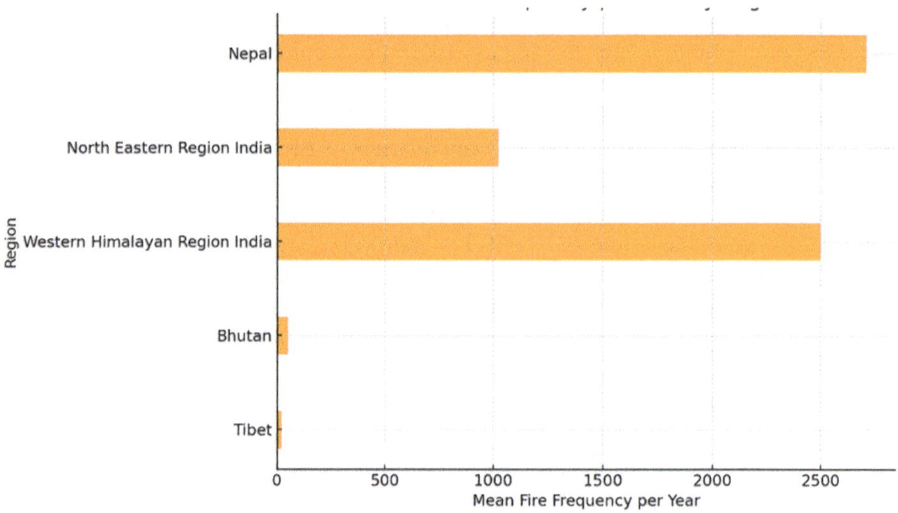

Fig. 14.2 Annual frequency of forest fires across Himalayan countries

Human carelessness, whether intentional or accidental, is a significant contributor to forest fires. Farmers' fires that spread, and carelessly discarded cigarette butts, pose serious risks, particularly in dry coniferous forests. The Eastern Himalayan region's biodiversity and human populations are vulnerable to forest fires (Fig. 14.2), which can be triggered by numerous causes. The region's steep, inaccessible terrain complicates fire suppression efforts. However, Forest fires, both those ignited by human activities and natural causes, have significant environmental consequences. These fires release substantial amounts of carbon dioxide and other greenhouse gases into the atmosphere, contributing to climate change (Hao and Liu 1994). Beyond climate change, forest fires lead to the loss of biodiversity. As these fires ravage ecosystems, they destroy habitats and threaten the survival of numerous plant and animal species (Brown and Davis 1973). Furthermore, forest fires alter atmospheric chemistry by releasing large quantities of trace gases, aerosol particles, and black carbon (Dwyer et al. 1998), which can have far-reaching effects on air quality and climate patterns. These fires can release hundreds of millions of tons of smoke aerosols, further exacerbating air pollution and contributing to climate change (Hao and Liu 1994).

14.3.3 Mega-Hydroelectric Power Projects Driven Habitat Fragmentation

In the context of the Eastern Himalayan ecosystem, habitat fragmentation has emerged as one of the most pressing ecological challenges, exacerbating the impacts of habitat loss. Over recent decades, rapid infrastructure expansion particularly road networks, mega hydropower projects, and urbanisation has transformed large, continuous tracts of forest into smaller, isolated patches (Sharma and Roy 2007; Xu et al. 2009). This spatial reconfiguration of habitats not only reduces the total available area but also alters the ecological connectivity essential for sustaining wildlife populations. Similar patterns of habitat fragmentation and their cascading ecological consequences are discussed in Chap. 9 (Sect. 9.1). Additionally, Chap. 10 (Sect. 10.3), explores how infrastructure projects and tourism-related development fragment critical habitats in alpine regions.

Infrastructure development remains a dominant driver of fragmentation. Large-scale hydroelectric power initiatives on river Teesta include four such big dam as Teesta Stage I–IV projects in various locations in Sikkim alter the hydrological footprints and impacts on associated vegetations. The Subansiri Lower Hydroelectric Project on the Brahmaputra River dam at Arunachal-Assam border and the proposed Yurlung Zangpo dam in Tibet Autonomous Region on the Chinese side of Tibet are reshaping entire riverine and forest landscapes. These projects, while contributing to regional energy demands, have inundated forested valleys, disrupted river flow regimes, and severed ecological corridors essential for terrestrial and aquatic species (Agarwal 2012; Pandit and Grumbine 2012; ICIMOD 2019a).

Hydropower development in the Himalayas results in a two-fold fragmentation effect:

- **Aquatic fragmentation**—Dams alter natural river connectivity, impeding migratory fish species such as *Tor putitora* and disrupting aquatic ecosystem processes (Dudgeon 2011; Winemiller et al. 2016).
- **Terrestrial fragmentation**—Reservoir creation submerges low-elevation forest tracts, while associated transmission lines and access roads dissect wildlife habitats, creating physical and ecological barriers for species movement (Chaturvedi et al. 2013).

The combined effect of these alterations is particularly acute in the Eastern Himalayas, where narrow valleys and steep gradients amplify the loss of contiguous forest cover. Studies indicate that cumulative hydropower projects on the Teesta and Subansiri basins could fragment up to 20–25% of riparian forest habitats within their influence zones (ICIMOD 2019a). While both processes are detrimental, research suggests that habitat loss has a more severe impact on species persistence than fragmentation alone (Brook et al. 2003). Species adapted to intact forest interiors often fail to survive in fragmented landscapes, where altered microclimates and increased edge effects prevail. Fragmentation can create new ecological edges that may temporarily benefit some generalist species, yet it also isolates populations, thereby reducing genetic exchange and increasing vulnerability to stochastic events.

14.3.4 Expansion of Roads and Cities on Eastern Himalayas

Urban and infrastructural expansion in the Eastern Himalayas has accelerated in recent decades, reshaping cities and towns such as Darjeeling, Kurseong, Gangtok, and Itanagar. Once small and compact, these hill settlements are now expanding onto fragile slopes under the combined pressures of economic development and tourism growth. Darjeeling and Kurseong, famous for their tea estates, colonial heritage, and scenic vistas, have witnessed a sharp surge in visitor numbers, leading to the rapid construction of multi-storey hotels, numerous homestays, and roadside resorts. Many of these developments occur without adequate adherence to environmental safeguards, altering traditional architectural landscapes and straining local resources (Singh 2019).

Large-scale connectivity projects are also transforming the region. The tunnel-based railway routes up to Rongpo and the upgrading of highways to 4–6 lanes have substantially improved access, boosting tourism and trade. However, such projects often require extensive hill-cutting, deforestation, and slope modification, which increase the risk of landslides and soil erosion (Tiwari and Joshi 2012).

The tourism boom has intensified vehicular pressure, with narrow mountain roads frequently congested during peak seasons. Increased emissions, noise, and road degradation are now common concerns (Chettri et al. 2018). While these developments contribute to economic vitality, the unchecked pace of urbanization and tourism threatens to erode the ecological stability and cultural heritage that form the core attractions of the Eastern Himalayas.

14.3.5 Ecological Consequences for Fauna

Both habitat loss and fragmentation across in-situ conservation areas of the Eastern Himalayan biogeographic zone have resulted in drastic biodiversity decline. Large, connected forest landscapes are critical for wide-ranging species such as the snow leopard (Uncia uncia) and various migratory bird species (Pandit et al. 2007). Endemic species like the Mishmi takin (*Budorcas taxicolor* ssp. *taxicolor*) and the Hoolock gibbon (*Bunopithecus hoolock* ssp. *hoolock*) face heightened extinction risk due to shrinking habitats. Other threatened large mammals include the black bear (*Ursus thibetanus*), musk deer (*Moschus* sp.), bharal (*Pseudois schaeferi*), Himalayan tahr (*Hemitragus jemlahicus*), and serow (*Capricornis sumatraensis*).

Among avian species, vulnerable and endangered taxa include the monal pheasant (*Lophophorus impeyanus*), koklas pheasant (*Pucrasia macrolopha*), western tragopan (*Tragopan melanocephalus*), Himalayan snowcock (*Tetraogallus himalayensis*), golden eagle (*Aquila chrysaetos*), steppe eagle (*Aquila nipalensis*), black eagle (*Ictinaetus malayensis*), and bearded vulture (*Gypaetus barbatus*).

14.3.6 Implications for Conservation

In the Himalayan context already under acute stress from climate change the distinction between habitat loss and fragmentation is critical for designing effective conservation strategies. Addressing fragmentation without simultaneously halting habitat conversion is unlikely to succeed. Conservation planning must therefore combine habitat area preservation with connectivity restoration, such as through wildlife corridors, transboundary ecological linkages, and expansion of protected areas (Lindenmayer and Fischer 2013). Strategic environmental assessments of large-scale infrastructure projects, particularly hydropower development, should be mandated to safeguard ecological integrity.

14.4 Impact of the Global Climate Crisis

The Eastern Himalayas rank among the most climate-sensitive ecosystems in the world. Anthropogenic greenhouse gas emissions, atmospheric pollutants, temperature rise, and altered hydrological cycles are driving profound environmental transformations, including glacier retreat, erratic weather patterns, and an upsurge in climate-related natural disasters. The Himalayan glaciers often termed the "Third Pole" are melting at an unprecedented pace due to global warming. Bolch et al. (2019) reported that glaciers in the Hindu Kush-Himalaya have shrunk by nearly 15% over the past century. This retreat directly affects major river systems such as the Ganges, Indus, Brahmaputra, Teesta, Gandak, and Sankosh, all of which depend heavily on glacial meltwater. Declining water availability poses serious threats to millions of people reliant on these rivers for agriculture, potable water, and hydroelectric energy.

Increased climatic instability has amplified the frequency and severity of extreme weather events such as landslides, floods, and cloudbursts resulting in heavy loss of life and property. Concurrently, altered temperature and precipitation regimes are causing prolonged droughts and irregular monsoon cycles, undermining agricultural productivity (Fig. 14.3). As detailed in Chap. 1 (Sect. 1.3.1), the region's altitudinal and climatic gradients heighten its vulnerability, while Chap. 12 (Sect. 12.2) explores the cascading socio-ecological effects of these climatic shifts.

14.4.1 Effects on Species Distribution and Community Structure

Climate change is triggering large-scale altitudinal shifts in species distribution. In the Eastern Himalayas, home to numerous endemic and medicinal plants such as *Saussurea obvallata*, *Rhododendron arboreum*, and *Aconitum heterophyllum* even

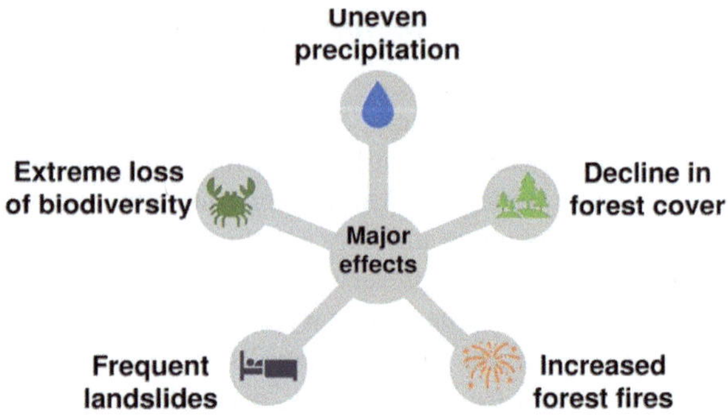

Fig. 14.3 Illustration representing major effects of climate change on the Himalayan ecosystem

subtle climatic variations can have substantial impacts (Telwala et al. 2013). Rising temperatures force cold-adapted species to migrate upslope, where they encounter new competitive pressures from lower-altitude taxa, leading to altered community composition and potential biodiversity loss (Telwala et al. 2013; Gottfried et al. 2012).

Alpine meadows, critical habitats for high-altitude flora such as *Primula denticulata*, *Meconopsis aculeata*, and *Delphinium brunonianum*, are shrinking due to glacial retreat and changing precipitation regimes (Pauli et al. 2015). Erratic monsoons and reduced snowfall disrupt phenological events as seed germination, flowering, and fruiting—affecting habitat stability for species like *Nardostachys jatamansi*, *Picrorhiza kurroa*, and *Dactylorhiza hatagirea* (Salick and Byg 2007).

14.4.2 Decline of High-Altitude and Endemic Flora

Glacial meltwater, a key moisture source for high-altitude plant communities, is diminishing rapidly, intensifying water stress for species such as *Rheum nobile*, *Bergenia ciliata*, and *Arnebia benthamii* (Gottfried et al. 2012). Extreme weather events, viz., landslides and droughts, further impair regeneration in species like *Taxus wallichiana*, *Betula utilis*, and *Juniperus indica* (Schickhoff et al. 2015). Changes in phenology threaten pollination success, affecting interdependent species such as *Fritillaria cirrhosa*, *Swertia chirayita*, and *Podophyllum hexandrum*.

The spread of invasive alien species, including *Lantana camara*, *Parthenium hysterophorus*, and *Eupatorium adenophorum*, is facilitated by climate change, intensifying competition for native species and degrading habitat quality (Salick and Ross 2009). Such invasions often lead to cascading ecological effects, disrupting pollinator networks, herbivory patterns, and nutrient cycling (Telwala et al. 2013).

14.4.3 Medicinal Plant Vulnerability

The alpine and subalpine belts host rare medicinal taxa, including *Ophiocordyceps sinensis*, which is highly sensitive to temperature anomalies and overharvesting (Salick and Konchar 2014). *Gentiana kurroo* and *Podophyllum hexandrum*, valued for pharmaceutical applications, are undergoing habitat fragmentation. *Ephedra gerardiana*, a drought-tolerant shrub of trans-Himalayan zones, faces population decline under intensifying aridity (Salick and Byg 2007).

Tree line species such as *Abies spectabilis* and *Quercus semecarpifolia* are shifting upward, altering forest structure and species composition (Schickhoff et al. 2015). Valuable timber species like *Cedrus deodara* and *Pinus wallichiana* are experiencing slower growth due to temperature stress and pest outbreaks exacerbated by climate change (Pauli et al. 2015).

14.4.4 Impacts on Shrublands and Grasslands

Shrubland species such as *Berberis aristata*, *Hippophae rhamnoides*, and *Rosa sericea*, important for erosion control, fodder, and traditional medicine are declining due to drought and land-use change. Medicinal herbs like *Valeriana jatamansi* and *Viola biflora* are unable to adapt to increasingly erratic moisture and temperature regimes (Telwala et al. 2013). Alpine grassland flora, including *Carex* spp., *Kobresia pygmaea*, and *Poa annua*, essential for soil stability and livestock forage, are diminishing due to warming trends and overgrazing (Gottfried et al. 2012).

14.4.5 Conservation Imperatives

Although rising CO_2 and warming may transiently enhance productivity in certain microhabitats, long-term climatic changes will likely result in net biodiversity loss (Salick and Ross 2009). Effective mitigation demands integrative strategies, including habitat restoration, ex-situ conservation, and community-led management (Salick and Byg 2007). The incorporation of traditional ecological knowledge with scientific approaches can bolster species resilience (Salick and Konchar 2014). Strengthening protected area networks, regulating medicinal plant harvesting, and prioritising afforestation with native species are essential to safeguarding Himalayan flora (Pauli et al. 2015).

14.5 Integrated Mitigation Strategies for Safeguarding the Eastern Himalayan Biodiversity

The Eastern Himalayan ecosystem, recognised as a globally significant biodiversity hotspot and a critical source of freshwater for millions of people, is increasingly threatened by the impacts of climate change. The region is experiencing accelerated warming, with the Trans-Himalayan zone heating at a rate higher than the global average, resulting in rapid glacial retreat, shifts in precipitation patterns, and a rising frequency of extreme weather events (Shrestha et al. 2012; Bolch et al. 2019). Addressing these challenges requires a holistic and integrated mitigation strategy that incorporates technological, ecological, policy, and socio-economic dimensions.

A transition to renewable and low-carbon energy systems is essential, as fossil fuel dependence remains a significant driver of greenhouse gas emissions. Decentralised renewable solutions such as micro-hydropower, solar photovoltaics, and biomass-based systems can simultaneously reduce emissions and improve energy access in remote Himalayan communities. Enhancing energy efficiency through innovations in building design, manufacturing processes, and energy management offers additional potential for emission reductions (IEA 2021). Sustainable land-use practices, particularly agroforestry, can strengthen ecosystem health by improving soil fertility, conserving biodiversity, and increasing carbon sequestration capacity. Conserving and restoring forested landscapes in the Eastern Himalayas is especially critical, given their role as major carbon sinks and reservoirs of unique flora and fauna. Community-based Forest management approaches have demonstrated notable success in safeguarding forest integrity while empowering local populations (Chaudhary et al. 2015).

Effective mitigation is also dependent on strong policy and governance mechanisms. Instruments such as carbon taxes and cap-and-trade systems can create economic incentives for emission reductions across multiple sectors (World Bank 2020). Complementary measures, including stringent emission standards and targeted subsidies for renewable energy projects, can further accelerate the shift towards a low-carbon economy (UNFCCC 2018). Given the transboundary nature of Himalayan ecosystems, enhanced regional cooperation among Himalayan nations is imperative for managing shared resources and addressing cross-border environmental challenges (ICIMOD 2019b).

Public engagement remains a vital pillar of successful climate action. Awareness campaigns that communicate the environmental, social, and economic implications of climate change, alongside the benefits of sustainable practices, can inspire community-driven initiatives (Singh 2019). Such efforts may include the adoption of energy-efficient appliances, waste reduction, sustainable tourism, and support for eco-certified products, collectively contributing to a lower regional carbon footprint (IPCC 2021). Integrating communities into decision-making processes ensures that strategies are locally relevant, culturally sensitive, and widely supported.

Financing these mitigation efforts represents a major challenge. The total estimated requirement for effective mitigation and adaptation in the six Himalayan countries stands at approximately USD 1085 billion (Asian Development Bank 2017). Mobilising this scale of funding necessitates the use of multiple channels, including international climate finance, public–private partnerships, and innovative instruments such as green bonds (World Economic Forum 2021). Strategic investments in climate-resilient infrastructure, renewable energy deployment, and sustainable agricultural systems can not only combat climate change but also drive economic development, create employment opportunities, and alleviate poverty (FAO 2019).

Overall, protecting the biodiversity and ecological integrity of the Eastern Himalayas demands an integrated approach that unites technological innovation, sustainable resource management, robust policy frameworks, active community participation, and substantial financial investment. If implemented cohesively, these measures can help mitigate the adverse effects of climate change, preserve the region's unique natural heritage, and secure the livelihoods of the millions who depend on its resources (IPCC 2021; UNDP 2020).

14.6 Conclusions

The Eastern Himalayan ecosystem is at a critical juncture due to the dual threats of climate change and human-induced environmental degradation. Urgent and comprehensive mitigation strategies are necessary to ensure the resilience and sustainability of this fragile region. By focusing on afforestation, sustainable land and water management, glacier conservation, renewable energy adoption, disaster preparedness, community involvement, and strong policies, we can safeguard the Himalayas for future generations. A combination of traditional ecological knowledge and modern scientific approaches, along with national and international cooperation, will be crucial in mitigating the impacts of climate change in the Himalayas. Immediate and sustained action is imperative to protect this vital ecological and cultural heritage.

Acknowledgments The authors express their sincere gratitude to the editors and reviewers for their valuable insights and constructive feedback that greatly improved the quality of this chapter. We also thank our respective institutions, Gour Banga University and Banaras Hindu University, for providing the necessary academic environment and support.

References

Agarwal A (2012) Dam building in the Himalayas: boon or bane? Econ Polit Wkly 47:18–21

Asian Development Bank (2017) Climate change mitigation and adaptation in Asia. ADB Report. https://www.adb.org/sites/default/files/institutional-document/358881/ccof-2017-2030.pdf

Bajracharya P, Sultana S (2022) Examining the use of urban growth boundary for future urban expansion of Chattogram, Bangladesh. Sustainability 14(9):5546. https://doi.org/10.3390/su14095546

Baloch QB, Shah SN, Iqbal N, Sheeraz M, Asadullah M, Mahar S, Khan AU (2023) Impact of tourism development upon environmental sustainability: a suggested framework for sustainable ecotourism. Environ Sci Pollut Res Int 30(3):5917–5930

Behera MD, Kushwaha SP, Roy PS (2002) High plant endemism in an Indian hotspot–eastern Himalaya. Biodivers Conserv 11:669–682. https://doi.org/10.1023/A:1015596309833

Bolch T et al (2019) Status and change of the cryosphere in the extended Hindu Kush Himalaya Region. In: Wester P, Mishra A, Mukherji A, Shrestha A (eds) The Hindu Kush Himalaya Assessment. Springer, Cham. https://doi.org/10.1007/978-3-319-92288-1_7

Brandt JS, Kuemmerle T, Li H, Ren G, Zhu J, Radeloff VC (2012) Using Landsat imagery to map forest change in southwest China in response to the national logging ban and ecotourism development. Remote Sens Environ 121:358–369. https://doi.org/10.1016/j.rse.2012.02.010

Brandt JS, Allendorf T, Radeloff V, Brooks J (2017) Effects of national forest-management regimes on unprotected forests of the Himalaya: management of Himalayan forests. Conserv Biol 31:1271–1282. https://doi.org/10.1111/cobi.12927

Brook BW, Sodhi NS, Ng PK (2003) Catastrophic extinctions follow deforestation in Singapore. Nature 424(6947):420–423. https://doi.org/10.1038/nature01795

Brown AA, Davis KP (1973) Forest fire: control and use. McGraw-Hill, New York

Chatterjee D (1940) Studies on the endemic flora of India and Burma. J Asiat Soc Bengal Sci 5(1):19–67

Chaturvedi R et al (2013) Fragmentation of Himalayan forests due to hydropower development. Curr Sci 104:250–256

Chaudhary R, Uprety Y, Rimal S (2015) Deforestation in Nepal: causes, consequences, and responses. In: Shroder JF, Sivanpillai R (eds) Biological and environmental hazards and disasters, 1st edn. Elsevier, pp 335–372

Chettri N, Sharma E, Shakya B, Bajracharya B (2018) Sustainable tourism and biodiversity conservation in the Himalaya. Mountain Res Develop 38(3):227–239

Choudhary RK, Srivastava RC, Das AK, Lee JG (2012) Floristic diversity assessment and vegetation analysis of Upper Siang district of eastern Himalaya in North East India. Korean J Pl Taxon 42(3):222–246

Dobriyal MJ, Bijalwan A (2017) Forest fire in western Himalayas of India: a review. N Y Sci J 10(6):39–46

Dorjee JSB, Buckmaster AJ, Downey PO (2020) Weeds in the land of Gross National Happiness: knowing what to manage by creating a baseline alien plant inventory for Bhutan. Biol Invasions 22(9):2899–2914. https://doi.org/10.1007/s10530-020-02306-5

DPR (2012) Plants of Nepal: fact sheet. Department of Plant Resources, Ministry of forests and soil conservation, Nepal, Kathmandu. https://dpr.gov.np/wp-content/uploads/2023/08/Plants-of-Nepal-Fact-sheet_2012-1.pdf

Dudgeon D (2011) Asian river fishes in the Anthropocene: threats and conservation challenges. J Fish Biol 79:1487–1524

Dwyer E, Grégoire JM, Malingreau JP (1998) A global analysis of vegetation fires using satellite images: spatial and temporal dynamics. Ambio 27(3):175–181

FAO (2019) Sustainable agriculture for climate change mitigation. Food and Agriculture Organization of the United Nations. https://www.fao.org/climate-change/en

Gottfried M, Pauli H, Futschik A, Akhalkatsi M, Barančok P, Benito Alonso JL, Coldea G, Dick J, Erschbamer B, Fernández Calzado MAR, Kazakis G (2012) Continent-wide response of mountain vegetation to climate change. Nat Clim Chang 2(2):111–115. https://doi.org/10.1038/nclimate1329

Hao WM, Liu MH (1994) Spatial and temporal distribution of tropical biomass burning. Glob Biogeochem Cycles 8(4):495–503. https://doi.org/10.1029/94GB02086

ICIMOD (2019a) Himalayan hydropower and ecological security. ICIMOD, Kathmandu

ICIMOD (2019b) Transboundary cooperation for climate resilience in the Himalayas. International Centre for Integrated Mountain Development. https://lib.icimod.org/record/34482/files/icimodBridgingBoundaries.pdf?type=primary

ICIMOD, WCD, GBPNIHESD, RECAST (2017) Kangchenjunga landscape conservation and development initiative feasibility assessment report - regional synthesis. ICIMOD Working Paper 2017/9. ICIMOD, Kathmandu. https://lib.icimod.org/record/32609/files/icimodWP_9_KLCDI2017.pdf

IEA (2021) Energy efficiency 2021. International Energy Agency. https://www.iea.org/reports/energy-efficiency-2021

IPCC (2021) Climate change 2021: the physical science basis. Intergovernmental Panel on Climate Change. https://www.ipcc.ch/report/ar6/wg1/

Ives JD, Messerli B (1989) The Himalayan dilemma: reconciling development and conservation. Routledge, New York. https://doi.org/10.4324/9780203169193

Kandel P, Chettri N, Chaudhary RP, Badola HK, Gaira KS, Wangchuk S, Bidha N, Uprety Y, Sharma E (2019) Plant diversity of the Kangchenjunga landscape, eastern Himalayas. Plant Divers 41(3):153–165. https://doi.org/10.1016/j.pld.2019.04.006

Khai TC, Mizoue N, Kajisa T, Ota T, Yoshida S (2016) Stand structure, composition and illegal logging in selectively logged production forests of Myanmar: comparison of two compartments subject to different cutting frequency. Glob Ecol Conserv 7:132–140. https://doi.org/10.1016/j.gecco.2016.06.001

Kumar D, Katoch SS (2016) Environmental sustainability of run of the river hydropower projects: a study from western Himalayan region of India. Renew Energy 93:599–607. https://doi.org/10.1016/j.renene.2016.03.032

Kunwar R (2006) Forest fire in the Terai, Nepal Causes and community management interventions. Int For Fire News 34:46–54

Lindenmayer DB, Fischer J (2013) Habitat fragmentation and landscape change: an ecological and conservation synthesis. Island Press

Maity D, Ghosch J, Midday M (2019) Spring flowers of Darjeeling and Sikkim Himalayas:(with a few additions of cryptogams). Citadel, Kolkata

Mehta P, Bisht K, Sekar KC (2021) Diversity of threatened medicinal plants of Indian Himalayan region. Plant Biosyst 155(6):1121–1132

Naithani HB (2020) Floristic diversity of North East India. Rain Forest Research Institute, (ICFRE), Jorhat

Padmawathe R, Qureshi Q, Rawat GS (2004) Effects of selective logging on vascular epiphyte diversity in a moist lowland forest of eastern Himalaya, India. Biol Conserv 119(1):81–92. https://doi.org/10.1016/j.biocon.2003.10.024

Pandey DK, Adhiguru P, Sah U, Devachandra N, Dubey SK, Deo C (2020) Is rubber monoculture banishing agrobiodiversity and happiness? Evidences from shifting cultivation landscape of Tripura, northeast India. Curr Sci 118(1):108–113. https://doi.org/10.18520/cs/v118/i1/108-113

Pandit MK, Grumbine RE (2012) Potential effects of ongoing and proposed hydropower development on terrestrial biological diversity in the Indian Himalaya. Conserv Biol 26:1061–1071

Pandit MK, Sodhi NS, Koh LP, Bhaskar A, Brook BW (2007) Unreported yet massive deforestation driving loss of endemic biodiversity in Indian Himalaya. Biodivers Conserv 16:153–163. https://doi.org/10.1007/s10531-006-9038-5

Pandit MK et al (2014a) Cumulative impacts of hydropower development on Himalayan biodiversity. Conserv Biol 28:1–12

Pandit MK, Manish K, Koh LP (2014b) Dancing on the roof of the world: ecological transformation of the Himalayan landscape. BioScience 64(11):980–992. https://doi.org/10.1093/biosci/biu152

Papworth S, Rao M, Oo MM, Latt KT, Tizard R, Pienkowski T, Carrasco LR (2017) The impact of gold mining and agricultural concessions on the tree cover and local communities in northern Myanmar. Sci Rep 7(1):46594. https://doi.org/10.1038/srep46594

Pauli H, Gottfried M, Lamprecht A, Nießner S, Rumpf S, Winkler M, Steinbauer K, Grabherr G (2015) The GLORIA field manual—standard Multi-Summit approach, supplementary methods and extra approaches. GLORIA-Coordination, Austrian Academy of Sciences and University of Natural Resources and Life Sciences, Viena

Prokop P (2018) Tea plantations as a driving force of long-term land use and population changes in the Eastern Himalayan piedmont. Land Use Policy 77:51–62. https://doi.org/10.1016/j.landusepol.2018.05.035

Rana SK, Rawat GS (2017) Database of Himalayan plants based on published floras during a century. Data 2(4):36. https://doi.org/10.3390/data2040036

Reddy CS, Satish KV, Jha CS, Diwakar PG, Murthy YK, Dadhwal VK (2016) Development of deforestation and land cover database for Bhutan (1930–2014). Environ Monit Assess 188:1–12. https://doi.org/10.1007/s10661-016-5676-6

Roy R, Karki S, Karky BS, Kotru R, Sohail M, Reinhardt S, Phuntsho K (2015) REDD+ in the Hindu Kush Himalayas: a stocktaking study from Bhutan, India, Myanmar, Nepal, and Pakistan. https://doi.org/10.13140/RG.2.1.3364.2964

Salick J, Byg A (2007) Indigenous peoples and climate change. Tyndall Centre for Climate Change Research. Int Indig Policy J 1. https://doi.org/10.18584/iipj.2010.1.1.2

Salick J, Konchar K (2014) Curating biocultural collections. Royal Botanic Gardens, Kew

Salick J, Ross N (2009) Traditional peoples and climate change. Glob Environ Change 19(2): 137–139

Samant SS (2015) Assessment, valuation and conservation prioritization of floristic diversity in trans, North Western and Western Himalaya. D.Sc. Thesis, Kumaun University, Nainital, Uttarakhand

Samant SS (2021) Floristic diversity of the Himalaya in relation to climate change: status, values and conservation. J Graph Era Univ 25:31–54

Schickhoff U, Bobrowski M, Böhner J, Bürzle B, Chaudhary RP, Gerlitz L, Heyken H, Lange J, Müller M, Scholten T, Schwab N (2015) Do Himalayan treelines respond to recent climate change? An evaluation of sensitivity indicators. Earth Syst Dynam 6(1):245–265. https://doi.org/10.5194/esd-6-245-2015

Sharma S, Roy PS (2007) Forest fragmentation in the Himalaya: a central Himalayan case study. Int J Sustain Dev World Ecol 14(2):201–210. https://doi.org/10.1080/13504500709469720

Sheth C, Datta A, Parashuram D (2019) Persistent loss of biologically-rich tropical forests in the Indian Eastern Himalaya. BioRxiv 827360

Shrestha AB, Wake CP, Mayewski PA, Dibb JE (2012) Maximum temperature trends in the Himalaya and its vicinity: an analysis based on temperature records from Nepal for the period 1971-94. J Clim 12(9):2775–2786

Singh S (2019) Tourism, urbanization and environmental challenges in Himalayan towns. J Mt Sci 16(5):1035–1047

Singh DK, Hajra PK (1996) Floristic diversity. Changing perspective of biodiversity status in the Himalaya. British Council, New Delhi, pp 23–38

Singh H, Kumar M (2022) Climate change and its impact on Indian Himalayan forests: current status and research needs. In: Climate change: impacts, responses and sustainability in the Indian Himalaya. Springer International Publishing, Cham, pp 223–242

Singh R, Kumar D, Gumber S, Tewari P, Singh SP (2016) Nature of forest fires in Uttarakhand: frequency, size, and seasonal patterns in relation to pre-monsoonal environment. Curr Sci 111: 398–403. https://doi.org/10.18520/cs/v111/i2/398-403

Tariq MAUR, Wangchuk K, Muttil N (2021) A critical review of water resources and their management in Bhutan. Hydrology 8(1):31. https://doi.org/10.3390/hydrology8010031

Telwala Y, Brook BW, Manish K, Pandit MK (2013) Climate-induced elevational range shifts and increase in plant species richness in a Himalayan biodiversity epicentre. PLoS One 8(2):e57103. https://doi.org/10.1371/journal.pone.0057103

Third P (2021) Fires ravage forests in Himalayas, threatening health and biodiversity. In: The Third Pole. https://dialogue.earth/en/forests/fires-ravage-forests-himalayas-threatening-health-biodiversity/#:~:text=Majority%20of%20forests%20face%20the%20fires&text=Indian%20magazine%20Outlook%20reported%20that,by%20fires%20in%20March%202021. Accessed 20 Apr 2021

Tiwari PC, Joshi B (2012) Urban growth in Himalaya: understanding the process and options for sustainable development. J Urban Reg Stud Contemp India 1(1):1–12

UNDP (2020) Climate change mitigation strategies in mountain regions. United Nations Development Programme. https://www.undp.org/events/climate-2020

UNFCCC (2018) The Paris agreement and its implementation in mountain ecosystems. United Nations Framework Convention on Climate Change. https://unfccc.int/process-and-meetings/the-paris-agreement#:~:text=The%20Paris%20Agreement%20speaks%20of,the%20well%2Dfunctioning%20Technology%20Mechanism

Vadrevu KP, Lasko K, Giglio L, Schroeder W, Biswas S, Justice C (2019) Trends in vegetation fires in south and southeast Asian countries. Sci Rep 9(1):7422

Van Den Hoek J, Smith AC, Hurni K, Saksena S, Fox J (2021) Shedding new light on mountainous forest growth: a cross-scale evaluation of the effects of topographic illumination correction on 25 years of forest cover change across Nepal. Remote Sens 13(11):2131. https://doi.org/10.3390/rs13112131

Ved DK, Kinhal GA, Ravikumar K, Prabhakaran V, Ghate U, Vijaya Shankar R, Indresha JH (2003) Conservation assessment and management prioritization for the medicinal plants of Jammu and Kashmir, Himachal Pradesh and Uttaranchal. Foundation for Revitalisation of Local Health Traditions, Bangalore, p 206

Ved DK, Kinhal GA, Ravikumar K, Vijaya Shankar R, Haridasan K (2005) Conservation assessment and management prioritization (CAMP) for the wild medicinal plants of North-East India. Med Plants Conserv 11:40–44

Wang C, Myint SW (2016) Environmental concerns of deforestation in Myanmar 2001–2010. Remote Sens 8(9):728. https://doi.org/10.3390/rs8090728

Watershed Management Division (2017) Drivers of deforestation and forest degradation in Bhutan. Department of Forests and Park Services (Thimphu), p 146. http://library.cnr.edu.bt/cgi-bin/koha/opac-detail.pl?biblionumber=7066&shelfbrowse_itemnumber=9996

Win RN, Reiji S, Shinya T (2009) Forest cover changes under selective logging in the Kabaung Reserved Forest, Bago Mountains, Myanmar. Mt Res Dev 29(4):328–338. https://doi.org/10.1659/mrd.00009

Winemiller KO et al (2016) Balancing hydropower and biodiversity in the Amazon, Congo, and Mekong. Science 351:128–129

World Bank (2020) State and trends of carbon pricing 2020. World Bank Group. https://openknowledge.worldbank.org/server/api/core/bitstreams/5afce9b5-03e0-5557-a550-6fac52275dde/content

World Economic Forum (2021) Financing a sustainable future: investment strategies for climate mitigation. World Economic Forum Report. https://initiatives.weforum.org/financing-transition-community/home

Xu J, Grumbine RE, Shrestha A, Eriksson M, Yang X, Wang Y, Wilkes A (2009) The melting Himalayas: cascading effects of climate change on water, biodiversity, and livelihoods. Conserv Biol 23(3):520–530. https://doi.org/10.1111/j.1523-1739.2009.01237.x

Chapter 15
Current Knowledge and Future Perspectives towards Spatiotemporal Maintenance of Flora in a Biodiversity Hotspot: A Case Study from the Western Himalayas

Balwant Rawat, Janhvi Mishra Rawat, Namini Joshi, Akshita Raj, Shivani Negi, Mrinalini Agarwal, Rowndel Khwairakpam, Rajneesh Bhardwaj, and Arvind Singh Negi

Abstract Among global mountain systems, which have emerged as a special thrust under the global change scenario, the Himalayas being the youngest and loftiest represent a highly diverse and complex biological and physical system. Mountain Biosphere Reserves (MBRs) have been identified among the most suitable areas for long-term Research & Development activities across the world. The globally selected MBRs for case studies include the Nanda Devi Biosphere Reserve (NDBR) as a potential site in the Indian Himalayan Region (IHR) from the Asia-Pacific Region. The intensive study sites include Pindari-Sunderdhunga-Kafni (PSK) in Kumaun region and Lata-Tolma-Phagti (LTP) in the Garhwal region in the buffer zone of NDBR. The present investigation deals with the assessment of LULC/compositional changes over the last 25 years in Nanda Devi Biosphere Reserve. Landsat TM-1990 and Landsat-8 SR 2015 were procured and Arc GIS 9.2 was used for spatial analysis. While considering the overall vegetation cover in the reserve, no significant change was recorded between 1990 and 2015. Most of the vegetation cover classes revealed negative change over the 25-year time interval, with a maximum decrease of 39.6 km^2 (5.6%) in the case of alpine meadows followed by alpine scrubs (15.8 km^2; 3.9%). However, in the case of forests, a noticeable increase was revealed for Evergreen broadleaf forests (15.4 km^2; 10.6%). The decline in forest cover in the Himalayan region is most often attributed to human disturbances whereas the increase is a consequence of favourable climatic

B. Rawat (✉) · R. Khwairakpam · R. Bhardwaj · A. S. Negi
School of Agriculture, Graphic Era Hill University, Dehradun, Uttarakhand, India

J. M. Rawat · N. Joshi · A. Raj · S. Negi · M. Agarwal
Department of Biotechnology, Graphic Era (Deemed to be University), Dehradun, Uttarakhand, India

D. Das et al. (eds.), *Plant Diversity and Conservation Strategies of the Himalayan Biodiversity Hotspot*, Topics in Biodiversity and Conservation 21,
https://doi.org/10.1007/978-3-032-21697-7_15

conditions. The outcomes of this study are indicative of potential LULC changes in the Himalayan ecosystems and need special attention.

Keywords Biodiversity · Forest vegetation · LULC patterns · Vegetation dynamics

15.1 Introduction

Himalayan forests are among the most biologically diverse ecosystems on Earth (Singh and Singh 1987; Singh et al. 1992), shaped by a unique geological history and vast altitudinal gradients, as described in Chap. 1 (Sect. 1.3.1) and based on the evidences it has been stated that these forests show variety and complexity when compared to tropical and temperate forests of the world (Zobel and Singh 1997). These forests are also deeply intertwined with the lives and traditions of local communities, whose dependence on forest resources for food, medicine, and livelihoods is well-documented in Chap. 2 (Sect. 2.1). Such socio-ecological linkages, however, make the forests vulnerable to human-induced pressures. Studies indicate that anthropogenic activities from resource extraction to infrastructure expansion can substantially alter forest composition and function (Silori 2001; Rawal et al. 2012; Rawat et al. 2021). In an investigation by Rawal et al. (2012) on representative oak forests of the region has reported differences in the sensitivity of diverse oak forests towards disturbance intensities. Another study (Gairola et al. 2009) from subalpine forests of west Himalaya has provided evidences of changing patterns of litterfall and nutrient cycling across levels of disturbances. Moreover, in recent decades global climate change has emerged as a major factor that is influencing the entire gamut of life on the Earth (IPCC 2007). Among others, investigations on world forest vegetation have generated evidence of changes in patterns and processes in response to changing climate and natural disturbances (Notaro et al. 2012; Rawat et al. 2022; Khali et al. 2023; Bisht et al. 2024). Evidences further indicate that Global Warming can also modify abiotic conditions that influence individual plant performances (Ise and Moorcroft 2006; Fissore et al. 2008; Xu et al. 2010). Moreover, it has been demonstrated that global species distribution changes are being driven by climate change at both latitudinal and altitudinal scales (Parmesan and Yohe 2003; Rosenzweig et al. 2007) leading to new species assemblages (Wing et al. 2005; Williams and Jackson 2007). Alpine treeline and high-altitude ecosystems and species exhibit trends of upward migration. It is anticipated that this well-observed phenomenon will persist with the predicted warming. Therefore, it is anticipated that alterations in the Earth's climate and atmospheric composition will affect both the natural and artificial environments (IPCC 2001). Alpine treelines and other climate-specific ecotones (Körner and Paulsen 2004) are thought to be especially vulnerable to changes in temperature (Theurillat and Guisan 2001).

The impacts of such structural dynamics may be paradoxical on a larger scale because high elevation forest changes might directly contribute to global warming in a variety of ways. In general, it is anticipated that plant communities and species compositions will fluctuate (Keller et al. 2000; Pauli et al. 2001; Walther et al. 2005) due to changing competition brought about by boundary alterations, as well as potential temperature change. Unfortunately, such evidences from Himalayan forests are not available to describe their responses, particularly as a consequence of climate change. This study is crucial for understanding how Himalayan forests, as biodiversity hotspots, are responding to these dual pressures of human-induced disturbances and climate change. By examining these responses, we can inform conservation strategies that aim to preserve the unique flora and fauna of the region, ensuring the continued provision of essential ecosystem services.

15.1.1 Mountains as Change Sensitive Ecosystems

The mountains have been recognised among the most change-sensitive ecosystems in the world. However, numerous human-caused changes, including altered land use and cover, excessive resource extraction, and growing effects of climate change, pose a danger to the integrity and sustainability of delicate alpine ecosystems. Around 33 years ago in 1992, during the Rio Earth Summit, Chapter 13 of Agenda 21, for the first time, brought mountains to global attention by way of providing convincing evidences that global mountains were facing tremendous degradation. Over the past 33 years, there have been notable advancements in our understanding of how global change affects mountain regions; nonetheless, in many cases, the negative trends in these settings persist. Realising the above, mountains are now being considered among the most fragile environments in the world (Diaz et al. 2003). The proven sensitivity of mountain ecosystems to climate change and the ongoing anthropogenic changes across the globe, make them ideal places for research towards understanding the impacts of such changes (Schaaf 2009).

15.1.2 Himalayas as Vulnerable Ecosystem

The susceptibility of the Himalayan Mountains to both natural and man-made disruptions is widely acknowledged. Given the region's abundance, uniqueness, and susceptibility to environmental changes, it has been identified as one of the 34 worldwide biodiversity hotspots (Anonymous 2009). The region's high endemism suggests the existence of numerous vital habitats and ecoregions of global significance (Samant et al. 1998), many of which are detailed in Chap. 4 (Sect. 4.1), which describes the diverse vegetation types and their ecological significance.

The Himalayan Mountains are a key candidate for rapid action due to their diversity of representative ecosystem features, their susceptibility to perturbations caused by humans and/or the climate, and—most importantly—their socioeconomic marginalisation and lack of livelihood possibilities. With a land area of more than 5.37 lakh km^2, the Indian Himalayan Region (IHR) makes up a sizeable chunk of the Himalayan Biodiversity Hotspot, encompassing 16.2% of the nation's total land area. Its temperature and physiography have varied significantly as a result of the temporal and spatial fluctuations brought about by different geological orogenies. This variation in local microclimate and physiography can be observed in the increased richness and representativeness of its biodiversity components (Anonymous 2009), including rare alpine and subalpine species discussed in Chap. 5 (Sect. 5.2).

The vulnerability of the IHR to various kinds of perturbations has been highlighted (Singh et al. 2010), and as outlined in Chap. 12 (Sect. 12.1), the region faces significant risks from climate-driven changes such as glacial retreat, altered precipitation regimes, and temperature fluctuations. These interacting stressors underscore the need for urgent, integrated conservation strategies that address both ecological fragility and socio-economic vulnerability.

15.1.3 Mountain Biosphere Reserves as Special Candidates

To address numerous intricate yet significant challenges related to global changes in mountain regions, several representative mountain landscapes have been designated as Biosphere Reserves (BRs). The fact that more than 40% of all Biosphere Reserves worldwide are located in mountainous regions and that these regions are widely dispersed over 40 nations, highlights the importance of mountain regions within UNESCO's World Network of Biosphere Reserves. The Mountain Biosphere Reserves (MBRs) have particularly received global attention as candidate areas to: (i) understand the response patterns, and (ii) implement multidisciplinary approaches in a changing environment. Therefore, several strategies have been considered to address the issues facing mountain ecosystems. Three goals have been established for these strategies as follows: (i) creating an integrative research strategy for identifying signals of environmental change on a global scale; (ii) defining the effects of these changes on both mountain and lowland areas; and (iii) assisting in the creation of sustainable resource management regimes for mountains (Price et al. 2006). In this context, the MBRs that promote and demonstrate a balanced relationship between man and nature, and are considered 'living laboratories' for integrated management of land, water and biodiversity; have emerged as global priority sites (UNESCO-MAB 2004) and are being used as 'early warning' systems (Rawal and Rawat 2012; Price et al. 2024).

The Nanda Devi Biosphere Reserve (NDBR) in the Indian Himalayan Region (IHR) has been highlighted as a possible site from the Asia-Pacific Region among the different MBRs that have been chosen for case studies around the globe (Reasoner et al. 2003; UNESCO-MAB 2004). Given the significance of NDBR, the target study location for this investigation is this BR. UNESCO recognised NDBR's distinctiveness by adding it to the World Network of Biosphere Reserves (WNBR) in 2004. Additionally, UNESCO has placed the Nanda Devi and Valley of Flowers National Parks, which comprise the reserve's core zone, on the World Heritage List under Natural Criteria vii and x. The reserve (30°05′–31°02′N

Latitude, 79°12′–80°19′E Longitude) is located in the northern part of the west Himalayas and includes parts of Chamoli district in Garhwal; and Bageshwar and Pithoragarh districts in Kumaun region of the Uttarakhand State, India. It falls in the West Himalayan Biogeographic province of zone Himalaya. The Reserve has a wide altitudinal range (1800–7817 m asl) and covers an area of 6407.03 km^2. Besides, the reserve represents unique landscapes with outstanding naturalness values (Rawal and Rawat 2012).

15.1.4 Forest Vegetation as the Specific Target

The preservation of forests has become the main goal as a result of awareness of the general benefits of forests and the extraordinary rate at which they are being lost. Therefore, it is well acknowledged that the loss of biodiversity is just one of the several ecological, social, and economic effects of forest decline (Jha et al. 2000). As the only renewable natural resource on Earth, forests hold a special place in the natural resource community because they sustain life in many ways and offer services that cannot be obtained through any other source. In addition to forming a variety of representative ecosystems, the forests in the NDBR provide a habitat for numerous uncommon and endangered species. About 27% of the buffer zones land is covered in woods, compared to 10% in the reserve's core zone. According to Rawal and Rawat (2012), these frosts contribute to the reserve's rich floral (fungi-128, lichens-77, bryophytes-146, pteridophytes-137 spp., gymnosperms-11, and angiosperms-699 spp.) and faunal (pisces-1, reptiles-3, amphibian-8, annelids-6, molluscs-14, insects-229, birds-243, and mammals-29 spp.) diversity.

15.1.5 RS & GIS as Monitoring Tools

The biodiversity of the Himalayan forests is changing at a rate never seen before as a complex reaction to both natural and human-induced environmental changes. This understanding calls for reliable information and data sets at a range of spatial scales (Alexander and Millington 2000). Therefore, optimisation of the use of effective monitoring methods is highly desired. Among others, the Remote Sensing (RS) data sets, which provide quick and objectively available information for decision-makers and managers, have emerged as an important and timely technological tools (Howard 1991; Millette et al. 1995; Raj et al. 2024). In India, natural resource mapping and monitoring, using space technology, has gained greater impetus with the launch of IRS-1A in 1988 followed by IRS-1B, IRS-1C and IRS-1D, which helped in providing satellite images of the entire country with enhanced capability to monitor and manage resources (Gopalan 1998; Udaya and Dutt 1998; Bisht and Rawat 2023). Moreover, remote sensing techniques have also gained worldwide popularity in the last few decades particularly w.r.t. conservation and management planning at species and community level. Particularly in mountain regions, due to the difficult access to most of these areas, remote sensing often remains the only way to investigate large sections of these mountains (Kaab 2005). In the case of NDBR, few studies have been conducted on mapping of NDBR resources using RS technology (Maikhuri et al. 2003; Nainwal et al. 2008; Bharti et al. 2011; Bali et al. 2011a, b; Pandey et al. 2024). However, there is a clear gap in integration between the traditional field-generated data sets and the RS-based spatial data. In this context, while considering global literature, some studies include direct long-term observations and measurements of changes (Sykora et al. 2002; Bush and Richter 2006; Notaro et al. 2012; Rinella et al. 2012). However, this approach requires a lot of effort, manpower and time. On the other hand, repeated satellite remote sensing over a range of spatial and temporal scales provides an affordable way to evaluate environmental characteristics including changes in land use and forest cover. As such, the importance of remote sensing and GIS has been recognised in the country for resource mapping (Amarnath et al. 2003; Goparaju et al. 2005), change detection/spatial analysis (Jha et al. 2000; Jayakumar et al. 2002) and decision-making (Prasad 1998; Lakshmi et al. 1998). As such, detection of changes through remote sensing now has a wide acceptability. Comparison of time series vegetation maps is considered a useful tool for analysing vegetation changes (Li et al. 2010; Zhao et al. 2022; Seyam et al. 2023; He et al. 2024). In the context of NDBR, very few studies (Nautiyal et al. 2005; Bali and Ali 2010) have provided information on changes in landuse/landcover.

15.2 Materials and Methods

15.2.1 Study Area

15.2.1.1 Background

The Government of India designated the Nanda Devi Biosphere Reserve (NDBR) as a biosphere reserve on January 18, 1988. The reserve hosts a rare assemblage of ecosystems, including alpine meadows, glaciers, diverse types of temperate forests, and traditional agroecosystems. As described in Chap. 5 (Sect. 5.2), alpine and subalpine zones of the Himalaya harbour unique and often endemic flora that contribute significantly to the region's biodiversity.

The NDBR lies between 30°06′ and 31°04′ North latitude and 79°13′ and 80°15′ East longitude (Fig. 15.1), encompassing a total area of 6407.03 km^2 (Core zone: 712.12 km^2; Buffer zone: 5148.57 km^2; Transition zone: 546.34 km^2). It falls within the West Himalayan Highland (2b) province of the Himalayan biogeographic zone, whose ecological characteristics are outlined in Chap. 4 (Sect. 4.1).

The two representative sites in the NDBR buffer zone Pindari-Sunderdhunga-Kafni (PSK) in Kumaun and Lata-Tolma-Phagti (LTP) in Garhwal, served as the intensive study locations. These areas, as highlighted in Chap. 12 (Sect. 12.1), are particularly vulnerable to climate-driven changes such as glacial retreat and altered

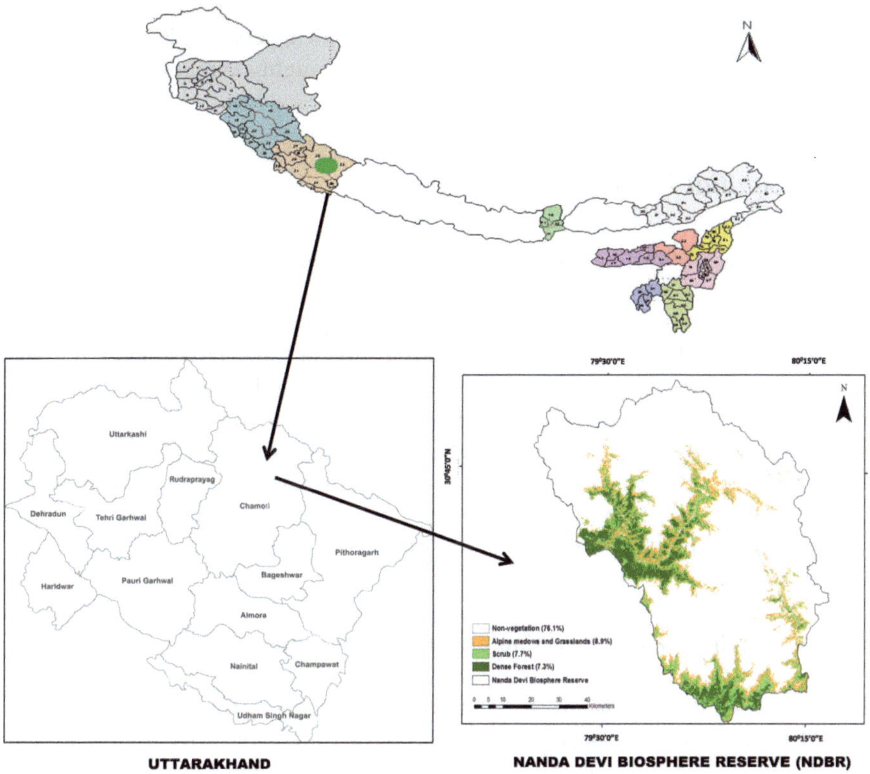

Fig. 15.1 Location map of study area—NDBR and target sites using ERDAS 9.1

alpine vegetation patterns, making them critical zones for long-term ecological monitoring and conservation planning.

15.2.1.2 Conservation Significance

The identified core zones (Nanda Devi and Valley of Flowers NPs) of the reserve represent unique landscapes with outstanding naturalness value. A sizeable glacial basin is represented by the Nanda Devi National Park (NDNP), which is the catchment area of the Rishi Ganga, an Eastern tributary of the Dhauli Ganga that feeds into the Alaknanda, a significant tributary of the Ganges. The basin displays rugged topography and is one of the deepest gorges in the world. Over a century, prominent mountaineers and botanists have recognised the Valley of Flower National Park (VoFNP) for its remarkable floral diversity and scenic beauty. The park has been set aside specifically for the preservation of Himalayan flora. Realising these values, both of the NPs have been inscribed in the list of World Heritage Sites (e.g. NDNP in 1988; VoFNP in 2005). The ecological integrity and authenticity of the core areas have been maintained thoroughly ever since they have been declared as National Parks. The reserve, besides fulfilling the livelihood needs of indigenous communities, is important for its unparalleled ecological services. For example, the glaciers in the reserve contribute significantly to the mighty Ganges River.

15.2.2 Spatial Database

Distribution maps of different forest communities at different altitudinal zones were developed using Remote Sensing (RS) and Geographical Information Systems (GIS). The main focus was to highlight the major physiognomic types present in the study area. Satellite data was used for delineation of forest cover, and topo-maps were used for topographic features. The digital data of Landsat TM (1990) and Landsat-8 SR (2015) with a spatial resolution of 28 m was acquired from the National Remote Sensing Centre (NRSC), Hyderabad, Telangana, India and www.landcover.org.T (free image downloader). The topo maps (1:50,000) of the study area were procured from the Survey of India (SOI) (Topo sheet numbers; 53N-01,05,06,09,10,11,13,14,15,16 and 62B-01,02,03,04,06,07).

15.2.2.1 Satellite Image Selection and Pre-processing

Landsat TM-1990 (path 145-row 39) and Landsat-8 SR 2015 (path 145-row 39) standard false colour composite imagery were procured for November; the years 1990 and 2015. The November imagery was chosen because of the lowest azimuth, sun angle, and cloud so that the shadow influence from the landform was efficiently avoided.

15.2.2.2 Software Used

The GIS lab of the Institute provided ERDAS software version 9.2 (which was utilised for digital image processing, georeferencing, and satellite image digital categorisation). Arc View 3.2 was used for plotting GPS locations on the image. Whereas, Arc GIS 9.2 was used for spatial analysis.

15.2.2.3 Radiometric Corrections

Unwanted artefacts were eliminated using a series of pre-processing or cleanup procedures, such as additive effects brought on by air scattering. Dark pixel removal was used to apply first-order radiometric adjustments. (Lillesand and Kiefer 1994). This method works under the assumption that there's a good chance certain pixels in an image should be black (0% reflectance). The imaging device does, however, record a non-zero Digital Number (DN) value at the purportedly dark-shadowed pixel location due to air dispersion. To eliminate the first-order scattering component, the DN value was deducted from the data.

15.2.2.4 Geometric Corrections

Ground Control Points (GCPs) were consistently distributed and identified with a root mean square error of less than one pixel in geometrically registered images. The closest neighbourhood method was used to resample the image. The research region was recovered using a boundary map after all of the scenes had been mosaiced.

15.2.2.5 Image Interpretation and Classification Accuracy

A visual interpretation scheme was used for the extraction of land cover and forest vegetation classes. The land cover features were visually digitised using ERDAS Software. Repetitive training areas and field sample locations assessed the accuracy of land cover and vegetation maps using the commission and omission error matrix, and as a result, the overall classification accuracy was computed.

15.2.2.6 Ground Truthing

Field surveys conducted from April 2015 to December 2019 provided a fair idea of broad vegetation types in the study area. Information was collected in the same season as the remotely sensed data, to achieve the best possible matching of information. Numerous observations regarding physiognomy and vegetation were made. Through contact and formal interviews, additional information about the history and present conditions of forests, fauna and its distribution, and land use and land cover were also obtained from local forest officials, other pertinent departments, and other stakeholders.

15.2.2.7 Post Classification

Thirteen land cover classes were determined for each image: Glacier, Snow and Moraine, Alpine meadow, Alpine scrub, Open rocks, Conifer dominated forest, Evergreen broadleaf forest, Mixed Forest, Forest blanks, River, High altitude lake, Grassland and Cropland. Comparison of maps for land cover changes followed post-classification technique and the identity of vegetation unit at each point for the years 1990 and 2015 was determined. Statistical measures, describing the landscape configuration and composition, were used for quantification of landscape structure and its changes. Towards achieving better clarity in comparison, forest communities were pooled into bigger units of physiognomy. After creating maps showing the density of forest cover, they were subjected to a geospatial change analysis, which produced change maps that showed the changes in forest cover over a 25-year span, from 1990 to 2015.

15.2.2.8 Prediction Maps

The whole approach using RS/GIS application in the study for change detection and identification of sensitive areas in the reserve has been depicted (Fig. 15.2). The lower resolution of imagery did not allow for reaching the individually marked forest stand. However, to overcome this problem and to highlight sensitive and rich areas, broad physiognomic types have been used.

15.3 Results and Discussion

15.3.1 Temporal Changes in the Forest Vegetation

Temporal changes in various landuse/landcover categories, with a particular focus on forest vegetation (Physiognomic types), have been investigated in the present study using Landsat TM 1990 and Landsat-8 SR 2015 data (Figs. 15.3 and 15.4). The present status and change statistics have been given (Table 15.1).

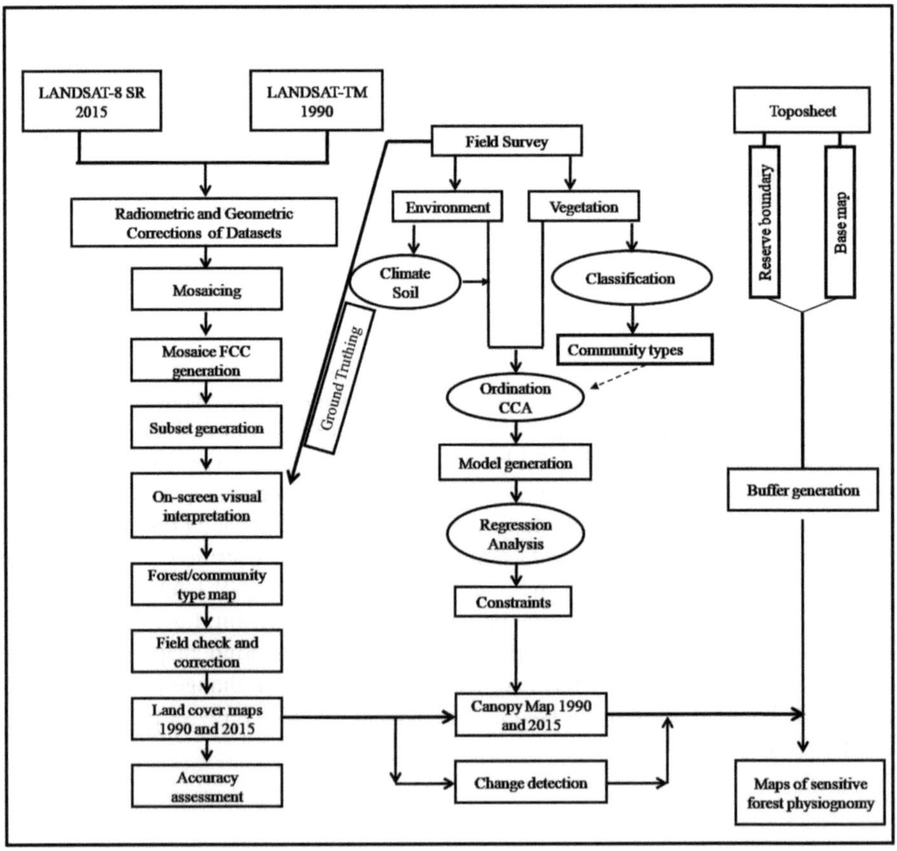

Fig. 15.2 Paradigm for change detection and assessment of rich and sensitive areas

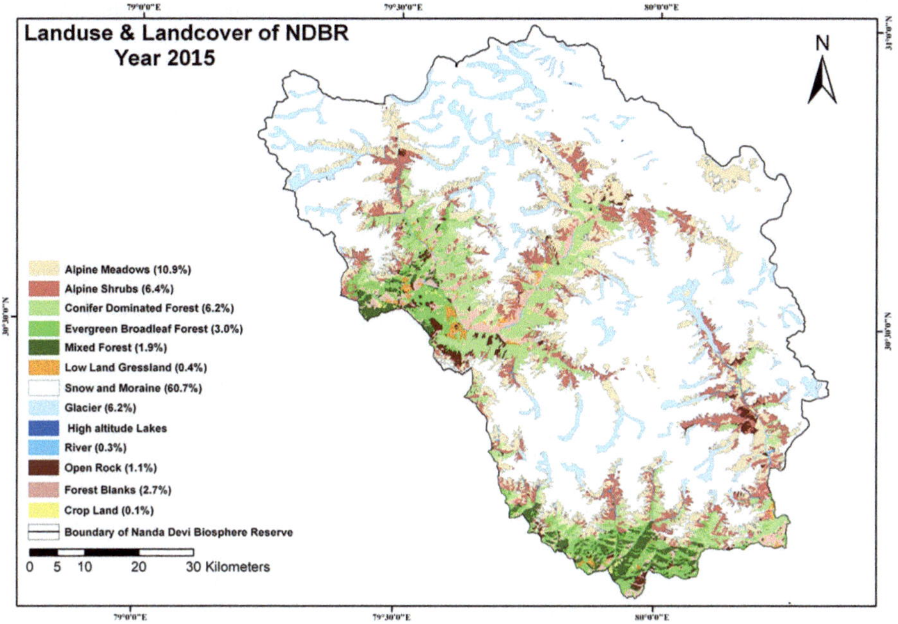

Fig. 15.3 Landuse and Land cover map of NDBR based on Landsat-8 SR 2015 using Arc GIS 9.2

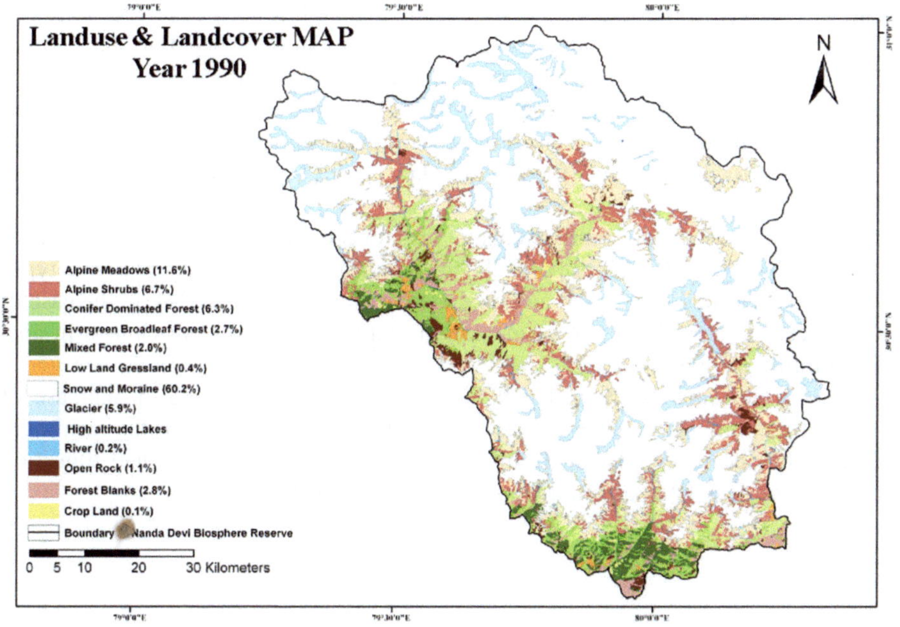

Fig. 15.4 Landuse and landcover map of NDBR based on Landsat-TM 1990 using Arc GIS 9.2

Table 15.1 Area statistics of different landuse/landcover classes of NDBR using Landsat TM 1990 and Landsat-8 SR 2015 data

Cover class	1990		2015	
	Area (km^2)	Area (%)	Area (km^2)	Area (%)
Vegetation				
Alpine meadows	708.5	11.6	668.9	10.9
Alpine scrubs	408.4	6.7	392.5	6.5
Conifer dominated forest	386.7	6.3	378.6	6.2
Evergreen broadleaf forest	163.5	2.7	180.9	3.0
Mixed broadleaf forest	122.6	2.0	115.8	1.9
Low land grassland	26.3	0.4	25.9	0.4
Non-vegetation				
Snow and moraine	3683.9	60.2	3715.9	60.7
Glacier	359.1	5.9	377.5	6.2
High altitude lake	0.3	0.0	0.2	0.0
River	11.3	0.2	17.4	0.3
Open rocks	69.5	1.1	68.6	1.1
Forest blanks	170.8	2.8	165.7	2.7
CropLand	7.3	0.1	8.1	0.1
Total	**6118.2**	**100.0**	**6118.0**	**100**

Investigations on RS/GIS-based mapping of vegetation cover have already been made in other parts of the world (Seyam et al. 2023). While considering the overall vegetation cover in the reserve, no significant change has been observed. Between 1990 and 2015. Most of the vegetation cover classes exhibited negative change over 25 years' time interval, with a maximum decrease of 39.6 km^2 (5.6%) in the case of alpine meadows followed by alpine scrubs (15.9 km^2; 3.9%). Among delineated forest physiognomic types in the reserve, conifer-dominated evergreen (8.1 km^2; 2.1%) and Mixed broadleaf forests (4.8 km^2; 3.9%) showed negative changes (Table 15.2).

However, a noticeable positive change was revealed for the Evergreen broadleaf forest (15.4 km^2; 10.6%). Temporal changes in different forest physiognomic types have been presented (Fig. 15.5) and predicted earlier (He et al. 2024). In the case of non-vegetation area River shows maximum increase w.r.t. percent change in area (54%) followed by Crop Land (11%).

Zone-based analysis showed that the core zone has registered an increase of 11 km^2 for conifer-dominated forests from 1990 till 2015. However, the maximum decrease has been recorded in Conifer-dominated forests in buffer and transition zones, which was 19.1 km^2 in the last 25 years (Table 15.3).

Table 15.2 Change in landuse/landcover of NDBR during the period 1990–2015 using Landsat TM 1990 and Landsat-8 SR 2015 data

Cover class	Area (km^2)	% change in area
Vegetation		
Alpine meadows	−39.6	−5.6
Alpine scrubs	−15.9	−3.9
Conifer dominated forest	−8.1	−2.1
Evergreen broadleaf forest	15.4	10.6
Mixed broadleaf forest	−4.8	−3.9
Low land grassland	−0.4	−1.5
Non-vegetation		
Snow and moraine	32	0.9
Glacier	18.4	5.1
High altitude lake	−0.1	−33.3
River	6.1	54.0
Open rocks	−0.9	−1.3
Forest blanks	−5.1	−3.0
Crop land	0.8	11.0

– Negative sign shows decrease; + Positive sign shows increase

15.3.2 *Evidence from LU-LC Changes*

On a broader spatial scale, a comparison of two time period data sets, using Landsat TM (1990) and Landsat-8 SR (2015) data, revealed an overall decline of 2.8% in forests in the reserve. Most of the vegetation cover classes exhibited a decline in cover over a 25-year time period. This trend does not agree with the indicated progression in forest density in different classes from the year 1986 to 1999 (Sahai and Kimothi 1994). Considering the forest physiognomic types, Conifer dominated forest (2.1%), and Mixed broadleaf forests (3.9%) exhibited an overall decline in cover. On the contrary, an increase in Evergreen Broadleaf Forests (10.6%) was noticed. The decline in forest cover in the Himalayan region is most often attributed to human disturbances (Singh et al. 1992; Gairola 2005) whereas the increase is a consequence of favourable climatic conditions (Rai et al. 2012). Vegetation Dynamics in Response to Climate Change and Human Activities Tibetan Plateau have already been reported (Zhao et al. 2022).

15.4 Conclusions

The observed decline in total forest cover by 2.8%, coupled with the specific decreases in Conifer-dominated (2.1%) and Mixed broadleaf (3.9%) forests, juxtaposed with a notable 10.6% increase in Evergreen broadleaf forests, underscores a dynamic and potentially transformative phase in the ecological trajectory of the region. These shifts, as delineated through rigorous Remote Sensing and GIS analyses spanning from 1990 to 2015, not only reflect the immediate impacts of

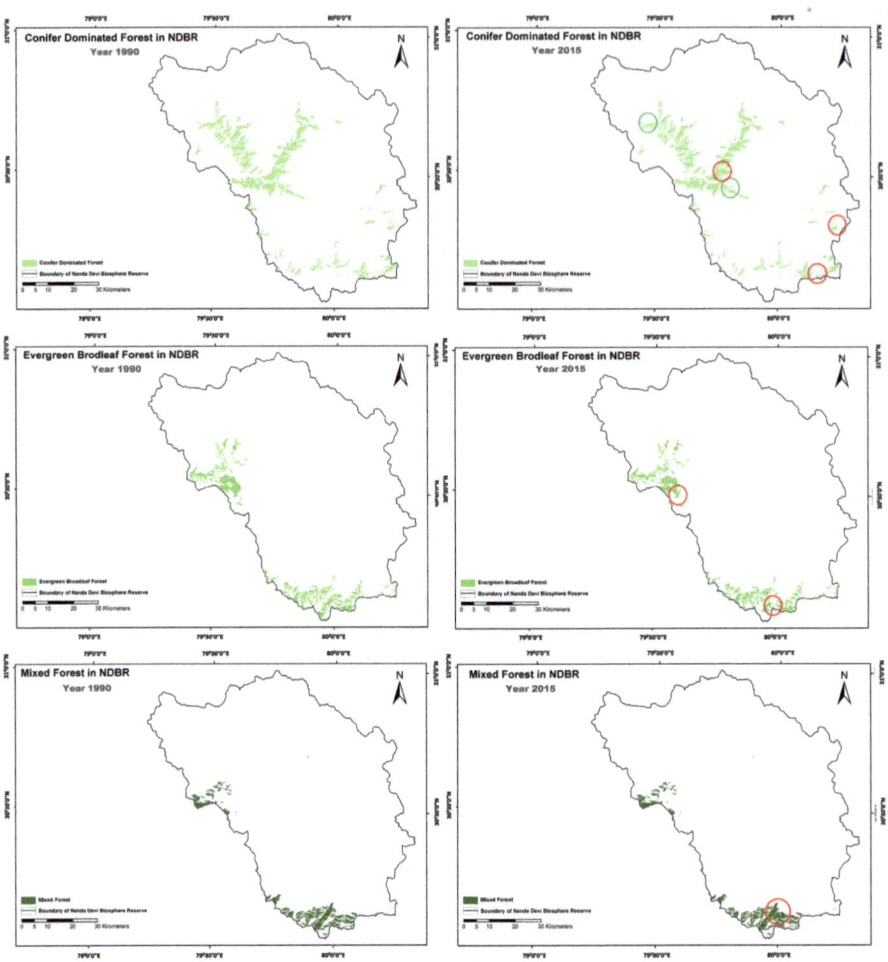

Fig. 15.5 Changes in different forest physiognomic types in NDBR (**a**, **b**: Conifer dominated forests (1990 & 2015); **c**, **d**: Evergreen broadleaf forests (1990 & 2015) and; **e**, **f**: Mixed broadleaf forests (1990 & 2015))

Table 15.3 Change statistics for forest physiognomy types across different zones of NDBR during the period 1990–2015 using Landsat TM 1990 and Landsat-8 SR 2015 data

Physiognomy	Core		Buffer & transition	
	1990	2015	1990	2015
Conifer dominated	13.2	24.2	373.5	354.4
Evergreen broadleaf	–	–	163.5	176.3
Mixed broadleaf forests	–	–	122.6	117.8

anthropogenic disturbances but also intimate the onset of broader ecological reconfigurations. Such alterations in forest composition and structure are indicative of underlying changes in ecosystem processes and biodiversity patterns.

Given the Himalayas' designation as a global biodiversity hotspot, harboring approximately 3160 rare, endemic, and sensitive plant species, the implications of these forest dynamics are profound. The observed compositional shifts may influence species interactions, habitat availability, and the resilience of ecosystems to climate-induced stresses. Moreover, the Himalayas' role as a significant carbon sink, storing about 62% of total forest carbon, highlights the critical need to monitor and understand these changes to inform conservation strategies.

Therefore, this study is pivotal in providing empirical evidence of ongoing ecological transformations, serving as a catalyst for comprehensive, long-term research initiatives. Such endeavors are essential to elucidate the multifaceted impacts of these changes on biodiversity, ecosystem services, and the overall ecological health of the region. Incorporating broader ecological attributes, including ecosystem processes, into future research will be instrumental in crafting informed, adaptive management strategies aimed at preserving the rich biodiversity and ecological integrity of the Himalayan forests.

Acknowledgements Vice Chancellor, Graphic Era Hill University and Graphic Era Deemed University are gratefully acknowledged for the necessary facility and support. Sincere acknowledgement goes to the Forest Department, Govt of Uttarakhand, India for providing necessary support during the field visits.

Funding JMR received financial support from SERB (TAR/2022/000018), Government of India.

References

Alexander RW, Millington AC (2000) Vegetation mapping: from patch to planet. John Wiley & Sons Ltd., Chichester, UK, pp. 339

Amarnath G, Murthy MSR, Britto SJ, Rajashekar G, Dutt CBS (2003) Diagnostic analysis of conservation zones using remote sensing and GIS techniques in wet evergreen forests of the Western Ghats—an ecological hotspot, Tamil Nadu, India. Biodivers Conserv 12:2331–2359

Anonymous (2009) Governance for Sustaining Himalayan Ecosystem (G-SHE)-guidelines & best practices. MoEF & GBPIHED, New Delhi, India, pp 1–51

Bali R, Ali SN (2010) Dynamics of Pindari glacier during the last 600 years. Curr Sci 99:1307

Bali R, Agrawal KK, Patil SK et al (2011a) Record of Neotectonic activity in the Pindari Glacier valley: a study based on Glacio-geomorphic and AMS Fabric evidences. Earth Sci India 4(1): 1–14

Bali R, Agarwal KK, Ali SN, Rastogi SK, Krishna K (2011b) Drainage morphometry of Himalayan Glacio-fluvial basin, India: hydrologic and neotectonic implications. Environ Earth Sci. https://doi.org/10.1007/s12665-011-1324-1

Bharti RR, Adhikari BS, Rawat GS (2011) Assessing vegetation changes in timberline ecotone of Nanda Devi National Park, Uttarakhand. Int J Appl Earth Obs Geoinf. https://doi.org/10.1016/j.jag.2011.09.018

Bisht S, Rawat KS (2023) A review of statistical approaches used for landslide susceptibility analysis with the help of remote sensing and GIS technology. Acta Sci Pol Formatio Circumiectus 22(3):83–96. https://doi.org/10.15576/ASP.FC/2023.22.3.13

Bisht S, Rawat KS, Singh SK (2024) Earth observation data and GIS-based landslide susceptibility analysis through frequency ratio model in lesser Himalayan region, India. Quat Sci Adv 1: 10014. https://doi.org/10.1016/j.qsa.2023.100141

Bush JK, Richter FA (2006) Two decades of vegetation change on terraces of a south Texas river. J Torrey Bot Soc 133(2):280–288

Diaz HF, Grosjean M, Graumlich L (2003) Climate variability and change in high elevation regions: past, present and future. Climate Change 59:1–4

Fissore C, Giardina CP, Kolka RK et al (2008) Temperature and vegetation effects on soil organic carbon quality along a forested mean annual temperature gradient in North America. Glob Chang Biol 14:193–205

Gairola S (2005) Assessment of diversity pattern in sub-alpine forests of west Himalaya: recruitment strategy, litterfall and nutrient return. Ph.D. Thesis(Forestry), HNB University, Srinagar (Garhwal)

Gairola S, Rawal RS, Dhar U (2009) Patterns of litterfall and return of nutrients across anthropogenic disturbance gradients in three subalpine forests of west Himalaya India. J For Res 14(2): 73–80. https://doi.org/10.1007/s10310-008-0104-6

Gopalan AKS (1998) IRS-1C/1D Mission, Keynote Paper, 8th user interaction workshop to commemorate one decade of IRS operations, 17–18 March, Hyderabad

Goparaju L, Tripathi A, Jha CS (2005) Forest fragmentation impacts on phytodiversity—an analysis using remote sensing and GIS. Curr Sci 88:1264–1274

He Q, Yang Q, Jiang S, Zhan C (2024) A comprehensive analysis of vegetation dynamics and their response to climate change in the loess plateau: insight from long-term kernel normalized difference vegetation index data. Forests 15(3):471. https://doi.org/10.3390/f15030471

Howard JA (1991) Remote sensing of forest resources-theory and application. University Press, Cambridge, pp 1–405

IPCC (2001) Climate change 2001. Impacts, adaptation, and vulnerability. Cambridge University Press, UK, Cambridge

IPCC (2007) Climate change 2007. The physical science basis. Summary for policymakers. Contribution of working group I to the fourth assessment report of the intergovernmental panel on climate change. Cambridge University Press, Cambridge, UK

Ise T, Moorcroft PR (2006) The global-scale temperature and moisture dependencies of soil organic carbon decomposition: an analysis using a mechanistic decomposition model. Biogeochem 12: 2370–2390

Jayakumar S, Arockiasamy DI, Britto SJ (2002) Conserving forests in the Eastern Ghats through remote sensing and GIS—a case study in Kolli hills. Curr Sci 82:1259–1267

Jha CS, Dutt CBS, Bawa KS (2000) Deforestation and land use changes in Western Ghats, India. Curr Sci 79:231–238

Kaab A (2005) Remote sensing of mountain environments. UNESCO Publication: Proceedings second and third GLOCHAMORE Workshop, pp. 92-99.

Keller F, Kienast F, Beniston M (2000) Evidence of response of vegetation to environmental change on high-elevation sites in the Swiss Alps. Reg Environ Change 1:70–77

Khali H, Rawat KS, Chatterjee RS (2023) Dynamic association of slope movements in the Uttarakhand Himalaya: a critical review on the landslide susceptibility assessment. Geomat Nat Haz Risk 14(1). https://doi.org/10.1080/19475705.2023.2273214

Körner C, Paulsen J (2004) A world-wide study of high altitude treeline temperatures. J Biogeogr 31(5):713–732. 10.1111/jbi.2004.31.issue-5 https://doi.org/10.1111/j.1365-2699.2003.01043.x

Lakshmi VU, Murthy MSR, Dutt CBS (1998) Efficient Forest resources management through GIS and remote sensing. Curr Sci 75:272–282

Li A, Deng W, Liang S, Huang C (2010) Investigation on the pattern of global vegetation change using a satellite-sensed vegetation index. Remote Sens 2:1530–1548

Lillesand TM, Kiefer RW (1994) Remote sensing and image interpretation, 3rd edn. Wiley, New York, Chichester, Brisbane, Toronto, Singapore, p 750

Maikhuri RK, Rao KS, Nautiyal S (2003) Land use land cover change impacts and strategies for rehabilitation of degraded land: a case study from the central(lesser) Himalaya (Uttaranchal). In: Rawat MSS (ed) Central Himalaya environment and development - potentials, actions and challenges. Print Media Publication, Srinagar, Garhwal, pp 223–229

Millette TL, Tuladhar AR, Kasperson RE, Turner BL (1995) The use and limits of remote sensing for analysing environmental and social change in the Himalayan Middle Mountains of Nepal. Glob Environ Change 5:367–380

Nainwal HC, Negi BDS, Chaudhary M, Sajwan KS, Gaurav A (2008) Temporal changes in rate of recession: evidence from Satopant and BhagirathKharak glaciers, Uttrakhand, using Total Station Survey. Curr Sci 94:653–660

Nautiyal S, Shibasaki R, Rajan KS, Maikhuri RK, Rao KS (2005) Impact of land use changes on subsidiary occupation: A case study from Himalayas of India. 4th International Conference on Environmental Informatics, ISEIS 2005. 14–23

Notaro M, Mauss A, Williams JW (2012) Projected vegetation changes for the American Southwest: combined dynamics modelling and bioclimatic-envelop approach. Ecol Appl 22:1365–1388

Pandey S, Bhandari MS, Shankhwar R et al (2024) Mortality mapping of *Pinuswallichiana* in Nanda Devi Biosphere Reserve: a UNESCO world heritage site in India. Earth Syst Environ 8: 561–586. https://doi.org/10.1007/s41748-024-00388-y

Parmesan C, Yohe G (2003) A globally coherent fingerprint of climate change impacts across natural systems. Nature 421:37–42

Pauli H, Gottfried M, Grabherr G (2001) High summits of the Alps in a changing climate. The oldest observation series on high mountain plant diversity in Europe. In: Walther GR, Burga CA, Edwards PJ (eds) Fingerprints of climate change. Adapted behaviour and shifting species ranges. Kluwer Academic/Plenum Publishers, New York

Prasad SN (1998) Conservation planning for the Western Ghats of Kerala: II. Assessment of habitat loss and degradation. Curr Sci 75:228–235

Price MF, Gurung AB, Dourojeanni P, Muselli D (2006) Social monitoring in Mountain Biosphere Reserves—conclusion from EU—GLOCHAMORE project. Mt Res Dev 26:174–180

Price MF, Schaaf T, Cárdenas M-R, Pisa PF, Köck G (2024) The world network of mountain biosphere reserves. In: Schneiderbauer S, Shroder JF, Pisa PF, Szarzynski J (eds) Safeguarding mountain social-ecological systems, Elsevier, New York, USA, pp 169–174. https://doi.org/10.1016/B978-0-12-822095-5.00025-5

Rai ID, Adhikari BS, Rawat GS, Basgali K (2012) Community structure along timberline ecotone in relation to micro-topography and disturbances in western Himalaya. Not Sci Biol 4:41–52

Raj S, Rawat KS, Tripathi VK (2024) Multi-temporal image processing for LULC classification and change detection. Environ Ecol 42(3B):1349–1357. https://doi.org/10.60151/envec/rodf5502issn0970-0420

Rawal RS, Rawat B (2012) Nanda Devi Biosphere Reserve-west Himalaya, India. In: Palni LMS, Rawal RS, Rai RK, Reddy SV (eds) Compendium on Indian Biosphere Reserve. Progression during two decades of conservation. Ministry of Environment and Forests, New Delhi, Government of India

Rawal RS, Gairola S, Dhar U (2012) Effect of disturbance intensities on vegetation patterns in oak forests of Kumaun, west Himalaya. J Mt Sci 9:157–165

Rawat B, Gaira KS, Gairola S, Tewari LM, Rawal RS (2021) Spatial prediction of plant species richness and density in high-altitude forests of Indian west Himalaya. Trees For People 6: 100132. https://doi.org/10.1016/j.tfp.2021.100132

Rawat KS, Sahu SR, Singh SK et al (2022) Cloudburst analysis in the Nainital district, Himalayan Region, 2021. Discov Water 2(12). https://doi.org/10.1007/s43832-022-00020-y

Reasoner M, Bugmann H, Schaaf T (2003) Background and concepts for collaborative work: global change research in mountain biosphere reserves. In: Carr CM (ed) Proceedings of the

international launching workshop. Entlebuch Biosphere Reserve, Switzerland 10–13 November. Curran Publishing Services. The United Nations Educational, Scientific and Cultural Organization, Paris
Rinella MJ, Mangold JM, Espeland EK, Shelly RL, Jacobs JS (2012) Long-term population dynamics of seeded plants in invaded grasslands. Ecol Appl 22:1320–1329
Rosenzweig C, Casassa DS, Karoly A et al (2007) Assessment of observed changes and responses in natural and managed system. In: Parry ML, Canziani, Palutid JP et al (eds) Climate change: impacts, adaptation and vulnerability. Cambridge University Press, Cambridge, UK, pp 79–131
Sahai B, Kimothi MM (1994) Remote sensing of Nanda Devi Biosphere Reserve for biodiversity conservation. In: Proceedings of seminar on biodiversity conservation. National Conservation Congress WWF, New Delhi, India, pp 131–137
Samant SS, Dhar U, Palni LMS (1998) Medicinal plants of Indian Himalaya: diversity distribution potential values. GyanodayaPrakashan, Nainital
Schaaf T (2009) Mountain biosphere reserves—a people centred approach that also links global knowledge. Sustain Mount Dev (ICIMOD) 55:13–15
Seyam MMH, Haque MR, Rahman MM (2023) Identifying the land use land cover (LULC) changes using remote sensing and GIS approach: a case study at Bhaluka in Mymensingh, Bangladesh. Case Stud Chem Environ Eng 7:100293. https://doi.org/10.1016/j.cscee.2022.100293
Silori CS (2001) Status and distribution of anthropogenic pressure in the buffer zone of Nanda Devi Biosphere in western Himalaya, India. Biodivers Conserv 10:1113–1130
Singh JS, Singh SP (1987) Forest vegetation of the Himalaya. Bot Rev 53(1):80–192
Singh SP, Singh RP, Rawat YS (1992) Patterns of soil and vegetation and factors determining their forms and hydrologic cycle in NDBR. Final technical report, Ministry of Environment and Forest, Government of India, New Delhi
Singh SP, Singh V, Skutsch M (2010) Rapid warming in the Himalayas: ecosystem response and development option. Clim Dev 2:221–232
Sykora KV, Kalwij JM, Peter-Jan K (2002) Phytosociological and floristic evaluation of 15-year ecological management of roadside verges in the Netherlands. Preslia-Praha 74:421–436
Theurillat JP, Guisan A (2001) Potential impact of climate change on vegetation in the European alps: a review. Climate Change 50:77–109
Udaya LV, Dutt CBS (1998) Paper presented in the Workshop on Experiences Sharing on GIS for Sustainable Forest Management, 16–17 March, Hyderabad
UNESCO-MAB (2004) Proceedings "Global change research in Mountain Biosphere Reserves, Switzerland, pp 10–13
Walther G-R, Beissner S, Burga CA (2005) Trends in the upward shift of alpine plants. J Veg Sci 16:541–548
Williams JW, Jackson ST (2007) Novel climates, no-analog communities, and ecological surprises. Front Ecol Environ 5:475–482
Wing SL, Harrington GJ, Smith FA et al (2005) Transient floral change and rapid global warming at the Paleocene-Eocene boundary. Science 310:993–996
Xu X, Zhou Y, Ruan H, Luo Y, Wang J (2010) Temperature sensitivity increased with soil organic carbon recalcitrance along an elevational gradient in the Wuyi Mountains, China. Soil Biol Biochem 42:1811–1815
Zhao G, Ren L, Ye Z (2022) Vegetation dynamics in response to climate change and human activities in a typical alpine region in the Tibetan plateau. Int J Environ Res Public Health 9(19):12359. https://doi.org/10.3390/ijerph191912359
Zobel DB, Singh SP (1997) Himalayan forests and ecological generalizations. BioScience 47:735–745

Part III
Conservation and Management of Himalayan Hotspot

Chapter 16
Richness, Traditional Use and Management of Non-timber Forest Plant Products in the Himalayan Communities

Rajib Biswas

Abstract The present study examines the Northeastern region of India, comprising the eight states of the Eastern Himalayas. Except for the plains of Assam and Tripura, the region is largely mountainous, with elevations ranging from 15 to 8500 m. Its diverse topography sustains a remarkable variety of vegetation, making it one of the richest biodiversity hotspots in the country. Alongside this ecological wealth, the region is inhabited by numerous indigenous communities, each with distinctive cultural traditions and practices. Rice forms the staple food for most tribes, supplemented by meat from domesticated animals as well as certain reptiles and birds. Fermented rice and millet beers hold cultural significance, being consumed both in daily life and during ritual ceremonies. Ethnobotanical surveys have documented 235 plant species used by these communities for food, medicine, and cultural practices. However, this knowledge system and its associated resources are increasingly at risk. Rapid population growth, anthropogenic pressure, rising economic pressures, habitat degradation, and the influence of global commercial markets threaten both the continuity of indigenous traditions and the biodiversity on which they depend. Safeguarding ethnobotanical knowledge and ensuring the sustainable management of natural resources are therefore of critical importance.

Keywords North East India · Ethnic community · Ethnic knowledge

16.1 Introduction

Ethnobotany is an interdisciplinary field that examines the intricate relationships between humans and plants, encompassing aspects of sociology, linguistics, anthropology, plant taxonomy, agriculture, medicine, phytochemistry, ecology, and archaeology. This discipline has expanded significantly over time, reflecting its growing importance in understanding human-plant interactions across various

R. Biswas (✉)
Department of Conservation Biology, Durgapur Government College, Durgapur, West Bengal, India

D. Das et al. (eds.), *Plant Diversity and Conservation Strategies of the Himalayan Biodiversity Hotspot*, Topics in Biodiversity and Conservation 21,
https://doi.org/10.1007/978-3-032-21697-7_16

cultures and environments. As discussed in Chap. 2 (Sect. 2.1), local communities in the Himalaya have traditionally depended on forest resources for their subsistence, livelihoods, and cultural practices, making ethnobotanical studies highly relevant for biodiversity conservation.

The term *ethnobotany* was first introduced by John William Harshberger in 1895, defining it as "the study of plants used by primitive and aboriginal people." Richard Evans Schultes, often regarded as the 'father of modern ethnobotany,' later refined this definition in 1962, describing ethnobotany as "the study of the relationships that exist between people of a primitive society and their natural resources." Schultes' work emphasised the significance of documenting traditional plant knowledge, particularly in the context of medicinal and psychoactive plants a theme further reflected in Chap. 5 (Sect. 5.2), where high-altitude medicinal flora of the Himalaya are described as central to both cultural traditions and modern pharmacological potential.

In 1987, S.K. Jain offered a more comprehensive perspective, characterising ethnobotany as "the total natural and traditional relationship and interactions between men, domesticated animals and their surrounding plant wealth." Jain's definition underscores the holistic nature of ethnobotany, integrating human, animal, and plant interactions within specific ecological contexts. This broader ecological lens resonates with the concerns raised in Chap. 13 (Sect. 13.1.2), where anthropogenic pressures on forest ecosystems are highlighted as threats not only to biodiversity but also to cultural traditions dependent on them.

These foundational definitions highlight the multifaceted nature of ethnobotany, illustrating its role in bridging cultural practices with scientific inquiry to foster a deeper understanding of plant utilisation and conservation in human societies. As emphasised in Chap. 12 (Sect. 12.1), ethnobotanical knowledge also plays an important role in building climate-resilient strategies, since communities adapt their resource use in response to environmental change.

16.2 Study Areas

The present study focuses on the North-Eastern Region (NER) of India, encompassing the states of Assam, Nagaland, Mizoram, Manipur, Meghalaya, Tripura, Arunachal Pradesh, Sikkim, and the hilly region of Darjeeling district in West Bengal. Geographically, the region spans latitudes between 22°19′N and 30°N, and longitudes between 89°42′E and 97°12′E (Singh 2000) (Fig. 16.1). Its geographical and biogeographic significance is also discussed in Chap. 1 (Sect. 1.3.1), which highlights how physiographic variation contributes to the immense biodiversity of the Himalaya.

This diverse region is predominantly mountainous, with the exception of the plains in Assam and Tripura. Elevations range from 15 m in the lower Assam plains to over 8500 m in northern Sikkim, where peaks such as Kangchenjunga rise to 8586 m. In Sikkim, areas like Lachen, Lachung, and Sakyong feature elevations

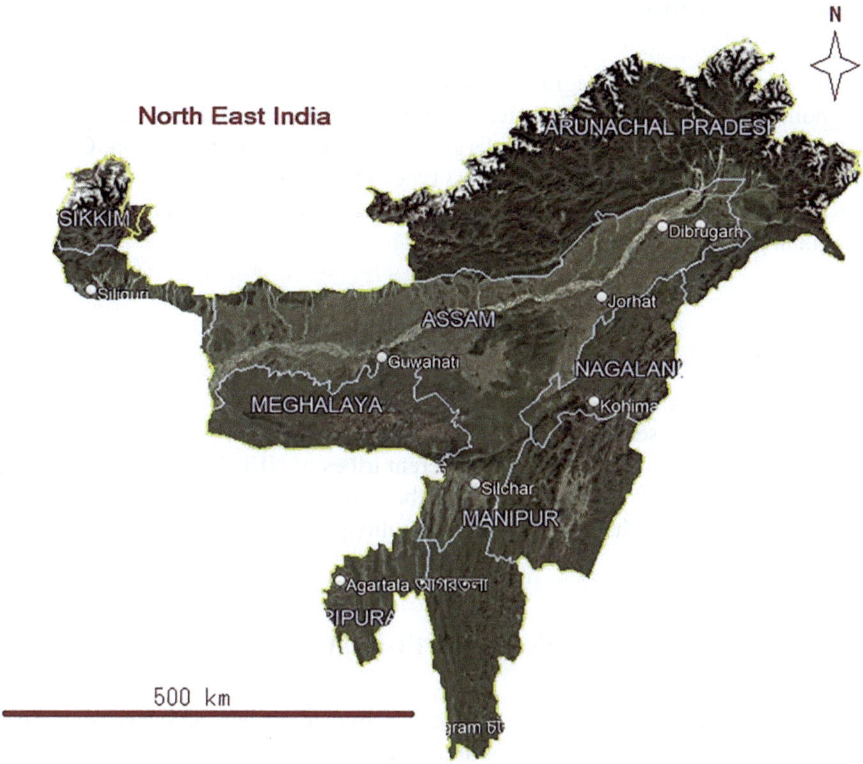

Fig. 16.1 Map of study area. (North East India extracted from Google Earth imagery)

exceeding 7000 m. Mizoram's steep hills, carved by rivers, create deep gorges, while Nagaland's Dhansiri, Doyong, Dikhu, and Jhanji rivers traverse its terrain. Such variation in altitude and terrain directly shapes the ecological distribution of species, an aspect further elaborated in Chap. 4 (Sect. 4.1), which describes vegetation zones across the Himalayan range.

The region's topography is further enriched by the perennial Teesta River, which originates in Sikkim and flows through the hilly parts of West Bengal. This river, along with others like the Brahmaputra and its tributaries, plays a crucial role in shaping the landscape and supporting the region's biodiversity, a theme closely related to the wetland and riverine ecosystems outlined in Chap. 6 (Sect. 6.1).

Climatically, the NER experiences a wide range of conditions due to its varied elevations. The combination of rolling mountains, enchanting valleys, flowing rivers, and tropical evergreen forests sustains its rich biodiversity. As emphasised in Chap. 12 (Sect. 12.1), these ecological treasures are particularly vulnerable to climate change, making the conservation of the North-Eastern Region's natural heritage a matter of urgent concern.

16.3 Vegetation

The vegetation of these regions in Eastern Himalaya is very rich and diverse abounding in tallest Trees, Tree ferns, Orchids, Primulas and Blue Poppies. The mountain slopes are covered with a large number of *Rhododendron* spp. Champion and Seth in 1968 classified Indian forest types (Table 16.1).

Rao and Panigrahi (1961) described the diverse forest ecosystems based on the altitudinal zones in five types (Table 16.2).

16.4 The Ethnic Groups of North-East India

North-East India presents great ethnic diversity and has nearly 400 ethnic communities. Dutta and Dutta 2005 describe different tribes of NEI. The areas are inhabited by a large number of aboriginals like Abar, Garo, Dafla, Khasi, Jaintia, Kuki, Mishiamiri, Mikir, Rabha, Naga, Apatani, Gallong, and Lushais (Table 16.3).

16.5 Life Style of People of North-East India

Each ethnic group possesses its own distinctive traditions and customs. These differences are reflected in their attire, marriage rituals, ceremonial songs and dances, and overall ways of life (Stoffle et al. 1990). Among the Angami Nagas, unmarried girls traditionally cropped their hair, while married women were permitted to grow it long. In contrast, other tribes regard hair as an essential element of feminine beauty and maintain it accordingly. Among the Lepchas, the day begins with worship of their deity, followed by the consumption of chi (a fermented millet drink), zo (boiled rice), and feeding of domestic animals and poultry (Singh 2000). Such cultural dimensions of human–nature interactions echo the socio-ecological linkages discussed in Chap. 2 (Sect. 2.1), where traditional lifestyles are shown to be tightly interwoven with natural resources.

The Naga community typically builds its settlements on hilltops. Village boundaries are demarcated by natural landmarks such as small streams, ditches, or hillocks. Their houses are generally elevated about 2.5 m above the ground using wooden posts. The walls are constructed from finely woven and joined split bamboo or are plastered with mud. Roofs are thatched using sungrass, leaves, or split bamboo. These resource-based settlement practices parallel the ethnobotanical traditions highlighted in Chap. 5 (Sect. 5.2), where local communities are shown to utilise available forest plants for construction and cultural purposes.

Rice is the staple food for most of these tribes. They also consume a variety of meats, including those from domesticated animals, birds, and even snakes. In Nagaland, Arunachal Pradesh, and Meghalaya, dog meat is consumed on special

Table 16.1 Forest types of North-East as classified by Champion and Seth (1968)

Forest types	Vegetation structure	Forest types	Vegetation structure
1B/C1.	Assam valley tropical wet evergreen forest	5/1S2	Khair-sissu forest
1B/C2(a, b).	Upper Assam valley tropical wet evergreen forest	8B/C2	Khasi sub-tropical hill forest
1B/C3.	Cachar tropical wet evergreen forest	8B/DS1	Assam sub-tropical hill savannah woodland
1/E1.	Cane brakes	9C1b	Himalayan chir pine forest
1/E2.	Wet bamboo brakes	9/C2/DS1	Assam sub-tropical pine forest
1/2S1.	Pioneer euphorbiaceous scrub	9/C2/DS1	Assam sub-tropical pine savannah
2B/C1(a, b).	Assam valley semi-evergreen forest	11B/C1 (a,b,c).	East Himalayan wet temperate forest
2B/C1/1S1.	Sub-Himalayan high alluvial semi-evergreen forest	11B/C2	Naga hills wet temperate forest
2B/C1/1S2	*Syzygium* parkland	12/C3(a, b)	East Himalayan moist temperate forest
2B/C1/2S2	Eastern alluvial secondary semi-evergreen forest	12/E1.	Cypress forest
2B/C/2S3.	Sub-Himalayan secondary wet mixed forest	12/DS1.	Montane bamboo brakes
2B/C2	Cachar semi-evergreen forest	12/1S1	Alder forest
2/E3	Moist bamboo brakes	12/2S1	Low level blue-pine forest
2/2S1	Secondary moist bamboo brakes	13/C6.	East Himalayan dry temperate coniferous forest
3C/C1(a, b,c)	Very moist sal forest	13/C6/E1	Larch forest
3C/C2(d)	Moist plains sal forest	13/C7	East Himalayan dry juniper/birch forest
3C/C2/DS1	Moist sal savannah	13/1S1	*Hippophae/Myricaria* scrub
3C/C3(b).	East Himalayan moist mixed deciduous forest	13/1S2	*Populus-Salix* forest
3C/C_3/2S1	Northern secondary moist mixed deciduous forest	14/C2.	East Himalayan sub-alpine birch/fir forest
3C/C_3/2S2	Secondary Euphorbiaceous scrub	14/C2/2S1	Sub-alpine blue pine forest
3/1S1	Low alluvial savannah woodland	14/DS1	Sub-alpine pasture
3/1S2(a, b).	Eastern hollock forest	15/C1	Birch/ *Rhododendron* scrub
4C/FS_2	Sub-montane hill valley swamp forest	15/C2	Deciduous alpine scrub
4C/FS_3.	Creepers swamp forest	15/C2/E1	Dwarf *Rhododendron* scrub
4D/SS1 to SS_5.	Tropical seasonal swamp forest	16/C1	Dry alpine scrub
4D/2S2.	Eastern wet alluvial grassland	16/E1	Dwarf juniper scrub
4E/RS1	Riparian fringing forest		

Table 16.2 Major ecosystem types and their dominated species

Forest types/Sub types	Altitude	Dominated species
Tropical forests (2 types)		
Tropical evergreen gorests	Up to 1000 m	*Shorea robusta, Tectona grandis, Terminalia myriocarpa, Dalbergia sissoo, Dipterocarpus* sp. *Artocarpus* sp., *Trewia nudiflora,* etc. with associates species *Gynocardia odorata, Saurauria nepalensis, Syzygium cumini, Ailanthus integrifolia, Knema angustifolia, Daubanga grandiflora, Careya arborea, Oroxylum indicum*
Tropical grasslands	Up to 1000 m	*Arundinella bengalensis, Setaria palmifolia, Phragmites karka, Arundo donax, lmperata cylindrica, Erianthus lingischorus, Saccharum spontaneum*
Subtropical forests	1000–2000 m	*Ficus - Castanopsis - Callicarpa* association with *Schima wallichii, Castanopsis indica, Saurauria fasciculata, Firmiana colorata, Exbuckalandia populnea, Stereospermum colais, Prunus cerasoides, Trevesia palmata, Wendlandia* sp. etc. Along the water courses *Alnus nepalensis* is quite common. On dry slopes *Rhododendron arboreum, Schima wallichii, Garuga pinnata* etc. The canes like *Calamus erectus, C. floribundus* etc. are also seen. The important shrubs are represented by species like *Rubus ellipticus, Artemisia nilagirica, Boehmeria macrophylla, Osbekia sikkimensis, O. nepalensis, Melastoma nepalensis* and *Maesa chisia.*
Temperate forests (3 types)		
Broad leaved forests	2500 m	*Acer campbellii, Corylopsis himalayana, Ilex* spp., *Persea* spp., *Symplocos* spp.
Evergreen oak forests	1800–2600 m	*Uthocarpus elegans, Utsea thomsonii, Rhododendron arboreum, R. lepidotum, Styrax serrulatum, Acer campbellii, Castanopsis tribuloides,* C. *purpurella, Juglans regia, Symplocos* spp., *Lyonia ovalifolia*
Temperate *Rhododendron* - conifer forests	2500–3500 m	*Cupressus torulosa, Tsuga dumosa, Abies delavayi, Cryptomeria japonica, Larix griffithii, Picea morinda,* etc. associated with *Prinsepia utilis, Daphne papyracea, Rhododendron lepidotum, Agapetes serpens, Lyonia ovalifolia*
Subalpine forests	3600–4200 m	*Abies densa* and *Betula utilis.* Associated with *Meconopsis* spp., *Cardamine* spp., *Potentilla* spp., *Astragalus* spp., *Corydalis* spp., *Primula* spp., *Pedicularis* spp., *Polygonum* spp., *Rheum* spp., *Saussurea* spp., *Anaphalis* spp. etc. At higher reaches *Rhododendron hodgsonii, R. anthopogon, R. campanulatum, R. thomsonii.*
Alpine vegetation	4200–5500 m	*Rhododendron anthopogon, R. campanulatum, R. nivale, R. pumilum, Primula sikkimensis, P. glabra, Meconopsis nepalensis, Sedum* spp. *Bergenia* spp. *Polygonum* spp., *Pedicularis* spp., *Ephedra* spp.

Table 16.3 State wise indigenous community distribution in North East India

State name	Important tribes
Assam	Barmons, Biata, Bodo, Changsan, Chakma, Choggloi, Deori, Dimasa, Doungel, Gamalhou, Gangte, Garo, Guite, Hojong, Hmar, Hojai, Haokip/ Houpit, Haolai, Hengna, Hongsungh, Jaintia, Jongbe, Kachari, Khasi, Hanneng, Hrangkhawl, Karbi, Khawehung, Khawathlang, Khelma, Kholhou, Kipgen, Kuki, Lakher, Lalung, Lengthang, Lhangum, Lhoujem, Lhouvun, Lupheng, Mangjel, Mech, Misao, Mishing, Mizo, Pawi, Rabha, Riang, Sairhem, Selnam, Singson, Sithou, Sonwal, Sukte, Synteng, Thado, Thangngen, Vibub, Vaiphei
Arunachal Pradesh	Adi, Aka, Apatani, Bagung, Bangni, Deuri, Kangbo, Khampti, Meyer, Miji, Miju, Miri, Mishing, Mishmi, Monpa, Na, Nishi, Nocte, Sherdukpen, Sinpho, Sulung, Tagin, Tangsa, Wanchu, Yabin
Manipur	Aimol, Anal, Angami, Chiru, Chothe, Gangte, Hmar, Kubui, Kacha Naga, Koirao, Koireng, Korn, Lamgang, Mao, Maram, Maring, Mizo, Monsang, Mayon, Paite, Purum, Ralte, Sabte, Serna, Simte, Tangkhul, Thodou, Vaiphei, Zou
Meghalaya	Banai, Baro, Bhoi, Biate, Dalu, Hajong, Jaintia and Garo, Koch, Lyngam, Man, Rabha, War
Mizoram	Chakma, Chowngthu, Chawhto, Hmar, Khawlhting, Khaingte, Kuki, Lakher, Lushai, Mora, Nagenta, Pautu, Paite, Pawi, Ralte, Rawite, Renthlet, Thode, Tlau, Vongchhia, Zawngte
Nagaland	Naga (16 Naga scheduled tribes viz. Angami, Ao, Chakhesang, Chang, Chirr, Konyak, Khei mnungan, Lotha, Makware, Phom, Rengma, Sangtam, Serna, Tikhir, Yimchungrel and Zeliang), Kuki, Kachari, Mikir and Garo tibes
Tripura	Chakma, Halam, Jamatia, Mag, Naotia, Riang, Tipra, Garo, Kuki, Lushai, Munda and Oraon
Sikkim and Darjeeling	Bhutia, Lepcha and Nepali

occasions. Indigenous communities have developed numerous tools made from bamboo and cane, such as fish traps, rods, lines, and hooks. To catch fish, they sometimes introduce plant-based poisons into rivers or streams. For bird-catching, they use the fruit of *Scurrula parasitica*. Crabs are lured using earthworms or chicken intestines. Additionally, some plant extracts are utilised to make poison for arrows. These traditional uses of plants and animals for subsistence reflect patterns also discussed in Chap. 13 (Sect. 13.1.2), which examines human dependence on forest ecosystems and the pressures this places on biodiversity.

These communities cultivate a range of traditional cereals, millets, pulses, and fruits, which constitute their primary diet (Singh 2000). Young bamboo shoots are a commonly consumed vegetable. Fermented rice or millet beer is integral to their social and cultural life. This drink is known by different names among various groups: *Apong* (Adi), *Bunkchung* (Monpa), *Chang* (Bhutia), *Chi* (Lepcha), *Jand* (Nepali), *Le Opani* (Deori and Aka), *Morpo* (Mikir), and *Zuh* (Naga), among others (Singh 2000). The Lepchas specifically grow *Eleusine coracana* (finger millet) for brewing beer, and the residue from this process is fed to pigs and chickens. During festivals, beer is traditionally served in wooden vessels, bowls, or bamboo tumblers. These practices align with the broader ethnobotanical perspectives discussed in Chap. 15 (Sect. 15.1), where traditional plant knowledge is linked with cultural continuity and ecosystem management.

Agricultural practices in these regions remain underdeveloped. Most tribal groups rely on shifting cultivation, locally known as *jhum*. In Arunachal Pradesh, this practice is called *Adi-arick*, while in Tripura, it is referred to as *Hooknis-mono*. Some communities, such as the Apatani and the Nagas, have adopted terrace or wet rice cultivation. The Khasi people of Meghalaya cultivate potatoes extensively on terraced fields. Terrace cultivation has been increasingly practiced in Arunachal Pradesh, Manipur, Nagaland, and Tripura (De et al. 2010). These shifting and settled agricultural systems mirror the land-use dynamics highlighted in Chap. 12 (Sect. 12.1), which detail the impact of climate and human activity on fragile Himalayan agroecosystems.

Despite the presence of multistructural economies, the region of Northeast India continues to experience economic underdevelopment due to its challenging physiography and socio-political factors. The agricultural economy primarily involves rice cultivation in the plains of Assam and Manipur, as well as terrace farming practiced by communities such as the Apatanis, Aos, Jaintias, and Nagas. In North Sikkim, cardamom is the principal cash crop, while in Meghalaya, potatoes and citrus fruits are widely cultivated. Assam also has a flourishing tea industry (Singh 2000). These agricultural adaptations connect with ecosystem services described in Chap. 6 (Sect. 6.1), especially in relation to wetland and riverine systems that support farming and food security.

Cane and bamboo play a vital role in the livelihoods of tribal communities across Northeast India. They are used to make mats, baskets of various types, containers, fish traps, furniture, ropes, headgear, and more (Singh 2000). Among the Nagas, wooden basketry is highly developed. The Lepchas utilise the bark of *Edgeworthia gardneri* (locally known as Argeli) for making straps. They also create ritual masks carved from single blocks of wood. The Nagas manufacture wooden and bamboo utensils, particularly for storing rice beer. In Assam, large traditional hats are made using the *Ucuala* leaf and bamboo (Tynsong et al. 2020). These cultural practices reinforce the sustainable use of non-timber forest products; a theme explored in Chap. 14 (Sect. 14.2).

Each tribal group possesses its own distinctive musical instruments. Among the Khasis, the drum is referred to as the *duitara*. The Mikirs play the *cheng*, *chang chickli*, and *thekleachang*. The Angami have a unique stringed instrument called the *upou*, a type of guitar with a wooden body and strings made from goat skin. The Mikirs also use a wooden flute (*muri*), a bamboo flute (*pangsi*), and the *achang*, an instrument similar to the *vina* (De et al. 2010).

Festivals in the North-Eastern region are closely associated with agricultural cycles, especially rice cultivation. The Mikirs celebrate three harvest festivals: *Chojun*, *Hachakkakan*, and *Horbong-amarm-kepu*. Among the Nagas, the *Heleikanebe* festival is held during March and April to mark the beginning of the sowing season, while *Nchangban* is observed in May (Chakraborty et al. 2012).

These festivals, grounded in ecological rhythms, reflect the intimate relationship between biodiversity and cultural identity, echoing themes discussed in Chap. 15 (Sect. 15.1) on the ethnobotanical basis of cultural traditions.

16.6 Methodology

A comprehensive list of Ethnobotanical important plant species was prepared by collecting published information from secondary sources, such as scientific studies, books, published floras, Ph.D. theses, and forest working plans. Search on different online databases such as Scopus, Google Scholar, PubMed, and Web of Science to collect data. The precision of species identification in this study was dependent on the source (Fig. 16.2).

16.7 Data Analysis

To evaluate the relative usefulness of different ethnobotanically important species, a quantitative valuation was also used by calculating various indices, such as Use report (UR), use value (UV), relative frequency of citation (RFC), relative importance index (RI), cultural importance index (CI), and cultural value (CV) as provided (Kushwaha et al. 2018) (Table 16.4).

16.8 Result and Discussion

Data compiled from 16 published papers, reviews, and a doctoral thesis focusing on the study area were systematically organised into an Excel sheet, encompassing extensive ethnobotanical knowledge. The information was analysed following the

Fig. 16.2 Hypothetical diagram of the methodology

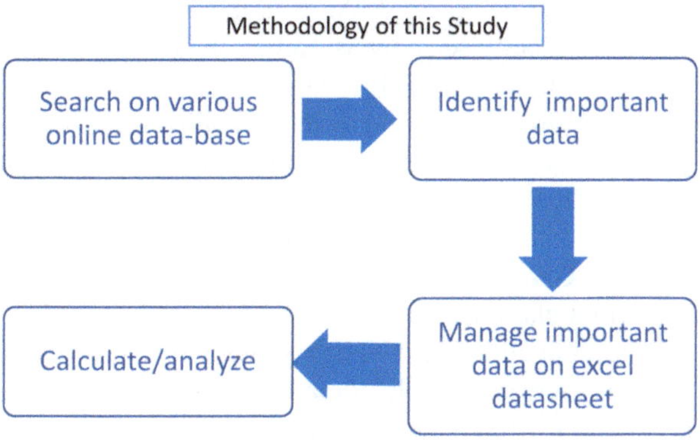

Table 16.4 Methodology of Analysis

Name	Calculation formulas	References
Use report (UR)	$UR_S = \sum_{u=u1}^{u_{NC}} \sum_{i=i1}^{i_N} UR_{ui}$ Informant "*i*" mentions the use of the species "*s*" in the use category "*u*." First, we sum the "UR" of all the literature sources (from "i_1" to "i_N"). Then, sum all the "UR" of each use-category (from "u_1" to "u_{NC}")	Tardìo and Pardo-De-Santayana (2008)
Use value (UV)	UVs = $\sum$ Us/N Where, UVs = Use Value for the species Σ Us = Sum of the uses mentioned for a species N = Total number of informants	Gómez-Beloz (2002), Martin (2004)
Relative frequency of citation (RFC)	RFC = FCs / N Where, RFC = Relative Frequency of Citation FCs = Number of informants who mentioned the use of species N = Total number of informants	Tardío et al. (2005, 2007), Tardìo and Pardo-De-Santayana (2008)
Family use values (FUVs)	FUVs = $\sum$ UVs/NS Where, FUVs = Family Use Value $\sum$UVs = Sum of the Use Values NS = Total number of species quoted from the family	Phillips (1996), Phillips and Gentry (1993), Prance et al. (1987)
Relative importance index (RI)	$RI_S = \frac{RFC_{S(max)} + RNU_{S(max)}}{2}$ $RFC_{s(max)} = FC_s/max\ (FC)$, $RNU_{s(max)}$ is the relative number of use-categories.	Tardío et al. (2005, 2007), Tardìo and Pardo-De-Santayana (2008)
Cultural significance index (CSI)	CSI = $\sum (I \times E \times C)\, CF$ I = species management, E = preference of use C = frequency of use, CF = correction factor.	Da Silva et al. (2006), Pelto and Pelto (1975)

prescribed methodology, and a comparative structural graph was generated. In total, 235 plant species were documented as being integral to the daily lives of local inhabitants (Devi et al. 2022; Jha and Smith-Hall 2023; Masoodi and Sundriyal 2020; Chatterjee et al. 2006; Sharma and Das 2018). The present study does not provide a tribe-wise analysis; rather, the overall uses of plants were calculated based on data extracted from the various published sources. A significant proportion of these species are recorded as medicinal plants with diverse therapeutic applications.

Most available studies have concentrated on the ethnobotanical practices of tribes such as the Bodo, Khasi and Jaintia, Apatani, Baro, Koch, Naga, Kuki, Kachari, Mikir, Angami, Chirr, Konyak, Chakma, Halam, Jamatia, Mag, Naotia, Riang, Tripura, Garo, Kuki, Lushai, Bhotia, and Lepcha, with particular emphasis on medicinal plant use among these indigenous groups.

The indigenous knowledge synthesised in this study demonstrates considerable expertise across multiple domains, including herbal medicine, traditional foods, fibre production, edible and essential oils, natural dyes, gums, resins, tannins, and tribal

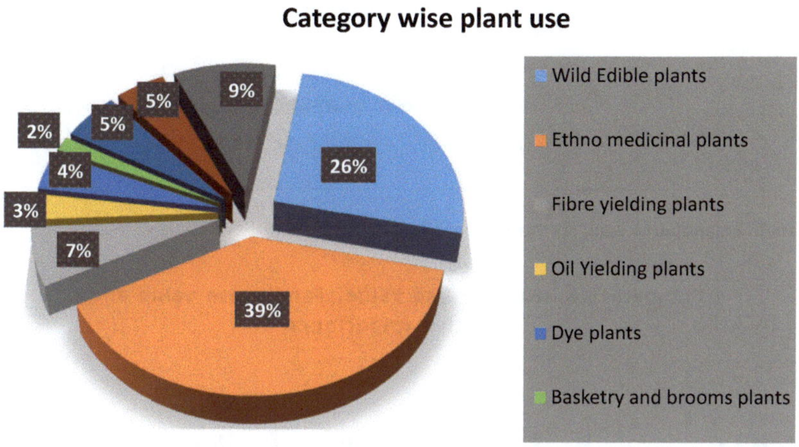

Fig. 16.3 Category wise ethnobotanical important plants used by indigenous people of NEI

crafts each contributing significantly to community welfare. The ethnobotanical wealth of North East India has been classified into ten major categories: wild edible plants (175 species), ethnomedicinal plants (264 species), fibre-yielding plants (45 species), oil-yielding plants (20 species), dye-producing plants (29 species), plants used in basketry and broom-making (11 species), gum and resin-yielding plants (13 species), tannin-producing plants (24 species), plants employed as fish and bird poisons (32 species), and timber and wood-yielding plants (65 species). Collectively, these categories illustrate the sustainable utilisation of biodiversity for human benefit, highlighting the remarkable economic plant wealth of the North East region (Fig. 16.3).

In this study, the highest Cultural Values (CV) were recorded for *Piper nigrum*, *Oryza sativa*, *Asparagus racemosus*, and *Centella asiatica* (CV = 0.36), whereas 81 species exhibited the lowest CV (0.04). The highest Relative Importance (RI) values were associated with *Justicia adhatoda*, *Curcuma longa*, *Centella asiatica*, *Asparagus racemosus*, *Oroxylum indicum*, *Terminalia chebula*, *Terminalia arjuna*, *Zanthoxylum oxyphyllum*, and *Zingiber officinale*. By contrast, the lowest RI values were noted for *Cuscuta reflexa*, *Rauvolfia serpentina*, *Carica papaya*, and *Morus alba*. With respect to Use Value (UV), the highest value (0.4) was recorded for *Centella asiatica*, *Azadirachta indica*, and *Ocimum tenuiflorum*, while 97 species showed the lowest UV (0.04). Similarly, the highest Family Use Values (FUV) were observed for *Azadirachta indica*, *Ocimum tenuiflorum*, *Andrographis paniculata*, *Justicia adhatoda*, *Acmella paniculata*, *Centella asiatica*, and *Aloe vera*. In contrast, the lowest FUVs were recorded for *Canavalia gladiata*, *Calotropis gigantea*, *Euphorbia neriifolia*, *Callicarpa arborea*, *Alpinia galangal*, *Cucumis sativus*, *Hibiscus macrophyllus*, and *Galinsoga parviflora* (Figs. 16.4 and 16.5).

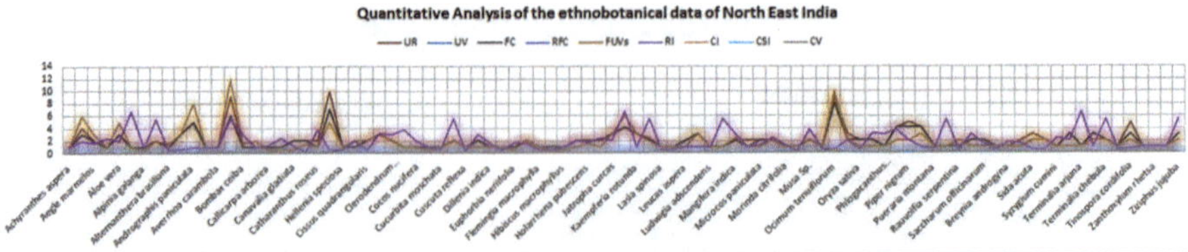

Fig. 16.4 Quantitative analysis of different Plant use indices

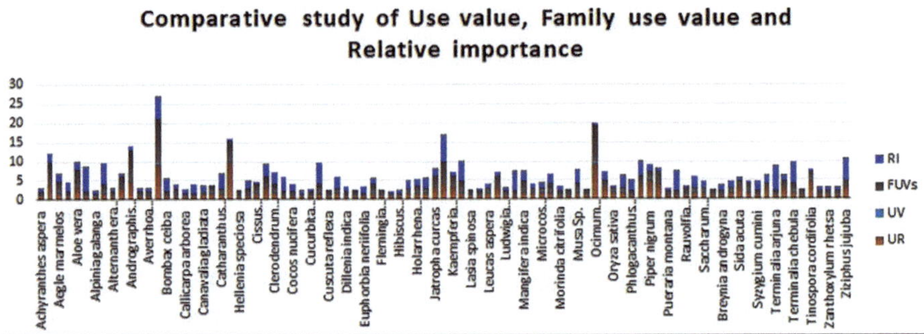

Fig. 16.5 Comparative analysis of RI, FUVs, UV and UR

16.9 Management

The management of ethnobotanical knowledge and natural resources is of paramount importance. Increasing population pressure, economic inflation, anthropogenic activities, habitat loss, and illegal exploitation driven by global commercial demand have placed both traditional knowledge and natural resources under severe threat. Sacred groves play a vital role in conserving not only biodiversity but also indigenous knowledge systems. In Meghalaya, these sacred groves, locally known as *Lawkyntang*, are prominent features of the landscape. One such sacred grove, Lungchuk, is situated in north Sikkim at an elevation of approximately 2000 m. The Khasi community believes that these groves are the dwelling places of deities who safeguard the welfare of the people. Within these groves, everything is protected by the reigning deity, and the removal of any material, including dead wood or twigs, is strictly taboo.

The Khasi sacred groves remain relatively undisturbed and are preserved due to longstanding traditional beliefs and customs. These groves serve as important repositories of wild germplasm and biological diversity, playing a crucial role in the conservation of indigenous flora, including many endangered species of orchids and medicinal plants. Notable endangered medicinal plants found in these groves include *Aconitum ferox* and *Aconitum heterophyllum* (*Ranunculaceae*), *Coptis teeta* (*Ranunculaceae*), *Dendrobium nobile* and *Flickingeria fugax* (*Orchidaceae*), *Nardostachys jatamansi* (*Valerianaceae*), *Panax assamicus* Banerjee (*Araliaceae*), and *Garcinia pedunculata* (*Clusiaceae*). These species are of significant value both ethnobotanically and commercially and are classified as Endangered (EN), necessitating urgent conservation measures (Mao et al. 2009). Similar concerns regarding the vulnerability of medicinal and alpine plant species are highlighted in Chap. 5 (Sect. 5.2), which documents the ecological and cultural importance of Himalayan high-altitude flora.

The socio-economic impacts on ethnobotanically important plants are considerable and warrant further detailed investigation. It is crucial to reassess the pressures on natural vegetation caused by overgrazing, excessive harvesting, and land-use changes such as urbanisation. The degradation of natural vegetation can be mitigated through effective management strategies that regulate anthropogenic activities and grazing. Moreover, there is an urgent need to document the invaluable traditional knowledge held by local communities across various parts of Northeast India. As emphasised in Chap. 15 (Sect. 15.1), the systematic documentation of ethnobotanical traditions is essential to ensure their preservation and sustainable use for future generations.

16.10 Conclusions

Traditional knowledge of plant use in Northeast India has been orally transmitted across generations, making it vulnerable to loss amid social and environmental changes. The high values of indices such as Cultural Value (CV), Relative Importance (RI), Use Value (UV), and Family Use Value (FUV) highlight the significant reliance of local communities on diverse plant species for medicinal, nutritional, and economic purposes.

It is crucial to systematically document this knowledge before it fades, as its loss would not only erode cultural heritage but also threaten biodiversity conservation. Indigenous knowledge encompasses detailed ecological insights essential for sustainable resource management, and its preservation supports both community well-being and ecosystem resilience.

Urgent conservation efforts combining ethnobotanical documentation and biodiversity protection are needed to safeguard these invaluable resources and ensure their continued benefit for future generations.

Acknowledgements The author, Rajib Biswas, gratefully acknowledges the support and resources provided by the Department of Conservation Biology, Durgapur Government College, Durgapur, West Bengal, India.

References

Chakraborty R, De B, Devanna N, Sen S (2012) North-East India an ethnic storehouse of unexplored medicinal plants. J Nat Prod Plant Resour 2:143–152

Champion GH, Seth SK (1968) A revised survey of the forest types of India. Govt. of India, New Delhi

Chatterjee S, Saikia A, Dutta P, Ghosh D, Pangging G, Goswami AK (2006) Biodiversity significance of North East India. WWF-India, New Delhi, pp 1–71. http://trpenvis.nic.in/test/doc_files/BiodiversitySignificanceNEIndia_WWF.pdf

Da Silva VA, Andrade LHC, de Albuquerque UP (2006) Revising the cultural significance index: the case of the Fulni-ô in Northeastern Brazil. Field Methods 18(1):98–108

De B, Debbarma T, Sen S, Chakraborty R (2010) Tribal life in the environment and biodiversity of Tripura, India. Curr World Environ 5:59–66

Devi S, Sinha R, Sibi G (2022) Northeast India: a treasury of medicinal plants. Review article. Emer Life Sci Res 8(1):5–21. https://doi.org/10.31783/elsr.2022.810521

Dutta BK, Dutta PK (2005) Potential of ethnobotanical studies in North East India: an overview. IJTK 4(1):7–14

Gómez-Beloz A (2002) Plant use knowledge of the *Winikina Warao*: the case for questionnaires in ethnobotany. Econ Bot 56:231–241

Jha KK, Smith-Hall C (2023) Three models to illustrate plant- people relationships in the medicinal plant hotspots of North East India. Ethnobot Res Appl 26(7):1–48. https://doi.org/10.32859/era.26.7

Kushwaha A, Jain S, Bhojwani K, Kalyani G (2018) Concise synopsis on quantitative ethnobotanical tools for medicinal plant analysis. Int J Pharm Sci Rev Res 48(1):128–132

Mao A, Hynniewta T, Sanjappa M (2009) Plant wealth of Northeast India with reference to ethnobotany. IJTK 8:96–103

Martin GJ (2004) Ethnobotany: a methods manual. Earthscan Publications, London

Masoodi HUR, Sundriyal RC (2020) Richness of non-timber forest products in Himalayan communities—diversity, distribution, use pattern and conservation status. J Ethnobiol Ethnomed 16: 56. https://doi.org/10.1186/s13002-020-00405-0

Pelto PJ, Pelto GH (1975) Intra-cultural diversity: some theoretical issues. Am Ethnol 2(1):1–18

Phillips O (1996) Some quantitative methods for Analyzing ethnobotanical knowledge. In: Alexiades MN (ed) Selected guidelines for ethnobotanical research: a field manual. New York Botanical Garden, New York, pp 171–197

Phillips O, Gentry AH (1993) The useful plants of Tambopata, Peru: statistical hypotheses tests with a new quantitative technique. Econ Bot 47:15–32

Prance GT, Balee W, Boom BM, Carneiro RL (1987) Quantitative ethnobotany and the case for conservation in Amazonia. Conserv Biol 1:296–310

Rao RS, Panigrahi G (1961) Distribution of vegetational types and their dominant species in Eastern India. J Ind Bot Soc 40:274–285

Sharma M, Das B (2018) Medicinal plants of northeast region of India: a small review. Int J Curr Pharm Res 10:11–12

Singh SP (2000) Studies on ethnobotany of North-Eastern Region of India. Phd Thesis. C. S. J. M. University, Kanpur

Stoffle RW, Halmo DB, Evans MJ, Olmsted JE (1990) Calculating the cultural significance of American Indian plants: Pauiute and Shoshone Ethnobotany at Yucca Mountain, Nevada. Am Anthropol 92(2):416–432

Tardìo J, Pardo-De-Santayana M (2008) Cultural importance indices: a comparative analysis based on the useful wild plants of southern Cantabria (Northern Spain). Econ Bot 62:24–39

Tardío J, Pascual H, Morales R (2005) Wild food plants traditionally used in the province of Madrid. Econ Bot 59(2):122–136

Tardío J, Blanco E, Carvalho AM, Lastra JJ, San Miguel E, Morales R (2007) Traditional knowledge on wild edible plants in the northwest of the Iberian Peninsula (Spain and Portugal): a comparative study. J Ethnobiol Ethnomed 3:27

Tynsong H, Dkhar M, Tiwari B (2020) Review: Traditional ecological knowledge of tribal communities of North East India. Biodiversitas J Biol Divers 21:3209–3224

Chapter 17
Trends, Gaps, and Policy Implications Towards Conservation of the Himalayan Biodiversity Hotspot

Kushankur Sarkar, Aaratrik Pal, Rajendra Yonzone, and Monoranjan Chowdhury

Abstract The Himalayan region, recognised as one of the world's most important biodiversity hotspots, harbors an unparalleled diversity of flora and fauna while providing vital ecosystem services. This chapter explores the current conservation trends shaping the region, including the establishment of protected areas, community-based conservation initiatives, climate adaptation strategies, technological advancements, and ecological restoration projects. Despite these efforts, significant gaps remain, such as fragmented ecosystems, policy incoherence across international borders, mass tourism, data deficiencies, and insufficient integration of indigenous knowledge systems. Case studies from Nepal, India, and Bhutan highlight successful conservation models, including community-led forest management, GIS-based monitoring, and watershed restoration. Policy implications emphasise the need for cross-border collaboration, climate-resilient policies, sustainable ecotourism regulations, and strengthened research initiatives. The chapter concludes by recommending holistic conservation strategies that balance ecological integrity with socio-economic development, ensuring long-term sustainability for both the environment and local communities.

Keywords Himalaya · Hotspot · Community conservation · Climate change adaptation · Protected areas · Policy implications

K. Sarkar · A. Pal
Taxonomy of Angiosperms and Biosystematics Laboratory, Department of Botany, University of North Bengal, Darjeeling, West Bengal, India

R. Yonzone
Department of Botany, Victoria Institution (College), Baithakkhana, Kolkata, West Bengal, India

M. Chowdhury (✉)
Department of Botany, University of North Bengal, Siliguri, West Bengal, India
e-mail: mchowdhury@nbu.ac.in

D. Das et al. (eds.), *Plant Diversity and Conservation Strategies of the Himalayan Biodiversity Hotspot*, Topics in Biodiversity and Conservation 21,
https://doi.org/10.1007/978-3-032-21697-7_17

17.1 Introduction

The Himalayan region, stretching across multiple countries and spanning over 7.5 lakh km^2, is one of the most significant biodiversity hotspots on Earth (Rawal et al. 2013; Premlata et al. 2024). It harbours an immense variety of flora and fauna, many of which are unique to this area. As one of the 36 global biodiversity hotspots, the Himalayas boast more than 10,000 plant species, 300 mammal species, and nearly 1000 bird species, playing a critical role in maintaining the ecological balance of the region (Apollo et al. 2021). Its ecosystems provide essential services such as water regulation, climate stabilisation, and carbon sequestration, all of which are crucial for the millions of people who rely on these resources for their survival. This region is not only ecologically but also culturally significant. The forests, wetlands, and grasslands of the Himalayas are crucial for regulating water systems and maintaining agricultural productivity in downstream areas (Rai et al. 2018). Additionally, local communities in the Himalayas, which include various indigenous groups, have preserved their cultural heritage through ethnomedicinal practices and spiritual connections to the land. Sacred groves, which are often preserved for religious purposes, serve as sanctuaries for biodiversity, protecting rare plant species and maintaining traditional ecological knowledge. These groves and community-based practices reflect the deep interconnection between culture, biodiversity, and conservation (Sarkar et al. 2023; Sarkar et al. 2025b). This perspective aligns with discussions in Chap. 1 (Sect. 1.3.1), which highlights Himalayan biodiversity patterns, and Chap. 2 (Sect. 2.1), which emphasises its global status as a biodiversity hotspot.

Despite its importance, the Himalayan hotspot faces growing environmental threats. Deforestation, infrastructure development, and agricultural expansion are leading to significant habitat destruction and fragmentation (Dhyani 2023). Climate change further exacerbates these threats, causing glacial melt, changing precipitation patterns, and altering ecosystems (Schickhoff et al. 2016). These environmental pressures not only threaten biodiversity but also destabilise local communities whose livelihoods depend on healthy ecosystems. Similar concerns are raised in Chap. 5 (Sect. 5.2), which discusses the vulnerability of alpine medicinal plants, and in Chap. 14 (Sect. 14.2), which highlights the dependence of local people on non-timber forest products. The geopolitical complexity of the Himalayas, spanning multiple countries, presents additional challenges to conservation efforts. Conservation policies are often fragmented, and transboundary issues like wildlife migration and shared river systems require cooperation between countries (Gurung et al. 2019). Yet, differing political priorities and economic pressures often hinder the effective implementation of cohesive conservation strategies. In this context, countries like Nepal India, and Bhutan have made notable progress in integrating conservation into national policies, particularly through community-based approaches, while other regions still struggle to protect their fragile ecosystems (Maraseni et al. 2019). Technological advancements, such as remote sensing and Geographic Information Systems (GIS), have revolutionised conservation efforts by enabling more accurate monitoring of land use changes, deforestation, and habitat loss. These innovations have improved the ability to track ecosystem health and predict future trends, allowing for more effective conservation strategies (Bhambri and Kautish 2024). Additionally, community involvement in conservation has become increasingly important. Local communities often have a deep understanding of their environment, and their participation in forest management, wildlife protection, and sustainable tourism is essential for long-term success (Nepal 1997). This emphasis on community engagement resonates with Chap. 15 (Sect. 15.1), which explores ethnobotanical traditions, and Chap. 13 (Sect. 13.1.2), which discusses community-forest interactions.

Given the growing environmental challenges, policy innovation is urgently needed to protect the Himalayan hotspot. Effective conservation strategies must balance environmental protection with socio-economic development, particularly in areas where local communities depend on natural resources. Cross-border collaboration, climate adaptation policies, and the inclusion of indigenous knowledge in formal conservation frameworks are essential steps for preserving the region's biodiversity. Without comprehensive and integrated conservation efforts, the future of the Himalayan hotspot and the millions of people who rely on its resources remain uncertain. This concern is consistent with Chap. 12 (Sect. 12.1), which addresses climate change impacts on ecosystems. The objective of this chapter is to analyse the pressing conservation issues in the Himalayan hotspot, assess the effectiveness of existing conservation strategies, and propose sustainable solutions for long-term ecological protection. It aims to provide a comprehensive understanding of biodiversity threats, conservation frameworks, and the role of scientific advancements in mitigating environmental challenges. It also emphasises the critical role of technology, local communities, and transboundary cooperation in safeguarding this ecologically and culturally significant region.

17.2 Geography and Climate of the Study Area

The Himalayan region is characterised by its extreme elevation gradient, ranging from low-lying foothills to some of the world's highest peaks, including Mount Everest (8849 m). This altitudinal variation profoundly influences the region's ecosystems, resulting in distinct ecological zones from subtropical forests at lower elevations to alpine meadows and barren high-altitude areas. Elevation plays a critical role in shaping the distribution of flora, fauna, and human settlements across the region. The elevation map illustrates the varying altitudinal features of the study area, highlighting key mountain ranges and valleys (Fig. 17.1).

The climate of the Himalayas is equally diverse, shaped by both elevation and seasonal monsoon patterns. Lower regions, especially in the eastern and central Himalayas, receive high annual rainfall during the monsoon, while higher elevations and western parts remain relatively arid. Temperature varies significantly with altitude, ranging from tropical climates in the foothills to freezing conditions at

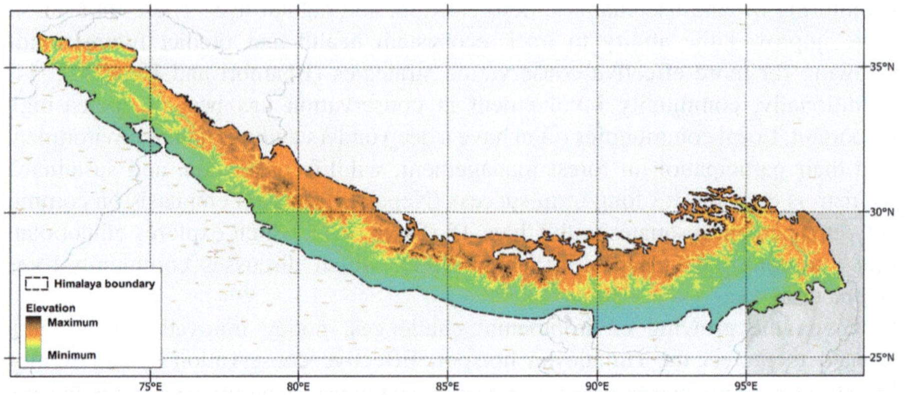

Fig. 17.1 Elevation map of the Himalayan region, depicting the altitudinal variations from low-lying foothills to the highest mountain peaks

higher altitudes. Annual maximum temperature for the region varies from 0 to 28 °C and annual minimum temperature varies from −25 to 7 °C. Whereas, annual rainfall varies from 238 mm to approx. 5000 mm per year. The spatial distribution of rainfall and temperature across the region is depicted in Fig. 17.2, providing a clear representation of climatic variability that impacts vegetation types, hydrology, and biodiversity (Zurick and Pacheco 2006). Chapter 12 (Sect. 12.1) also addresses climate change impacts on ecosystems.

17.3 Current Conservation Trends in the Himalayan Hotspot

The Himalayan region, known for its unique biodiversity, faces significant environmental pressures. Recent conservation efforts in the region reflect a multidimensional approach, focusing on protected area networks, community involvement, climate change mitigation, technological integration, and ecological restoration.

17.3.1 Protected Area Networks

The Himalayan hotspot encompasses numerous protected areas like biosphere reserves, national parks, wildlife sanctuaries along with several patches of reserve forests. The prominent UNESCO word heritage biosphere reserves are Nanda Devi Biosphere Reserve, Khangchendzonga Biosphere Reserve and Manas Biosphere Reserve located at unique altitudinal ranges (Chettri et al. 2008; Xu et al. 2019).

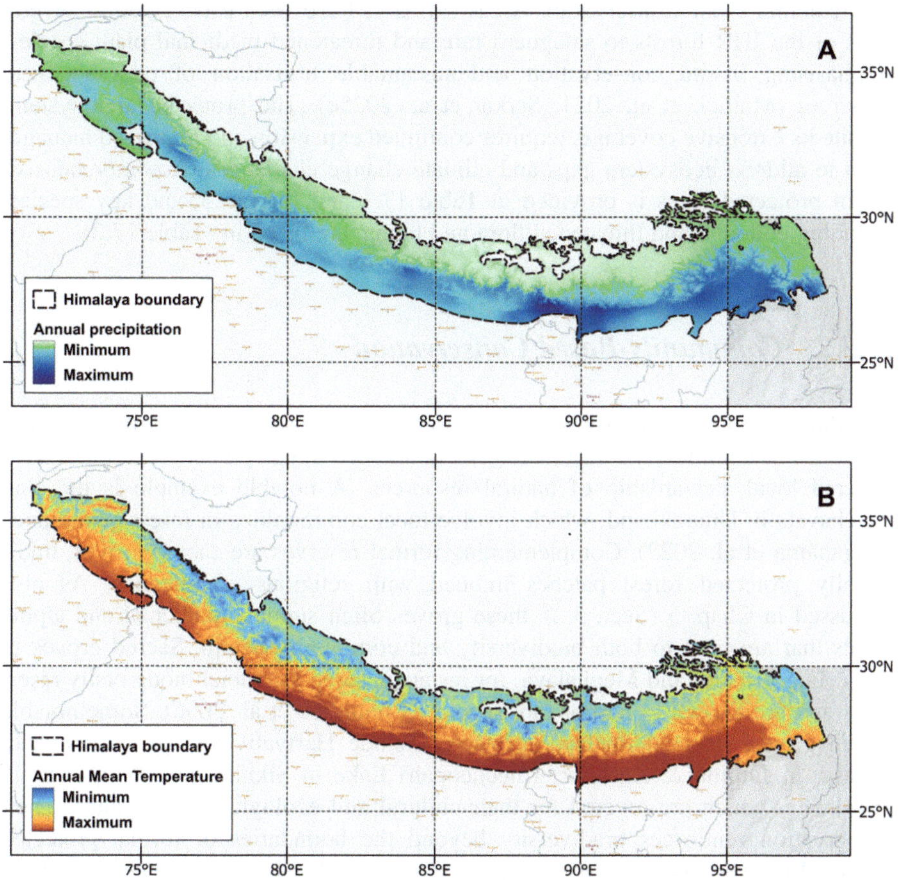

Fig. 17.2 Rainfall (**a**) and temperature (**b**) distribution map of the Himalayan region, illustrating the spatial variation in annual precipitation and temperature

These protected areas play a pivotal role in preserving endemic species and fragile ecosystems (Pringle 2017). The protected area network (PAN) in the Indian, Nepalese, and Bhutanese Himalayas covers a significant portion of the region, ensuring the conservation of diverse ecosystems and species. The Indian Himalayan Region (IHR) alone comprises numerous national parks and wildlife sanctuaries, covering approximately 9.6% of the geographical area, which is higher than the national average of 4.66% (Rawal and Dhar 2001). In Nepal, 22.5% of the country's land is designated as protected areas, including national parks, wildlife reserves, and conservation areas, but gap analyses reveal underrepresentation of mid-altitude ecosystems (Shrestha et al. 2010). Bhutan has one of the highest proportions of protected land in the world, with 51% of the country under protection, yet research suggests that additional conservation efforts are needed outside existing PAs to maintain habitat connectivity for multiple species (Penjor et al. 2021).

Medicinal Plant Conservation Areas (MPCA) have been established in several parts of the IHR forests to safeguard rare and threatened medicinal plant species, emphasising in-situ conservation and sustainable utilisation of ethnobotanical resources (Mallick et al. 2021; Sarkar et al. 2025c). The protected area system, despite its extensive coverage, requires continued expansion and improved management to address ecosystem gaps and climate change challenges. A comprehensive list of protected areas is provided in Table 17.1 and Fig. 17.3, and key species, including endemic and threatened flora and fauna, are listed in (Table 17.2).

17.3.2 Community-Based Conservation

Community-led conservation is gaining momentum in the Himalayan region. Initiatives such as community forest management and eco-development committees have fostered local stewardship of natural resources. A notable example is the Van Panchayats in Uttarakhand, which involve local communities in forest governance (Nagahama et al. 2022). Complementing formal reserves are sacred groves, traditionally protected forest patches imbued with religious significance. As also discussed in Chap. 5 (Sect. 5.2), these groves often support medicinal and alpine plants that are vital to both biodiversity and community health. Sacred groves in Darjeeling, Sikkim and Meghalaya, for instance, serve as crucial biodiversity reservoirs and are known to harbor rare species (Chatterjee et al. 2006). Some notable sacred groves in the Himalayan Hotspot include Hariyali Devi in Uttarakhand, Sanasar in Jammu & Kashmir, Khecheopalri Lake in Sikkim, and Kalipokhri in Darjeeling, which are revered for their cultural and ecological significance. Their conservation reinforces biodiversity beyond the boundaries of formal protected areas. Moreover, the revival of sacred groves by local communities' aids both cultural preservation and biodiversity protection (Mandal et al. 2021). The integration of traditional ecological knowledge with modern conservation practices is a critical aspect of community-based initiatives (O'Neill et al. 2017), and this linkage to livelihoods resonates with the themes of Chap. 14 (Sect. 14.2), which emphasises the importance of non-timber forest products.

17.3.3 Climate Change Mitigation

High-altitude ecosystems in the Himalayas are highly vulnerable to climate change, with visible impacts such as glacial retreat and altered precipitation patterns (Negi et al. 2022a, b). Consequently, climate adaptation and mitigation strategies are central to conservation efforts. Projects such as the "Glacier Lake Outburst Flood (GLOF) monitoring system" aim to reduce climate-related hazards. Additionally, agroforestry and reforestation programmes focusing on carbon sequestration are being promoted (Nair et al. 2009).

Table 17.1 Protected areas and ramsar sites in the Himalayan region and foothills of India, Nepal, and Bhutan

Country	Protected areas	Province/State	Area (km^2)
Bhutan	Jigme Singye Wangchuck National Park	Central Bhutan	1730
	Wangchuck Centennial National Park	Central Bhutan	4914
	Phrumsengla National Park	Central-Eastern Bhutan	905.5
	Bumdeling Wildlife Sanctuary	Eastern Bhutan	1520.61
	Jomotsangkha Wildlife Sanctuary	Eastern Bhutan	334.73
	Sakteng Wildlife Sanctuary	Eastern Bhutan	740.6
	Royal Manas National Park	Southern Bhutan	1057
	Phibsoo Wildlife Sanctuary	Southern Bhutan	268.93
	Jigme Dorji National Park	Western Bhutan	4316
	Jigme Khesar Strict Nature Reserve	Western Bhutan	609.51
	Motithang Takin Preserve	Western Bhutan	0.47
	Royal Botanical Park, Lamperi	Western Bhutan	47
India	Namdapha National Park	Arunachal Pradesh	1985
	Manas National Park	Assam	950
	Nameri National Park	Assam	200
	Dibru-Saikhowa National Park	Assam	765
	Great Himalayan National Park	Himachal Pradesh	1171
	Pin Valley National Park	Himachal Pradesh	675
	Bandli Wildlife Sanctuary	Himachal Pradesh	41
	Chail Wildlife Sanctuary	Himachal Pradesh	109
	Churdhar Wildlife Sanctuary	Himachal Pradesh	66
	Daranghati Wildlife Sanctuary	Himachal Pradesh	167
	Darlaghat Wildlife Sanctuary	Himachal Pradesh	6
	Dhauladhar Wildlife Sanctuary	Himachal Pradesh	944
	Gamgul Siahbehi Wildlife Sanctuary	Himachal Pradesh	109
	Gobind Sagar Bird Sanctuary	Himachal Pradesh	100
	Kalatop Khajjiar Wildlife Sanctuary	Himachal Pradesh	69
	Kanwar Wildlife Sanctuary	Himachal Pradesh	61
	Khokhan Wildlife Sanctuary	Himachal Pradesh	14
	Kias Wildlife Sanctuary	Himachal Pradesh	14
	Kibber Wildlife Sanctuary	Himachal Pradesh	1400
	Kugti Wildlife Sanctuary	Himachal Pradesh	379
	Lippa Asrang Wildlife Sanctuary	Himachal Pradesh	349
	Majathal Wildlife Sanctuary	Himachal Pradesh	57.55
	Manali Wildlife Sanctuary	Himachal Pradesh	32
	Naina Devi Wildlife Sanctuary	Himachal Pradesh	123
	Nargu Wildlife Sanctuary	Himachal Pradesh	278
	Pong Dam Lake Wildlife Sanctuary	Himachal Pradesh	307
	Rakcham Chitkul Wildlife Sanctuary	Himachal Pradesh	304
	Chandertal Wetland	Himachal Pradesh	0.49

(continued)

Table 17.1 (continued)

Country	Protected areas	Province/State	Area (km^2)
	Pong Dam Lake	Himachal Pradesh	156.6
	Renuka Wetland	Himachal Pradesh	0.20
	Renuka Wildlife Sanctuary	Himachal Pradesh	4
	Rupi Bhaba Wildlife Sanctuary	Himachal Pradesh	738
	Sainj Wildlife Sanctuary	Himachal Pradesh	90
	Sechu Tuan Nala Wildlife Sanctuary	Himachal Pradesh	103
	Shikari Devi Wildlife Sanctuary	Himachal Pradesh	72
	Shilli Wildlife Sanctuary	Himachal Pradesh	2
	Shimla Water Catchment Wildlife Sanctuary	Himachal Pradesh	10
	Simbalbara Wildlife Sanctuary	Himachal Pradesh	19
	Talra Wildlife Sanctuary	Himachal Pradesh	40
	Tirthan Wildlife Sanctuary	Himachal Pradesh	61
	Tundah Wildlife Sanctuary	Himachal Pradesh	64
	Kishtwar High Altitude National Park	Jammu & Kashmir	400
	Dachigam National Park	Jammu & Kashmir	141
	Hokera Wetland	Jammu & Kashmir	13.76
	Surinsar-Mansar Lakes	Jammu & Kashmir	3.5
	Wular Lake	Jammu & Kashmir	189
	Hemis National Park	Ladakh	4400
	Tso Moriri	Ladakh	120
	Tso Kar Wetland Complex	Ladakh	95.77
	Barsey Rhododendron Sanctuary	Sikkim	104
	Fambonglho Wildlife Sanctuary	Sikkim	51.76
	Khangchendzonga National Park	Sikkim	1784
	Kitam Bird Sanctuary	Sikkim	6
	Kyongnosla Alpine Sanctuary	Sikkim	31
	Maenam Wildlife Sanctuary	Sikkim	35.34
	Pangolakha Wildlife Sanctuary	Sikkim	128
	Shingba Rhododendron Sanctuary	Sikkim	43
	Valley of Flowers National Park	Uttarakhand	87.5
	Nanda Devi National Park	Uttarakhand	630
	Gangotri National Park	Uttarakhand	2390
	Govind Pashu Vihar National Park	Uttarakhand	958
	Askot Musk Deer Sanctuary	Uttarakhand	599.93
	Binsar Wildlife Sanctuary	Uttarakhand	45.59
	Kedarnath Wildlife Sanctuary	Uttarakhand	975.24
	Mussoorie Wildlife Sanctuary	Uttarakhand	10.82
	Nandhaur Wildlife Sanctuary	Uttarakhand	269.95
	Sonanadi Wildlife Sanctuary	Uttarakhand	301.18
	Asan Conservation Reserve	Uttarakhand	4.44
	Neora Valley National Park	West Bengal	88

(continued)

Table 17.1 (continued)

Country	Protected areas	Province/State	Area (km^2)
	Singalila National Park	West Bengal	78
	Mahananda Wildlife Sanctuary	West Bengal	159
	Buxa Tiger Reserve	West Bengal	760
	Senchal Wildlife Sanctuary	West Bengal	39
	Jaldapara National Park	West Bengal	216
	Gorumara National Park	West Bengal	80
	Chapramari Wildlife Sanctuary	West Bengal	9.6
	Jore Pokhri Wildlife Sanctuary	West Bengal	2
Nepal	Chitwan National Park	Bagmati & Madhesh Province	932
	Langtang National Park	Bagmati Province	1710
	Shivapuri Nagarjun National Park	Bagmati Province (Kathmandu Valley)	159
	Annapurna Conservation Area	Gandaki Province	7629
	Manaslu Conservation Area	Gandaki Province	1663
	Rara National Park	Karnali Province	106
	Shey Phoksundo National Park	Karnali Province	3555
	Bardiya National Park	Lumbini Province	968
	Banke National Park	Lumbini Province	550
	Parsa National Park	Madhesh & Bagmati Province	637
	Sagarmatha National Park	Province No. 1	1148
	Koshi Tappu Wildlife Reserve	Province No. 1	175
	Makalu Barun National Park	Province No. 1	1500
	Kanchenjunga Conservation Area	Province No. 1	2035
	Khaptad National Park	Sudurpashchim & Karnali Province	225
	Shuklaphanta National Park	Sudurpashchim Province	305

17.3.4 Restoration Projects

Restoration projects in the Himalayan Hotspot are focused on degraded forest landscapes, rivers, and watersheds. Initiatives like reforestation in the Indian Himalayan region and wetland restoration aim to revive ecosystems and enhance biodiversity (Mallick and Chakraborty 2018). Sacred groves are also undergoing restoration, with efforts to restore their traditional ecological functions, and as highlighted in Chap. 5 (Sect. 5.2), these groves often harbour medicinal and alpine plants that make their protection ecologically significant. These restoration efforts are often community-driven, involving active participation from local populations (Verma et al. 2021), a theme that also connects to Chap. 14 (Sect. 14.2), which discusses how non-timber forest products link biodiversity conservation with community livelihoods.

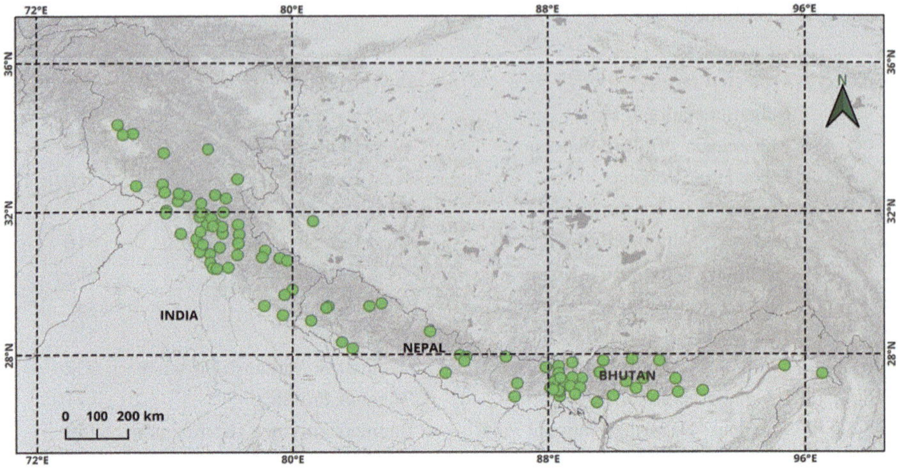

Fig. 17.3 Map showing the protected area network across the Himalayan region

17.4 Major Gaps in Conservation in the Himalayan Hotspot

Despite various conservation initiatives, significant gaps persist in the effective management of the Himalayan ecosystem. These gaps hinder long-term sustainability and highlight the need for more integrated and strategic conservation efforts.

17.4.1 Fragmented Ecosystems

Forest patches occur as a result of natural processes, e.g., forest expansion, succession, etc. Modifying the land surface is a part of the human survival process, and it can also result in the creation of forest patches (Foley et al. 2005; Matte et al. 2015). Unfortunately, several human activities, including clearing for agricultural practices and urbanisation, caused increased forest fragmentation (Haddad et al. 2015). Anthropogenic activities, coupled with natural factors, contribute to ecosystem fragmentation in the Himalayan landscapes (Sharma and Roy 2007). Rapid deforestation and forest fragmentation not only lead to the reduction of species-rich communities in dry tropical forest regions, but they also become species-poor and less diverse. Eventually, increased anthropogenic disturbance changes community organisation; for example, invasive *Lantana camera* is changing forest structure heavily, despite its shrubby nature; it changes the understorey of large tree ecosystems (Raghubanshi and Tripathi 2009). Even forest fragmentation is not an exception for Khangchendzonga Biosphere Reserve, Sikkim, India. Sharma et al. (2016) reported a decrease in large core (5.93%), edge (3.64%), and patches (0.66%) areas

Table 17.2 Some key species that contribute to the rich biodiversity of the Himalayan hotspot

Common name	Scientific name	Primary locations
Assam Roofed Turtle	*Pangshura sylhetensis*	Brahmaputra-Meghna drainage systems in India and parts of Bhutan.
Bengal Florican	*Houbaropsis bengalensis*	Grasslands of the Terai in Nepal and parts of Assam (India).
Blue Sheep (Bharal)	*Pseudois nayaur*	High-altitude terrains across the Himalayas, including India, Nepal, and Bhutan.
Brahma Kamal	*Saussurea obvallata*	Alpine meadows of the Himalayas
Himalayan Cucurbit	*Herpetospermum darjeelingense*	Eastern Himalayas, Nepal, Tibet, Western Himalayas
Golden Langur	*Trachypithecus geei*	Foothills of the Eastern Himalayas, particularly in Bhutan and Assam (India).
Himalayan Blue Poppy	*Meconopsis baileyi*	Alpine meadows in Arunachal Pradesh (India) and adjacent regions.
Himalayan Brown Bear	*Ursus arctos isabellinus*	Northern India and Nepal.
Himalayan Calamus	*Calamus pseudoerectus*	Eastern Himalayas
Himalayan Griffon Vulture	*Gyps himalayensis*	High-altitude regions across the Himalayas.
Himalayan Knotweed	*Koenigia polystachya*	Alpine meadows and forest edges in the Himalayas.
Himalayan Mayapple	*Podophyllum hexandrum*	Forest floors in the Himalayas, including India, Nepal, and Bhutan.
Himalayan Monal	*Lophophorus impejanus*	Forested areas of the Himalayas in India and Nepal.
Himalayan Musk Deer	*Moschus leucogaster*	Alpine forests and meadows in Nepal, Bhutan, and northern India.
Himalayan Quail	*Ophrysia superciliosa*	Western Himalayas in Uttarakhand (India); possibly extinct.
Himalayan Salamander	*Tylototriton verrucosus*	Eastern Himalayas, particularly in Darjeeling (India) and surrounding areas.
Himalayan Wolf	*Canis lupus chanco*	High-altitude regions of India and Nepal.
Himalayan Yew	*Taxus wallichiana*	Temperate forests in the Himalayas, including India, Nepal, and Bhutan.
Mahananda Calamus	*Calamus mahanandensis*	Eastern Himalayas
Red Panda	*Ailurus fulgens*	Eastern Himalayas, notably in Sikkim (India), Bhutan, and Nepal.
Rhododendron (various spp.)	*Rhododendron spp.*	Sub-alpine and alpine regions throughout the Himalayas.
Snow Leopard	*Panthera uncia*	High-altitude regions across the Himalayas, including India, Nepal, and Bhutan.
Takin	*Budorcas taxicolor*	Forested valleys and rocky terrains in Bhutan, northeastern India.

in the protected landscape during 2000–2009, while an increase in non-forest (6.59%) and the perforated regions (4.01%). That includes a land cover reduction in cases of open forest (2.81%), alpine scrub (0.39%), alpine meadows (8.81%), snow (3.46%), and hill shadow areas (0.60%).

17.4.2 Megaprojects

Despite the reputation of hydroelectric projects (HEPs) as cheap, green, and renewable, recent studies showed that HEPs do impact negatively in multiple ways, including socio-economic, environmental and geopolitical (Choudhury and Dey Choudhury 2021). Most of the proposed as well as constructed dams are planned to be in major river valleys of the Indian Himalayan landscapes, and therefore correlate with the biodiversity, including angiosperms, birds, insects, fishes, and reptiles. Among these dams, several are in relatively undisturbed, dense forests. Submergence of these forests due to dam building contribute directly to the habitat degradation (Grumbine and Pandit 2013). In addition to dams, the construction of large-scale rail and road tunnels in the Himalayas has significantly altered fragile mountain ecosystems. The excavation of tunnels for highway expansions and railway corridors, such as the Char Dham Highway Project and the Rishikesh-Karnaprayag Railway, has triggered landslides, deforestation, and groundwater depletion. Blasting and drilling for tunnel construction disturb geological stability, increasing the risk of land subsidence, as seen in Joshimath, where unregulated tunneling has been linked to widespread land sinking (Awasthi et al. 2024). However, these tunnels also serve as critical infrastructure, enhancing connectivity, reducing travel time, and facilitating economic growth in remote and strategically significant regions. Improved road and rail networks are essential for disaster response, military mobilisation, and regional development, particularly in border states like Sikkim, Arunachal Pradesh and Ladakh (Mishra et al. 2024). Additionally, tunnels help mitigate extreme weather disruptions, providing year-round access to areas prone to landslides and heavy snowfall. Nevertheless, the long-term ecological consequences of these megaprojects pose serious challenges for conservation efforts in the Himalayas. Balancing infrastructure development with environmental sustainability remains a key priority for policymakers and conservationists (Awasthi et al. 2024; Yadawa 2024).

17.4.3 Policy Gaps

A significant gap in conservation policy lies in the lack of cohesive cross-border strategies among the Himalayan nations—India, Nepal, Bhutan, China, and Pakistan (Mishra et al. 2019). Since the Himalayan ecosystem spans multiple countries, transboundary ecosystems require coordinated management. However, fragmented national policies and limited collaboration hinder efforts to preserve shared biodiversity (Sharma et al. 2007). For example, the management of species with large home ranges, such as the snow leopard, is often compromised by inconsistent policies across borders.

17.4.4 Biodiversity Loss and Data Deficiency

Biodiversity loss in the Himalayas remains a critical issue, further complicated by inadequate data on less-studied and endemic species (Sharma et al. 2010). Many taxa, particularly invertebrates, fungi, lichens, and alpine flora, are under-researched, resulting in insufficient information for effective conservation planning. For instance, recent work highlights how lichens, which serve as sensitive bioindicators of air pollution, show significant vulnerability to environmental stressors in the region, yet remain largely excluded from mainstream biodiversity assessments (Sarkar et al. 2025a). Additionally, existing biodiversity databases often lack updated and region-specific records, reducing their effectiveness in tracking species distributions and population trends (Wani et al. 2023). Addressing these gaps requires enhanced investment in systematic biodiversity surveys, incorporation of indicator taxa like lichens, and broader ecological research to inform evidence-based conservation strategies.

17.4.5 Funding and Resource Allocation

Funding shortages pose a significant barrier to the successful implementation of conservation projects in the region. Many initiatives are reliant on short-term grants, which limit their scope and fail to ensure long-term sustainability. Additionally, the lack of adequate financial resources results in poorly maintained protected areas and insufficient staffing for monitoring and enforcement (Uprety et al. 2021). Furthermore, there is a significant underutilisation of resource persons and specialists across different segments such as biodiversity research, ecosystem management, climate adaptation, and community engagement, which further weakens conservation effectiveness. Increasing both domestic and international funding, as well as encouraging public-private partnerships and interdisciplinary collaborations, is critical to addressing this gap and ensuring a more robust and scientifically informed conservation framework.

17.4.6 Human-Wildlife Conflict

Human-wildlife conflict has intensified in the Himalayan region due to habitat encroachment and fragmentation (Sharma et al. 2020). The expansion of human settlements and agricultural lands into wildlife habitats has increased encounters with species such as leopards, elephants, and bears. These interactions often result in crop losses, livestock predation, and even human fatalities, leading to retaliatory killings of wildlife (Lamichhane et al. 2018). Although some mitigation measures, such as fencing and compensation schemes, are in place, they remain inadequate due to limited policy support and funding.

17.4.7 Mass Tourism Pressure

Mass tourism in the Himalayas has caused extensive ecological degradation, resource stress, and socio-economic challenges (Ziegler et al. 2021). The Joshimath disaster, marked by land subsidence, is directly linked to unregulated tourism and rapid infrastructural expansion in fragile terrains (Awasthi et al. 2024). Popular sites like Everest Base Camp, Kedarnath, and Mount Kailash face soil erosion, deforestation, and biodiversity loss due to heavy tourist footfall (Apollo 2017; Huddart et al. 2020). High-altitude towns such as Leh and Namche Bazaar suffer acute water shortages, with tourist water use exceeding local household consumption by threefold during peak seasons (KC and Thapa Parajuli 2015; Das et al. 2022). Waste pollution has intensified, with plastics and trekking litter accumulating in sensitive pilgrimage sites (Nyaupane and Timothy 2022). Microplastics in glaciers like Gangotri are accelerating melt and disrupting hydrology (Ashraf et al. 2024). Urban centers in the Indian Himalayas generate 6346 tonnes of solid waste daily, led by Jammu & Kashmir (1792 t/day) and Uttarakhand (1528 t/day) (Tiwari and Rani 2024). Additionally, the carbon footprint from transport and black carbon deposits is accelerating glacial retreat and water insecurity (Riaz et al. 2019).

17.5 Policy Implications and Recommendations for Conservation

To address the pressing environmental challenges faced by the Himalayan Hotspot, strategic policy measures are essential. These policies must emphasise regional cooperation, climate resilience, sustainable development, ecotourism regulation, indigenous knowledge integration, and research enhancement to ensure effective conservation.

17.5.1 Climate Action Policies

The Himalayas, being highly sensitive to climate change, require robust adaptation policies to mitigate the impacts of rising temperatures, glacial melt, and altered precipitation patterns. Policies must focus on developing ecosystem-based adaptation strategies and supporting local communities in building climate resilience. Strengthening early warning systems for climate-induced disasters such as glacial lake outburst floods (GLOFs) is also crucial (Sedai 2021). Incentivising carbon sequestration through reforestation and agroforestry programmes can further contribute to climate action (Sarkar et al. 2024).

17.5.2 Linking Conservation to Sustainable Development Goals (SDGs)

Conservation strategies in the Himalayan region should be closely aligned with the global Sustainable Development Goals (NITI Aayog 2018a). Integrating biodiversity conservation with poverty reduction (SDG 1), promoting gender equality in conservation leadership (SDG 5), and climate action (SDG 13) can foster sustainable livelihoods and enhance local participation. Ecotourism can be linked with SDG 8 (decent work and economic growth), provided it is well-regulated to prevent environmental degradation.

17.5.3 Strengthening Ecotourism Regulations

While ecotourism has the potential to support local economies and conservation, unregulated tourism can harm fragile ecosystems (Tiwari and Rani 2024). Stronger policies are needed to regulate ecotourism, focusing on limiting visitor numbers, enforcing waste management protocols, and promoting low-impact tourism models. Certification schemes for sustainable tourism operators should be introduced to ensure adherence to conservation standards (NITI Aayog 2018b).

17.5.4 Strengthening Research and Data Collection

Enhancing biodiversity research and establishing comprehensive ecological databases are essential for evidence-based policy formulation. Policy initiatives should prioritise funding for research on lesser-known species, forest ecosystems, and water resources. Improved data collection will support habitat restoration, species protection, and sustainable development efforts. Partnerships with academic institutions and international research organisations can accelerate data-driven conservation outcomes (Paygude et al. 2023). By implementing these policy recommendations, the Himalayan countries can foster a holistic conservation approach that addresses ecological, social, and economic dimensions, ensuring the long-term sustainability of this critical global biodiversity hotspot.

17.6 Case Studies and Best Practices in Conservation

Efforts in the Himalayan Hotspot have yielded numerous successful case studies and best practices, particularly in community-led conservation, technology-driven monitoring, and ecological restoration. These examples demonstrate effective strategies that can be replicated across the region.

17.6.1 Community-Led Conservation Success Stories

Bhutan's Gross National Happiness (GNH)-aligned conservation policies emphasise community stewardship. Community forests, covering over 2% of the country's land, are managed by local groups, ensuring sustainable resource use while enhancing forest cover (Sears et al. 2017). In Nepal, the Annapurna Conservation Area Project (ACAP) has empowered local communities to manage natural resources, resulting in increased wildlife populations and improved livelihoods through regulated ecotourism (Bajracharya et al. 2006).

17.6.2 Application of Remote Sensing and GIS in Conservation

Technological advancements have played a crucial role in conservation by providing accurate and real-time data for monitoring ecosystems. In the Indian Himalayan Region, remote sensing and Geographic Information System (GIS) tools have been used to monitor deforestation, habitat degradation, and land-use changes (Bera et al. 2021). Various projects employed Normalised Difference Vegetation Index (NDVI) analysis and Land Use Land Cover (LULC) models to assess changes in forest cover (Chakraborty et al. 2016; Husain et al. 2023). These results were used to guide reforestation efforts and protect critical wildlife corridors.

17.6.3 Successful Ecological Restoration Projects

Successful ecological restoration projects have been implemented across the Himalayas, focusing on degraded forests, grasslands, and watersheds. In Uttarakhand, a land restoration initiative was taken to improve degraded ecosystems by integrating scientific baseline data on soil and land attributes into decision-making, adopting context-specific strategies, promoting traditional knowledge alongside scientific approaches, supporting community-driven reforestation and sustainable agriculture, and fostering collaboration among various departments and institutions for effective implementation and resource management (Negi et al. 2022a, b). The Sikkim Springshed Rejuvenation Initiative, also known as "Dhara Vikas programme", is another notable example. By restoring natural springs and their catchments, the project has ensured year-round water availability for local communities, benefiting both agriculture and biodiversity. This initiative has become a model for spring revival across the Himalayan region (Tambe et al. 2012).

17.7 Conclusions

The Himalayan region, with its unparalleled biodiversity and vital ecosystem services, faces complex conservation challenges. Balancing ecological preservation with socio-economic development requires innovative, collaborative, and adaptive approaches. Effective conservation must transcend national boundaries, promote the participation of local communities, and harness both traditional knowledge and advanced technologies to safeguard the region's ecological integrity. Cross-border cooperation is pivotal for the sustainable management of shared resources, emphasising the need for cohesive policies across nations. Community-led initiatives demonstrate the potential for local stewardship to enhance conservation outcomes, highlighting the importance of integrating indigenous knowledge into formal strategies. Moreover, the growing use of advanced monitoring tools such as remote sensing and GIS underscores the role of technology in tracking environmental changes and guiding restoration efforts. While policy frameworks are evolving to address climate adaptation, ecological restoration, and sustainable development, persistent gaps particularly in funding, research, and data collection hinder long-term effectiveness. Addressing these gaps through increased investment, policy coherence, and stakeholder engagement is essential for ensuring the sustainability of conservation efforts. The way forward involves a holistic approach that incorporates ecological, social, and economic dimensions into conservation planning. By building on successful models and addressing existing gaps, stakeholders can work towards a future where both biodiversity and human well-being are preserved. Strengthening partnerships, enhancing research, and fostering long-term commitment from all sectors will be key to sustaining the ecological richness and cultural heritage of the Himalayan Hotspot.

Acknowledgements The authors express sincere gratitude to the Department of Botany, University of North Bengal, for providing the academic support, resources, and encouragement necessary for the completion of this work. They also recognise the use of open-source databases, remote sensing products, and published literature that have significantly enriched this study.

References

Apollo M (2017) The true accessibility of mountaineering: the case of the high Himalaya. J Outdoor Recreat Tour 17:29–43

Apollo M, Andreychouk V, Mostowska J, Jin Z, Maciuk K, Rettinger R, Jones TE (2021) Himalayan nature-based tourism. Potential, regional diversity, nature conservation and touristic load. Int J Conserv Sci 12(4):1527–1546

Ashraf MV, Misri S, Pant S, Ahmad S (2024) Unorganized tourism practices and the allied challenges in the Himalayas. In: Borthakur A, Singh P (eds) The resilience of traditional knowledge systems for a sustainable future, Advances in Asian human-environmental research. Springer, Cham. https://doi.org/10.1007/978-3-031-56858-9_9

Awasthi S, Jain K, Sahoo S, Kumar R, Goswami A, Joshi GC, Kulkarni AV, Srivastava DC (2024) Analyzing Joshimath's sinking: causes, consequences, and future prospects with remote sensing techniques. Sci Rep 14(1):10876

Bajracharya SB, Furley PA, Newton AC (2006) Impacts of community-based conservation on local communities in the Annapurna Conservation Area, Nepal. Biodivers Conserv 15:2765–2786

Bera A, Mukhopadhyay BP, Chowdhury P, Ghosh A, Biswas S (2021) Groundwater vulnerability assessment using GIS-based DRASTIC model in Nangasai River Basin, India with special emphasis on agricultural contamination. Ecotoxicology and Environmental Safety 214:112085

Bhambri P, Kautish S (2024) Technological advancements in promoting ecosystem health. In: Leal Filho W, Kautish S, Wall T, Rewhorn S, Paul SK (eds) Digital technologies to implement the UN sustainable development goals, World sustainability series. Springer, Cham. https://doi.org/10.1007/978-3-031-68427-2_21

Chakraborty A, Sachdeva K, Joshi PK (2016) Mapping long-term land use and land cover change in the central Himalayan region using a tree-based ensemble classification approach. Appl Geogr 74:136–150

Chatterjee S, Saikia A, Dutta P, Ghosh D, Pangging G, Goswami AK (2006) Biodiversity significance of North East India, vol 30. WWF-India, New Delhi, pp 1–71

Chettri N, Shakya B, Thapa R, Sharma E (2008) Status of a protected area system in the Hindu Kush-Himalayas: an analysis of PA coverage. Int J Biodivers Sci Manag 4(3):164–178

Choudhury NB, Dey Choudhury SR (2021) Implications for planning of hydroelectric projects in Northeast India: an analysis of the impacts of the Tipaimukh project. GeoJournal 86:1999–2019. https://doi.org/10.1007/s10708-020-10158-8

Das MM, Barfal SS, Joshi M, Joshi R, Kumar K, Kumar D (2022) Review on water scarcity across Indian Himalayan Region. In: Facets of a Smart City: computational and experimental techniques for sustainable urban development. Bentham Sci Pub, Sharjah, pp 92–102

Dhyani S (2023) Are Himalayan ecosystems facing hidden collapse? Assessing the drivers and impacts of change to aid conservation, restoration and conflict resolution challenges. Biodivers Conserv 32:3731–3764. https://doi.org/10.1007/s10531-023-02692-x

Foley JA, DeFries R, Asner GP, Barford C, Bonan G, Carpenter SR et al (2005) Global consequences of land use. Science 309(5734):570–574

Grumbine RE, Pandit MK (2013) Threats from India's Himalaya dams. Science 339(6115):36–37

Gurung J, Chettri N, Sharma E, Ning W, Chaudhary RP, Badola HK, Wangchuk S, Uprety Y, Gaira KS, Bidha N, Phuntsho K (2019) Evolution of a transboundary landscape approach in the Hindu

Kush Himalaya: key learnings from the Kangchenjunga Landscape. Glob Ecol Conserv 17: e00599
Haddad NM, Brudvig LA, Clobert J, Davies KF, Gonzalez A, Holt RD et al (2015) Habitat fragmentation and its lasting impact on Earth's ecosystems. Sci Adv 1(2):e1500052
Huddart D, Stott T, Huddart D, Stott T (2020) Adventure tourism in the Himalayas. In: Adventure tourism: environmental impacts and management. Springer International Publishing, Cham, pp 241–289
Husain MA, Kumar P, Gonencgil B (2023) Assessment of spatio-temporal land use/cover change and its effect on land surface temperature in Lahaul and Spiti, India. Land 12(7):1294
KC A, Thapa Parajuli RB (2015) Climate change and its impact on tourism in the Manaslu conservation area, Nepal. Tour Plan Dev 12(2):225–237
Lamichhane BR, Persoon GA, Leirs H, Poudel S, Subedi N, Pokheral CP, Bhattarai S, Thapaliya BP, De Iongh HH (2018) Spatio-temporal patterns of attacks on human and economic losses from wildlife in Chitwan National Park, Nepal. PLoS One 13(4):e0195373
Mallick PH, Chakraborty SK (2018) Forest, wetland and biodiversity: revealing multi-faceted ecological services from ecorestoration of a degraded tropical landscape. Ecohydrol Hydrobiol 18(3):278–296
Mallick D, Dasgupta S, Paul P, Mondal S, Pal A, Chowdhury M (2021) Diversity of arboreal spermatophytes in three MPCAs of India: with special reference to West Bengal. Indian Forester 147(12):1145–1154
Mandal P, Roy P, Sarkar K (2021) Pristine treasure trove: exploring sacred groves of Uttar Dinajpur, West Bengal, India. Lambert Academic Publishing, Maldova
Maraseni TN, Bhattarai N, Karky BS, Cadman T, Timalsina N, Bhandari TS, Apan A, Ma HO, Rawat RS, Verma N, San SM (2019) An assessment of governance quality for community-based forest management systems in Asia: prioritisation of governance indicators at various scales. Land Use Policy 81:750–761
Matte ALL, Müller SC, Becker FG (2015) Forest expansion or fragmentation? Discriminating forest fragments from natural forest patches through patch structure and spatial context metrics. Austral Ecol 40(1):21–31
Mishra A, Appadurai AN, Choudhury D, Regmi BR, Kelkar U, Alam M, Chaudhary P, Mu SS, Ahmed AU, Lotia H, Fu C (2019) Adaptation to climate change in the Hindu Kush Himalaya: stronger action urgently needed. In: Wester P, Mishra A, Mukherji A, Shrestha A (eds) The Hindu Kush Himalaya assessment. Springer, Cham. https://doi.org/10.1007/978-3-319-92288-1_13
Mishra V, Thakur A, Nautiyal D, Sharma A, Bhardwaj S, Prakash S, Rajput B, Sharma RS (2024) Himalayan dam projects in India and their implications: a deep dive into the Dhauliganga hydropower initiative. In: Borthakur A, Singh P (eds) The Himalayas in the anthropocene. Springer, Cham. https://doi.org/10.1007/978-3-031-50101-2_3
Nagahama K, Tachibana S, Rakwal R (2022) Critical aspects of people's participation in community-based forest management from the case of Van Panchayat in Indian Himalaya. Forests 13(10):1667
Nair PKR, Mohan Kumar B, Nair VD (2009) Agroforestry as a strategy for carbon sequestration. J Plant Nutr Soil Sci 172(1):10–23
Negi VS, Pathak R, Dhyani V, Durgapal M, Joshi RK, Bhatt ID (2022a) Land restoration in the Himalayan Region: steps towards biosphere integrity. Land Use Policy 121:106317
Negi VS, Tiwari DC, Singh L, Thakur S, Bhatt ID (2022b) Review and synthesis of climate change studies in the Himalayan region. Environ Dev Sustain 24(9):10471–10502
Nepal SK (1997) Sustainable tourism, protected areas and livelihood needs of local communities in developing countries. Int J Sustain Dev World Ecol 4(2):123–135
NITI Aayog (2018a) Contributing to sustainable development in the Indian Himalayan region. NITI Aayog, Government of India, New Delhi
NITI Aayog (2018b) Sustainable tourism in the Indian Himalayan region. Report of working group II. NITI Aayog, Government of India, New Delhi

Nyaupane G, Timothy DJ (2022) Tourism and development in the Himalaya. Routledge, London
O'Neill AR, Badola HK, Dhyani PP, Rana SK (2017) Integrating ethnobiological knowledge into biodiversity conservation in the Eastern Himalayas. J Ethnobiol Ethnomed 13:1–4
Paygude A, Jose VS, Shamoon A (2023) Leveraging on technology-driven information systems for conservation through informed decisions in the Hindu Kush Himalayas. In: Kumar A, De Jong W, Kumar M, Pandey R (eds) Climate change in the Himalayas. Academic. https://doi.org/10.1016/B978-0-443-19415-3.00003-7
Penjor U, Kaszta Ż, Macdonald DW, Cushman SA (2021) Prioritizing areas for conservation outside the existing protected area network in Bhutan: the use of multi-species, multi-scale habitat suitability models. Landsc Ecol 36:1281–1309
Premlata KR, Hajam YA, Giri A (2024) Threats, challenges, and conservation strategies of Himalayan faunal biodiversity. In: Sobti RC (ed) Role of science and technology for sustainable future. Springer, Singapore. https://doi.org/10.1007/978-981-97-0710-2_19
Pringle RM (2017) Upgrading protected areas to conserve wild biodiversity. Nature 546(7656): 91–99
Raghubanshi AS, Tripathi A (2009) Effect of disturbance, habitat fragmentation and alien invasive plants on floral diversity in dry tropical forests of Vindhyan highland: a review. Trop Ecol 50(1): 57
Rai R, Zhang Y, Paudel B, Acharya BK, Basnet L (2018) Land use and land cover dynamics and assessing the ecosystem service values in the trans-boundary Gandaki River Basin, Central Himalayas. Sustainability 10(9):3052
Rawal RS, Dhar U (2001) Protected area network in Indian Himalayan region: need for recognizing values of low profile protected areas. Curr Sci 25:175–184
Rawal RS, Bhatt ID, Chandra Sekar K, Nandi SK (2013) The Himalayan biodiversity: richness, representativeness, uniqueness and life support values. G.B. Pant Institute of Himalayan Environment and Development (GBPIHED), Kosi-Katarmal, Almora, p 84
Riaz R, Ali U, Li J, Zhang G, Alam K, Sweetman AJ, Jones KC, Malik RN (2019) Assessing the level and sources of Polycyclic Aromatic Hydrocarbons (PAHs) in soil and sediments along Jhelum riverine system of lesser Himalayan region of Pakistan. Chemosphere 216:640–652
Sarkar K, Roy P, Panda S, Choudhuri C, Chowdhury M (2023) Ethnomedicinal study on plant resources from sacred groves of Dakshin Dinajpur district, West Bengal, India. Ethnobot Res Appl 25:1–35
Sarkar K, Barman T, Choudhuri C, Chowdhury M (2024) Carbon storage in sacred groves of Uttar Dinajpur: implications for climate change mitigation and conservation strategies. Environ Dev Sustain. https://doi.org/10.1007/s10668-024-05454-z
Sarkar K, Ishore A, Chowdhury M, Bhattacharya A (2025a) Assessing the impacts of air pollution on lichens of Darjeeling through air pollution tolerance index. Biol Bull 52(2):45
Sarkar K, Dey S, Farhat Z, Kumar A, Choudhuri C, Chowdhury M (2025b) Ag/ZnO nanodots from Artemisia austroyunnanensis as antibacterial and anticancer agents against human neuroblastoma cells. Scientific Reports 15(1):22598.
Sarkar K, Mondal S, Mallick D, Chowdhury M (2025c) Ecological Impact of Tropical Invasive Alien Species on the Indigenous Vegetation of Sursuti MPCA of Gorumara National Park, India. In: Roy S, Mathur P, Chowdhury M (eds) Recent Trends and Applications in Plants, Microbes and Agricultural Sciences. APMAS 2023. Sustainable Landscape Planning and Natural Resources Management. Springer, Cham. https://doi.org/10.1007/978-3-031-90139-3_6
Schickhoff U, Singh RB, Mal S (2016) Climate change and dynamics of glaciers and vegetation in the Himalaya: an overview. In: Singh R, Schickhoff U, Mal S (eds) Climate change, glacier response, and vegetation dynamics in the Himalaya. Springer, Cham. https://doi.org/10.1007/978-3-319-28977-9_1
Sears R, Phuntsho S, Dorji T, Choden K, Norbu N, Baral H (2017) Forest ecosystem services and the pillars of Bhutan's gross national happiness. Center for International Forestry Research CIFOR, Bogor

Sharma S, Roy PS (2007) Forest fragmentation in the Himalaya: a central Himalayan case study. Int J Sustain Dev World Ecol 14(2):201–210
Sharma E, Chettri N, Gurung J, Shakya B (2007) Landscape approach in biodiversity conservation: a regional cooperation framework for implementation of the convention on biological diversity in Kangchenjunga landscape. ICIMOD, Kathmandu, p 29
Sharma E, Chettri N, Oli KP (2010) Mountain biodiversity conservation and management: a paradigm shift in policies and practices in the Hindu Kush-Himalayas. Ecol Res 25:909–923
Sharma M, Areendran G, Raj K, Sharma A, Joshi PK (2016) Multitemporal analysis of forest fragmentation in Hindu Kush Himalaya - a case study from Khangchendzonga Biosphere Reserve, Sikkim, India. Environ Monit Assess 188(10):596
Sharma P, Chettri N, Uddin K, Wangchuk K, Joshi R, Tandin T, Pandey A, Gaira KS, Basnet K, Wangdi S, Dorji T (2020) Mapping human-wildlife conflict hotspots in a transboundary landscape, Eastern Himalaya. Glob Ecol Conserv 24:e01284
Shrestha UB, Shrestha S, Chaudhary P, Chaudhary RP (2010) How representative is the protected areas system of Nepal? Mt Res Dev 30(3):282–294
Tambe S, Kharel G, Arrawatia ML, Kulkarni H, Mahamuni K, Ganeriwala AK (2012) Reviving dying springs: climate change adaptation experiments from the Sikkim Himalaya. Mt Res Dev 32(1):62–72
Tiwari C, Rani M (2024) Trouble in paradise: evaluating the effects of unorganized tourism on the Himalayan ecology. In: Borthakur A, Singh P (eds) The Himalayas in the anthropocene. Springer, Cham. https://doi.org/10.1007/978-3-031-50101-2_7
Uprety Y, Chettri N, Dhakal M, Asselin H, Chand R, Chaudhary RP (2021) Illegal wildlife trade is threatening conservation in the transboundary landscape of Western Himalaya. J Nat Conserv 59:125952
Verma A, Schmidt-Vogt D, De Alban JD, Lim CL, Webb EL (2021) Drivers and mechanisms of forest change in the Himalayas. Glob Environ Chang 68:102244
Wani SA, Mugal MA, Dar FA, Reddy CS, Rashid I, Khuroo AA (2023) Biodiversity data synthesis on trees of Indian Himalayan region: policy and management implications. Glob Ecol Conserv 48:e02698
Xu J, Badola R, Chettri N, Chaudhary RP, Zomer R, Pokhrel B, Hussain SA, Pradhan S, Pradhan R (2019) Sustaining biodiversity and ecosystem services in the Hindu Kush Himalaya. In: Wester P, Mishra A, Mukherji A, Shrestha A (eds) The Hindu Kush Himalaya assessment. Springer, Cham. https://doi.org/10.1007/978-3-319-92288-1_5
Yadawa SK (2024) Landslide mitigation and sustainable management and policies. In: Chatterjee U, Lalmalsawmzauva K, Biswas B, Pal SC (eds) Landslides in the Himalayan region. Disaster risk reduction. Springer, Singapore. https://doi.org/10.1007/978-981-97-4680-4_19
Ziegler DA, Wasson RJ, Sundriyal Y, Srivastava P, Sasges G, Ramchunder SJ, Ong CE, Nepal SK, McAdoo BG, Gillen J, Bishwokarma D, Bhardwaj A, Apollo M (2021) A call for reducing tourism risk to environmental hazards in the Himalaya. Environ Hazards 22(1):1–28. https://doi.org/10.1080/17477891.2021.1984196
Zurick D, Pacheco J (2006) Illustrated atlas of the Himalaya. University Press of Kentucky, Kentucky

Chapter 18
Protected Areas of the Himalayan Hotspot: A Strategy Towards Biodiversity Conservation

Debahooti Baruah, Bishmita Boruah, Himamani Kalita, Arjun Adhikari, Sukanya Sarma, and Partha Pratim Baruah

Abstract The Himalayan Hotspot is recognised as one of the world's most significant biodiversity zones, home to a wide variety of plants and animals, many of which are endemic and endangered species. Protected areas in Himalaya serve as crucial sanctuaries for safeguarding ecosystems, preventing habitat loss, and mitigating the impacts of climate change and human activities such as deforestation and land conversion. These regions hence, are essential to preserving the natural resources that local residents rely on for their livelihoods and the ecological balance. The present study highlights the challenges that threaten the effectiveness of protected areas, including illegal poaching, habitat fragmentation, and limited resources for enforcement and management. To achieve long-term success, it is also highlighted that better conservation legislation, local community involvement, and trans-boundary cooperation are the need of the hour. This chapter examines the role of protected areas in the conservation of biodiversity within this region.

Keywords Endemic species · Ecosystem management · Illegal poaching · Endangered species

---

D. Baruah
Department of Botany, Tezpur College, Tezpur, Assam, India

B. Boruah · H. Kalita · P. P. Baruah (✉)
Department of Botany, Gauhati University, Guwahati, Assam, India
e-mail: ppbaruah@gauhati.ac.in

A. Adhikari
Department of Botany, Madhab Choudhury College, Barpeta, Assam, India

S. Sarma
Department of Botany, Salbari College, Baksa, Assam, India

D. Das et al. (eds.), *Plant Diversity and Conservation Strategies of the Himalayan Biodiversity Hotspot*, Topics in Biodiversity and Conservation 21,
https://doi.org/10.1007/978-3-032-21697-7_18

18.1 Introduction

The Himalayan hotspot is home to a complex mountain ecosystem which is characterised by significant topographic as well as climatic variation. The entire Himalayan hotspot which is about 7,50,000 km^2 divided into two geographical regions: Eastern and Western Himalaya. Eastern Himalaya covers parts of Assam, Arunachal Pradesh, Bhutan, West Bengal, Sikkim, Nepal, Southeast Tibet and Northern Myanmar. Whereas Western Himalaya covers the Kumaon- Garhwal, Northern west Kashmir and Northern Pakistan. Furthermore, the area is composed of four separate tectonic units: the Greater Himalaya, the Lesser Himalaya, the Outer Himalaya, and the Trans-Himalaya (Valdiya 2002). The region is also considered as one of the most important global biodiversity hotspots which host a wide variety of flora and fauna including a high number of endemic species. This discussion on floral and faunal endemism aligns with the detailed ecological and geological stratification described in Chap. 1 (Sect. 1.2) and further supported by biogeographic regionalisation insights from Chap. 4 (Sect. 4.1). Over 10,000 different species of vascular plants are thought to be found in the Himalayan hotspot, with nearly a third which is about 31.6% being endemic (Wambulwa et al. 2021). However, the exact number of species is still up for debate, and numerous new species are constantly being discovered in the area till today (Borah et al. 2020; Maity 2014). A significant number of orchid species, many of which are relatively young endemic species, have recently been reported from the hotspot, suggesting that more research will likely reveal a much higher degree of plant endemism (Wambulwa et al. 2021). Additionally, several widely dispersed plant taxa, including *Primula, Pedicularis*, and *Rhododendron* distributed in the Eastern Himalaya including Arunachal, Sikkim and Bhutan. Whereas, western part of Himalaya is dominated by trees such as *Abies pindrow*, *Pinus wallichiana*, *Cedrus deodara*, *Olea cuspidate* and *Shorea robusta* (Shaheen et al. 2012; De Jong 2023).

Each of the major ecosystems that make up the Himalayan hotspot is home to a distinct collection of flora and wildlife that have adapted to the particular ecological conditions of the area (Xu et al. 2019). It is also home to an astounding variety of animal species, many of which are rare or endangered because of the region's distinctive and varied environments. Important populations of several large mammals and birds, such as tigers, Himalayan tahr, snow leopard, elephants, rhinos, red panda, wild water buffalo and vultures are also found in the hotspot (Behera et al. 2002). The region is home to a wide range of important habitats, including tropical rainforests, temperate forests, and alpine meadows (Das and Bera 2018). High altitudes above the treeline, alpine meadows are home to resilient grasses and wildflowers that flourish in the severe weather (Singh et al. 2024). These areas are vital habitat for animals like the Himalayan tahr and snow leopard. A wide variety of species, including the red panda and clouded leopard, can be found living in the deep undergrowth and rich tapestry of broadleaf and coniferous trees which make the mid-elevation temperate forests (Das and Bera 2018). On the other hand, the lush and diverse tropical rainforests of the eastern Himalayan foothills are characterised by regular rainfall and high humidity levels. These rainforests are home to animals like the Indian elephant and Bengal tiger. Complex web of ecological interactions between various habitats are thus seen in the hotspot. Further, through affecting water cycles and nutrient flows, forest areas improve the health of nearby wetlands and alpine meadows. Strong ecological processes and a variety of food webs are supported by this connectivity. The habitat diversity is important from a cultural and economic standpoint (Hens and Nath 2003). The area's natural settings promote ecotourism, indigenous livelihoods, and customs (Hawkins 2003).

The Himalaya hotspot region is particularly vulnerable to rapid climate change which poses several threats to its unique biodiversity (Negi et al. 2012). Numerous factors, such as the melting of glaciers due to global warming, habitat alteration, floods, landslides, invasive hydrothermal projects, road construction, deforestation and the destruction of trees for fuel and domestic purposes, growth in tourism, and much more, are putting it in danger of causing irreversible damage (Banerjee et al. 2019). The distinctive biodiversity of the Himalayan hotspot needs to be effectively protected and conserved, which requires addressing all these challenges.

18.2 Protected Areas of the Himalayan Hotspot

Protected areas are essential for preserving the unique ecosystems and rich biodiversity of the Himalayan hotspot, particularly given its exceptionally high level of endemism. Spanning multiple countries including Nepal, India, Bhutan, and China these regions encompass a diverse array of ecosystems, from subtropical rainforests to alpine tundra. They serve as ecological strongholds, safeguarding habitats of critical ecological and scientific importance. The description of these protected areas aligns with the conservation challenges outlined in Chap. 11 (Sect. 11.3), while their biodiversity mirrors the species-specific conservation concerns discussed in Chap. 9 (Sect. 9.2). Notable protected areas across the Himalayan hotspot (Chettri et al. 2008; Banerjee et al. 2019) are presented in Table 18.1, which highlights their role in conserving numerous endemics and threatened species, including the red panda, snow leopard, wild yak, marmots, musk deer, and various pheasant species. In the Indian Himalayan region alone covering approximately 45,774 km^2 there are 133 designated protected areas (Rawal et al. 2013). These landscapes play a crucial role in sustaining vital ecosystem services such as water regulation, soil stabilisation, and climate buffering, which are indispensable to both wildlife and local human communities. To ensure the long-term effectiveness of these conservation zones, proactive management is imperative. This includes addressing emerging threats such as climate change, habitat encroachment, and unsustainable human activities that jeopardise the ecological integrity and natural heritage of the Himalayan hotspot.

Table 18.1 Some of the important protected areas of the Himalaya hotspot

Sl. No.	Country	Protected areas	Flora	Fauna	Importance
1	India	Khangchendzonga National Park	*Rhododendron*, Oak, Fir, Alpine meadows	*Panthera unica*, *Ailurus fulgens*, *Hemitragus jemlahicus*	UNESCO World Heritage Site
2		Namdapha National Park	Hollong, Bamboo, Dipterocarps	*Neofelis nebulosa, Hoolock leuconedys, Panthera tigris*	Easternmost Tiger Reserve of India
3		Singalila National Park	*Rhododendron*, Magnolia, Bamboo	*Ailurus fulgens, Ursus thibetanus laniger, Muntiacus muntjak*	Famous for trekking trails
4		Barsey *Rhododendron* Sanctuary	*Rhododendron*, Oak, Bamboo	*Ailurus fulgens*, *Paguma larvata, Panthera pardus*	Famous for its *Rhododendron* blooms
5		Mouling National Park	Hollong, Pine, Tropical Rainforest	*Budorcas taxicolor*, *Neofelisnebulosa*, *Hoolock hoolock*	Part of the Dihang-Dibang Biosphere Reserve
6		Pakke Wildlife Sanctuary and Tiger Reserve	Hollong, Bamboo, Evergreen Forests	*Panthera tigris tigris, Rhyticeros undulatus, Panthera pardus*	Known for Hornbill Conservation Programme
7		Eaglenest Wildlife Sanctuary	*Rhododendron*, Oak, Bamboo	*Ailurus fulgens*, *Bugun liocichla*, *Neofelis nebulosa*	Biodiversity hotspot for birdwatchers
8		Neora Valley National Park	*Rhododendron*, Ferns, Bamboo	*Neofelis nebulosa*, *Ailurus fulgens*, *Naemorhedus goral*	One of the least explored areas in India
9		Nameri National Park and Tiger Reserve	Evergreen and Semi-evergreen Forests	*Elephas maximus*, *Panthera tigris*, *Asarcorniss cutulata*	Located along the Jia-Bhoreli River
10		Manas National Park	Grasslands, Semi-evergreen Forests	*Panthera tigris tigris, Trachypithecus geei, Porcula salvania*	UNESCO World Heritage Site
11		Shingba *Rhododendron* Sanctuary	*Rhododendron*, Birch, Alpine Forests	*Moschus moschiferus, Ailurus fulgens, Pseudois nayaur*	Known for diverse *Rhododendron* species
12		The Great Himalayan National Park	Coniferous & Deciduous Forests	*Ursus arctos isabellinus, Panthera uncia, Capra ibex*	UNESCO World Heritage Site
13	Bhutan	JigmeDorji National Park	*Meconopsisbetonicifolia*, Fir, Spruce	*Budorcas taxicolor*, *Panthera tigris tigris*, *Ursus thibetanus laniger*	Largest National Park in Bhutan

14		Sakteng Wildlife Sanctuary	Oak, *Rhododendron*, Alpine Grasslands	*Panthera uncial, Ailurus fulgens, Moschus chrysogaster*	Home to the mythical Yeti
15		Royal Manas National Park	Sal Forests, Bamboo, Tropical Grasslands	*Trachypithecus geei*, *Neofelis nebulosa, Panthera tigirs tigris*	Connected to India's Manas National Park
16		Jigme Singye Wangchuck National Park	Subtropical to Alpine Forests	*Panthera tigris, Ursus thibetanus laniger, Ailurus fulgens*	Second-largest park in Bhutan
17		Bumdeling Wildlife Sanctuary	Hemlock, Spruce, *Rhododendron*	*Panthera uncia, Grus nigricollis, Budorcas taxicolor*	Important Bird Area (IBA)
18	Nepal	Makalu Barun National Park	Pine, Oak, *Rhododendron*	*Ailurus fulgens*, *Panthera uncia, Hemitragus jemlahicus*	Part of the Everest region
19		The Sagarmatha National Park	Blue Pine, Birch, Juniper	*Lophophorusimpejanus, Moschus moschiferus, Panthera uncial*	Home to Mount Everest
20	China	Qomolangma National Nature Preserve	Alpine Shrubs, *Rhododendron*	*Pantholops hodgsonii, Lynx lynx, Ursus arctos*	Protects Everest (Qomolangma) region
21		Medog Nature Reserve	Tropical Rainforest, Pine, Oak	*Neofelis nebulosa*, *Ailurus fulgens*, *Elephas maximus*	One of the most remote reserves in Tibet
22		Nujiang Langcang Gorge Alpine Nature Reserve	Conifers, *Rhododendron*	*Budorcas taxicolor*, *Grus nigricollis, Panthera uncial*	Located near the Three Parallel Rivers area

18.3 Threats to Biodiversity in the Himalayas

Ecosystems and species across the globe are facing unprecedented threats due to escalating human pressures on the environment. Key drivers such as overhunting, invasive species, pollution, habitat fragmentation, and land-use change have already jeopardised approximately 25% of mammal species and 13% of bird species, alongside over 21,000 other plant and animal taxa currently at risk of extinction (Steffen et al. 2007). These multidimensional threats to biodiversity closely mirror the human-induced ecological pressures outlined in Chap. 5 (Sect. 5.2) and are further contextualised through community and governance responses described in Chap. 11 (Sect. 11.3). Likewise, climate-driven species redistributions are consistent with projections presented in Chap. 12 (Sect. 12.1).

Among these threats, land-use change is widely regarded as the leading cause of global biodiversity decline. Habitat loss and degradation primarily driven by agricultural expansion, logging, urbanisation, mining, and infrastructure development remain the most direct threats to terrestrial fauna. These pressures reduce the total area available to species, diminishing population sizes and fragmenting their ranges into small, isolated patches (Vitousek et al. 1997). Approximately 80% of all vulnerable terrestrial birds and mammals are imperiled by agricultural habitat degradation (Visconti et al. 2011). The governance and institutional responses to these drivers are further analysed in Chap. 11 (Sect. 11.2) with additional insights on species-specific conservation needs in Chap. 5 (Sect. 5.1).

Human-induced mortality particularly from hunting and poaching continues to pose a critical threat. An estimated 40–50% of at-risk bird and mammal species, along with a disproportionately high percentage of large herbivores, face extinction risks due to direct human exploitation (Newbold et al. 2015). Despite increasing awareness and conservation efforts, poaching remains a pressing concern, with high-value body parts such as elephant ivory, rhinoceros' horn, tiger and lion skins, bones, claws, and pangolin scales being trafficked at alarming rates. In addition, the consumption of bush meat has led to drastic declines in wildlife populations and continues to endanger numerous species (Ripple et al. 2016). These issues also intersect with livelihood strategies and enforcement challenges discussed in Chap. 11 (Sect. 11.4). Invasive predators, particularly on islands, have caused catastrophic collapses in bird populations and continue to affect terrestrial fauna elsewhere.

Species redistributions due to climate change have profound consequences for ecosystem functionality, human well-being, and even global climate regulation (Pecl et al. 2017). Current projections suggest that the planet is experiencing the largest species redistribution event since the last glacial maximum. These shifts are occurring alongside surging human demands for natural resources, placing immense pressure on ecosystems to continue delivering services. As with other climate change impacts, such redistributions are expected to create stark disparities "winners" and "losers" in both biodiversity and socio-economic sectors, potentially fuelling regional conflicts. The static nature of most protected area boundaries makes them ill-suited for safeguarding dynamic, shifting species ranges (Negi et al. 2012), a concern echoed in Chap. 5 (Sect. 5.3).

The Himalayan region referred to as the "Third Pole" due to its vast glacial reserves is particularly vulnerable. Climate change is already altering hydrological cycles, ecosystem integrity, and the livelihoods of millions. Its acute sensitivity to warming necessitates the urgent adoption of adaptive strategies and resilient development models. These needs are explored in Chap. 12 (Sect. 12.2) along with practical examples of ecosystem-based adaptation and scientific monitoring.

Recent research using MaxEnt modelling and field data indicates that by 2050 and 2070, roughly 16% and 18% of endemic angiosperm species in the Sikkim Himalaya may lose suitable habitat. Likewise, alpine meadows are expected to contract by 1–3%, being gradually replaced by shrublands (Manish et al. 2016). A broader climate vulnerability assessment across the Indian Himalayan Region has emphasised the increasing threats posed by glacial lakes and landslides, as well as the growing strain on traditional subsistence farming systems (Morton 2007). Furthermore, poor land-use decisions such as unplanned road-cutting and deforestation continue to exacerbate these risks.

The five principal human-induced threats to biodiversity remain: invasive species, pollution, overexploitation of resources, deforestation and habitat loss, and climate change. Additional pressures include hunting, poaching, overgrazing, urban expansion, and the creation of transport corridors, all of which fragment habitats and weaken ecological resilience (Ogunkunle et al. 2019). These themes are interwoven with socio-political dimensions explored in Chap. 11 (Sect. 11.1), reinforcing the call for integrative and inclusive conservation strategies across the Himalayas.

18.4 Strategies for the Conservation of Himalayan Biodiversity

The Himalayan region, recognised as a global biodiversity hotspot, faces mounting threats from climate change, deforestation, and habitat degradation. Addressing these challenges requires a comprehensive conservation strategy that includes the establishment and adaptive management of protected areas, promotion of ecologically responsible tourism (Chap. 10, Sect. 10.4), afforestation, and community-based conservation initiatives. Technological innovations such as GIS-based monitoring, ecological modelling, and habitat restoration tools detailed in Chap. 10 (Sect. 10.3) play a crucial role in supporting these efforts. In parallel, effective governance, robust legislative frameworks, and decentralised policy implementation are essential, as highlighted in (Chap. 13, Sect. 13.2). Strengthening institutional coordination, advancing scientific research, and fostering cross-border cooperation remain pivotal to preserving the Himalayan region's unique biodiversity for future generations (Fig. 18.1).

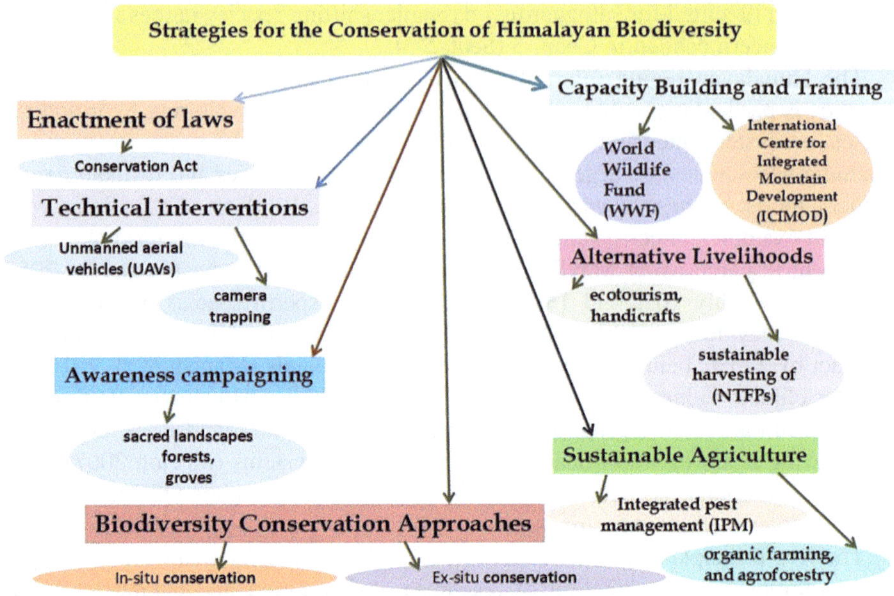

Fig. 18.1 Conservation strategies for Himalayan biodiversity

18.4.1 Enactment of Laws

Globally, numerous acts have been introduced in order to protect endangered wildlife and to prevent human-wildlife conflicts and preserve biodiversity. For instance, recently India enforces the Wildlife (Protection) Amendment Act, 2022 that persuasively argues for examination of specific situations, such as the use of elephants for religious and other purposes. A few practical measures, such as controlling invasive alien species and releasing caged animals without facing consequences, have been included in this Act (Gautam and Kaur 2023). Similarly, one of the stringent and restrictive enactment was the Yellowstone model, which placed an emphasis on preserving a pristine area and forbade people from residing within the park, was followed by the National Parks and Wildlife Conservation Act, 1973 (NPWCA) of Nepal (Bhattarai et al. 2017).

18.4.2 Technical Interventions

Unmanned aerial vehicles (UAVs) and camera trapping are two technical approaches for wildlife monitoring that are remotely operated and controlled by embedded autonomous computer programmes (Mishra and Natalizio 2020; Keystone Foundation 2021). In Nepal's Banke National Park, Bardia National Park,

Chitwan National Park, Parsa Wildlife Reserve and Suklaphanta Wildlife Reserve drones have been deployed to locate Royal Bengal tigers and reduce human-wildlife conflict by monitoring their proximity to villages (Karmacharya et al. 2018). UAVs equipped with high-resolution cameras are used in China's Xinglong mountain nature reserve to monitor musk deer populations and analyse the impact of climate change (Zhang et al. 2022).

A contemporary technology for precisely locating and monitoring wildlife is the GPS (Global Positioning System). An added advantage of this method is that it helps in identification and tracking of animals in diverse habitats without imposing harm to them (Joshi et al. 2021). In Bhutan, GPS tracking showed that cranes migrate between the Phobjikha Valley (Bhutan) and the Tibetan Plateau, leading to policies protecting key wetland habitats along these routes (Lhamo 2019). In Uttarakhand, India, GPS-monitored bears were tracked to predict crop-raiding behavior, allowing timely interventions by forest departments (Habib et al. 2015). In Nepal, researchers used GPS collars on Himalayan tahr to understand how they navigate shrinking alpine habitats caused by rising temperatures (Dhakal et al. 2023).

18.4.3 Awareness Campaigning

Besides National Parks and Wildlife Sanctuaries, sacred landscapes, forests, and groves, which are areas of vegetation traditionally protected by the local communities, serve as outstanding examples of in-situ biodiversity conservation (Gupta and Sharma 2008). In Nepal, traditional systems of resource governance have been well-developed and effective, playing a significant role in biodiversity conservation over the centuries (Bhattarai and Kunwar 2015). For example, in Byas, Darchula district of Nepal, communities set a specific date during the last week of September (around the full moon) for harvesting grasses and herbs. Restrictions are placed on harvesting to ensure fair and equitable access to fodder during times of scarcity, particularly in the winter, while also conserving the resources. (Bajracharya and Malla 2012). Similarly, some freshwater lakes also play a role in protecting aquatic flora and fauna. For instance, Khecheopalri Lake in Sikkim has been designated as sacred by the local people to prevent the degradation of aquatic life (Gurung and Thapa 2017).

18.5 Biodiversity Conservation Approaches

Biodiversity conservation is practiced at two levels viz., in-situ conservation and ex-situ conservation (Mir et al. 2015).

18.5.1 In-Situ Conservation

The Great Himalayan National Park in Himachal Pradesh helps protect species such as the Himalayan monal (*Lophophorus impejanus*), the state bird of Himachal Pradesh, the vulnerable snow leopard (*Panthera uncia*), and the endangered Himalayan brown bear (*Ursus arctos isabellinus*) under in-situ conservation (Singh and Ranjitsinh 2016). The Kangchenjunga National Park in Sikkim is a haven for species like the vulnerable common leopard (*Panthera pardus*) (Sharma and Pradhan 2017). Similarly, Kaziranga National Park in Assam is dedicated to conserving the Indian rhinoceros (*Rhinoceros unicornis*), Bengal tiger (*Panthera tigris tigris*) and wild buffalo (*Bos gaurus*) (Sarma and Jha 2015). The *Bugun liocichla* (*Liocichla bugunorum*) is a critically endangered bird species endemic to the eastern Himalayas, primarily found in the forests of Arunachal Pradesh, India. Indigenous and local communities are extending their efforts to conserve this bird, as it faces threats from habitat destruction, particularly from logging and infrastructure development (Ghosh and Mishra 2015).

18.5.2 Ex-Situ Conservation

The Central Zoo, Elephant Conservation Breeding Center, Vulture Conservation Breeding Center and Crocodile Conservation Breeding Center are prominent examples of ex-situ conservation efforts in Nepal (Sapkota and Sharma 2018). The Sikkim Himalayan Botanical Garden, for instance, specialises in conserving endemic plants such as *Primula sikkimensis* and *Rhododendron*, as well as medicinal plants like *Sanguisorba* and *Berberis*, supporting biodiversity in the eastern Himalayas (Sharma and Bhatt 2017). In 1983, a collaborative mission between the Royal Government of Bhutan and International Rice Research Institute of Bhutan gathered 184 traditional rice varieties from high and medium-altitude rice-growing regions, although many remote areas were not included in the survey (Pradhan 1996). The collection and conservation of germplasm from endemic, rare, endangered, threatened, and other valuable plants is an important effort. For instance, the National Botanical Garden (NBG) in Godavari, Lalitpur, Nepal has ex-situ conserved plants such as *Hypericum cordifolium* (endemic), *Alstonia scholaris* (rare), and *Podocarpus neriifolius* (endangered). In total, NBG conserves 860 plant species, with 395 of these being preserved ex-situ. (Adhikari and Lama 2014; Bhatta and Gurung 2015). Additionally, the Northeast Regional Agricultural Research Station of Indian Agricultural Research Institute in Assam, India focuses on conserving the genetic resources of economically important crops like rice and tea (Sarma et al. 2015).

18.6 Sustainable Agriculture and Biodiversity Conservation

Sustainable agriculture aims to meet current demands for food and fibre without compromising the ability of future generations to fulfil their own needs (Velten et al. 2015). In the Himalayan hotspot characterised by rugged terrain and ecologically fragile landscapes adopting environmentally responsible agricultural practices is imperative. In these biodiversity-rich but delicate ecosystems, techniques such as Integrated Pest Management (IPM), organic farming, and agroforestry are essential for minimising the ecological footprint of agriculture (Saxena et al. 2005). Notably, agroforestry has been instrumental in restoring degraded land and improving soil fertility through the integration of trees into farming systems (Sharma et al. 2007), building upon foundational insights discussed in Chap. 2 (Sect. 2.2).

Organic farming supports long-term soil health by reducing dependence on synthetic chemicals, while IPM promotes crop protection through ecological methods, thereby conserving native biodiversity (Kumar et al. 2023). A successful example is the state-wide organic agriculture initiative launched by the government of Sikkim, India, which not only enhanced crop yields but also contributed to the conservation of species-rich areas adjacent to protected natural zones (Kumar et al. 2018). Similarly, Bhutan's approach to community-based natural resource management anchored in sustainable agriculture has yielded both social and economic benefits while supporting biodiversity conservation (Neuhoff et al. 2014), echoing the community livelihood strategies outlined in Chap. 11 (Sect. 11.4).

18.7 Alternative Livelihoods for Conservation

In places where native communities mostly depend heavily on natural resources for their livelihoods, the risk of overexploitation of the resources poses a significant threat to biodiversity. For the protected area to stay protected, alternative income-gathering options that lessen reliance on natural resources are therefore essential (Pandey 2011; Badola et al. 2018). Alternative livelihoods including ecotourism, handicrafts, and sustainable harvesting of non-timber forest products (NTFPs) have shown great promise as income generating sources in the Himalayan Hotspot (Pandey et al. 2016). In Nepal, the Annapurna Conservation Area Project (ACAP) has successfully combined conservation efforts with the promotion of ecotourism, leading to both improved livelihoods for local communities and enhanced protection for the region's rich biodiversity (KC et al. 2015). In Himalayan regions, weaving and basketry provides a wide range of economic benefits to local artisans. This eventually reduces their reliance on activities that harm nature (Singh et al. 2021).

One of the successful projects in this regard includes the Phobjikha Valley Community-Based Ecotourism (CBET) initiative in Bhutan, which shows how alternative livelihoods can aid in the protection of biodiversity. The project has helped in empowering local communities by providing training in hospitality and guiding services. This has enabled them to benefit financially from the growing number of tourists while maintaining the valley's ecological integrity (Thinley 2010; Gogoi and Maheshwari 2024).

18.8 Education and Awareness: Building a Conservation Ethic

In the Himalayan Hotspot, raising awareness about the values of conservation ethics among the next generation is crucial for ensuring the sustainability of conservation efforts. To ensure this, environmental education programmes in schools and communities are crucial (Dhar et al. 2002; Thoker et al. 2025). Activities like plantation, wildlife monitoring and waste management included at school level education not only educate participants but also contribute directly to conservation efforts. Awareness campaigns are equally important in promoting conservation.

Moreover, to protect ecologically diverse and sensitive regions, capacity building and training programmes are critical. Such initiatives equipe local communities and conservation staff with the knowledge and skills needed to manage protected areas effectively. To facilitate this, organisations like World Wildlife Fund (WWF) and the International Centre for Integrated Mountain Development (ICIMOD) have been instrumental in Himalayan regions, by providing training on sustainable resource management, biodiversity monitoring, and climate change adaptation (WWF 2000; Acharya and Sellmyer 2016; Sher et al. 2017).

18.9 Conclusions

The Himalayan Hotspot's protected areas are essential to maintaining one of the planet's most biologically varied and ecologically critical locations. These regions provide as an essential barrier against habitat destruction, climate change, and human expansion by preserving habitats that are home to thousands of endemic and endangered species. The implementation of conservation techniques inside these zones serves to safeguard not only the local residents' livelihoods but also the species, thereby guaranteeing the ecological and socio-economic viability of conservation initiatives. Nonetheless, persistent issues like poaching, deforestation, and environmental degradation call for more robust legislation, cross-national collaboration, and community involvement. The Himalayan region can preserve its biodiversity richness and guarantee the long-term survival of its distinctive species and ecosystems by growing and managing these protected zones well. Ultimately, the accomplishment of these projects is critical to the health of future generations as well as the biodiversity of the planet.

Acknowledgments Authors are grateful to MoEF and CC (Govt. of India), DST-FIST, UGC-SAP, OIL, ASTEC for providing facilities for research infrastructure in the Department of Botany, Gauhati University.

References

Acharya S, Sellmyer A (2016) ICIMOD Annual Report 2015

Adhikari M, Lama S (2014) Plant conservation efforts in Nepal: the role of the National Botanical Garden in ex-situ preservation of native flora. J Himal Pl Stud 25(3):12–19

Badola R, Hussain SA, Dobriyal P, Manral U, Barthwal S, Rastogi A, Gill AK (2018) Institutional arrangements for managing tourism in the Indian Himalayan protected areas. Tour Manag 66:1–12

Bajracharya B, Malla Y (2012) The role of local communities in biodiversity conservation in Nepal: the case of Byas, Darchula. Environ Manag Policy 19(4):130–145

Banerjee A, Chakraborty P, Bandopadhyay R (2019) Urgent conservation needs in the Sikkim Himalaya biodiversity hotspot. Biodivers J 20(2–3):88–97

Behera MD, Kushwaha SPS, Roy PS (2002) High plant endemism in an Indian hotspot–eastern Himalaya. Biodivers Conserv 11:669–682

Bhatta S, Gurung K (2015) Conservation of endemic and endangered plants in Nepal: a case study from the National Botanical Garden, Godavari. J Environ Sci Manag 17(4):102–110

Bhattarai N, Kunwar R (2015) Sacred landscapes and their role in the conservation of biodiversity in the Himalayan region of Nepal. Himal Ecol Dev 24(3):125–134

Bhattarai BR, Wright W, Poudel BS, Aryal A, Yadav BP, Wagle R (2017) Shifting paradigms for Nepal's protected areas: history, challenges and relationships. J Mt Sci 14:964–979

Borah D, Gap N, Singh RK (2020) *Pediculariskhoiyangii* (Orobanchaceae), a new species from the Eastern Himalaya, India. Phytotaxa 430:287–293

Chettri N, Shakya B, Thapa R, Sharma E (2008) Status of a protected area system in the Hindu Kush-Himalayas: an analysis of PA coverage. Int J Biodivers Sci Manag 4(3):164–178

Das AP, Bera S (2018) Plant diversity in the Himalaya hotspot region. Bishen Singh Mahendra Pal Singh, Dehradun. isbn:978-81-211-0946-8

De Jong W (2023) An overview of the Himalaya forests and their drivers of change. Clim Change Himalayas 1:11

Dhakal P, Sharma RK, Rajak BR, Pandey N, Khanal L (2023) Population and distribution of Himalayan tahr (*Hemitragusjemlahicus*) in Lamtang National Park of the Nepal Himalaya. MARKHOR (J Zool) 4(01):2–10

Dhar U, Rawal RS, Airi S, Bhatt ID, Samant SS (2002) Promoting outreach through conservation education programmes–case study from Indian Himalayan region. Curr Sci 82(7):808–815

Gautam A, Kaur S (2023) Management and conservation of wildlife in India: with special reference to Wildlife Protection (Amendment) Act, 2022. Indian J Integr Res Law 3:1

Ghosh S, Mishra C (2015) Involving local communities in the conservation of critically endangered species: a case study of Bugun Liocichla in Arunachal Pradesh. Environ Conserv 42(3):254–262

Gogoi P, Maheshwari A (2024) CABI, Community-Based Tourism as a Sustainable Approach to Tourism Development in Jibhi Valley, Himachal Pradesh, India. Tourism Cases https://doi.org/10.1079/tourism.2024.0066

Gupta A, Sharma R (2008) Sacred groves and in-situ conservation: traditional ecological knowledge and biodiversity conservation in India. Environ Conserv 35(1):36–44

Gurung A, Thapa S (2017) Sacred Lakes and biodiversity conservation in the eastern Himalayas: the case of Khecheopalri Lake, Sikkim. Int J Environ Sci Dev 8(6):426–431

Habib B, Shrotriya S, Mahar N, Lyngdoh S, Rawat GS, Mohan D, Mondal I (2015) Field sampling protocol–mammalian fauna in trans-Himalayan landscape, Uttarakhand, India. Wildlife Institute of India and Uttarakhand Forest Department, Dehradun, Uttarakhand, India. p 22

Hawkins BA (2003) Energy, water and broad-scale geographic patterns of species richness. Ecology 84:3105–3117

Hens L, Nath B (2003) The Johannesburg conference. Environ Develop Sustain 5:7–39

Joshi VC, Pande PP, Dhabekar AK, Frank GS, Rabra NS, Mahajan H, Mehra D (2021) 5G technology: a futuristic solution for wildlife monitoring. Int J Res Publ Rev 2(12):1661–1665

Karmacharya D, Sherchan AM, Dulal S, Manandhar P, Manandhar S, Joshi J, Hughes J (2018) Species, sex and geo-location identification of seized tiger (*Pantheratigristigris*) parts in Nepal—a molecular forensic approach. PLoS One 13(8):e0201639

KC A, Rijal K, Sapkota RP (2015) Role of ecotourism in environmental conservation and socioeconomic development in Annapurna conservation area, Nepal. Int J Sustain Dev World Ecol 22(3):251–258

Keystone Foundation, "Wildlife Monitoring - Observing bovines with verve", [Online]. Available: https://keystone-foundation.org/wildlife-monitoring/. [Accessed: 07-Sep.-2021]

Kumar J, Pradhan M, Singh N (2018) Sustainable organic farming in Sikkim: an inclusive perspective. In: Advances in smart grid and renewable energy: proceedings of ETAEERE-2016. Springer, Singapore, pp 367–378

Kumar A, Saha S, Layek J, Babu S, Kumar R, Choudhary AK, Das A (2023) Organic farming in Indian Himalayan region: innovations for sustainability. Indian J Agron 68:36–51

Lhamo P (2019) Assessing the influence of human settlements on the plant diversity in wetlands of Phobji and Gangtey, Bhutan. http://hdl.handle.net/123456789/80

Maity D (2014) A new species of Gentiana (Gentianaceae) from the Sikkim Himalaya. Edinb J Bot 71(3):289–296

Manish K, Telwala Y, Nautiyal DC, Pandit MK (2016) Modelling the impacts of future climate change on plant communities in the Himalaya: a case study from eastern Himalaya, India. Model Earth Syst Environ 2:1–12

Mir ZR, Noor A, Habib B, Veeraswami GG. (2015) Attitudes of local people toward wildlife conservation: A case study from the Kashmir Valley. Mt Res Dev 35(4):392–400. https://doi.org/10.1659/MRD-JOURNAL-D-15-00030.1

Mishra D, Natalizio E (2020) A survey on cellular-connected UAVs: design challenges, enabling 5G/B5G innovations, and experimental advancements. Comput Netw 182:107451

Morton JF (2007) The impact of climate change on smallholder and subsistence agriculture. Proc Natl Acad Sci USA 104(50):19680–19685. https://doi.org/10.1073/pnas.0701855104

Negi G, Samal P, Kuniyal J, Kothyari B, Sharma R, Dhyani P (2012) Impact of climate change on the Western Himalayan Mountain ecosystems: an overview. Trop Ecol 53(3):345–356

Nepal WWF (2000) WWF in Nepal. Three decades of partnership in conservation (1967–2000). Kathmandu

Neuhoff D, Tashi S, Rahmann G, Denich M (2014) Organic agriculture in Bhutan: potential and challenges. Org Agric 4:209–221

Newbold T, Hudson LN, Hill SL, Contu S, Lysenko I, Senior RA, Purvis A (2015) Global effects of land use on local terrestrial biodiversity. Nature 520(7545):45–50

Ogunkunle T, Adewumi A, Adepoju A (2019) Biodiversity: over-exploited but underutilized natural resource for human existence and economic development. Environ Ecosyst Sci 3(1):26–34

Pandey R (2011) Consumption and valuation of livestock fodder under different forest types of Himalayas, India. Silva Lus 19(2):195–207

Pandey AK, Tripathi YC, Kumar A (2016) Non timber forest products (NTFPs) for sustained livelihood: challenges and strategies. J For Res 10(1):1–7

Pecl GT, Araújo MB, Bell JD, Blanchard J, Bonebrake TC, Chen IC, Williams SE (2017) Biodiversity redistribution under climate change: impacts on ecosystems and human well-being. Science 355(6332):9214

Rawal RS, Bhatt ID, Chandra Sekar K, Nandi SK (2013) The Himalayan biodiversity: richness, representativeness, uniqueness and life-support values. G.B. Pant Institute of Himalayan Environment and Development, Almora, p 84

Ripple WJ et al (2016) Bushmeat hunting and extinction risk to the world's mammals. R Soc Open Sci 3(10):160–498
Sapkota K, Sharma H (2018) Ex-situ conservation efforts in Nepal: a review of zoos and breeding centers. Nepal J Biodivers 7(1):54–62
Sarma MK, Bhattacharyee M, Sharma AK (2015) Crop genetic resources of North East India: Current status and future strategies. Progressive Agriculture 15(1):28–34
Sarma S, Jha B (2015) Conservation and Management of the Indian Rhinoceros in the Kaziranga National Park, Assam. J Wildl Biodivers 9(4):45–55
Saxena KG, Maikhuri RK, Rao KS (2005) Changes in agricultural biodiversity: implications for sustainable livelihood in the Himalaya. J Mt Sci 2:23–31
Shaheen H, Ullah Z, Khan SM, Harper DM (2012) Species composition and community structure of western Himalayan moist temperate forests in Kashmir. For Ecol Manag 278:138–145
Sharma R, Bhatt M (2017) Conservation of endemic plants of Sikkim Himalayas at the Sikkim Himalayan botanical garden. J Himalaya Bot 7(2):45–52
Sharma R, Pradhan K (2017) Conservation of red panda and snow leopard in Kangchenjunga Conservation Area. Himal Biodivers Conserv 5(2):120–125
Sharma R, Xu J, Sharma G (2007) Traditional agroforestry in the eastern Himalayan region: land management system supporting ecosystem services. Trop Ecol 48(2):189
Sher H, Bussmann RW, Hart R (2017) Promoting sustainable use of medicinal and aromatic plants for livelihood improvement and biodiversity conservation under global climate change, through capacity building in the Himalaya Mountains, Swat District, Pakistan. Ann Missouri Bot Gard 102(2):309–315
Singh R, Ranjitsinh MK (2016) Great Himalayan National Park: Conservation and Management. In: Himalayan Biodiversity: Conservation and Management Strategies. Himalayan Forest Research Institute, Shimla
Singh BP, Kumar B, Devi A (2021) Traditional handloom weaving: a cultural heritage in jeopardy among Gaddi scheduled tribe of Bhaderwah (J & K), North West Himalayas. Int J Adv Res 7(12):24–31
Steffen W, Crutzen PJ, McNeill JR (2007) The Anthropocene: are humans now overwhelming the great forces of nature? Ambio 36:614–621
Thinley P (2010) Empowering people, enhancing livelihood, and conserving nature. University of New Brunswick. Fredericton, New Brunswick, Canada
Thoker IA, Bhat MS, Shah SA, Lone FA, Parrey HA, Mir S (2025) Towards sustainable forest management: an examination of environmental knowledge among forest dwellers in the Kashmir Himalayas. Environ Dev Sustain 27(11):27095–27123. https://doi.org/10.1007/s10668-024-04823-y
Valdiya KS (2002) Emergence and evolution of Himalaya: reconstructing history in the light of recent studies. Prog Phys Geogr 26:360–399
Velten S, Leventon J, Jager N, Newig J (2015) What is sustainable agriculture? A systematic review. Sustainability 7(6):7833–7865
Visconti P, Pressey RL, Giorgini D et al (2011) Future hotspots of terrestrial mammal loss. Philos Trans R Soc B Biol Sci 366(1578):2693–2702
Vitousek PM, Mooney HA, Lubchenco J et al (1997) Human domination of Earth's ecosystems. Science 277:494–499
Wambulwa MC, Milne R, Wu ZY et al (2021) Spatiotemporal maintenance of flora in the Himalaya biodiversity hotspot: current knowledge and future perspectives. Ecol Evol 11(16): 10794–10812
Wester A, Mishra A, Mukherji AB,. Shrestha (Eds.) The Hindu Kush Himalaya assessment: Mountains, climate change, sustainability and people (pp. 127–165), Springer, Cham
Xu J, Badola R, Chettri N, Chaudhary RP, Zomer R, Pokhrel B, Pradhan, R (2019) Sustaining biodiversity and ecosystem services in the Hindu Kush Himalaya. In P. Wester, A. Mishra, A. Mukherji, & A. B. Shrestha (Eds.), The Hindu Kush Himalaya assessment: Mountains, climate change, sustainability and people (pp. 127–165), Springer, Cham
Zhang L, Sun Z, An B et al (2022) Alpine musk deer (*Moschus chrysogaster*) adjusts to a human-dominated semi-arid mountain ecosystem. Animals 12(21):3061

Chapter 19
Diversity and Conservation Strategies of *Rhododendron* in Sikkim Himalayas

Kingsuk Sarkar, Anindya Sundar Ray, Sandeep Rawat, Achintya Mitra, and Animesh Sen

Abstract The Sikkim Himalayas are home to an exceptional diversity of *Rhododendron*, with 44 taxa, including 36 species, 1 subspecies, 1 variety, and 6 natural hybrids. These species exhibit a remarkable range of morphological adaptations, from mat-forming shrubs to towering trees, thriving in varied habitats from rocky slopes to moss-covered cliffs. The genetic and population structure of *Rhododendron* species in the region reflects significant variability, indicating their evolutionary adaptability and ecological importance. Many species hold ethnomedicinal and cultural significance, being used in traditional remedies and local rituals, further highlighting their importance to the indigenous communities. However, several *Rhododendron* species in Sikkim are under threat due to habitat loss, overexploitation, climate change, and invasive species, with some listed under various categories of the IUCN Red List. Conservation strategies include habitat restoration, ex-situ propagation, establishment of protected areas, and community-based conservation programmes. Sustainable utilisation, combined with advanced techniques like molecular markers for genetic assessment and in vitro propagation, can aid in safeguarding these iconic plants. This chapter emphasises the need for integrated approaches to conserve *Rhododendron* diversity while addressing ecological and socio-cultural dynamics in the Sikkim Himalayas.

Keywords IUCN Red List · Threats · Conservation · Habitat loss · Ethnomedicinal

K. Sarkar · A. Mitra · A. Sen (✉)
Regional Ayurveda Research Institute, Gangtok, Sikkim, India

A. S. Ray
Ethnopharmacology Laboratory, Department of Biological Sciences and Microbiology, Biswa Bangla Biswabidyalay (University), Bolpur, Birbhum, West Bengal, India

S. Rawat
Sikkim Regional Centre, G.B. Pant National Institute of Himalayan Environment, Pangthang, Gangtok, Sikkim, India

D. Das et al. (eds.), *Plant Diversity and Conservation Strategies of the Himalayan Biodiversity Hotspot*, Topics in Biodiversity and Conservation 21,
https://doi.org/10.1007/978-3-032-21697-7_19

19.1 Introduction

Rhododendron, a genus of woody plants belonging to the family *Ericaceae*, holds profound ecological, cultural, and economic significance in the Himalayan region. The name originates from the Greek words rhodo (rose) and dendron (tree), and the genus was first formally classified by Carl Linnaeus in 1837 (Tiwari and Chauhan 2006). These plants range from ground-hugging shrubs to small trees, embellishing mountain landscapes with brilliant floral displays and playing a vital role in sustaining ecological balance. Popular in landscaping for their diverse forms, evergreen nature, and prolonged blooming periods (Eşen et al. 2004), *Rhododendron* exhibit a spectrum of colours, white, pink, red, magenta, yellow, orange, and blue determined by carotenoids and anthocyanins within petal cells (Xia et al. 2022).

Globally, the genus encompasses approximately 1200 species (Clinton and Vose 1996), with *Rhododendron* arboreum being the most widely distributed species in the Indian Himalaya (Srivastava 2012). Beyond their ornamental appeal, the species of *Rhododendron* are valued in traditional medicine for treating ailments such as heart disease, inflammation, bronchitis, and asthma. Their antioxidant-rich young leaves are commonly used to relieve headaches, while the wood is utilised in crafting tools and cultural artefacts such as khukri handles and pack saddles (Kumar et al. 2019).

Ecologically, *Rhododendron* serve as keystone species in subalpine and alpine ecosystems, stabilising slopes and watersheds, and offering vital nectar resources to pollinators such as bees, butterflies, and birds. Their dense canopies also provide nesting habitats for high-altitude fauna. However, these species face growing threats from habitat degradation, overharvesting, and climate change, which are contributing to shifts in their distribution, a phenomenon consistent with trends noted in Chap. 12 (Sect. 12.1) regarding climate-induced range shifts. Some species have extensive ranges across the Himalayas, while others are restricted to niche habitats, such as rocky alpine zones (Gibbs et al. 2011; Kumar 2012), highlighting species-specific vulnerabilities discussed in Chap. 9 (Sect. 9.2).

In regions like Sikkim, where rare and endemic *Rhododendron* species are concentrated, conservation is paramount (Vetaas 2002; Basnett and Ganesan 2022). Successful conservation will require strategies that are both ecologically informed and socially inclusive integrating traditional knowledge and local stewardship practices as outlined in Chap. 11 (Sect. 11.4). Balancing ecological preservation with community needs is essential to ensure that these iconic mountain floras continue to thrive amid increasing environmental pressures.

19.2 Diversity of *Rhododendron* in Sikkim Himalayas

19.2.1 Habitat, Morpho-Taxonomy and Distribution of Rhododendron

Worldwide, approximately 1200 species of *Rhododendron* have been identified, with China hosting the highest diversity, boasting 571 species, of which 409 are endemic. This remarkable representation underscores China as a global hub for *Rhododendron* diversity. India, too, harbours significant *Rhododendron* diversity, with around 80 species, 10 sub-species, and 14 varieties reported. Notably, the Himalayan region is home to 98% of India's *Rhododendron* species, highlighting the region's crucial role in sustaining this genus. Among these, Sikkim stands out as a biodiversity hotspot, supporting 72% of the *Rhododendron* species found in India (Sekar and Srivastava 2010).

Rhododendron species are highly adaptable to various habitats, though they predominantly thrive in open, well-aerated, and moist environments with humus-rich, acidic soils that are low in nutrients like nitrogen, calcium, phosphorus, and magnesium. They are commonly found in forest bases, glades, marshes, along stream sides, in open meadows, on rocks and boulders, or as epiphytes associated with mosses. The roots of *Rhododendron* plants form symbiotic relationships with ericoid mycorrhizal fungi, which enhance the availability of nutrients for the plants. Their evergreen nature helps them adapt to nutrient-deficient soils, promoting consistent photosynthesis and minimising nutrient loss (Sekar and Srivastava 2010).

Rhododendron plays a significant role in biodiversity, especially in the timberline-alpine transition (ecotone), and is regarded as a keystone species in extreme cold climates where temperatures range from +20 to −20 °C. Its ability to adapt to sub-freezing temperatures and tolerate frost or hailstorms is a key factor driving its distribution toward higher altitudes and northern latitudes (Vetaas 2002). Several *Rhododendron* species are classified under various threat categories by the International Union for Conservation of Nature (IUCN), including one species as extinct in the wild, two as critically endangered, five as endangered, 18 as vulnerable, and three as near threatened (IUCN 2022). *Rhododendron* is classified into eight subgenera, each with distinct characteristics. The subgenus *Hymenanthes* (elepidotes) includes evergreen shrubs and large trees (up to 40 m) with varying leaf and flower sizes. *Rhododendron* (Lepidotes), found in regions like the sub-Himalaya and Southeast Asia, consists of smaller bushes with scaly hairs on the stems, smaller leaves, and fragrant flowers. The azaleas, classified under the subgenera *Pentanthera* (30 species), *Tsutsutsi* (117 species), and *Azaleastrum* (34 species), are characterised by scaleless, soft leaves and single terminal flowers. The remaining subgenera *Mumeazalea*, *Therorhodion*, and *Candidastrum* contain only four species (Chamberlain et al. 1996).

The Sikkim Himalaya is a treasure trove of *Rhododendron* diversity, with a total of 44 taxa found in the region. This includes 36 species, along with 1 subspecies, 1 variety, and 6 naturally occurring hybrids, reflecting the incredible richness and beauty of the local flora (Gogoi et al. 2022). Figure 19.1 shows the major

Fig. 19.1 Different taxa of *Rhododendron* from Sikkim Himalaya: (**a**) *R. anthopogon* (**b**) *R. arboreum* subsp. *arboreum*, (**c**) *R. baileyi*, (**d**) *R. barbatum*, (**e**) *R. ciliatum*, (**f**) *R. dalhousieae*, (**g**) *R. grande*, (**h**) *R. glaucophyllum*, (**i**) *R. maddenii*, (**j**) *R. setosum*, (**k**) *R. thomsonii*, (**l**) *R arboreum* subsp. *cinnamomeum*

Rhododendron species of Sikkim Himalaya and Table 19.1 shows the diversity of *Rhododendron* in Sikkim Himalaya along with its distribution and IUCN status.

Rhododendrons in the Sikkim Himalayas display a remarkable range of morphological features. These plants vary in growth forms, from miniature, mat-forming shrubs like *R. pumilum* to dense, scrub-forming shrubs such as *R. aeruginosum* and *R. fulgens*, to large shrubs like *R. campanulatum* and *R. thomsonii*. Some species, like *R. arboreum*, *R. falconeri*, and *R. grande*, grow into towering trees, while others, like *R. camelliiflorum* and *R. vaccinioides*, adopt an epiphytic lifestyle, thriving on moss-covered cliffs and rocky slopes. The bark of *Rhododendron* varies widely, ranging from smooth and thin to coarse and multi-layered. It peels annually, revealing distinct textures and colours sooty grey, reddish-brown, pale yellow, or dark brown depending on the species. This peeling phenomenon has ecological significance, supporting mosses and other life forms on thicker bark while discouraging lichens on smooth bark. Leaves are evergreen across all regional species, displaying diversity in size, shape, and instrumentation, which hold taxonomic importance. Leaf blades can be tiny, as in *R. nivale* (0.4 cm), or large, as in *R. hodgsonii* (up to 38 cm). Surfaces vary from smooth to leathery or bullate, with some species possessing dense or wooly hairs and glandular scales of diverse colours and forms. Flowers are arranged in terminal racemes, varying from compact, globular clusters to elongated or laxly distributed arrangements. Corolla shapes include campanulate, funnelform, tubular, and ventricose, with colours ranging

Table 19.1 Taxa of *Rhododendron* from Sikkim Himalaya

Sl. No.	Name of species	Local name	Altitude (m)	Habitat	Distribution in Sikkim	Flowering season	IUCN status
1	*Rhododendron abhayae*	–	2300–2500	Temperate zone in open areas close to streams	Shingba *Rhododendron* Sanctuary	May	NE
2	*Rhododendron aeruginosum*	Nilo-patay Chimal (Nepali)	4200–5000	Alpine zone in open and rocky places	Kupuk, Nathu-L, Tamze, Tembao, Tukla, Yumesamdong, Yumthang	April to June	NE
3	*Rhododendron anthopogon*	Balu (Bhutia); Tshoo Loo (Lepcha); Sunpati, Dhupi Gurans (Nepali)	3000–5500	Sub-alpine to alpine zone in open grasslands, sandy and rocky places	Dzongri, Kupup, Thangu, Yumesamdong, Yumthang	May to June	NE
4	*Rhododendron arboreum* Sm. subsp. *arboreum*	Etok (Bhutia), Etok-Kung (Lepcha), Lali Gurans (Nepali)	1500–3600	Sub-tropical to sub-alpine zone in open and rocky slopes	Barsey *Rhododendron* Sanctuary, Lachen, Lachung, Ribdi, Yumthang	March to May	NE
5	*Rhododendron arboreum* Sm. subsp. *cinnamomeum* var. *cinnamomeum*	Gurans (Nepali)	2100–3500	Sub-tropical to sub-alpine zone in open and rocky slopes	Kyongnosla Alpine Sanctuary, Shingba *Rhododendron* Sanctuary	March to June	NE
6	*Rhododendron arboreum* Sm. subsp. *cinnamomeum* var. *roseum*	Gurans (Nepali)	2500–3500	Temperate to sub-alpine zone in mixed broad-leaved forest, and open and rocky slopes of coniferous forest	Barsey *Rhododendron* Sanctuary, Kyongnosla Alpine Sanctuary, Lachung, Lachen, Maenam Wildlife Sanctuary	February to May	NE
7	*Rhododendron baileyi*	Bailey ko Chimal (Nepali)	3100–3900	Temperate to sub-alpine zone, open slopes in coniferous forest and *Rhododendron*-forest margins	Kyongnosla Alpine Sanctuary, Shingba *Rhododendron* Sanctuary	March to April	NE
8	*Rhododendron barbatum*	Lal Chimal (Nepali)	2400–3700	Temperate to sub-alpine zone in edges of mixed broad-leaved forest, and open and rocky slopes in coniferous forest.	Barsey *Rhododendron* Sanctuary, Lachung, Lachen, Maenam, Kyongnosla Alpine Sanctuary, Shingba *Rhododendron* Sanctuary, Dzongri	March to April	NE

(continued)

Table 19.1 (continued)

Sl. No.	Name of species	Local name	Altitude (m)	Habitat	Distribution in Sikkim	Flowering season	IUCN status
9	*Rhododendron camelliiflorum*	Chia-phulay Gurans (Nepali)	2400–3600	Temperate to sub-alpine zone in mixed broad-leaved, and coniferous forest, epiphytic or growing on open and moss laden rocky slopes	Lachung, Maenam Wildlife Sanctuary, Ratey Chu, Shingba *Rhododendron* Sanctuary, Zuluk	May to July	NE
10	*Rhododendron campanulatum*	Nile Chimal (Nepali)	3000–4500	Temperate to alpine zone in coniferous forest margins, and rocky slopes in *Rhododendron* thickets.	Kyongnosla Alpine Sanctuary, Lachung, Tsomgo, Yumthang	April to May	NE
11	*Rhododendron campylocarpum*	Bongo-fale Garuns (Nepali)	1200–4300	Temperate to sub-alpine zone in rocky slopes of coniferous forest and edges of *Rhododendron* thickets	Kyongnosla Alpine Sanctuary, Shingba *Rhododendron* Sanctuary, Shiv-mandir, Yumthang	April to May	NE
12	*Rhododendron candelabrum*	Candle-tree *Rhododendron*	3400–4300	Temperate to sub-alpine zone in mixed *Rhododendron* forests and edges of coniferous forest	Shingba *Rhododendron* Sanctuary, Yumthang	April to May	EN
13	*Rhododendron ciliatum*	Jungay Chimal (Nepali)	2400–3800	Temperate to sub-alpine zone in open rocky slopes in coniferous forest, and gravelly to marshy edges of *Rhododendron* thickets	Karponang, Shingba *Rhododendron* Sanctuary, Yumthang, Zuluk	April to May	NE
14	*Rhododendron cinnabarinum* Hook. f. subsp. *cinnabarinum*	Kechung, Kema (Lepcha); Sano Chimal (Nepali)	2900–4000	Temperate to sub-alpine zone, rocky hillsides in coniferous forest, and edges of *Rhododendron* thickets	Dzongu, Kyongnosla Alpine Sanctuary, Shingba *Rhododendron* Sanctuary, Yumthang, Zuluk	April to May	NT

15	*Rhododendron dalhousieae* Hook.f. var. *dalhousieae*	Lahare Chimal (Nepali)	1700–2700	Temperate zone, epiphytic or lithophyte in broad-leaved forest, rocky cliffs and hillsides	Barsey *Rhododendron* Sanctuary, Barapathing, Chungthang, Lachung, Maenam, Pangthang	April to May	VU
16	*Rhododendron decipiens*	Hukkaune Korlingo (Nepali)	3100–3400	Temperate zone in mixed *Rhododendron* forests and edges of coniferous forest	Chiabhanjyang-Singalila	April to May	EN
17	*Rhododendron edgeworthii*	Edgeworth ko Chimal (Nepali)	2100–2800	Temperate zone, epiphytic or lithophyte in broad-leaved forest, rocky cliffs and hillsides	Lachen, Ratey Chu, Upper Dzongu	April to June	NE
18	*Rhododendron falconeri* subsp. *falconeri*	Kalma (Bhutia); Korlinga (Nepali)	2700–3500	Temperate zone, in broad-leaved forests interior and hill slopes forming dense patches	Barsey *Rhododendron* Sanctuary, Dzongri, Lachen, Menla, Maenam Wildlife Sanctuary, Pangolakha Wildlife Sanctuary	April to June	NE
19	*Rhododendron formosum* var. *formosum*	–	1300–2100	–	Assam Lingzey	March to May	NT
20	*Rhododendron fulgens*	Chimal (Nepali)	3000–4200	Temperate to alpine zone in coniferous forest margins, and rocky slopes in *Rhododendron* thickets.	Kyongnosla Alpine Sanctuary. Shingba *Rhododendron* Sanctuary, Yumthang, Thangu, Tsomgo, Upper Dzongu	April to May	NE
21	*Rhododendron glaucophyllum*	Takma Chimal (Nepali)	2900–3900	Temperate to sub-alpine zone, open slopes, rocky slopes and cliffs in coniferous forest margins.	Karponang, Kyongnosla Alpine Sanctuary, Menla, Shingba *Rhododendron* Sanctuary, Yumthang, Thangu, Yakchey, Yakla	April to May	NE
22	*Rhododendron grande*	Etok Amal (Lepcha); Patle Korlinga (Nepali)	2100–3200	Temperate zone, in broad-leaved forests interior, margins and hill slopes.	Barsey *Rhododendron* Sanctuary, Menla, Ratey Chu, Maenam Wildlife Sanctuary, Karponang	March to April	NE

(continued)

Table 19.1 (continued)

Sl. No.	Name of species	Local name	Altitude (m)	Habitat	Distribution in Sikkim	Flowering season	IUCN status
23	*Rhododendron griffithianum*	Seto Chimal (Nepali)	1800–2900	In moist temperate forest mixed with bamboo, Oaks, Magnolias and other *Rhododendron*.	Bakhim, Hilley, Chewbhanjyang, Lachen, Lachung, Ratey Chu, Tholung	April to May	NE
24	*Rhododendron hodgsonii*	Gulabi Korlinga (Nepali)	2900–3900	In temperate to sub-alpine areas, in forest ridges, and steep slopes mixed with bamboo, conifers	Bega, Dzongri, Kishong Lachen, Lachung. Menla, Shingba *Rhododendron* Sanctuary, Yumthang, Zuluk	April to June	NE
25	*Rhododendron lanatum*	Bhutle Gurans (Nepali)	3500–3900	Temperate to sub-alpine zone, steep slopes, exposed boulders and ridges of coniferous forest	Baba mandir, Dzongri, Lachung, Menmechu lake area, Nathula, Phedang. Sherathang	May to July	NE
26	*Rhododendron lepidotum*	Tsuma (Bhutia); Bhule Sunpati (Nepali)	2400–5000	Temperate to alpine zone, on steep slopes, exposed boulders, ridges of coniferous forest and open meadows mixed in *Rhododendron* scrub	Chungthang, Karponang, Lachung Lachen. Shingba *Rhododendron* Sanctuary, Yumesamdong, Yumthang, Zuluk	May to July	NE
27	*Rhododendron leptocarpum*	Jhino-phale Gurans (Nepali)	2400–4000	Temperate zone, epiphytic or lithophytic on steep slopes, and dead tree stumps in coniferous forest	Tshoka-Dzongri, Maenam Wildlife Sanctuary	April to June	EN
28	*Rhododendron lindleyi*	Sano Lahare Chimal (Nepali)	2500–3000	Temperate zone, epiphytic or lithophyte in broad-leaved forest, rocky cliffs and hillsides	Chitrey, Ratey Chu, Ravangla, Rimbi-Deorali Dara	April to May	NE
29	*Rhododendron maddenii* Hook.f. subsp. *maddenii*	Chimal (Nepali)	1500–3000	Sub-tropical to temperate zone, epiphytic or lithophytic, on steep rocky cliffs and hillside thickets	Chungthang, Mensithang, Ratey Chu, Upper Dzongu	April to June	EN

30	*Rhododendron mekongense* Franch. var. *mekongense*	–	2900–4400	Temperate to alpine zone, in open rocky slopes	Shingba *Rhododendron* Sanctuary, Lum area in Upper Dzongu	May to June	NE
31	*Rhododendron nivale*	Hiun Gurans (Nepali)	4500–5800	Open-mountain slopes and screes in alpine zone	Tembao, Yumesamdong	June to July	NE
32	*Rhododendron niveum*	Hiun-pate Gurans (Nepali)	2900–3700	Temperate zone in rocky valleys and ridges of coniferous forests	Kyongnosla Alpine Sanctuary, Shingba *Rhododendron* Sanctuary, Upper Dzongu, Yakchey	April to May	EN
33	*Rhododendron pendulum*	Jhundinae Chimal (Nepali)	2900–3700	Temperate to sub-alpine zone in coniferous forest, epiphytic on Abies densa tree trunks, and moss laden rocky cliffs	Kyongnosla Alpine Sanctuary, Lachen, Menla, Shingba *Rhododendron* Sanctuary, Tshoka-Dzongri	April to May	NE
34	*Rhododendron pumilum*	Purke Gurans (Nepali)	3400–4300	Sub-alpine to alpine zone in open places and stream banks, often on moss laden boulders	Kupup, Lhonak, Yakchey-Yumthang, Tshoka-Dzongri, Zemu	May to June	EN
35	*Rhododendron setosum*	Tsallu (Bhutia, Nepali)	3600–4900	Open-mountain slopes and screes in sub-alpine to alpine zone	Dzongri, Katao, Kupup, Lhonak, Muguthang, Tembao, Thangu, Yumesamdong, Yumthang	May to July	NE
36	*Rhododendron sikkimense*	Sikkimae Gurans (Nepali)	3000–3700	Temperate to sub-alpine zone in mixed *Rhododendron* forests and edges of coniferous forest	Kyongnosla Alpine Sanctuary, Menla, Menmechu Lake area, Phuni, Shingba *Rhododendron* Sanctuary	April to June	NE
37	*Rhododendron sundarraianum*	–	3300–3600	Temperate to sub-alpine zone in mixed *Rhododendron* forests and edges of coniferous forest	Shingba *Rhododendron* Sanctuary	April to May	DD
38	*Rhododendron thomsonii* subsp. *thomsonii*	Dr. Thomson ko Gurans (Nepali)	3200–4300	Temperate to sub-alpine zone in mixed *Rhododendron* forests and edges of coniferous forest.	Dzongri, Jakthang, Kyongnosla Alpine Sanctuary, Lungthung, Shingba *Rhododendron* Sanctuary, Thangu, Tsomgo, Yumthang	April to May	NE

(continued)

Table 19.1 (continued)

Sl. No.	Name of species	Local name	Altitude (m)	Habitat	Distribution in Sikkim	Flowering season	IUCN status
39	*Rhododendron thupdenii*	Thupden Tsering's *Rhododendron*	3500–3900	Temperate to sub-alpine zone in coniferous forest	Upper Yumthang valley	May	NE
40	*Rhododendron triflorum*	Pahele Chimal (Nepali)	2200–3200	Temperate zone, rocky hill-sides and cliffs in broad-leaved and coniferous forest margins.	Barsey *Rhododendron* Sanctuary, Lachung, Lachen, Yakchey, Zuluk	May to June	LC
41	*Rhododendron vaccinioides*	Khiaune-patey Gurans (Nepali)	2000–3000	Temperate zone in moist shady places along mixed broad-leaved forest fringe, epiphytic or growing on open and moss laden rocky slopes.	Barsey *Rhododendron* Sanctuary, Chungthang, Fambonglho Wildlife Sanctuary, Lachung, Maenam, Ratey Chu, Tendong, Zuluk	May to June	NE
42	*Rhododendron virgatum*	Flanginae Gurans (Nepali)	2200–3200	Temperate zone, rocky hill-sides and steep slopes along broad-leaved, and coniferous forest margins	Bichu, Chungthang, Karponang, Lachung, Lachen-Chaten	March to April	NE
43	*Rhododendron wallichii*	Wallich ke Chimal, Nilo Chimal (Nepali)	3000–4300	Temperate to alpine zone along coniferous forest margins, slopes, and in *Rhododendron* thickets	Shingba *Rhododendron* Sanctuary, Shiv-mandir, Yumthang	April to May	NE
44	*Rhododendron wightii*	Dr. Wight ko Gurans (Nepali)	3500–4500	Sub-alpine to alpine zone in valleys, river-side slopes and open rocky places	Dzongri, Kupup, Menmechu Lake area, Shiv-mandir, Tamze, Tukla, Tsomgo, Yumesamdong, Yumthang	April to May	NE

DD data deficient, *EN* endangered, *NE* not evaluated, *NT* near threatened, *VU* vulnerable from plain red, pink, white, and yellow to intricate patterns of blotches or freckles. Stamens number 10–25 and vary in length and instrumentation, contributing to species identification. The calyx is often reduced but is distinct and variable in some species, from cup-shaped to irregularly lobed. Fruits are septicidal capsules that vary from ovoid or oblong to linear-cylindrical, with some, like *R. campylocarpum*, curving upon maturity. These traits, combined with their ecological adaptations, underline the extraordinary diversity of *Rhododendron* in this Himalayan region (Gogoi et al. 2022).

19.2.2 Population Structure and Genetic Diversity

Population Structure and Genetic diversity is a fundamental aspect of biodiversity, essential for the sustainable use of germplasm resources and the enhancement of genetic traits. It also serves as a foundation for devising effective conservation strategies. The resilience and evolutionary potential of a species in the face of adverse environmental conditions are intrinsically linked to its genetic diversity and population genetic structure. High genetic diversity equips species with a broader range of adaptive traits, enabling them to survive and thrive in changing environments. Similarly, a well-structured population facilitates gene flow and reduces the risk of inbreeding, ensuring long-term survival and evolutionary adaptability. Advances in molecular biology have significantly facilitated the analysis of genetic diversity, breeding, and evolutionary studies through the use of molecular markers. Among these, simple sequence repeat (SSR) markers, such as expressed sequence tag SSR (EST-SSR) and genomic SSR (gSSR), are widely valued for their high polymorphism, reliability, and ease of application (Stebbins 1950; Frankham 2003; Wei et al. 2011; Liu et al. 2012, 2019; Mable 2019). Dar et al. (2024) studied the population structure and genetic diversity of *R. anthopogon* across the altitudinal gradient in the Kashmir Himalaya using SCoT (Start Codon Targeted) markers. They found genetic diversity decreased from 3200 to 3600 m, then increased from 3601 to 4000 m. The highest diversity was at higher altitudes, with low gene flow ($Nm = 0.1804$) and significant genetic differentiation ($Gst = 0.7349$). Here 'Nm' stands for the number of migrants per generation, and 'Gst' is a measure of genetic differentiation between populations. The study revealed that altitude drives genetic variation and that higher altitude populations may be important for migration because of climate change (Dar et al. 2024). Sharma et al. (2020) developed 249 SSR markers from *R. catawbiense* and *R. mucronatum* EST data to study genetic diversity in *R.* arboreum. 168 SSR sequences were used to design primer pairs, and 26 out of 30 primers successfully amplified 36 genotypes of *R. arboreum* from the western Himalayas. They found di-nucleotide repeats were the most prevalent, followed by tri- and tetra-nucleotide repeats. The genetic diversity analysis revealed a range of heterozygosity and PIC (Polymorphism Information Content) values, with the markers effectively distinguishing geographic differences. These EST-SSRs are valuable for micro-evolutionary studies and population genetics in *Rhododendron* species (Sharma et al. 2020).

19.3 Traditional and Cultural Significance of *Rhododendron*

Rhododendron species have gained significant value in the international market, reaching a total of 110 million United States dollars (USD) for perfumes, essential oils, foods, and medicines. This market is largely dominated by countries such as Belgium (44.1 million USD), the Netherlands (37.4 million USD), Germany (16.0 million USD), France (14.8 million USD), Finland (11.4 million USD), and China (10.98 million USD), with notable contributions from Denmark, Britain, Japan, the USA, and others (OEC 2024). The red-flowered *Rhododendron arboreum* holds deep cultural and ecological significance across the Himalayan region. It is the national flower of Nepal, where it is known as *lali gurans*. During its blooming season, women often adorn their hair with its vibrant blossoms, which are also offered at Buddhist temples. In Nepal, the fresh petals are used to flavour sweet drinks, consumed with salt and chilies, and even incorporated into fish curries, where they are traditionally believed to help dissolve fish bones lodged in the throat. In Uttarakhand, India, *R. arboreum* is honoured as the state tree and is widely used to prepare the popular local beverage, "buransh squash".

The ritual and medicinal uses of *Rhododendron* extend to other species as well. For example, the fragrant leaves of *R. anthopogon* are commonly used as incense in Buddhist religious practices, and its essential oils are valued for their calming properties. The wood of various *Rhododendron* species is utilised for making utensils, furniture, and fences, and is also used as firewood and for charcoal production demonstrating the plant's wide-ranging utility in rural Himalayan livelihoods (Srivastava 2012). These cultural and practical uses are in line with the broader ethnobotanical patterns described in Chap. 4 (Sect. 4.3), which highlights the multipurpose nature of medicinal and ritual plants in the Himalayan region.

Many *Rhododendron* species are also employed in traditional medicine systems, reflecting their pharmacological relevance. *R. arboreum*, in particular, is used to treat diarrhoea and dysentery (Idrisi et al. 2010; Lepcha et al. 2019), as well as heart ailments and stomach disorders (Kumar et al. 2019). Fresh or dried petals are chewed to relieve throat pain, blood dysentery, and gastrointestinal issues, while young leaves are traditionally used to ease headaches (Badola and Pradhan 2003). These medicinal applications align with the broader theme of plant-based traditional healthcare explored in Chap. 13 (Sect. 13.2) and support the conservation-focused discussions in Chap. 16 (Sect. 16.4), where the sustainable use and cultural value of threatened Himalayan species are highlighted as integral to biodiversity preservation and community well-being.

Its petals, rich in phytochemicals, have been found to inhibit severe acute respiratory syndrome coronavirus 2 (SARS-CoV-2) infection in vitro, showcasing its promising antiviral properties (Lingwan et al. 2023). The root bark is used to treat pneumonia, dried flowers are used for dysentery and menstrual disorders, and the flowers are also fermented to produce a local wine, 'Guranse,' which serves as an antidote to altitude sickness (Mandal et al. 2013). *R. anthopogon* is valued for its medicinal properties, with leaves, twigs, and flowers being used to treat indigestion, vomiting, respiratory disorders, stomach aches, and dysentery (Tamang et al. 2017). It is also employed in treating headaches, rheumatism, and inflammation (Nand and Naithani 2018). *R. setosum* is traditionally used to heal wounds, with young leaves being applied directly (Tamang et al. 2017). *R. campanulatum* is known for its effectiveness in relieving cough, where leaves are chewed, and the juice from crushed leaves is used as a remedy (Pradhan and Badola 2008). These biological activities of *Rhododendron* are attributed to biologically active compounds such as rhodojaponin III (which has antihypertensive effects), daurichromenic acid (known for its anti-inflammatory and anti-HIV properties), and andromedotoxin (which induces hypotension and bradycardia) (Popescu and Kopp 2013).

Flower juice of some *Rhododendron*, such as *R. arboreum*, is a popular refreshing drink, valued for its potential medicinal benefits. Beyond juice and beverages, the flowers hold promises for innovative marketing strategies and product development. They can be processed into dried products such as powders, chutney mixes, and pickles. Traditionally, *Rhododendron* flowers are used to make chutney by crushing them and mixing them with mint, salt, and souring agents like tamarind or pomegranate seeds (Kumar et al. 2019). *Rhododendron* honey, sometimes referred to as wild, mad, or toxic honey, has been recognised since ancient times for its unique properties. Produced by bees collecting nectar from *Rhododendron* flowers, it has traditionally been used as a medicine for various ailments. The honey contains active compounds such as grayanotoxin and andromedotoxin, which, when consumed in large quantities, can cause adverse effects like severe vertigo, low blood pressure, and bradycardia (Jansen et al. 2012; Rasgele and Kekecoglu 2013). However, research has shown its potential therapeutic benefits, including lowering blood glucose and lipid levels in diabetic rats. Additionally, *Rhododendron* honey exhibits antibacterial, antifungal, antioxidant, and antimicrobial properties (Silici et al. 2010).

19.4 Threats to *Rhododendron* Diversity

Rhododendron in their natural habitats face significant environmental pressures, particularly in the fragile mountainous regions that are highly vulnerable to natural calamities and anthropogenic threats. These species, vital for delivering ecosystem services in high-altitude areas, are under increasing threat due to habitat degradation, unsustainable exploitation, and developmental activities. There are 1157 taxa of *Rhododendron* and 39 identified as endangered and 36 as critically endangered (Gibbs et al. 2011). Sikkim Himalaya is home to many ethnic groups whose traditional practices significantly influence *Rhododendron* survival. Habitat degradation caused by landslides, flash floods, and forest clearing poses a major threat in this region. For example, a 1980 landslide near Yakchey in Lachung, North Sikkim, wiped out an entire population of *R. niveum*. Landslide-prone areas, such as the Shingba *Rhododendron* Sanctuary, also face similar threats (Gogoi et al. 2022).

The regeneration of *Rhododendron* species is significantly hindered by various environmental and biological factors, which is another major threat. One major challenge is the allelopathic effect exerted by their leaves or litter layer, which can inhibit the growth and development of nearby plants, including their own saplings. Additionally, adverse environmental conditions, both edaphic (soil-related) and climatic, further exacerbate the issue, creating unfavourable habitats for seedling establishment. Poor seed germination is another critical factor contributing to the decline of these exquisite species. Once germinated, saplings often face early mortality due to limited sunlight penetration on the forest floor, as dense canopy cover restricts light availability. The rapid accumulation of leaf litter further complicates their survival by creating physical and chemical barriers to growth. These combined factors severely impact the natural regeneration dynamics of *Rhododendron* species, posing a significant challenge to their conservation (Facelli and Pickett 1991; Sundriyal and Sharma 1996; Singh et al. 2010).

Rapid climate change has caused shifts in species distributions across various taxa, with many moving poleward or to higher elevations as global temperatures rise. If greenhouse gas emissions remain unchecked, narrowly distributed species with small populations face a high risk of extinction within the next century, though extinction estimates vary based on taxa, scale, and methodologies (Parmesan 2006; Chen et al. 2011; Pearson et al. 2014; Lenoir and Svenning 2015; Frishkoff et al. 2016). Yu et al. (2019) predict the impacts of future climate and land use changes on the distribution of 10 narrow- and 10 wide-ranging *Rhododendron* species in China by 2070, using MaxEnt modeling. Their findings reveal that narrow-ranging species are more vulnerable, with potential extinctions and range shifts, while wide-ranging species face overall range reductions, highlighting species-specific responses to environmental changes (Yu et al. 2019).

In addition to climate change, human-induced habitat loss and fragmentation from activities like deforestation, agriculture, urbanisation, and infrastructure development are major drivers of biodiversity loss. The combined effects of climate change and land use alterations are expected to amplify impacts on ecosystems, making it crucial to predict species' responses for effective biodiversity conservation (Sala et al. 2000; Foley et al. 2005; Brook et al. 2008; Oliver and Morecroft 2014). Overharvesting, particularly of species like *R. anthopogon* for incense production, exacerbates the problem. The species of *Rhododendron* are often exploited for their fine-grained wood and resin, which are used as firewood and charcoal, threatening species such as *R. barbatum, R. nivale, R. setosum, R. wallichii, and R. wightii.* Shrinking habitats due to development projects further strain these populations. Activities like road construction and the expansion of military establishments, especially in sensitive areas like Yakchey, have resulted in significant habitat loss and uprooting of plants, including epiphytic species and seedlings crucial for biodiversity. Additionally, tourism-driven practices, such as roadside cleanings, inadvertently damage *Rhododendron* populations and associated biodiversity. A lack of awareness and basic ecological knowledge among authorities and the public amplifies the damage, posing an ongoing threat to these iconic species and their ecosystems (Gogoi et al. 2022). Species such as *R edgeworthii, R. fulgens,*

R. niveum, R. leptocarpum, R. lindleyi, R. pumilum, and *R. sikkimense* are under significant threat in the Sikkim and Darjeeling Himalayas and require immediate conservation efforts (Gogoi et al. 2022). These species are crucial to the region's biodiversity and play an essential role in maintaining ecological balance. The increasing environmental pressures, habitat degradation, and unsustainable practices have placed these *Rhododendron* species at risk. Therefore, targeted conservation strategies are needed to protect their populations and ensure their survival in this fragile ecosystem.

19.5 Conservation Strategies

Rhododendron, known for their ecological, medicinal, and cultural significance, are a vital component of Himalayan biodiversity. However, increasing anthropogenic pressures, habitat loss, and climate change have threatened their survival, necessitating effective conservation strategies. These strategies aim to protect their natural habitats, preserve genetic diversity, and ensure sustainable utilisation. Conservation efforts also focus on involving local communities and leveraging modern scientific techniques to balance ecological preservation with socio-economic benefits.

Protected Areas (PAs) are globally recognised as essential tools for in-situ conservation and safeguarding natural communities and ecosystems. As of May 2025, India encompasses a total of 1022 Protected Areas (covering 178,640.69 km^2), which represents 5.43% of the country's geographical area of 3,287,263 km^2. This includes a total of 106 National Parks (44,402.94 km^2, 1.35%), 573 Wildlife Sanctuaries (127,197.55 km^2, 3.87%), 220 Community Reserves (1455.15 km^2, 0.04%), and 123 Conservation Reserves (5585.05 km^2, 0.17%). Additionally, India's forest and tree cover, as per the Forest Survey of India (2021), spans 809,537 km^2, accounting for 24.62% of its geographical area, with forest cover contributing 21.71% and tree cover 2.91%. Khangchendzonga National Park, established in 1977, was the first PA in Sikkim, and then the Kyongnosla Alpine Wildlife Sanctuary was established in the same year. Over time, PAs gradually increased to eight, with the latest addition being Kitam Wildlife Sanctuary on February third, 2005. With approximately 31% of its geographical area under protection, Sikkim has the highest percentage of land dedicated to protected areas among Indian states (Manish and Pandit 2019). Table 19.2 shows the major protected areas of Sikkim Himalaya having rich diversity of *Rhododendron*.

Botanic gardens and arboreta serve as essential repositories for the ex-situ conservation of plant diversity. However, global analyses by Botanic Gardens Conservation International (BGCI) reveal challenges such as the disproportionate presence of species in temperate regions, their cultivation outside native habitats, and limited representation across institutions. To address these issues, The Global Conservation Consortium for *Rhododendron* was founded in 2018 to facilitate and coordinate conservation efforts on a global scale. Although 78% of threatened *Rhododendron* taxa currently conserved in botanic gardens and ongoing projects

Table 19.2 Protected areas of the Sikkim Himalaya having rich *Rhododendron* diversity

Sl. No.	Name of the protected areas	Area (Sq km)	Altitude range	*Rhododendron* species
1	Barsey *Rhododendron* Sanctuary	104.00	2200–4100	*R. arboreum* Sm. Exot. subsp. *arboreum*, *R. arboreumsm* subsp. *cinnamomeum* var. *roseum*, *R dalhousieae*, *R. barbatum*, *R. falconeri*, *R. grande*, *R. triflorum*, *R vaccinioides*
2	Fambonglho Wildlife Sanctuary	51.76	1524–2749	*R. vaccinioides*, *R dalhousieae*
3	Kyongnosla Alpine Sanctuary	31.00	3292–4116	*R. sikkimense*, *R. arboreumsm* subsp. *cinnamomeum*, *R. baileyi*, *R. barbatum*, *R. campanulatum*, *R. campylocarpum*, *R. cinnabarinum*, *R. fulgens*, *R. glaucophyllum*, *R. niveum*, *R. pendulum*, *R. thomsonii*
4	Maenam Wildlife Sanctuary	53.34	2400–3263	*R. arboreumsm* subsp. *cinnamomeum*, *R. barbatum*, *R. camelliiflorum. Dalhousieae*, *R. falconeri*, *R. grande*, *R. leptocarpum*, *R. vaccinioides*
5	Shingba *Rhododendron* Sanctuary	43.00	3048–4575	This sanctuary contains 25 taxa of *Rhododendron*, includes *R. arboreumsm* subsp. *cinnamomeum*, *R. baileyi*, *R. barbatum*, *R. camelliiflorum*, *R. campylocarpum*, *R. ciliatum*, *R. cinnabarinum*, *R. fulgens*, *R. glaucophyllum*, *R. abhayae*, *R. sikkimense*, etc.

targeting their preservation, genetic representation in these ex-situ collections remains underexplored (Hu et al. 2024). The ex-situ conservation of *Rhododendron* subsection Maddenia highlights the efforts of several botanical gardens worldwide, which collectively hold 55 of the 65 taxa in cultivation. Leading in this endeavour are the *Rhododendron* Species Botanical Garden (USA), which conserves 48 taxa, and the Royal Botanic Garden Edinburgh (UK), including its sites at Edinburgh, Logan, and Benmore, with 37 taxa. The University of California Botanical Garden at Berkeley (USA) follows with 33 taxa, while the Dandenong Ranges Botanic Garden (Australia) and Dunedin Botanic Garden (New Zealand) host 31 and 27 taxa, respectively. Other significant contributors include RhodoBG_02 (Australia), with 25 taxa; Pukeiti Garden (New Zealand), with 23 taxa; and Royal Botanic Gardens Kew (UK), including Richmond and Wakehurst, with 19 taxa. The Blue Mountains Botanic Garden (Australia) and the Germplasm Bank of Wild Species (China) also play pivotal roles, conserving 19 and 14 taxa, respectively. Smaller but notable contributions come from the St Andrews Botanic Garden (UK) and the University of British Columbia Botanical Garden (Canada), which house 14 and 11 taxa. These gardens are critical repositories for subsection Maddenia's conservation, with 86%

of taxa, including all the threatened taxa, represented in cultivation. However, gaps in genetic diversity representation remain, particularly in wild accessions, emphasising the need for expanded collections and international collaboration (Hu et al. 2024).

The Botanical Survey of India (BSI) at its Gangtok regional centre has successfully introduced several species of *Rhododendron*, including *R. arboreum* subsp. *arboretum*, *R. Dalhousie*, *R. formosum*, and *R. madden*, as part of their ex-situ conservation efforts. Additionally, the Sikkim Forest Department has made notable progress in introducing *Rhododendron* species, including the successful establishment of several species at the State Conservatory in Bulbulay near Gangtok. Of particular note is the successful introduction of *R. niveum* around the periphery of Kyongnosla Alpine Sanctuary, a significant achievement in conserving the region's unique *Rhododendron* diversity (Gogoi et al. 2022).

Tissue culture is vital for conserving *Rhododendron* species in Sikkim, a region rich in biodiversity. Many *Rhododendron* species are threatened by habitat loss, climate change, and over-exploitation. Tissue culture techniques, such as micropropagation, enable large-scale propagation in controlled environments, preserving genetic diversity. This approach helps overcome slow natural regeneration, supporting the restoration and reintroduction of these species into their native habitats and contributing to long-term conservation efforts in Sikkim. Singh et al. (2016) developed an in vitro propagation protocol for *R. griffithianum* (Seto Chimal), an endangered species of the Sikkim Himalaya, to address its decline caused by habitat loss, exploitation, and environmental disturbances. Seeds from the Barsey *Rhododendron* Sanctuary were cultured on Anderson's medium with antioxidants and GA3, while shoot proliferation was optimised using Anderson's medium enhanced with 5 Micromolar (μM) N6-(2-isopentenyl) adenine (2-iP) and 1.0 μM Indole-3-Butyric Acid (IBA), achieving 75% proliferation with an average of 11 shoots per explant. Rooting on Anderson's liquid medium containing 1.5 μM IBA showed 72% success, and hardened plantlets exhibited an 84% survival rate, offering a practical approach to conservation (Singh et al. 2016). The Sikkim Regional Centre of the GB Pant Institute of Himalayan Environment and Development, in collaboration with the Department of Forest, Government of Sikkim, the institute has established a two-hectare conservation area within the Himalayan Zoological Park, Bulbulay, Gangtok. This initiative includes planting tissue-culture-raised *R. maddeni* and traditionally propagated six *Rhododendron* species (*R. griffithianum*, *R. dalhousieae*, *R. grande*, *R. maddeni*, *R. baileyii*, and *R. ciliatum*) in both the park and the institute's arboretum, where they are thriving (Kumar et al. 2004; Singh et al. 2008a, b, 2009). The conservation of *Rhododendron* species in Sikkim requires a combination of traditional and advanced methods. In-situ conservation, habitat restoration, and tissue culture, along with seed banking and ex-situ techniques, play vital roles in preserving genetic diversity and ensuring long-term sustainability. These integrated strategies help protect *Rhododendron* species and their unique diversity in the face of environmental challenges.

19.6 Conclusions

The diverse *Rhododendron* species of the Sikkim Himalayas represent not only ecological and genetic wealth but also cultural and ethnomedicinal heritage. Their wide-ranging adaptability and significance underscore the need for urgent conservation efforts. Despite their resilience, these species face mounting threats from habitat degradation, climate change, and human activity. Effective conservation strategies, including habitat protection, ex-situ propagation, and community involvement, are vital for their sustainable management. Integrating modern genetic tools with traditional knowledge offers a promising pathway to preserve this invaluable floral diversity. Ensuring the survival of *Rhododendron* in the Sikkim Himalayas is crucial, not just for biodiversity conservation but also for maintaining the region's ecological balance and cultural identity.

19.7 Future Prospects

The conservation and sustainable management of *Rhododendron* species in the Sikkim Himalayas present several future opportunities. Advancements in genetic research, such as DNA barcoding and genome sequencing, can enhance our understanding of population dynamics, genetic diversity, and adaptive traits, aiding in the identification of conservation priorities. In vitro propagation techniques and cryopreservation can further support ex-situ conservation efforts. Ecotourism centered on *Rhododendron* blooms can promote regional economic growth while fostering awareness of their ecological significance. Additionally, exploring the bioactive compounds in *Rhododendron* for pharmacological applications could provide innovative solutions in medicine, generating incentives for their preservation. Engaging local communities through participatory conservation programmes and incorporating traditional knowledge into sustainable use practices will ensure the long-term survival of these species. Climate-resilient restoration strategies, such as assisted migration and habitat augmentation, should be developed to mitigate the impacts of climate change. By integrating scientific advancements with policy frameworks and community involvement, the future of *Rhododendron* conservation in the Sikkim Himalayas holds immense potential for success.

Acknowledgments We sincerely express our gratitude to Assistant Director In-Charge, Regional Ayurveda Research Institute (RARI), Gangtok, for his unwavering support, encouragement, and valuable guidance throughout the preparation of this book chapter. We also thank the National Medicinal Plants Board (NMPB) for their generous funding and financial assistance.

References

Badola HK, Pradhan BK (2003) Plants used in healthcare practices by limboo tribe in south--west of khangchendzonga biosphere reserve, Sikkim, India. IJTK 12:355–369

Basnett S, Ganesan R (2022) A comprehensive review on the taxonomy, ecology, reproductive biology, economic importance and conservation status of Indian Himalayan *Rhododendrons*. Bot Rev 88:505–544. https://doi.org/10.1007/s12229-021-09273-z

Brook B, Sodhi N, Bradshaw C (2008) Synergies among extinction drivers under global change. Trends Ecol Evol 23:453–460. https://doi.org/10.1016/j.tree.2008.03.011

Chamberlain D, Hyam R, Argent G et al (1996) The genus *Rhododendron*: its classification and synonymy. Royal Botanic Garden, Edinburgh

Chen I-C, Hill JK, Ohlemüller R et al (2011) Rapid range shifts of species associated with high levels of climate warming. Science (1979) 333:1024–1026. https://doi.org/10.1126/science.1206432

Clinton BD, Vose JM (1996) Effects of *Rhododendron maximum* L. on *Acer rubrum* L. seedling establishment. Castanea 61:38–45

Dar TUH, Mangral ZA, Islam SU et al (2024) Genetic variation and population structure of *Rhododendron anthopogon* along an altitudinal gradient: a case study from Himalaya. Plant Mol Biol Report 42:650–664. https://doi.org/10.1007/s11105-024-01438-5

Eşen D, Zedaker SM, Kirwan JL, Mou P (2004) Soil and site factors influencing purple-flowered rhododendron (*Rhododendron ponticum* L.) and eastern beech forests (Fagus orientalis Lipsky) in Turkey. For Ecol Manag 203:229–240. https://doi.org/10.1016/j.foreco.2004.07.052

Facelli JM, Pickett STA (1991) Plant litter: light interception and effects on an old-field plant community. Ecology 72:1024–1031. https://doi.org/10.2307/1940602

Foley JA, DeFries R, Asner GP et al (2005) Global consequences of land use. Science (1979) 309: 570–574. https://doi.org/10.1126/science.1111772

Frankham R (2003) Genetics and conservation biology. C R Biol 326:22–29. https://doi.org/10.1016/S1631-0691(03)00023-4

Frishkoff LO, Karp DS, Flanders JR et al (2016) Climate change and habitat conversion favour the same species. Ecol Lett 19:1081–1090. https://doi.org/10.1111/ele.12645

Gibbs D, Chamberlain D, Argent G (2011) The red list of *Rhododendrons*. Botanic Gardens Conservation International, Richmond

Gogoi R, Sherpa N, Mao AA et al (2022) Rhododendrons of Sikkim and Darjeeling Himalaya: an illustrated account. Botanical Survey of India, Kolkata and Directorate of Cinchona and Other Medicinal Plants, West Bengal

Hu L, MacKay M, Gardiner SE, Tate JA (2024) *Rhododendron* diversity conservation in global botanic gardens: a case study of *Maddenia* species. Oryx:1–11. https://doi.org/10.1017/S0030605324000759

Idrisi MS, Badola HK, Singh R (2010) Indigenous knowledge and medicinal use of plants by local communities in Rangit Valley, South Sikkim, India. NeBIO 1:34–45

International Union for Conservation of Nature (2023) IUCN 2022: International Union for Conservation of Nature annual report. IUCN, Gland

Jansen SA, Kleerekooper I, Hofman ZLM et al (2012) Grayanotoxin poisoning: 'mad honey disease' and beyond. Cardiovasc Toxicol 12:208–215. https://doi.org/10.1007/s12012-012-9162-2

Kumar P (2012) Assessment of impact of climate change on Rhododendrons in Sikkim Himalayas using Maxent modelling: limitations and challenges. Biodivers Conserv 21:1251–1266. https://doi.org/10.1007/s10531-012-0279-1

Kumar S, Singh K, Rai L (2004) In vitro propagation of an endangered Sikkim Himalayan rhododendron (*R. maddenii*) from cotyledonary nodal segments. Digital Library and Archives of the Virginia Tech University Libraries

Kumar V, Suri S, Prasad R et al (2019) Bioactive compounds, health benefits and utilization of *Rhododendron*: a comprehensive review. Agric Food Secur 8:6. https://doi.org/10.1186/s40066-019-0251-3

Lenoir J, Svenning J-C (2015) Climate-related range shifts – a global multidimensional synthesis and new research directions. Ecography 38:15–28. https://doi.org/10.1111/ecog.00967

Lepcha TT, Pradhan P, Gaira KS et al (2019) Ethnomedicinal use of plants by Bhutia tribe in Sikkim Himalaya. In: Proceedings of the Ist Himalayan researchers consortium, Dehradun, pp 71–78

Lingwan M, Shagun S, Pahwa F et al (2023) Phytochemical rich Himalayan *Rhododendron arboreum* petals inhibit SARS-CoV-2 infection *in vitro*. J Biomol Struct Dyn 41:1403–1413. https://doi.org/10.1080/07391102.2021.2021287

Liu Y-F, Xing M, Zhao W et al (2012) Genetic diversity analysis of *Rhododendron aureum* Georgi (Ericaceae) located on Changbai Mountain using ISSR and RAPD markers. Plant Syst Evol 298:921–930. https://doi.org/10.1007/s00606-012-0601-0

Liu Y, Geng Y, Song M et al (2019) Genetic structure and diversity of *Glycyrrhiza* populations based on transcriptome SSR markers. Plant Mol Biol Report 37:401–412. https://doi.org/10.1007/s11105-019-01165-2

Mable BK (2019) Conservation of adaptive potential and functional diversity: integrating old and new approaches. Conserv Genet 20:89–100. https://doi.org/10.1007/s10592-018-1129-9

Mandal D, Panda AK, Rana M (2013) Medicinal plants used in folk medicinal practices available in rich biodiversity of Sikkim. Environ Ecol 31:1445–1449

Manish K, Pandit MK (2019) Identifying conservation priorities for plant species in the Himalaya in current and future climates: a case study from Sikkim Himalaya, India. Biol Conserv 233:176–184. https://doi.org/10.1016/j.biocon.2019.02.036

Nand K, Naithani S (2018) Ethnobotanical uses of wild medicinal plants by the local community in the Asi Ganga Sub-basin, Western Himalaya. J Complement Med Res 9:34. https://doi.org/10.5455/jcmr.20180507034822

Oliver TH, Morecroft MD (2014) Interactions between climate change and land use change on biodiversity: attribution problems, risks, and opportunities. WIREs Clim Change 5:317–335. https://doi.org/10.1002/wcc.271

Parmesan C (2006) Ecological and evolutionary responses to recent climate change. Annu Rev Ecol Evol Syst 37:637–669. https://doi.org/10.1146/annurev.ecolsys.37.091305.110100

Pearson RG, Stanton JC, Shoemaker KT et al (2014) Life history and spatial traits predict extinction risk due to climate change. Nat Clim Chang 4:217–221. https://doi.org/10.1038/nclimate2113

Popescu R, Kopp B (2013) The genus *Rhododendron*: an ethnopharmacological and toxicological review. J Ethnopharmacol 147:42–62. https://doi.org/10.1016/j.jep.2013.02.022

Pradhan BK, Badola HK (2008) Ethnomedicinal plant use by Lepcha tribe of Dzongu valley, bordering Khangchendzonga Biosphere Reserve, in North Sikkim, India. J Ethnobiol Ethnomed 4:22. https://doi.org/10.1186/1746-4269-4-22

Rasgele PG, Kekecoglu M (2013) Physico-chemical properties of *Rhododendron* honey produced in Turkey. Herba Pol 59:88–97. https://doi.org/10.2478/hepo-2013-0019

Sala OE, Stuart Chapin F, Armesto JJ et al (2000) Global biodiversity scenarios for the year 2100. Science (1979) 287:1770–1774. https://doi.org/10.1126/science.287.5459.1770

Sekar KC, Srivastava SK (2010) Rhododendrons in Indian Himalayan region: diversity and conservation. Am J Plant Sci 01:131–137. https://doi.org/10.4236/ajps.2010.12017

Sharma H, Kumar P, Singh A et al (2020) Development of polymorphic EST-SSR markers and their applicability in genetic diversity evaluation in *Rhododendron arboreum*. Mol Biol Rep 47:2447–2457. https://doi.org/10.1007/s11033-020-05300-1

Silici S, Sagdic O, Ekici L (2010) Total phenolic content, antiradical, antioxidant and antimicrobial activities of *Rhododendron* honeys. Food Chem 121:238–243. https://doi.org/10.1016/j.foodchem.2009.11.078

Singh K, Kumar S, Rai S (2008a) Raising planting materials of Sikkim Himalayan *Rhododendron* through vegetative propagation using 'Air-wet technique'. Digital Library and Archives of the Virginia Tech University Libraries

Singh KK, Kumar S, Pandey A (2008b) Soil treatments for improving seed germination of rare and endangered Sikkim Himalayan rhododendrons. World J Agric Sci 4:288–296

Singh KK, Rai LK, Gurung B (2009) Conservation of rhododendrons in Sikkim Himalaya: an overview. World J Agric Sci 5:284–296

Singh KK, Gurung B, Rai LK, Nepal LH (2010) The influence of temperature, light and pre-treatment on the seed germination of critically endangered Sikkim Himalayan rhododendron (*R. niveum* Hook *f.*). J Am Sci 6:172–177

Singh KK, Singh M, Chettri A (2016) In vitro propagation of *Rhododendron griffithianum* Wt.: an endangered *Rhododendron* species of Sikkim Himalaya. J Appl Biol Biotechnol. https://doi.org/10.7324/JABB.2016.40211

Srivastava P (2012) *Rhododendron arboreum*: an overview. J Appl Pharm Sci 2:158–162

Stebbins GL (1950) Chapter VI. Isolation and the origin of species. In: Variation and evolution in plants. Columbia University Press, New York, pp 189–250

Sundriyal RC, Sharma E (1996) Anthropogenic pressure on tree structure and biomass in the temperate forest of Mamlay watershed in Sikkim. For Ecol Manag 81:113–134. https://doi.org/10.1016/0378-1127(95)03657-1

Tamang M, Pal K, Kumar Rai S et al (2017) Ethnobotanical survey of threatened medicinal plants of West Sikkim. Int J Bot Stud 2:116–125

Tiwari ON, Chauhan UK (2006) *Rhododendron* conservation in Sikkim Himalaya. Curr Sci 90: 532–541

Vetaas OR (2002) Realized and potential climate niches: a comparison of four *Rhododendron* tree species. J Biogeogr 29:545–554. https://doi.org/10.1046/j.1365-2699.2002.00694.x

Wei W, Qi X, Wang L et al (2011) Characterization of the sesame (*Sesamum indicum* L.) global transcriptome using Illumina paired-end sequencing and development of EST-SSR markers. BMC Genomics 12:451. https://doi.org/10.1186/1471-2164-12-451

Xia X, Gong R, Zhang C (2022) Integrative analysis of transcriptome and metabolome reveals flavonoid biosynthesis regulation in *Rhododendron pulchrum* petals. BMC Plant Biol 22:401. https://doi.org/10.1186/s12870-022-03762-y

Yu F, Wang T, Groen TA et al (2019) Climate and land use changes will degrade the distribution of Rhododendrons in China. Sci Total Environ 659:515–528. https://doi.org/10.1016/j.scitotenv.2018.12.223

Chapter 20
Himalayan Plant Biodiversity: A Concluding Overview

Dipayan Das, Payel Paul, and Monoranjan Chowdhury

Abstract The Himalayan mountain system is among the world's youngest and richest biodiversity regions. The rich plant diversity is due to the diverse geological landscape, variable environmental gradients, and the overlapping of multiple biogeographic zones. The present summary brings together findings from 19 chapters that examine Himalayan plant biodiversity from taxonomic, ecological, and conservation perspectives. The Himalayan hotspot is enriched with approximately 10,500–11,900 vascular seed plant species, distributed across 240 families and 2322 genera. The Indian Himalayan Region is a rich pool of 18,940 plant species, of which nearly 8500 angiosperms, around 40% of which are endemic. These species are observed in different elevations, from roughly 200 m to above 8000 m. Species-rich and ecologically important groups such as orchids (about 1036 taxa), *Rhododendron* (121 taxa), and medicinal plants (1748 species) are particularly vulnerable. Projections suggest that climate change could lead to a 16–18% loss of suitable habitat for endemic species between 2050 and 2070. Anthropogenic pressures such as deforestation, mega-structures, overharvesting, and invasive species continue to accelerate the process of biodiversity loss. Although conservation efforts like protected areas, community-based initiatives, ex-situ measures, and policy interventions are playing their part but requires further strict initiatives. The outcome of the 19 chapters suggests the need for coordinated conservation strategies that safeguard ecological integrity while supporting sustainable livelihoods across the Western, Central, and Eastern Himalaya.

D. Das (✉)
Department of Microbiology, Royal School of Biosciences, The Assam Royal Global University, Guwahati, Assam, India

P. Paul
Department of Botany, North Bengal St. Xavier's College, Jalpaiguri, West Bengal, India

M. Chowdhury
Department of Botany, University of North Bengal, Siliguri, West Bengal, India

D. Das et al. (eds.), *Plant Diversity and Conservation Strategies of the Himalayan Biodiversity Hotspot*, Topics in Biodiversity and Conservation 21,
https://doi.org/10.1007/978-3-032-21697-7_20

Keywords Himalayan biodiversity hotspot · Endemic species · Orchid diversity · Medicinal plants · Climate change vulnerability · Anthropogenic impacts · Conservation policy · Protected areas

20.1 Introduction

The Himalayan mountain system that stretches nearly 3000 km between 27–36° N latitude and 72–91° E longitude, represents one of the most biologically rich regions of the world (Samant 2021). The Himalayas are recognised as a major biodiversity hotspot due to their plant diversity, high levels of endemism, and intricate biogeographic patterns (Anthal et al. 2025), a theme comprehensively introduced through the biogeographic and historical synthesis presented in Chap. 1. The Himalaya is a geologically young mountain chain, formed by the continuing collision of the Indian and Eurasian tectonic plates (Singh and Pusalkar 2020), and the ongoing uplift has resulted in pronounced altitudinal variation and the development of complex environmental gradients, as elaborated in Chaps. 1 and 3, giving rise to a wide range of habitats that support diverse plant life (Ahmad et al. 2025).

The Indian Himalayan Region (IHR), covering about 419,873 km^2, nearly 18% of India's land area, encompasses 18,940 plant species (Samant 2021). These include approximately 8500 angiosperms, 44 gymnosperms, and around 600 pteridophytes, with angiosperm endemism estimated at nearly 40% (Samant 2021), reinforcing the Himalayan hotspot status discussed in Chap. 2. Such richness indicates both the region's multifaceted geological condition and its wide climatic variability. Himalayan flora shows a distinctive biogeographic character resulting from the overlap of Palaeotropical and Palaearctic elements, along with extensive local speciation originating through topographic isolation and diverse microclimates (Singh and Pusalkar 2020), a pattern further illustrated through floristic and endemic diversity assessments in Chaps. 5 and 9, and synthesised at a broader scale in Chap. 16. Distinct elevational zonation ranging from tropical and subtropical foothills (<1000 m) through temperate forests (2000–3500 m) to alpine meadows (3500–5000 m) and the nival zone (>5000 m) results in the origin of well-defined vegetation belts (Rawat et al. 2017), a pattern analysed in detail for the Central Himalaya in Chap. 3. Each belt supports plant communities adapted to specific temperature regimes, precipitation patterns, and soil conditions, as demonstrated in regional case studies presented in Chaps. 6, 7, and 15. The contrasting environmental conditions across the east to west Himalaya, viz., drier conditions in the Western Himalaya and high rainfall in the Eastern Himalaya, are also responsible for species richness and community composition across the region (Saikia et al. 2017), as documented across regional case studies in Chaps. 2 and 4, and integrated at a landscape scale in Chap. 16.

The Himalayan plant communities also play a critical role in regulating water flow, stabilising fragile slopes, sequestering carbon, and sustaining soil fertility (Manzoor et al. 2025). The region serves as an important source of non-timber forest products, medicinal plants, and wild edible species that support traditional healthcare systems and rural livelihoods (Masoodi et al. 2020), themes explored extensively in Chap. 5 and revisited from a socio-economic perspective in Chap. 17. The Himalayan local communities regularly uses 1748 medicinal plant species and 675 wild edible species (Samant 2021). Taxonomic groups such as *Orchidaceae*, comprising approximately 1036 taxa, and *Rhododendron*, with about 121 taxa, are particularly important for their species richness and high conservation value. Orchid diversity is studied in detail in Chap. 4, while members of *Polygonaceae* and other endemic lineages are studied in Chaps. 8, 9, and 14. The chapter indicates their ecological importance, cultural relevance, and conservation significance across the Himalayan region. Although the Himalayan plant biodiversity is rich, but is increasingly under severe threat. of The Himalayan belt is gradual warming, currently estimated at around 0.13 °C per decade and projected to rise to 0.25–0.48 °C per decade by the end of the century. This warming is altering species distributions, shifting phenology, and reducing suitable habitats, particularly at higher elevations (Manish et al. 2016). These climate-driven impacts are comprehensively studied in Chaps. 11 and 12, also integrated into future risk scenarios in Chap. 19. Human-made pressures, including deforestation, infrastructure expansion, unregulated tourism, overharvesting of medicinal plants, and the spread of invasive species, further intensify these impacts (Bisht et al. 2022), as discussed in Chaps. 6 and 7 and synthesised from a governance perspective in Chap. 18. Endemic and geographically restricted taxa, along with species of significant economic importance, are particularly vulnerable to ongoing pressures, and a substantial proportion of these remain inefficiently protected within present conservation frameworks (Thapliyal et al. 2024), an issue emphasised across Chaps. 9, 13, and 15 and addressed through strategic recommendations in Chaps. 16 and 19.

This summary chapter draws together insights from 19 chapters that collectively examine Himalayan plant biodiversity from multiple perspectives. These include analyses of biogeographic patterns and diversity metrics, taxonomic assessments of important groups such as orchids, medicinal plants, rhododendrons, and endemic taxa, evaluations of major threats, and discussions on conservation and management strategies. By integrating findings across the Western, Central, and Eastern Himalaya, the chapter provides a consolidated foundation for evidence-based conservation planning and sustainable management of plant resources in this globally significant mountain system.

20.2 Himalayan Biodiversity Hotspot: Biogeographic Foundations and Global Significance

20.2.1 Orogenic History and Biogeographic Position

The plant diversity of the Himalayan mountain system is associated with the tectonic evolution and geographic location. The collision between the Indian and Eurasian plates, which began around 50 million years ago, led to the formation of the world's youngest and highest mountain range. This process resulted in the formation of the extensive altitudinal variation and allowed speciation through geographic isolation and habitat diversification (Thapliyal et al. 2024). Elevational gradients exceeding 8000 m occur over short horizontal distances, resulting in sharp climatic transitions comparable to shifts from tropical to polar regions within restricted spatial scales (Samant 2021). The ecological consequences of these gradients for plant adaptation, community assembly, and ecosystem resilience are reflected in taxon- and region-specific assessments presented in Chaps. 8, 9, and 15.

The Himalaya lies at the interface of two major biogeographic realms, viz., the Palaeotropical (Oriental) region to the south and the Palaearctic region to the north. This transitional position allows the coexistence of tropical, temperate, and alpine floristic elements, contributing significantly to regional species richness (Thapliyal et al. 2024). The longitudinal variation in monsoon influence has led to distinct floristic differentiation across the Western, Central, and Eastern Himalaya, each exhibiting characteristic species composition and patterns of endemism (Saikia et al. 2017). Regional variations were examined across multiple case studies in Chaps. 2 and 4. These east–west diversity differences and their implications for endemic and threatened flora are studied in Chaps. 9 and 14. While the influence of tourism-based land use change on regional plant diversity is studied in Chap. 10.

20.2.2 Species Richness and Endemism Patterns

The Himalayan hotspot is enriched with approximately 10,500–11,900 vascular seed plant species, distributed across 240 families and 2322 genera. This region is dominated by herbaceous taxa (66.7%), followed by shrubs (21.8%) and trees (11.7%). This pattern reflects the presence of widespread alpine and subalpine zones and reduced forest cover at higher elevations (Rana et al. 2017), a life-form composition strongly influenced by extensive alpine and subalpine habitats discussed in Chap. 3. Patterns of tree species richness are strongly associated with mean annual temperature and elevational range, indicating the influence of climatic variability and topographic density on woody plant diversity (Wani et al. 2023). Approximately 12% of Himalayan angiosperms are endemic, and a similar proportion is categorised as threatened or vulnerable, reflecting the high conservation value of this flora (Singh and Pusalkar 2020). The vulnerability of endemic taxa is explored in depth in Chap. 9, while the interacting roles of climate change (Chap. 12), anthropogenic pressures (Chap. 13), and regional land-use dynamics (Chap. 14) provide a broader context for understanding ongoing diversity loss. These findings are synthesised in Chap. 16 and extended into forward-looking conservation strategies in Chap. 19, reinforcing the need for integrated, spatiotemporal approaches to Himalayan plant conservation (Table 20.1).

The Eastern Himalaya represents the maximum percentage of species richness and endemism, with endemic taxa accounting for approximately 30–40% of the

Table 20.1 Overall species statistics for the Himalayan hotspot

Category	Count	References
Total vascular seed plants (Pan-Himalaya)	11,886	Lu et al. (2023)
Total species (compiled floras)	10,503	Rana et al. (2017)
Total species (Indian Himalaya)	11,743	Wani et al. (2024)
Families	240	Rana et al. (2017)
Genera	2322	Rana et al. (2017)
Endemic species (estimated)	3500-4000	Manish (2017)
Herbs	7918	Lu et al. (2023)
Shrubs	2587	Lu et al. (2023)
Trees	1388	Lu et al. (2023)

regional flora. Within this zone, Arunachal Pradesh itself supports nearly 6000 species of flowering plants, indicating its floristic diversity (Behera et al. 2002). These patterns of richness and endemism, and their conservation implications, are studied in Chaps. 9 and 14. These chapters mostly focus on endemic and regionally restricted taxa of the Eastern Himalaya. On the other side, the Western Himalaya is characterised by lower species richness, but bears distinct localised endemism, particularly in high-altitude landscapes such as the Rama Valley, where endemic species constitute approximately 41% of the recorded flora (Shaheen et al. 2011). A detailed assessment of endemic flora in Sikkim indicates climate-induced range dynamics. A total of 124 endemic plant species were analysed, of which 108 species (87%) exhibited upward elevational shifts, with dislocation magnitudes ranging from 23 to 998 m. The mean rate of upward migration was estimated at 27.53 ± 22.04 m per decade, indicating a rapid biological response to regional warming trends (Telwala et al. 2013). These climate-induced distributional shifts and their implications for the long-term persistence of endemic taxa are discussed in detail in Chaps. 11 and 12. While conservation requirements under changing climates are addressed in Chap. 15.

20.2.3 Geoclimatic Conditions and Ecosystem Services

The wide range of climatic zones in the Himalaya, ranging from tropical lowlands to alpine environments, supports diverse plant communities that participate in essential ecosystem processes (Ahmad et al. 2025). These include regulation of hydrological processes through watershed protection for major Asian river systems such as the Indus, Ganges, and Brahmaputra; carbon sequestration in forest biomass and soils; stabilisation of fragile slopes; and local climate regulation (Ahmed 2025). The region also provides timber, fuelwood, and fodder, along with approximately 1748 medicinal plant species and 675 wild edible species that sustain traditional livelihoods and healthcare practices (Samant 2021). These ecosystem services and their socio-ecological importance are examined in Chaps. 8, 9, and 13, while emerging threats from climate change, land-use transformation, and unsustainable resource extraction are synthesised in Chaps. 10, 11, and 12.

20.3 Regional Floristic Diversity and Elevational Patterns

20.3.1 Western Himalayan Floristic Composition

The Western Himalaya, encompassing Jammu and Kashmir, Himachal Pradesh, and Uttarakhand, is characterised by comparatively lower precipitation and colder winters than the eastern Himalaya. Floristic list from Himachal Pradesh documents more than 5000 vascular plant species, nearly one-third of which possess medicinal properties, indicating the region's ethnobotanical importance (Bargali et al. 2022). These medicinal plant resources and associated conservation aspects are studied in Chaps. 8 and 9. Endemic diversity is particularly restricted in alpine and subalpine habitats. Studies from the Pir Panjal Range recorded 328 plant species belonging to 228 genera and 78 families, with native taxa comprising 68.51% and non-native species accounting for 31.49%, indicating considerable invasion pressure (Surmal et al. 2022). Such invasion dynamics and their impacts on native plant communities are discussed in Chap. 13. Life-form spectrum analysis revealed dominance of chamaephytes (29%), therophytes (28%), and phanerophytes (21%), reflecting adaptive strategies to cold and moisture-based environments (Surmal et al. 2022). Medicinal plant assessments in protected areas such as the Khokhan Wildlife Sanctuary documented 363–377 medicinal species, including 179 native taxa, 84 near-endemics, and 6 endemics (Samant et al. 2024).

20.3.2 Central Himalayan Vegetation Gradients

The Central Himalaya shows distinct vegetation that changes through altitude-mediated environmental gradients. Floristic surveys conducted in the Nanda Devi Biosphere Reserve documented 451 plant species across forest transects, where mid-elevation communities were dominated by *Quercus* forests and mixed *Abies–Quercus* associations (Rawat and Negi 2021). Patterns of species richness and phylogenetic diversity along elevational gradients show peak values near 2700 m, corresponding to optimal climatic conditions at intermediate elevations and consistent with mid-domain effects (Liang et al. 2023). These gradient-driven diversity patterns are quantitatively analysed in Chap. 15. Transect-based quantitative analyses indicate that herbaceous taxa dominate species richness (79.04%), followed by trees (11.4%) and shrubs (9.52%). Total plant richness declines significantly with increasing elevation ($R^2 = 0.74$, $p < 0.001$) (Wani et al. 2022b). Beta diversity analyses further reveal very high species turnover along elevational gradients

(Sørensen dissimilarity = 0.92), with species replacement (0.88) contributing substantially more than notedness (0.04). This indicates distinct community groupings across elevation zones rather than simple species loss (Wani et al. 2022b).

20.3.3 Eastern Himalayan Diversity Hotspots

The Eastern Himalaya, particularly Sikkim and Arunachal Pradesh, is known for its floristic diversity within the Himalayan region. This vegetational richness is driven by high monsoonal precipitation, complex topography, and strong biogeographic connectivity with Southeast Asia. Sikkim alone supports approximately 584 endemic angiosperm species distributed across an elevational range of 300–5300 m (Manish et al. 2016), patterns that are analysed in Chaps. 9 and 14. The eastern Himalaya shows a diverse representation of moisture-dependent plant families, particularly *Orchidaceae* and *Ericaceae* (notably *Rhododendron*), reflecting humid climatic conditions (Singh and Pusalkar 2020). These taxonomic groups receive focused treatment in Chaps. 4, 9, and 14. Forest diversity assessments in Arunachal Pradesh reveal heterogeneous plant communities. Comprehensive and standardised quantitative records remain less extensive than those available for the western and central Himalayan regions (Saikia et al. 2017). Understanding these data gaps and strengthening regional floristic inventories form an important component of future research priorities outlined in Chap. 15. The close biogeographic affinity of this area with Southeast Asian floristic provinces also enhances its floristic uniqueness and contributes substantially to the high levels of regional endemism (Singh and Pusalkar 2020).

20.4 Taxonomic Diversity of Critical Plant Groups

20.4.1 Orchid Diversity and Distribution

Orchidaceae is one of the species-rich plant families within the Himalayan flora, comprising approximately 1036 taxa distributed across habitats ranging from tropical lowlands to alpine zones (Singh and Pusalkar 2020), with its diversity patterns synthesised in the taxonomic overview presented in Chap. 4, while their conservation relevance and prospects are revisited in the integrative discussions of Chap. 16. Field-based investigations revealed that the diversity and distribution of orchids depends on environmental gradients. A study in Jammu and Kashmir across 57 sampling sites documented 32 orchid species belonging to 18 genera across. The study observed peak species richness between 1500 and 3500 m elevation. Several taxa, including *Herminium monorchis*, *Habenaria furcifera*, and *Malaxis muscifera*, were recorded as single-individual occurrences, indicating their rarity and potential vulnerability within the region (Hussain et al. 2024). Both epiphytic and terrestrial or geophytic orchids are present across the elevational gradient. The terrestrial forms are mostly dominant at higher elevations (Acharya et al. 2011), indicating habitat transitions described in Chaps. 3 and 4.

Several medicinally important orchid species are under threat because of unsustainable and destructive harvesting practices. A pattern characteristic of high-value Himalayan medicinal taxa was studied in Chap. 5 and revisited under sustainability frameworks in Chap. 17. *Dactylorhiza hatagirea*, a critically endangered orchid collected primarily for its tubers, occurs at very low population densities (0.70–2.43 individuals m^{-2}) in alpine meadows between 2500 and 4500 m elevation. Its distribution shows a curvilinear relationship between altitude and population density ($r^2 = 0.666$). The limited distribution of this species is because of short life cycle, low seed germination rates, narrow microhabitat specificity, sparse populations, and high commercial demand (Singh et al. 2021). As studied in Chaps. 15, 16 and 17, effective conservation of threatened orchid taxa requires systematic documentation of species richness, population structure, and spatial distribution to establish reliable outcome for evidence-based conservation planning (Hussain et al. 2024).

20.4.2 Medicinal Plant Diversity and Utilisation Pressures

The Himalayan region documents around 1748 medicinal plant species, and is considered a reservoir of traditional medicinal resources (Samant 2021). Survey reports also support medicinal plant diversity, with 363–377 species reported from individual protected areas in Himachal Pradesh (Samant et al. 2024). These patterns are studied in depth in Chap. 8 and expanded into livelihood and value-chain perspectives in Chap. 17. Nepal records approximately 2331 medicinal and aromatic plant species, of which nearly 300 are actively traded (Shrestha et al. 2022). Medicinal plants are mostly represented by the families *Asteraceae*, *Rosaceae*, and *Lamiaceae*. Leaves, roots, or tubers are the most commonly harvested plant parts (Samant et al. 2024). The deep root extraction creates significant pressure on geophytic species. Such practices often involve destructive harvesting and restrict the process of natural regeneration (Singh et al. 2021).

20.4.3 Wetland Flora and Aquatic Plant Diversity

Himalayan wetlands, including alpine lakes, marshes, riparian corridors, and seasonal pools, are another Himalayan repository of distinct plant groupings adapted to prolonged soil saturation and variable hydrological regimes. They perform ecological functions such as water purification, flood moderation, sediment trapping, and provide habitat for specialised plant and animal taxa (Samant 2021). Functions evaluated from an ecosystem services perspective are described in Chap. 18. Climate projections suggest that alpine meadows, which frequently exhibit wetland characteristics, are likely to decline by approximately 1–3% by the mid-to-late twenty-first century due to temperature-mediated shrub encroachment. Such structural shifts can reduce suitable habitat for wetland-associated species. Diminishing the ecosystem services sustained by these landscapes (Manish et al. 2016), reinforcing concerns discussed in Chaps. 12 and 19.

20.4.4 IUCN Threat Categories and Market Pressures

Conservation assessments indicate a high concentration of extinction risk among the Himalayan medicinal plants. A Conservation Priority Index (CPI) analysis conducted in Gulmarg Wildlife Sanctuary evaluated 361 species and recorded 6 as critically endangered, 20 as endangered, 28 as vulnerable, and 98 as near threatened (Wani et al. 2022a). These categories account for approximately 42% of the assessed flora, a proportion much higher than global averages and indicative of acute regional vulnerability. These findings align with broader threat assessments presented in Chaps. 9 and 11.

Intensive market demand has led to unsustainable extraction of commercially valuable species. Overharvesting, habitat degradation, and environmental change have contributed to rapid population declines. Management practice requires ecological monitoring and increased awareness among local communities and forest personnel regarding the sustainable use of forest products (Samant et al. 2024). Conservation action remains restricted due to limited information on standing stock, inadequate cultivation protocols for threatened taxa, and insufficient documentation of trade volumes, challenges discussed in Chaps. 17 and 18 (Bargali et al. 2022).

20.4.5 Endemic Flora Status and Conservation Priorities

Endemic species in the Himalayan region are facing extinction risk due to limited geographic ranges and specialised habitat requirements (Mishra et al. 2023). The IHR reports approximately 3400 endemic angiosperms (40% of the angiosperm flora), with the highest endemism in the Eastern Himalaya (Samant 2021). Threat assessments of endemic alpine species reveal concerning patterns like *Aconitum novoluridum* (EOO = 60.46 km^2, AOO = 24 km^2) and *Bistorta longispicata* (EOO = 51.484 km^2, AOO = 12 km^2), both of which qualify as Endangered under IUCN criteria due to restricted ranges and ongoing threats (Das and Lahiri 2023). The climate change vulnerability assessments project indicates severe impacts on endemic species. In Sikkim, approximately 16% and 18% of endemic angiosperms are reported to lose climatically suitable habitat by 2050 and 2070, respectively (Manish et al. 2016). These projections are integrated into adaptive conservation strategies proposed in Chaps. 12, 16, and 19, which emphasise climate-resilient planning, assisted regeneration, and ex-situ conservation through seed banking and cultivation (Das and Lahiri 2023).

20.5 Alien Invasive Plant Species: Ecological Impacts and Management

Biological invasions represent an increasing threat to the Himalayan plant diversity, particularly in disturbed environments. The issues associated with settlements, transport corridors, grazing lands, and tourist infrastructure are addressed in Chap. 7. Floristic investigations in the Pir Panjal range have recorded non-native species comprising 31.49% of the total flora (328 species), indicating invasion pressure in accessible landscapes (Surmal et al. 2022). It was reported that the invasive taxa like *Polygonum polystachyum*, *Impatiens sulcata*, and *Cuscuta europaea* are mostly found in high-altitude disturbed sites (Bisht et al. 2022). Invasion rate shows a strong positive relationship with anthropogenic disturbance, as modified habitats support higher densities of invasive species (Bisht et al. 2022). The invasion dynamics were discussed in Chaps. 7 and 16 and addressed through management measures in Chap. 19.

20.6 Fungal Diversity in Maintaining the Himalayan Floral Diversity

The Himalayan region has a wide variation in fungal diversity due to altitudinal gradients, climatic zones, and vegetation types. Despite its ecological significance, fungal diversity in this region remains less studied compared to the flora and fauna (Yadav et al. 2023). Recent surveys documented widespread macrofungal richness across Himalayan landscapes. Analysis of fungarium data in the Western Himalayas (last 150 years) showed high Shannon diversity in Polyporaceae, with species distributed from 0 to 4500 m, while families like *Thelephoraceae* showed restricted distributions above 2000 m (Yadav et al. 2023). High-altitude surveys in Kishtwar National Park and Kashmir documented diversity in mushroom species with altitude-dependent distribution patterns (Sharma et al. 2022; Ullah et al. 2022). The Far Eastern Himalayan landscape shows distinct fungal communities adapted to extreme environmental conditions (Fang et al. 2023).

Fungal species play important roles in maintaining the Himalayan ecosystems. Ectomycorrhizal fungi dominate many forest types, forming symbiotic associations with oak, pine, deodar, and spruce that facilitate nutrient cycling and support forest health (Han et al. 2023; Yadav et al. 2023). Complementing the forest ecology perspectives was discussed in Chap. 4. Saprotrophic fungi-mediated decomposition processes are essential for nutrient recycling in these mountain ecosystems. Many species also serve as a food source and traditional medicine for local communities (Ullah et al. 2022). The ethnobiological resource utilisation in the Himalayan region was discussed in Chap. 5.

Conservation is required in relation to climate change and anthropogenic pressures that threaten Himalayan fungal diversity. Species richness correlates positively with rainfall and negatively with high solar radiation, making fungi vulnerable to shifting climate patterns (Yadav et al. 2023), a vulnerability pattern consistent with climate impacts described for alpine and wetland ecosystems in Chaps. 3 and 6. The limited altitudinal ranges of certain endemic taxa increase the extinction risks. Establishing germplasm banks, promoting sustainable harvesting practices, and conducting comprehensive biodiversity surveys are essential to establish sustainable conservation practice (Ullah et al. 2022). Understanding and protecting Himalayan fungal diversity is crucial for maintaining ecosystem functioning and supporting mountain communities.

20.7 Conservation Trends and Policy Frameworks

Protected areas such as national parks, wildlife sanctuaries, biosphere reserves, etc., are important government-initiated projects towards conserving the Himalayan biodiversity (Wani et al. 2022a). Details related to protected areas are discussed in Chap. 6 and revisited under future conservation scenarios in Chap. 18. Although the Indian Himalayan Region contains an extensive protected area, notable gaps persist, particularly in climate-vulnerable landscapes and centres of endemism (Wei et al. 2025). Common problems that restricts conservation process include shortages of staff and funding, conflicts between people and wildlife, land encroachment, illegal use of biological resources, and poor involvement of local communities. (Wani et al. 2022a). These limitations indicate the better management of underrepresented ecosystems, improvements in governance and management effectiveness, and strengthened ecological connectivity through corridor protection (Wei et al. 2025). Himalayan plant conservation operates within interconnected local, national, and international policy frameworks, including forest and biodiversity legislation and global agreements such as the Convention on Biological Diversity, CITES, and the Nagoya Protocol (Mosoh et al. 2024). Ongoing gaps in implementation due to weak enforcement, limited institutional capacity, fragmented policies, and poor integration of climate change concerns continue to deteriorate conservation efforts. This indicates the need for flexible, coordinated governance approaches at multiple levels, as discussed in Chaps. 12 and 15.

20.8 Conclusion

This edited volume examines the versatility of Himalayan plant diversity, which has emerged through millions of years of tectonic uplift, climatic oscillations, and evolutionary diversification across one of the world's most dynamic mountain systems. Climate change caused 16–18% habitat loss for endemic species by mid-century, along with ongoing land-use change, unsustainable resource extraction, and biological invasions. Effective conservation of Himalayan plant biodiversity requires integrated approaches from local community management to transboundary cooperation. The Himalayan belt requires regulation over sustainable harvest for medicinal plants and NTFPs. Conservation strategies must align with sustainable livelihoods, traditional knowledge systems, and climate change adaptation. The synthesis presented across these 19 chapters reports the scientific and survey-based outcomes for the Himalayan plant conservation. The study indicates that limitations persist in taxonomy, distribution, population dynamics, threat assessments, and management effectiveness. Thus, it is recommended that the Himalayan belt requires sustained research investment, particularly in underexplored regions and for understudied taxonomic groups, to enhance evidence-based conservation planning.

References

Acharya K, Vetaas O, Birks H (2011) Orchid species richness along Himalayan elevational gradients. J Biogeogr 38:1821–1833. https://doi.org/10.1111/j.1365-2699.2011.02511.x

Ahmad M, Luo Y-H, Rathee S, Spicer RA, Zhang J, Wambulwa MC, Zhu G-F, Cadotte MW, Wu Z-Y, Khan SM, Maity D, Li D-Z, Liu J (2025) Multifaceted plant diversity patterns across the Himalaya: status and outlook. Plant Divers 47(4):529–543. https://doi.org/10.1016/j.pld.2025.04.003

Ahmed R (2025) Decoding the fate of Himalayan glaciers under climate change: impacts, challenges, research gaps, and policy pathways. Evol Earth 3:100082. https://doi.org/10.1016/j.eve.2025.100082

Anthal R, Jasrotia R, Dhaka R (2025) Chapter 2 The Himalayas: a hotspot of biodiversity, pp. 21–36. https://doi.org/10.1201/9781774919965-2

Bargali H, Kumar A, Singh P (2022) Plant studies in Uttarakhand, Western Himalaya–a comprehensive review. Trees For People 8:100203. https://doi.org/10.1016/j.tfp.2022.100203

Behera MD, Kushwaha SPS, Roy PS (2002) High plant endemism in an Indian hotspot—eastern Himalaya. Biodivers Conserv 11(4):669–682. https://doi.org/10.1023/A:1015596309833

Bisht M, Chandra Sekar K, Mukherjee S, Thapliyal N, Bahukhandi A, Singh D, Bhojak P, Mehta P, Upadhyay S, Dey D (2022) Influence of anthropogenic pressure on the plant species richness and diversity along the elevation gradients of Indian Himalayan high-altitude protected areas. Front Ecol Evol 10. https://doi.org/10.3389/fevo.2022.751989

Das M, Lahiri S (2023) Threat assessment of two Himalayan endemic alpine plant species and conservation implications. Int J Bioresour Sci 9:139–144. https://doi.org/10.30954/2347-9655.0.2022.11

Fang W, Devkota S, Arunachalam K, Phyo KMM, Shakya B (2023) Systematic review of fungi, their diversity and role in ecosystem services from the Far Eastern Himalayan Landscape (FHL). Heliyon 9(1). https://doi.org/10.1016/j.heliyon.2022.e12756

Han X, Liu D, Zhang M, He M, Li J, Zhu X, Wang M, Thongklang N, Zhao R, Cao B, Han X, Liu D, Zhang M, He M, Li J, Zhu X, Wang M, Thongklang N, Zhao R, Cao B (2023) Macrofungal diversity and distribution patterns in the primary forests of the Shaluli Mountains. J Fungi 9(4). https://doi.org/10.3390/jof9040491

Hussain K, Dar ME-U-I, Khan AM, Iqbal T, Mehmood A, Habib T, Moussa IM, Casini R, Elansary HO (2024) Temperature, topography, woody vegetation cover and anthropogenic disturbance shape the orchids distribution in the western Himalaya. S Afr J Bot 166:344–359. https://doi.org/10.1016/j.sajb.2024.01.042

Liang J, Ding Z, Lie G, Zhou Z, Zhang Z, Hu H (2023) Patterns and drivers of phylogenetic diversity of seed plants along an elevational gradient in the central Himalayas. Glob Ecol Conserv 47:e02661. https://doi.org/10.1016/j.gecco.2023.e02661

Lu Z-X, Yang M, Li B, Hu J-J, Yu H-B (2023) Elevational patterns of seed plants and the driving mechanisms in the Himalaya. Ying Yong Sheng Tai Xue Bao = J Appl Ecol 34(7):1787–1796. https://doi.org/10.13287/j.1001-9332.202307.008

Manish K, Pandit MK, Telwala Y, Nautiyal DC, Koh LP, Tiwari S (2017) Elevational plant species richness patterns and their drivers across non-endemics, endemics and growth forms in the Eastern Himalaya. J Plant Res 130:829–844

Manish K, Telwala Y, Nautiyal DC, Pandit MK (2016) Modelling the impacts of future climate change on plant communities in the Himalaya: a case study from Eastern Himalaya, India. Model Earth Syst Environ 2(2):92. https://doi.org/10.1007/s40808-016-0163-1

Manzoor M, Ahmad M, Gillani SW, Waheed M, Grießinger J, Shaheen H, Meier WJ-H (2025) Biodiversity patterns and environmental drivers of fragile alpine glacial lake ecosystems in the Himalaya: implications for conservation and management. Ecol Indic 178:114028. https://doi.org/10.1016/j.ecolind.2025.114028

Masoodi HUR, Sundriyal RC (2020) Richness of non-timber forest products in Himalayan communities—diversity, distribution, use pattern and conservation status. J Ethnobiology Ethnomedicine 16:56. https://doi.org/10.1186/s13002-020-00405-0

Mishra AP, Kumar A, Yadav SN (2023) Ecology and conservation of threatened medicinal plants in the Trans-Himalayan region of Nanda Devi Biosphere Reserve, Western Himalaya. Trees For People 14:100451. https://doi.org/10.1016/j.tfp.2023.100451

Mosoh DA, Prakash O, Khandel AK, Vendrame WA (2024) Preserving earth's flora in the 21st century: climate, biodiversity, and global change factors since the mid-1940s. Front Conserv Sci 5. https://doi.org/10.3389/fcosc.2024.1383370

Rana SK, Rawat GS, Rana SK, Rawat GS (2017) Database of Himalayan plants based on published floras during a century. Data 2(4). https://doi.org/10.3390/data2040036

Rawat B, Negi AS (2021) Plant diversity patterns along environmental gradients in Nanda Devi Biosphere Reserve, West Himalaya. Trop Ecol 62(1):61–70. https://doi.org/10.1007/s42965-020-00122-5

Rawat GS (2017) The Himalayan Vegetation along Horizontal and Vertical Gradients. In: Prins HHT, Namgail T, eds. Bird Migration across the Himalayas: Wetland Functioning amidst Mountains and Glaciers. Cambridge University Press. pp. 189–204

Saikia P, Deka J, Bharali S, Kumar A, Tripathi OP, Singha LB, Dayanandan S, Khan ML (2017) Plant diversity patterns and conservation status of eastern Himalayan forests in Arunachal Pradesh, Northeast India. For Ecosyst 4(1):28. https://doi.org/10.1186/s40663-017-0117-8

Samant S (2021) Floristic diversity of the Himalaya in relation to climate change: status, values and conservation. J Graph Era Univ. https://doi.org/10.13052/jgeu0975-1416.913

Samant S, Tewari L, Prakash O, Kundra R, Singh A, Dutt S (2024) Diversity, distribution pattern and indigenous uses of medicinal plants in Khokhan Wildlife Sanctuary of Himachal Pradesh, North Western Himalaya, India. J NonTimber For Prod 31:75–94. https://doi.org/10.54207/bsmps2000-2024-PD95HL

Shaheen H, Ahmad N, Alam N, Ahmad K, Ullah Z (2011) Phytodiversity and endemic richness in high altitude Rama Valley, Western Himalayas, Northern Pakistan. J Med Plant Res 5:1489–1493

Sharma Y, Hashmi S, Sharma R, Koul S, Manhas R (2022) Macrofungal diversity and distribution in Kishtwar High Altitude National Park, Jammu and Kashmir, India. Curr Sci 122:1415. https://doi.org/10.18520/cs/v122/i12/1415-1425

Shrestha UB, Lamsal P, Ghimire SK, Shrestha BB, Dhakal S, Shrestha S, Atreya K (2022) Climate change-induced distributional change of medicinal and aromatic plants in the Nepal Himalaya. Ecol Evol 12(8):e9204. https://doi.org/10.1002/ece3.9204

Singh D, Pusalkar P (2020) Floristic diversity of the Indian Himalaya, pp 93–126. https://doi.org/10.1007/978-981-32-9174-4_5

Singh L, Bhatt ID, Negi VS, Nandi SK, Rawal RS, Bisht AK (2021) Population status, threats, and conservation options of the orchid Dactylorhiza hatagirea in Indian Western Himalaya. Reg Environ Chang 21(2):40. https://doi.org/10.1007/s10113-021-01762-6

Surmal O, Singh B, Musarella CM, Surmal O, Singh B, Musarella CM (2022) Understanding species diversity, phenology and environmental implications of different life forms in coniferous forests: a case study from Bhallesa Hills of Pir Panjal Mountain, Western Himalaya, India. Forests 13(12). https://doi.org/10.3390/f13122050

Telwala Y, Brook BW, Manish K, Pandit MK (2013) Climate-induced elevational range shifts and increase in plant species richness in a Himalayan biodiversity epicentre. PLoS One 8(2):e57103. https://doi.org/10.1371/journal.pone.0057103

Thapliyal N, Bhojak P, Mehta P, Sekar KC (2024) Threats and conservation of plant diversity in the Himalayan biodiversity hotspot. In T. Pullaiah (Ed.), Biodiversity hotspot of the Himalaya. Oakville, Ontario, Canada: Apple Academic Press

Ullah TS, Firdous SS, Shier WT, Hussain J, Shaheen H, Usman M, Akram M, Khalid AN (2022) Diversity and ethnomycological importance of mushrooms from Western Himalayas, Kashmir. J Ethnobiol Ethnomed 18(1):32. https://doi.org/10.1186/s13002-022-00527-7

Wani ZA, Bhat JA, Negi VS, Satish KV, Siddiqui S, Pant S (2022a) Conservation Priority Index of species, communities, and habitats for biodiversity conservation and their management planning: a case study in Gulmarg Wildlife Sanctuary, Kashmir Himalaya. Front For Glob Change 5. https://doi.org/10.3389/ffgc.2022.995427

Wani ZA, Khan S, Bhat JA, Malik AH, Alyas T, Pant S, Siddiqui S, Moustafa M, Ahmad AE, Wani ZA, Khan S, Bhat JA, Malik AH, Alyas T, Pant S, Siddiqui S, Moustafa M, Ahmad AE (2022b) Pattern of β-diversity and plant species richness along vertical gradient in northwest Himalaya, India. Biology 11(7). https://doi.org/10.3390/biology11071064

Wani SA, Mugal MA, Dar FA, Sudhakar Reddy C, Rashid I, Khuroo AA (2023) Biodiversity data synthesis on trees of Indian Himalayan Region: policy and management implications. Glob Ecol Conserv 48:e02698. https://doi.org/10.1016/j.gecco.2023.e02698

Wani SA, Khuroo AA, Zaffar N, Rafiqi S, Farooq I, Afzal S, Rashid I (2024) Data synthesis for biodiversity science: a database on plant diversity of the Indian Himalayan region. Biodivers Conserv 33(12):3377–3397. https://doi.org/10.1007/s10531-024-02784-2

Wei J-D, Wang W-T, Wei J-D, Wang W-T (2025) Vulnerability assessment of six endemic Tibetan-Himalayan plants under climate change and human activities. Plants (Basel) 14(15). https://doi.org/10.3390/plants14152424

Yadav N, Rakholia S, Ali N, Yosef R, Yadav N, Rakholia S, Ali N, Yosef R (2023) Fungal diversity and distribution in the biodiversity hotspots of the Western Himalayas. Diversity 15(11). https://doi.org/10.3390/d15111106

MIX
Papier aus verantwortungsvollen Quellen
Paper from responsible sources
FSC® C105338

If you have any concerns about our products,
you can contact us on
ProductSafety@springernature.com

In case Publisher is established outside the EU,
the EU authorized representative is:
Springer Nature Customer Service Center GmbH
Europaplatz 3, 69115 Heidelberg, Germany

Printed by Libri Plureos GmbH
in Hamburg, Germany